Crop Protection Strategies under Climate Change Scenarios

By

P. PARVATHA REDDY

Former Director

Indian Institute of Horticultural Research, Bengaluru

Published by
SCIENTIFIC PUBLISHERS (INDIA)

Jodhpur –
5 A, New Pali Road
P.O. Box 91
Jodhpur - 342 001 INDIA

Delhi –
4806/24, Ansari Road
Daryaganj
New Delhi - 110 002 INDIA

ISBN: 978-93-87307-08-7

Visit the Scientific Publishers (India) website at
http://www.scientificpub.com
Printed in India

Preface

Agriculture in the twenty-first century faces the challenge of meeting food demands while satisfying sustainability goals. Global food production must increase by 70% to meet the projected demand of the world's population which is likely to reach 9.5 billion by 2050. The challenge is further complicated by climate change, which affects the distribution of crop pests (insect pests, disease and nematode pathogens and weeds) and the severity of their outbreaks. More than 30 per cent of crops worldwide are blemished, damaged or destroyed by agricultural pests. The crop losses due to pests viewed in terms of food security would represent the equivalent of food required to feed over one billion people. Most existing pest management strategies have been based on the indiscriminate use of pesticides in agriculture. Increasing concerns over health and the environment, as well as new legislation on pesticide use urge us to find sustainable alternatives to pesticide-based pest management. Crop damage could occur more rapidly than expected due to climate change, because many pests develop more rapidly in response to rising temperatures. In addition, as temperatures increase, the frequency of spring frosts will decline and the resulting extended frost-free periods will increase the duration and intensity of pest outbreaks. This has major implication for the intensification of yield losses due to potential changes in crop diversity and increased incidence of insect-pests in the context of impending climate change. Agriculture needs to meet food demands, satisfy sustainability goals, and adapt to climate change. Indeed, maintaining, or even increasing, crop yield while simultaneously reducing reliance on pesticides represents a challenging task that is further complicated against the background of climate change scenarios.

Current agricultural practices may need to be redesigned because of climate change. Crop protection per se has an important role to play in mitigating climate change. Crop protection practices can contribute to the reduction of greenhouse gases (GHG) emissions. The conventional crop production systems coupled to reduced tillage is generally the best for producing high yields to minimize GHG emissions and to contribute to global food security. Optimum control of pests is insurance for optimum exploitation of all other investments in crop production. IPM is a widely used strategy that integrates cultural, biological, and chemical controls to reduce harmful insect populations below a threshold which otherwise will cause economic losses. Hence, time has come to elaborate new crop protection strategies that are robust and reliable enough for the challenges of changing climatic conditions and more stringent human and environmental protection.

In the view of these developments, the present book on "**Crop Protection Strategies under Climate Change Scenarios**" is timely and is need of the hour. The book provides the following information on insect and mite pests, disease pathogens (fungi, bacteria, viruses and nematodes), and weeds in a very systematic manner:

1. Overview on the crop pest scenarios in changing climate.
2. Effects of abiotic stresses due to climate change [increased temperatures, elevated carbon dioxide levels, varying precipitation patterns and frequency and magnitude of extreme weather events (drought, cyclones, floods, etc.), and elevated levels of atmospheric pollutants [ozone, acid rain, and elevated ultraviolet B] on crop pests.
3. Impacts of climate change induced consequences (expansion of geographical distribution, increase in number of generations, increased overwintering survival, pest population dynamics and outbreaks, risk of introducing invasive alien species, crop-pest interactions, loss of ecological biodiversity, changes in phenology, increased incidence of insect vectored plant diseases, disruption of plant-pollinator interactions, reduced effectiveness of pest management strategies) on crop pests.
4. Development of modelling to predict future pest change scenarios.
5. Formulation of sustainable adaptation and mitigation pest management strategies including physical, cultural, chemical, biological, host resistance, and integrated methods under climate change scenarios.

This book will be of immense value to scientific community involved in teaching, research and extension activities related to crop protection under changing climate. The material can be used for teaching post-graduate courses. The book can also serve as a very useful reference to policy makers and practicing farmers. Suggestions to improve the contents of the book are most welcome (E-mail: reddypp42@gmail.com). The publisher, Scientific Publishers, Jodhpur, India, deserves commendation for their professional contribution.

P. Parvatha Reddy

34, UAS Layout, Ist Main, 7th Cross

Sanjay Nagar, Bengaluru-560 094

Bengaluru

January 26, 2018

Contents

About Author

Dr. P. Parvatha Reddy obtained his Ph. D. degree jointly from the University of Florida, USA and the University of Agricultural Sciences, Bengaluru.

Dr. Reddy served as the Director of the prestigious, Indian Institute of Horticultural Research (IIHR) at Bangalore from 1999 to 2002 and during this period the Institute was honoured with "ICAR Best Institution Award". He also served as the Head, Division of Entomology and Nematology at IIHR and gave tremendous impetus and direction to research, extension and education in developing bio-intensive integrated pest management strategies in horticultural crops. These technologies are being practiced widely by the farmers across the country, since they are effective, economical, eco-friendly and residue-free. Dr. Reddy has about 34 years of experience working with horticultural crops and involved in developing an F1 tomato hybrid "Arka Varadan" resistant to root-knot nematodes. He has over 250 scientific publications to his credit, which also includes 30 books. He has guided a few Ph.D. students at the University of Agricultural Sciences, Bengaluru.

Dr. Reddy served as Chairman, Research Advisory Committee (RAC), Indian Institute of Vegetable Research, Varanasi; Senior Scientific Advisor, Dr. Prem Nath Agricultural Science Foundation, Bengaluru; Member, RAC of National Centre for Integrated Pest Management, New Delhi; Member of the Expert Panel for monitoring the research program of National Initiative on Climate Resilient Agriculture (NICRA) in the theme of Horticulture including Pest Dynamics and Pollinators; Member of the RAC of the National Research Centre for Citrus, Nagpur and the Project Directorate of Biological Control, Bengaluru. He served as a Member, QRT to review the progress of the Central Tuber Crops Research Institute, Thiruvananthapuram; AICRP on Tuber Crops; AICRP on Nematodes and AINRP on Betel vine. He is the Honorary Fellow of the Society for Plant Protection Sciences, New Delhi; Fellow of the Indian Phytopathological Society, New Delhi and Founder President of the Association for Advancement of Pest Management in Horticultural Ecosystems (AAPMHE), Bengaluru.

Dr. Reddy has been awarded with the prestigious "Association for Advancement Pest Management in Horticultural Ecosystems Award", "Dr. G.I. D'souza Memorial Lecture Award", "Prof. H.M. Shah Memorial Award" and "Hexamar Agricultural Research and Development Foundation Award" for his unstinted efforts in developing sustainable, bio-intensive and eco-friendly integrated pest management strategies in horticultural crops.

He has organized "Fourth International Workshop on Biological Control and Management of *Chromolaena odorata*", "National Seminar on Hi-tech Horticulture", "First National Symposium on Pest Management in Horticultural Crops: Environmental Implications and Thrusts", and "Second National Symposium on Pest Management in Horticultural Crops: New Molecules and Bio-pesticides".

Section I
INTRODUCTION

1

Climate Change – An Overview

1.1. INTRODUCTION

The human population is ever increasing, with conservative estimates predicting that the population will rise to approximately 9.5 billion by 2050. The United Nations Food and Agriculture Organization predicted that food production needs to increase by 70 per cent globally in order to feed the increased population. This demand has to be met primarily through yield increases on existing cultivated lands in order to be environmentally sustainable and cost effective (James, 2009). Thus, the major challenges facing the world are to feed and provide shelter for a world population that is increasing at an exponential rate. Agriculture must play a major role in achieving these goals both by providing ever-increasing food yields and an ever-increasing supply of natural products required by industries. An immediate priority for agriculture is to achieve maximum production of food and other products in a manner that is environmentally sustainable and cost effective, against the backdrop of climate change which not only is predicted to result in the loss of agricultural land as a consequence of rising sea levels, but is also likely to have a major impact on the dynamics of pest populations.

1.2 WHAT IS CLIMATE CHANGE?

Climate change is one of the most important global environmental problem today. It is a term used to describe a gradual increase in the average temperature of earth's atmosphere and its oceans, a change that is believed to be changing the earth's climate for ever. It is a long-term change in the statistical distribution of weather pattern over periods of time that range from decades to millions of years. It may be limited to a specific region, or may occur across the whole world.

The terms 'weather' and 'climate' are frequently considered to be interchangeable, but weather and climate refer to different things. Weather is the brief, rapidly changing condition of the atmosphere at a given place and time, influenced by the movement of air masses. Climate, on the other hand, should more accurately be the term applied to the average weather conditions over longer period of years to decades.

UNFCCC defines climate change as "a change of climate that is attributed directly or indirectly to human activity that alters the composition of the global atmosphere and that is in addition to natural climate variability observed over comparable time periods".

The IPCC defines climate change, as a statistically significant variation in either the mean state of the climate or in its variability, persisting for an extended period. Climate change may be due to natural internal processes or external forcing, or persistent anthropogenic changes in the composition of the atmosphere or in the land use.

The threat of climate change poses a challenge for sustainable agricultural growth, on account of its close linkages with other economic sectors such as sugar, textiles, jute, food and milk processing etc. This threat is compounded due to accumulated greenhouse gas emissions in the atmosphere, anthropogenically generated through long-term intensive industrial growth and high consumption lifestyles and preferences.

1.3. GLOBAL WARMING/GREENHOUSE EFFECT

The atmosphere is full of invisible gases, some of which are greenhouse gases (GHGs) which insulate the earth. They trap the sun's heat and keep our planet warm enough to sustain life. Some GHGs in the atmosphere do exist naturally. But a large portion of the GHGs in the atmosphere today have been, and continue to be, created by humans. This means that more of the sun's heat is being trapped than the earth's actual need. As a result, too much heat is being trapped, and the planet is warming too much. This is what's known as *'global warming'* or *'greenhouse effect'* (Fig. 1.1). Global warming is affecting weather patterns all over the world and this effect is what's known as *'climate change'*.

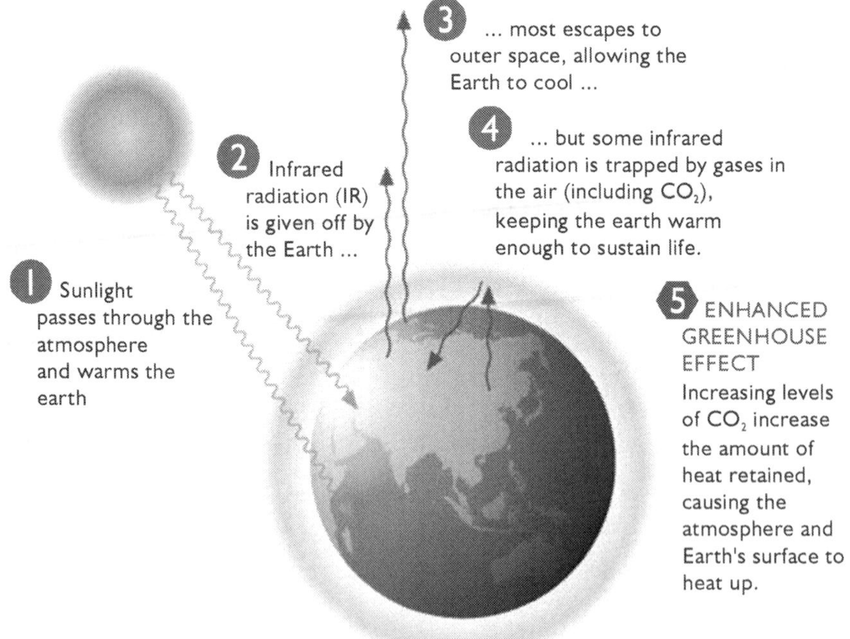

Fig. 1.1. Global warming or the greenhouse effect
(*Source:* CO_2 Cooperative Research Centre)

1.4. MAIN INDICATORS OF CLIMATE CHANGE

There are seven indicators that would be expected to increase in a warming world (Fig. 1.2).

- Temperature over land
- Ocean heat content
- Sea level
- Sea surface temperature
- Temperature over ocean
- Humidity
- Tropospheric temperature

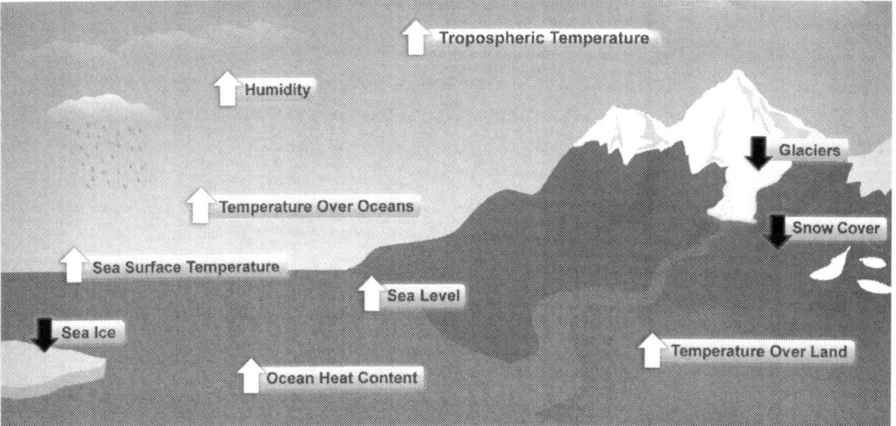

Fig. 1.2. Indicators for global warming

There are three indicators that would be expected to decrease in a warming world (Fig. 1.2).

- Sea ice
- Snow cover
- Glaciers

1.5. CAUSES OF CLIMATE CHANGE

Planet earth comprises of several layers. It starts with the three layers of the earth called as crust, mantle, and then five layers of atmosphere namely, core, troposphere, stratosphere, mesosphere, and exosphere. The atmosphere provides air, water, warmth and protects the earth from the harmful rays (UV) of the sun. The earth partially absorbs energy from the sun and simultaneously releases it in the form of infrared waves (as heat we feel). This heat is not completely lost to space, but is partially absorbed by some gases present in very small quantities in the atmosphere called as greenhouse gases (GHGs) and they remit some of this heat to the earth surface.

1.5.1. Natural causes

The earth's climate can be affected by natural factors that are external to the climate system, such as changes in volcanic activity, solar output, and the earth orbit around the Sun. Of these, the two factors relevant on timescales of contemporary climate change are changes in volcanic activity and changes in solar radiation. In terms of the earth energy balance, these factors primarily influence the amount of incoming energy. Volcanic eruptions are episodic and have relatively short-term effects on climate. Changes in solar irradiance have contributed to climate trends over the past century but since the industrial revolution, the effect of additions of GHGs to the atmosphere has been about ten times that of changes in the Sun's output.

1.5.2. Human causes

Climate change can be caused by human activities, such as the burning of fossil fuels (coal and oil for homes, factories and transportation) and the conversion of land from forestry to agriculture. Since the beginning of the industrial revolution, these human influences on the climate system have enhanced substantially, resulting in increased concentration of greenhouse gases (carbon dioxide, methane, nitrous oxide, ozone and chlorofluorocarbons) in the atmosphere that is making the earth warmer day by day. Three main causes of the increase in greenhouse gases observed over the past 250 years have been fossil fuels, land use, and agriculture (Fig. 1.3) (IPCC, 2007a).

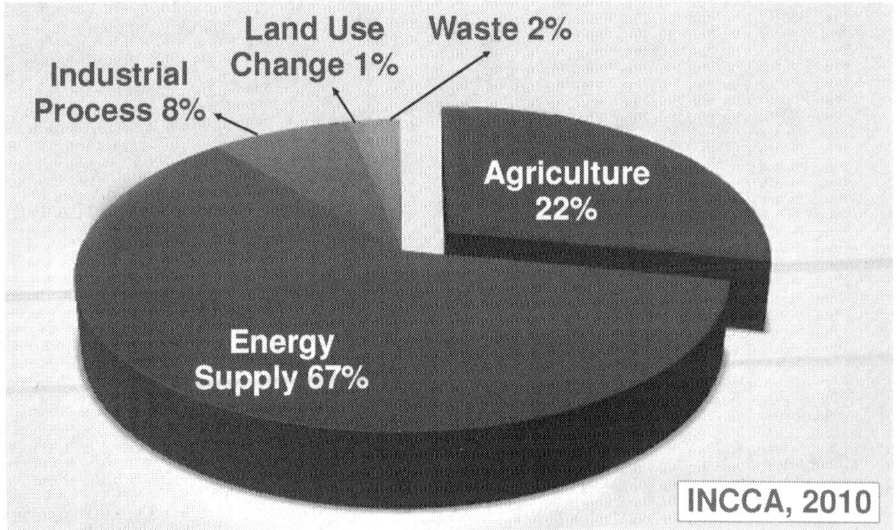

Fig. 1.3. Contribution of different sectors to greenhouse gas emissions in India

1.6. GREENHOUSE GASES

Six main greenhouse gases are carbon dioxide (CO_2), methane (CH_4) and nitrous oxide (N_2O), plus three fluorinated industrial gases: hydro-fluorocarbons (HFCs), perfluorocarbons (PFCs) and sulphur hexafluoride (SF_6). Water vapour (H_2O) is also considered a greenhouse gas.

1.6.1. Carbon dioxide (CO_2)

Carbon dioxide is undoubtedly, the most important greenhouse gas in the atmosphere. A minor but very important component of the atmosphere, CO_2 is released through natural processes such as respiration, volcano eruptions and through human activities such as deforestation, land use changes, land clearing, agriculture, burning fossil fuels and other activities.

1.6.2. Methane (CH_4)

Methane is another important greenhouse gas in the atmosphere. Wetlands, organic matter decay, cattle and refuse landfills, termites and natural gas and oil extraction are the main sources of methane. The primary sources of methane emission in agriculture includes rice cultivation, biomass burning, animal digestive processes and manure storage and handling. Methane has 25 times the Global Warming Potential (GWP) of carbon dioxide and is therefore 25 CO_2e.

1.6.3. Nitrous oxide (N_2O)

Nitrous oxide is a powerful greenhouse gas produced by soil cultivation practices, especially the use of commercial and organic fertilizers, fossil fuel combustion, nitric acid production, and biomass burning. Forests, grasslands, oceans, soils, nitrogenous fertilizers, burning of biomass and fossil fuels are the sources of N_2O. Nitrous oxide has the capacity to absorb and re-emit approximately 310 times the amount of heat compared to carbon dioxide – it has 310 times the GWP of CO_2.

1.6.4. Chlorofluorocarbons (CFCs)

CFCs are synthetic compounds, entirely of industrial origin used in a number of applications, which contribute to destruction of the ozone layer.

1.6.5. Water vapour (H_2O)

Water vapour is the most abundant greenhouse gas, but importantly, it acts as a feedback to the climate. It increases as the earth's atmosphere warms, but so does the possibility of clouds and precipitation, making these some of the most important feedback mechanisms to the greenhouse effect. Much of the heat that escapes from the oceans is in the form of water vapour, the most abundant greenhouse gas on earth.

1.7. CLIMATE CHANGE PROJECTIONS

- Over the next two decades, a warming of 0.2–0.4 °C per decade.
- Warming and sea-level rise to continue for centuries.
- A 1.8° C rise above 1980–1999 levels, in global average surface air temperatures by the end of the twenty-first century for a low emission scenario (IPCC, B1).
- A 4.0° C rise above 1980–1999 levels, in global average surface air temperatures by the end of the twenty-first century for a high emission fossil-fuel intensive scenario (IPCC, A1Fl).
- Warming expected to be greatest over land and at most high northern latitudes.

- By the end of the twenty-first century, rainfall very likely to have increased in high latitudes and East Africa; and decreased in most subtropical regions.
- Increase in extreme weather events (droughts, heat waves, heavy precipitation, and intense tropical cyclones).
- Temperature increases - An increase in global mean annual temperatures of 1° C by 2025 and 3° C by the end of the next century.
- Sea level rises - Global mean sea level is estimated – risen 10-25 cm over the last 100 years. In the next 100 years, the average sea level rise is projected to be about 50 cm.
- Increased variability of weather events.
- Carbon dioxide level increases – concentration of CO_2, the predominant greenhouse gas have increased from 280 ppm to 400 ppm over the last 150 years.
- Agriculture – Up to 50% reduction in maize yields – 4-35% reduction in rice yields (with some exceptions) – Rise in coconut yields (with some exceptions); reduced apple production – Negative impacts on livestock in all regions.
- Fresh water supply – High variability predicted in water yields (from 50% increase to 40-50% fall) – 10-30% increased risk of floods; increased risks of droughts.
- Forests and natural ecosystems – Shifting forest borders; species mix; negative impact on livelihoods and biodiversity.
- Human health – Higher morbidity and mortality from heat stress and vector/water-borne diseases – Expanded transmission window for malaria.

1.7.1. Projected impacts of climate change on agriculture

- Increased crop and pasture yields in colder environments and decreased yields in warmer and seasonally dry environments (virtually certain).
- Increased insect outbreaks (virtually certain).
- Reduction in yields in warmer regions due to heat stress (very likely).
- Increased heavy precipitation events, causing damage to crops, soil erosion and difficulty in land cultivation (very likely).
- Increased area affected by drought, leading to land degradation, lower yields/crop damage and failure, and more livestock deaths (likely).
- Storm intensity increased, leading to damaged crops and uprooting of trees (likely).
- Increased incidence of extreme high seas, causing salinization of irrigation water and freshwater systems (likely).

1.8. AGRICULTURE AS A SOURCE OF GHGs

Agriculture releases to the atmosphere significant amounts of CO_2, CH_4, and N_2O (IPCC, 2001). CO_2 is released largely from microbial decay or burning of plant litter and soil organic matter (Smith, 2004). CH_4 is produced when organic materials decompose in oxygen-deprived conditions, notably from fermentative digestion by ruminant livestock, from stored manures, and from rice grown under flooded conditions (Mosier *et al.*, 1998). N_2O is generated by the microbial transformation of nitrogen in soils and manures, and is often enhanced where available nitrogen (N) exceeds plant requirements, especially under wet conditions (Oenema *et al.*, 2005).

Major agricultural activities which contribute to elevated GHG emissions include:

- Agricultural soil management covers a broad array of practices including fertilization with synthetic fertilizer and animal manures; manure deposition by grazing animals, soil cultivation; production on N fixing crops and forages; irrigation and other practices. The category covers GHG emissions from both cropland and grasslands.
- Enteric fermentation is primarily CH_4 produced by the digestive processes of agricultural animals which are emitted from the animals as gas.
- Manure management emissions are CH_4 and N_2O released from manure during storage and handling.
- Rice cultivation done under anaerobic conditions in flooded fields, results in CH_4 emissions.
- Field burning of agricultural residues results mostly in CO_2 emissions, which are not counted because it is assumed that CO_2 will be reabsorbed by plants in the next growing season. Field burning, however, also results in release of CH_4, N_2O and other minor GHGs.

Agriculture accounted for an estimated emission of 5.1 to 6.1 Gt CO_2e/yr in 2005 (10-12% of total global anthropogenic emissions of GHGs). CH_4 contributes 3.3 Gt CO_2e/yr and N_2O 2.8 Gt CO_2e/yr. Despite large annual exchanges of CO_2 between the atmosphere and agricultural lands, the net flux is estimated to be approximately balanced, with CO_2 emissions around 0.04 Gt CO_2/yr only (low agreement, limited evidence). Together, these agricultural processes comprise 54% of methane emissions, roughly 80% of nitrous oxide emissions, and virtually all carbon dioxide emissions tied to land use (Table 1.1) (IPCC, 2007).

Table 1.1. Sources of direct and indirect agricultural GHGs

Sources of agricultural GHGs	CO_2e (Million tons)
Methane from cattle enteric fermentation (CH_4)	1792
Manures (CH_4 + N_2O)	413
Nitrous oxide from fertilized soils (N_2O)	2128
Fertilizer production (CO_2 + N_2O)	410
Biomass burning (CH_4 + N_2O)	672
Rice production (CH_4)	616
Farm machinery (seeding, tilling, spraying, harvest) (CO_2)	158
Irrigation (CO_2)	369
Pesticide production	72
Land conversion to agriculture	5900

1.9. IMPACTS OF CLIMATE CHANGE

The possible impacts of climate change include:

- Unprecedented heat waves.
- Cyclone - Intensity of storm would increase by at least 10%.
- Flood- precipitation may increase by 5-10% due to intensification of hydrological cycle.
- Draught.
- Decreased snow cover.
- Erratic monsoon.
- Rising sea level causing displacement along coast lines.
- Hot summers and warmer winters.
- Increased water salinity.
- More soil salinization- decreased soil fertility.
- Less cultivated land- less crop productivity with low quality agricultural produce.
- Increased incidence of insect pests and insect borne diseases in human beings.
- Effect on rain-fed agriculture, fisheries, livestock, and health.

Already over the last three decades, warming has had a discernible influence at the global scale on observed changes:

- Global warming
- Shifts in precipitation patterns
- Rising global mean sea level
- The retreat of glaciers
- Decline in the extent of Arctic sea ice coverage
- River run-off
- Increase in global mean ocean temperatures
- Widespread melting of snow and ice sheets
- Increased flood risk for urban areas and ecosystems
- Ocean acidification
- Extreme climatic events including heat waves

The impacts of climate change on agriculture have significant repercussions on livelihoods, food production, and the overall economy of countries, particularly those with agriculture-based economies in the developing world. At the same time, the agricultural sector holds significant climate-change mitigation potential through reductions of greenhouse gas (GHG) emissions as well as enhancement of agricultural sequestration (Fig. 1.4).

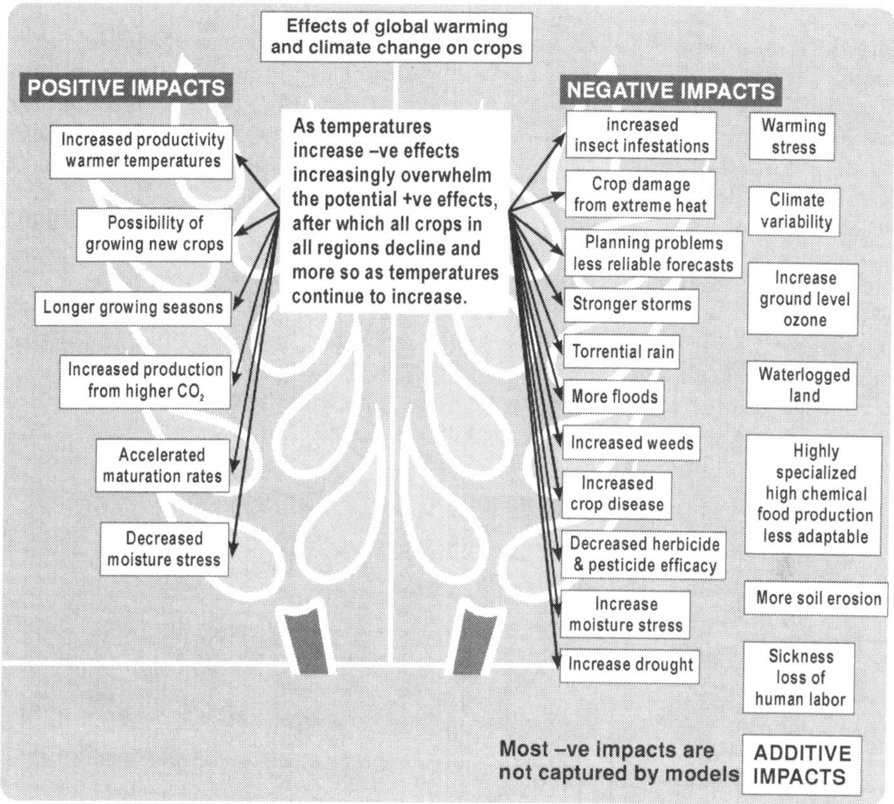

Fig. 1.4. Impacts of climate change on crops

1.9.1. Impacts on crop pests

Climate change scenario raises questions on the sustainability of current cropping systems which are likely to be impacted. Concomitantly, global food production must increase by 70 % to meet the projected demand of the world's population by 2050 (Chakraborty and Newton, 2011; Foley *et al.*, 2011). This difficult challenge is further complicated by climate change which influences the distribution of crop pests (i.e., insect and mite pests, disease pathogens, nematodes, and weeds) and their severity. Although pest management has played an important role in the marked rise in food production in the last 50 years (Oerke, 2006), most existing management strategies have been based on the indiscriminate use of pesticides in agriculture. Increasing awareness of health and environmental issues as well as increasingly stringent legislation on pesticide use, poses a particular compounded challenge to twenty-first century agriculture. Agriculture needs to meet food demands, satisfy sustainability goals, and adapt to climate change. Indeed, maintaining, or even increasing, crop yield while simultaneously reducing reliance on pesticides represents a challenging task (Hossard *et al.*, 2014) that is further complicated against the background of climate change scenarios.

In the current situation, two factors increase the risk of potential pest pressure in agriculture. While increasing anthropogenic impact, in particular increasing global movement of people and trade of plant commodities, continues to lead to the introduction of exotic pests into new regions. For example, South American tomato moth (*Tuta absoluta*) arrived in Europe (Spain) in 2006 and rapidly spread in Afro-Eurasia (Desneux *et al.*, 2011) (Fig. 1.5). It is currently widespread throughout most European countries. Climate change provides suitable conditions for such pests to adapt across the areas which were previously detrimental for their survival (Chakraborty, 2013; Fletcher *et al.*, 2010). In this way, climate change further accelerates pest pressure both in space and time and may have dramatic consequences throughout specific regions of the globe. In some regions, excessively dry, wet, or warm conditions may directly hamper crop production and consequently current crop protection methods may become ineffective (Chakraborty and Newton, 2011). Recent examples of pest adaptation to warmer temperatures have given rise to escalating disease epidemics in new areas where, in retrospect, the current cropping systems are excessively vulnerable to unforeseen threats. Even less direct effects, such as changes in crop or pest phenology can also have significant consequences in plant production.

Fig. 1.5. South American tomato moth (*Tuta absoluta*) and its damage on tomato.

Crop losses due to pests can account for up to or more than 40 % worldwide (Oerke, 2006). These values might significantly rise under changing climatic conditions whereby new and more aggressive pests affect levels and stability of crop yield thereby threatening food security. However, it is not possible to make estimates on overall potential crop losses due to climate change. This is in part because of the unpredictable adaptation of certain pests to a changing environment which creates situations of uncertainty (Chakraborty and Newton, 2011). Indeed, the main consequence of climate change and accelerated globalization is a heightened level of unpredictability regarding spatial and temporal interactions between weather, cropping systems, and pests (Chakraborty, 2013). This is the on-going crop protection challenge that farmers, advisers, researchers, and policy makers face.

In light of this, the reduction of pest impact has become more relevant than ever not only to produce sufficient food and agricultural commodities but also to reduce excessive input and unnecessary CO_2 emissions. Interactions between the changing environment, crops, cropping systems, and pests demand a new multi-disciplinary and integrative knowledge base for the development of viable adaptive strategies. The capacity to adapt is needed to reduce the impact of both invasive alien species as well as rapidly evolving resident pest populations

(Chakraborty, 2013; Roy *et al.*, 2014). Hence, there is a need for anticipatory strategies, enhanced adaptation capacity, and the development of more resilient cropping systems which require new insights and innovative organizational change. This book will mainly focus on scenarios of climate change and pest evolution. The objective is to identify relevant research needs and policy implications in order to understand how cropping systems and integrated pest management (IPM) strategies can be designed to better deal with invasive as well as existing but evolving pest populations under climate change. The emerging invasive alien pests evade biosecurity measures, and are likely to become established. There is a need for development of strategies to reduce their impact. For indigenous pests, there is a need for development of prevention, escape, or control strategies to manage plant-pest evolutionary processes.

1.10. CLIMATE CHANGE THREATS TO AGRICULTURE

- Increased temperature drastically affected the rice production due to decrease in crop duration in Philippines (10% reduction in yield in rice per 1.0° C rise in temperature).
- Reduction in yield due to warm days and nights: leads to less elongation of rice grain and lower quality of rice grain and higher chaffyness.
- An increase of 6° C in temperature and precipitation deficit of 300 mm reduced the maize yield by 36% in European Union. It will lead to a shift in cultivation of crops in non-traditional area and a change in cultural practices.
- Increase in temperature and carbon dioxide causes availability of less irrigated water and more pests and diseases of crops.
- IARI study on "Vulnerability of Indian agriculture to climate change" reveals - an increase in temperature by 1° C may reduce the yield of wheat, soybean, mustard, ground nut and potato by 3- 10%.
- If the trend of climate change continues in the same pace, the losses will be 10-40% by 2100.
- For 1° C rise in temperature in areas like Uttar Pradesh, Punjab and Haryana could amount to a loss of about 6 million tonnes of wheat annually.
- 0.5° C rise in winter temperature would reduce wheat yield by 0.45 t/ha.

1.11. CLIMATE CHANGE IMPOSED CHALLENGES FOR AGRICULTURE

The Indian climate has undergone significant changes showing increasing trends in annual temperature with an average of 0.56 °C rise over last 100 years (IPCC, 2007; Rao *et al.* 2009; IMD, 2010). Warming was more pronounced during post-monsoon and winter season with increase in number of hotter days in a year (IMD, 2010). Even though, there was slight increase in total rainfall received, number of rainy days decreased. The rain-fed zone of the country showed significant negative trends in annual rainfall (De and Mukhopadhyay, 1998; Lal, 2003; Rao *et al.* 2009). The semiarid regions of the country had maximum probability of prevalence of droughts of varying magnitudes (20-30%), leading to sharp decline in water tables and crop failures (Lal, 2003; IMD, 2010; Rao *et al.* 2009; Samra, 2003). By the end of next century (2100), the temperature in India is likely to increase by 1-5 °C (De and Mukhopadhyay, 1998; Lal, 2003; IPCC, 2007; IMD, 2010). According to the estimates of NATCOM (2004), there will be 15-40% increase in rainfall with high degree of variation in its distribution. Apart from this, the country is likely to experience frequently occurring extreme events like heat and cold waves, heavy tropical cyclones, frosts, droughts and floods (NATCOM, 2004; IPCC, 2007).

Being a tropical country, India is more challenged with impacts of looming climate change (Chahal *et al.*, 2008). Already, the productivity of Indian agriculture is limited by its high dependency on monsoon rainfall which is most often erratic and inadequate in its distribution (Chand and Raju, 2009). The country is experiencing declining trend of agricultural productivity due to fluctuating temperatures (Samra and Singh, 2004, Aggarwal, 2008; Joshi and Viraktamath, 2004), frequently occurring droughts and floods (Samra, 2003), problem soils, and increased outbreaks of insect-pests (Joshi and Viraktamath, 2004; Srikanth, 2007; Dhawan *et al.* 2007; IARI News, 2008; IRRI News, 2009) and diseases. These problems are likely to be aggravated further by changing climate which put forth major challenge to attain the goal of food security.

1.12. CLIMATE CHANGE SCENARIOS AND PREDICTION MODELS

A climate scenario is a reasonable description of the future climate based on a range of climatological relationships and assumptions of radioactive forcing (IPCC, 2007a). It can be visualized by global climate models (GCMs) and regional climate models (RCMs), which are complicated three dimensional mathematical representations to show the processes of interactions between the atmosphere, land surface, oceans and sea ice which resulted from the climate (Mearns, 2000). Climate projections should be considered as efficient methods to figure out the possible futures under the particular given emission scenarios rather than a forecasting tool (Suppiah *et al.*, 2007). GCMs are useful tools to simulate the important aspects of current and future climates although they still have significant errors (IPCC, 2007b). The GCMs with higher spatial resolutions can perform reasonable regional climate simulations, consequently, they provide climate scientists with the ability to acquire better insights into climate change impacts on a regional scale and estimate the impacts of the climate change on crop production (Mearns, 2000). Blenkinsop and Fowler (2007) suggested that RCMs have some trouble when they are used to reproduce the observed duration of low sensitivity monthly rainfall amounts.

Climate models need to be integrated with other modelling approaches to predict climate vulnerability and climate parameters such as rainfall and temperature. Metzger *et al.* (2005) provided an advanced terrestrial ecosystem analysis and modelling (ATEAM) framework to qualify the vulnerability of climate change using the GCM, HadCM3 (Hadley centre climate model 3) to predict climate change in 2080. Suppiah *et al.* (2007) used the statistical method to select 15 best models to derive the annual and seasonal average projections of rainfall and temperature in the whole of Australia.

Khan (2008) discussed climate variability and droughts in Australia according to historical climate data and suggested some possible policy approaches to deal with the extreme climate change variability, such as to adjust water allocations of surface and groundwater using prediction models, to improve water use efficiency in agriculture and to build a national legal framework to manage water resources in accordance with anticipated climate change impacts on water resources. Groves and Lempert (2007) provided a new analytical method based on Robust Decision Making to quantify the SRES scenarios (Special Report on Emissions Scenarios) developed using the scenario-axes method for decision makers. Fowler *et al.* (2005) studied the Neyman–Scott Rectangular Pulse (NSRP) rainfall scenario generation model to derive rainfall climate information using the stochastic simulation method. Xu *et al.* (2006) used the PRECIS climate model to predict mean temperature and rainfall in China for 2080s under B2 scenarios, the results show that the mean temperature and rainfall will increase across China.

GCMs have been used to predict climate scenarios and impacts in many cases using the downscaling approach (Thomas, 2008). GCM data typically have a low resolution of several degrees, lack the spatial and temporal precision necessary for detailed regional analysis and in the many cases have errors to simulate even the present-day climate (Bonan *et al.*, 2002). GCMs have uncertainties in predicting future climate data, while they can provide a reasonable accuracy about large-scale features and other variations due to climate forcing (Salinger et al., 2000). Since each climate model has its own uncertainty, more than one climate model would be better to deal with the accurate projection problem (Cuculeanu *et al.*, 2002). Thus, further study on GCMs will be focused on how to improve the sensitivity of GCMs and to evaluate feedback of the factors influencing climate model projection results. Meanwhile, future climate models will consider more detailed factors and have higher precise latitude and longitude in order to reduce the spatial and temporal errors for the accurate regional climate study.

1.13. CLIMATE-RESILIENT AGRICULTURE

Since climate change poses complex challenges like multiple abiotic stresses on crops and livestock, shortage of water, land degradation and loss of bio-diversity, a focused and long term research is required to find solutions to the problems. The necessary infrastructure to carry out basic and strategic research has to be put in place. At the same time, there is a scope to improve the resilience of agriculture by application of existing knowledge and technology on farmers' field as a holistic package. Hence, there is a need to develop improved technologies through short term and long term research and also demonstrate the existing technologies on farmers' fields for enhancing the resilience.

Sustainable agriculture seeks to transform agriculture into an ecologically sustainable climate resilient production system while at the same time, exploiting its fullest potential and thereby ensuring food security, equitable access to food resources, enhancing livelihood opportunities and contributing to economic stability. Climate-resilient agriculture (CRA) contributes to the achievement of sustainable development goals. It integrates the three dimensions of sustainable development (economic, social and environmental) by jointly addressing food security and climate challenges. It is composed of three main pillars:

- Sustainably increasing agricultural productivity and incomes.
- Adapting and building resilience to climate change.
- Reducing and/or removing greenhouse gases emissions, where possible.

CRA is an approach in developing the technical, policy and investment conditions to achieve sustainable agricultural development for food security under climate change. The magnitude, urgency and broad scope of the effects of climate change on agricultural systems create a compelling need to ensure comprehensive integration of these effects into national agricultural planning, investments and programmes. The CRA approach is designed to identify and operationalize sustainable agricultural development within the explicit parameters of climate change.

Achieving the transformations required for CRA and meeting these multiple objectives requires an integrated approach that is responsive to specific local conditions. Coordination across agricultural sectors (e.g. crops, livestock, forestry and fisheries) as well as other sectors, such as with energy and water sector, development is essential to capitalize on potential synergies, reduce trade-offs and optimize the use of natural resources and ecosystem services.

This approach also aims to strengthen the livelihoods and food security, especially of smallholders, by improving the management and use of natural resources and adopting appropriate methods and technologies for the production, processing and marketing of agricultural goods. To maximize the benefits and minimize the trade-offs, CRA takes into consideration the social, economic, and environmental context where it will be applied. Repercussions on energy and local resources are also assessed. A key component is the integrated landscape approach that follows the principles of ecosystem management and sustainable land and water use.

1.13.1. Adaptation

Climate change adaptation involves taking action to adjust to, or respond to, the effects of changes in climate, such as reduced rainfall or rising sea level. Adaptations to climate change exist at various levels of agricultural organization. In temperate regions, farm-level adaptations include changes in planting and harvest dates, tillage and rotation practices, substitution of crop varieties or species more appropriate to the changing climate regime, increased fertilizer or pesticide applications, and improved irrigation and drainage systems. Governments can facilitate adaptation to climate change through water development projects, agricultural extension activities, incentives, subsidies, regulations, and provision of crop insurance. There is considerable scope for agricultural adaptation throughout the food chain, for example, better post-harvest storage and distribution of food, to ameliorate the gap between good and poor years.

Easterling *et al.* (2007) describe a range of options, at the level of autonomous adaptation, for cropping and livestock systems:

- Different varieties or species with greater resistance to heat or water stress, or adapted phenology (maturation times and responses).
- New cropping practices, including adjustments in timing and locality of crop production, and changed water and fertilizer management to maintain yield quality and quantity.
- Greater use of water conservation technologies, including those to harvest water and conserve soil moisture, or, in flood-prone areas, water management to prevent water logging, erosion, and nutrient leaching.
- Diversification of on-farm activities and enhancement of agro-biodiversity, with greater integration between livestock and cropping systems.

- Adapted livestock and pasture management, including re-matching stocking rates and timing with pasture production, new varieties and species of forage and livestock, updated fertilizer applications, and using supplementary feeds and concentrates.
- Improved management of pests, diseases and weeds, for example, through integrated pest management, new crop and livestock varieties, improved quarantine, and sentinel monitoring programmes.
- Better use of short-term and seasonal climate forecasting to reduce production risk.

1.13.2. Mitigation

Climate change mitigation refers to actions that aim to reduce the amount of climate change, typically by limiting the future increases in concentrations of greenhouse gases in the atmosphere - either by reducing emissions from a wide range of our industrial and agricultural activities, or by increasing the amount of CO_2 taken up and stored in natural 'sinks' such as forests and soils. Actions that reduce greenhouse gas emissions in many cases also improve our preparedness for future climate change.

Smith *et al.* (2007) distinguish seven broad sets of options for mitigating GHG emissions from agricultural ecosystems:

- Cropland management, including nutrient management; tillage and residue management; water management (irrigation, drainage); rice paddy management; agroforestry; set-asides; crop rotations; and land-use change.
- Grazing land management and pasture improvement, including grazing intensity, increased productivity (e.g., fertilization), nutrient management, fire management, and species introduction (including legumes).
- Management of organic soils, avoiding drainage of wetlands included.
- Restoration of degraded lands, including erosion control, organic amendments, and nutrient amendments.
- Livestock management, including improved feeding practices, dietary additives, longer term structural and management changes, and animal breeding.
- Manure management, including improved storage and handling, anaerobic digestion, and more efficient use as a nutrient source.
- Bioenergy, including energy crops (solid, liquid, biogas, and residues).

The potential for synergies among these land-based agricultural mitigation actions that promote food security is particularly high for specific practices such as adopting improved crop varieties (e.g., with higher water-use efficiency), breeding livestocks to increase sustainable productivity of meat or milk, avoiding bare fallow lands and changing crop rotations to incorporate food producing cover crops and legumes, adopting precision fertilizer management, improving forage quality and quantity on pastures, expanding energy-efficient irrigation and water conservation techniques (e.g., in rice systems), and implementing agroforestry that does not take significant amounts of land out of food production.

Technical options for mitigation in agriculture needs to be locally appropriate. For example, although land management presents the major opportunity for mitigation in the agricultural sector globally, other interventions, such as improved livestock breeding and feeding, or manure management, may be more effective for particular countries, farming systems, or agro-ecological zones.

1.14. CONCLUSIONS

In the last decade, an overwhelming consensus has emerged among scientists that the world has entered an era of rapid global climate change, much of which is attributable to greenhouse gas (GHG) emissions from human activities. Rapid global climate change is expected to impact agriculture by causing shifts in temperature, precipitation, soil quality, pest regimes, and seasonal growth patterns. The exact nature and degree of these changes for any given region will be difficult to predict. At the same time that the agricultural sector is impacted by climate change. Research indicates that current agricultural activities are a significant source of greenhouse gases that aggravate climate disruption. The amount of GHGs emitted from an agricultural operation depends on its system and management.

To cope with climate change that is likely to be both rapid and unpredictable, agricultural systems must be resilient and able to adapt to change. Resilient agriculture systems are those that are more likely to maintain economic, ecological and social benefits in the face of dramatic exogenous changes such as climate change and price swings. In the face of uncertainty, food production systems should be established which are diverse and relatively flexible, with integration and coordination of livestock and crop production.

Sustainable and organic agricultural systems can help reduce agricultural GHG emissions through energy conservation, lower levels of carbon-based inputs, lower use of synthetic fertilizer and other features that minimize GHG emissions and sequester carbon in the soil. Agricultural land can serve as a sink for GHG emissions, especially through soil carbon sequestration, which could help moderate climate change. But agricultural land can serve as an effective GHG sink over the long-term only if agricultural systems are adopted which improve overall soil quality and provide for relatively stable GHG reduction or sequestration. Agricultural crop and forage production system features include, among others, fertilizer use and efficiency, nitrogen sequestration and overall GHG emissions of associated livestock production systems.

2

Climate Change, Crop Pests and Food Security

"Crop protection is much more important than production technology"

- Late Prof. S. Pradhan, an eminent Entomologist from India

2.1. INTRODUCTION

The effect of climate change on potential biotic stresses is well known. Climate change can induce elevated temperatures, droughts, rising sea levels, floods, changes in wind patterns and abnormal weather. These events can cause changes in ecosystem balance that can disrupt ecosystem services and favour crop pest development resulting in heavy crop losses, heavy/ injudicious pesticide use and in extreme cases, poverty, due to severe disruption of livelihoods. Changes in CO_2 concentration, increased temperature, increased drought, changing on rainfall patterns have all been documented in the IPCC reports and in several recent publications. Plants will be affected because their capacity to reproduce, grow and produce enough yields is linked to the surrounding environment. The effects of climate change will be different in various parts of the world and there exist several studies from different angles and perspectives that describe how climate changes will impact on agriculture and some of the proposed solutions to help adapting to climate changes. What is much less known and investigated is how pests and diseases, and therefore the biotic stress, will change under climate change scenario and how this will, in turn, affect agriculture. All in all the whole issue of adapting to climate changes has to do with food security which has a very strong social component, including education, land rights, food prices, poor markets, poor market access among others. Nonetheless, having the capacity of producing enough food is a prerequisite to achieve food security and pests and diseases can reduce productivity of major crops by up to 50% (Oerke, 2006).

To put things on right perspective, food production needs to increase by 70% by 2050 due to population growth (9.5 billion by 2050) and increasing food demand. This needs to be achieved under increased land degradation and land conversion, under reduced water availability. Increase in oil prices will result in higher cost of fertilizers thus making them less affordable for farmers. It is evident that pests and diseases will also change and mutate under climate change. Like plants, they will be subject to natural selection and mutation to adapt to the new conditions. How this will impact on agriculture?

Crop plants live in a very complex ecosystem in competition with neighbouring plants including weeds. Both are supported and/or attacked by viruses, bacteria, fungi, insects, mites, spiders, nematodes, birds etc. All of these species interact with each other. It is estimated that globally 70,000 pest species, including 9,000 insect and mites, 50,000 plant pathogens and 8,000 species of weeds exist. About 10% of these 70,000 are considered major pests. The climate change will increase the challenges from pests.

2.2. CROP LOSSES

There have been several efforts to provide a measure of global crop losses by weeds, insects, and diseases. The most recent and comprehensive of these estimates are those made by Oerke (2006) (Table 2.1). Estimates on potential and actual losses despite the current crop protection practices are given for wheat, rice, maize, potatoes, soybeans, and cotton for the period 2001–03 on a regional basis (19 regions) as well as for the global total. Among crops, the total global potential loss due to pests varied from about 50% in wheat to more than 80% in cotton production. The responses are estimated as losses of 26–29% for soybean, wheat and cotton, and 31, 37 and 40% for maize, rice and potatoes, respectively. Overall, weeds produced the highest potential loss (34%), with animal pests and pathogens being less important (losses of 18 and 16%). The effiacy of crop protection was higher in cash crops than in food crops. Weed control can be managed mechanically or chemically, therefore worldwide effiacy was considerably higher than for the control of animal pests or diseases, which rely heavily on synthetic chemicals. Despite a clear increase in pesticide use, crop losses have not significantly decreased during the last 40 years. However, pesticide use has enabled farmers to modify production systems and to increase crop productivity without sustaining the higher losses likely to occur from an increased susceptibility to the damaging effect of pests.

Table 2.1. Estimated crop loss potential of weeds, animal pests (arthropods, nematodes, rodents, birds, snails and slugs), disease pathogens (fungi, bacteria and viruses) due to pest groups in six major crops worldwide, in 2001-03

Crop	Attainable production (Million tons)	Crop losses (%) due to				
		Weeds	Animal pests	Pathogens	Viruses	Total
Wheat	785.0	23.0	8.7	15.6	2.5	49.8
Rice	933.1	37.1	24.7	13.5	1.7	77.0
Maize	890.8	40.3	15.9	9.4	2.9	68.5
Potatoes	517.7	30.2	15.3	21.2	8.1	74.9
Soybeans	244.8	37.0	10.7	11.0	1.4	60.0
Cotton*	78.5*	35.9	36.8	8.5	0.8	82.0

Source: (Oerke, 2006).

* Seed cotton

Nationally, pests are estimated to destroy about one third of our crops and are an increasingly serious constraint to crop production, in spite of the advances in pest control technology over the last half century. Agricultural trends are influencing the incidence and importance of pests. First, the expansion of worldwide trade in food and plant products is spreading the impact of weeds, insects, and diseases. Second, changes in cultural techniques, particularly intensification of cropping, reduction in crop rotations, and increase in monocultures, encourages the activity of pests.

2.3. IMPACT OF CLIMATE CHANGE ON CROP PESTS

During the last four decades, there have been major advances in food grain production in India due to the adoption of green revolution technology, which included high yielding varieties and intensive use of chemical fertilizers and pesticides. Use of high yielding varieties, chemical fertilizers and pesticides have invited several new pest problems along with deterioration of soil health and environment, which are currently being added with climate change/ global warming. In major agricultural crops, pests (insects, mites, etc.), diseases (fungal, bacterial, viral and nematode diseases) and weeds cause considerable yield losses (30-40%) every year in India. Climate change in terms of increased temperature, elevated CO_2 and erratic rainfall leading to draught, flood, submergence, salinity situations, has direct and indirect effects on the crop pests.

Increased temperature and rainfall directly affect the crop pests, whereas elevated CO_2 has indirect effect on these pests through hosts/ quality of the food materials produced by the host plants. Under both the circumstances, life activities of the crop pests are affected. In direct effects of climate change on crop pests, pest growth and development, reproduction and survival and indirectly affects breakdown in temperature sensitive host plant resistance. Plant defensive systems are lowered under stress, making them more susceptible to pest attacks. Changes in temperature, rainfall, and wind speed pattern influence their migratory behaviour. Natural enemy and host insect-pest population may respond differently. Parasitism could be reduced, if host population emerges and passes through vulnerable life stages more quickly at higher temperature. However, such pest problems will vary with sub-ecologies/ locations of the crop fields, which need to be tackled using appropriate IPM strategies having holistic approach and eco-friendly interventions based on the climate resilient suitable crop varieties tolerant to biotic stresses. Many people believe that global warming as predicted would increase pressure from weeds, pests and diseases. Higher temperatures and longer growing seasons could result in increased pest populations in temperate regions of Asia. Warmer winter temperatures would reduce winter kill, favouring the increase of insect populations. Overall temperature increases may influence crop pathogen interactions by speeding up pathogen growth rates which increases reproductive generations per crop cycle, by decreasing pathogen mortality due to warmer winter temperatures, and by making the crop more vulnerable (Cruz *et al.*, 2007).

Climate change will affect crop protection challenges. In cooler latitudes, global warming brings new species but others may disappear. Whether or not new species translate into pest problems is uncertain. Invasive species are often brought to other places by global trade of food and goods.

Climate change might have an influence on pesticide use due to presence of weeds, diseases, pests and their natural enemies. The latter factors are influenced by the weather and in the midterm by climatic changes (Goudriaan and Zadoks, 1995). Tilman *et al.* (2001) foresee a 2.4 to 2.7-fold increase in pesticide use by 2050. Chen and McCarl (2001) investigated the relationship of temperature, precipitation and pesticide costs for several crops in the USA and concluded that increases in rainfall leads to increases in average pesticide costs for corn, cotton, potatoes, soybeans, and wheat; while hotter weather increases pesticide costs for corn, cotton, potatoes, and soybeans but decreases the cost for wheat. In addition, increased atmospheric carbon dioxide is expected to alter the nutritional makeup of crops, thereby affecting the severity of attack from insects and disease organisms (Coviella and Trumble, 1999).

In general, however, most pest species are favoured with warm and humid conditions. Pest infestations often coincide with changes in climatic conditions, such as early or late rains, drought, or increases in humidity, which, in themselves, can reduce yields (Table 2.2).

Table 2.2. Serious crop pest epidemics critically influenced by climate change

Event effects	Pest damage to crops
Floods and heavy rains	Increased moisture benefits epidemics and prevalence of fungal leaf pathogens. • Rice leaf blight caused great famine in Bengal (1942), 2 million people died. • Wheat stripe rust outbreak in major production regions of China contributed to the 1960s famine. • Fungal epidemics in corn, soybean, alfalfa, and wheat in the US Midwest (1993). • Mycotoxin produced by wheat scab (*Fusarium* sp.) reached a record high in the US Great Plains (1993). • Humid summers drive epidemics of gray leaf spot of maize in Iowa and Illinois (1996). Water-induced soil transport increases dissemination of soil-borne pathogens to non-infected areas. • Outbreaks of soybean sudden death syndrome in the north central US states (1993). Continuous soil saturation causes long-term problems related to rot development and increase damage by diseases. • Crazy top and common smut in maize.
Drought	Water stress diminishes plant vigour and alters carbon-to-nitrogen ratios, lowering plant resistance to nematodes and insects. Attack by fungal pathogens of stems and roots are favoured by weakened plant conditions. Drought promotes insect outbreaks. • Outbreak of soybean cyst nematode correlated to drought conditions in the north central US states (1990). • Summer locust outbreak correlated to drought in Mexico (1999).

Event effects	Pest damage to crops
	Dry and warm conditions promote growth of insect vector populations, increasing viral epidemics.
Storms and air currents	Air currents provide large scale transportation for disease agents (e.g., spores of fungi) or insects from overwintering areas to attacking areas. • The spread of the stem rust fungus that overwinters in Mexico and Texas is always favoured by moist southern air currents. • The southern leaf blight of corn spread from Mississippi to the Midwest by air currents of a tropical storm in the Gulf of Mexico during 1970.
Warm winters	Warm winters increase overwintering populations of all pests. • Data reported for the European corn borer; wheat scab, and wheat rust. Increase overwintering populations of insect vectors. • Increase population of aphids that carry the soybean mosaic virus.

2.4. PEST EVOLUTION UNDER CLIMATE CHANGE

Climate change is generally associated with increased CO_2 levels, higher seasonal temperature profiles, and more precipitation (Pachauri and Reisinger, 2007). These effects of global climate change will affect the distribution of pests, although whether such changes will cause more frequent and more severe outbreaks cannot be answered in general terms. The fertilization effect of higher CO_2 levels might increase photosynthesis and crop yields (Ainsworth and Long, 2005). However, it is widely considered to also affect plant morphology, canopy structure, and hence micro-climate and the resident micro-floral populations in such environments (Eastburn *et al.*, 2011). Thus far, different studies have given conflicting results showing that elevated CO_2 levels may have negative, neutral, or positive effects on pests (Luck *et al.*, 2011). Indeed, the Climapest project, covering 20 different crops throughout Brazil, concluded that climate change can "decrease, increase, or have no impact on plant diseases, pests, or weeds depending on the region and the time period". For example, experiments showed that in an environment with higher CO_2 level, some rust-caused diseases seemed to increase while others decreased in severity (Ghini *et al.*, 2011).

The effects of temperature change are manifold and will affect crops and plant diseases in many different direct and indirect ways. For example, a 5.8 °C increase in temperature is expected to reduce suitable coffee-growing areas by more than 95% in several regions of Brazil (Assad *et al.*, 2004). Higher temperatures are often associated with increased rates of development, growth, and reproduction and hence the fear that plant diseases will become more severe makes intuitive sense. Warmer climate crops, such as those found in the tropics, tend to suffer greater pest problems than crops in more temperate regions. Yet, it is not easy to predict which combinations of crops and pests will tend to become problematic in the future and in which areas (Shaw and Osborne, 2011). On the other hand, higher levels of precipitation in certain areas could be useful for plant growth, although rainfall and moisture are also vital for the occurrence of fungal and bacterial species (Huber and Gillespie, 1992). Infection, sporulation, and spread of many fungi are moisture-dependent, and outbreaks are often triggered by long periods of wetness (Juroszek and von Tiedemann, 2011).

The body of knowledge on the effects of climate change, new exotic pest incursions, and globalization on crop protection points to a higher level of unpredictability regarding future crop-pest-climate interactions and increased frequency and amplitude of climatic fluctuations. Overall, there will be accelerations in the rate of exotic pests entering and establishing in new areas as well as an increased rate of evolution of existing pest populations. In such situations, the priority may shift from maximizing yields to achieving more stable levels of production and avoiding total crop failures. Hence, IPM strategies that are dynamic and internally diversified to locally adapted cropping systems, and more resilient to fluctuating weather conditions and new and evolving pests, should be developed accordingly.

A wide range of pests including insects, fungal and bacterial pathogens, and weeds may benefit from changing climate. The way in which climate change impacts their evolution can be summarized as below.

2.4.1. Weeds

Worldwide, weeds have been estimated to cause annual crop production losses of about 12% (Oerke *et al.*, 1994). In the US, annual losses in crop production due to weeds have been valued at approximately US$ 12 billion, amounting to some 10% of potential production. Large efforts are made to limit these damages through a variety of weed control measures.

Around the world, more human labour is expended in hand weeding than in any other agricultural task, and most cultivation and tillage practices are designed to aid in weed control. The chemical industry manufactures herbicides, which, next to fertilizers account for the largest volume of chemicals applied to crops. Over US$ 9 billion are spent on weed control every year in the US (USDA, 1999).

Weeds, which are better adapted to arid conditions than crops, will provide increased competition for moisture, nutrients, and light. Herbicidal controls are less effective under hot and dry conditions, but mechanical cultivation is more effective (Pimentel *et al.*, 1991). Another problem with herbicides applied under arid conditions is that they accumulate in the soil, which can lead to serious environmental problems.

Overall, US crop losses due to weeds are projected to rise from 5 to 50 per cent for selected crops. Similarly, warm and moist conditions projected for Africa are expected to increase crop losses due to weeds (Pimentel *et al.*, 1993).

Elevated CO_2, changes in temperature and precipitation patterns may affect weeds as much as crops. Higher CO_2 will stimulate photosynthesis and growth in C_3 weeds and C_3 crops, and reduce transpiration and increase water use efficiency in both C_3 and C_4 weeds and crops. Higher temperatures can possibly offset some of the benefits of elevated CO_2 for both, weeds and crops. High temperatures sometimes limit reproductive development and global warming may decrease reproductive output in such situations despite an increase in CO_2. It is unclear whether this is more likely to occur in C_3 than C_4 species, but if it were, it could alter weed community compositions and affect crop/weed interactions (Bunce and Ziska, 2000).

This would imply that weeds and crops both benefit or lose on the same scale. However, weeds are usually already very competitive due to greater genetic variation and physiological plasticity, otherwise they would not cause yield losses. Hence they may gain more advantages from climate change than crops (Bunce and Ziska, 2000).

In temperate regions, global warming will affect the growth and marginally affect phenology, and influence the geographical distribution of weeds. Weed species of tropical and subtropical origins, currently restricted to the southern regions, may expand northwards (Patterson, 1999).

Overall, invasive plants have been shown to severely affect plant communities. A recent meta-analysis estimated that invasive plants reduced fitness of native plant communities on average by 41.7 %, growth by 22.1 %, species abundance by 43.5 %, and species diversity by 50.7 % (Vilà *et al.*, 2011). More specifically to crops, weeds compete with them for light, nutrients, and water. It is likely that the effect of environmental disturbances such as rising CO_2 or increasing temperatures, will be manifested as a change in the competitiveness between crops and weeds. Herbicide performance is also expected to reduce at elevated CO_2 levels (Ziska, 2010). Being inherently adapted to the prevailing conditions, it is likely that weeds will respond more favourably to climate change than crops. This has been confirmed in the majority of studies examining the response of crops and weeds to increasing levels of CO_2. However, the physiological characteristics of crops and weeds being either a C_3 or C_4 plant will also determine their respective responses to CO_2 (Hatfield *et al.*, 2011; Ziska, 2011). While CO_2 is mainly expected to influence crop-weed competition, the most likely effect of a rising temperature is the northwards expansion of native and invasive weed species (Hatfield *et al.*, 2011).

In comparison to many diseases and insect pests, weeds are relatively non-mobile, and their spread and establishment into new regions is expected to take longer than for other pest groups. However, as projected, warming may exceed maximum rates of plant migration in post-glacial periods (Malcolm *et al.*, 2002), it could favor the most mobile weed species. Many agricultural weeds have characteristics associated with long-distance dissemination and a wide geographic range such as small seeds, phenotypic plasticity, and a short juvenile period that suggests that agricultural weeds may be among the fastest to spread (Dukes and Mooney, 2000).

Most of the current information on the impact of climate changes on future weed distribution is based on predictions performed using bioclimatic envelope models (Hyvönen *et al.*, 2012). However, these models have a limited value for predicting the long-term changes in the composition of weed communities (Thuiller *et al.*, 2008). Firstly, climatic conditions alone cannot predict range shifts without considering biotic and management practices such as soil type, tillage practice, and cropping sequence. Secondly, the bioclimatic envelope models assume that weed species will only gain foothold in areas that have a climate similar to its native range. However, this assumption has recently been challenged by studies revealing that the range expansion of alien species exceeded that predicted by the models used (Clements and DiTommaso, 2011). Rapid genetic evolution and/or phenotypic plasticity may explain these observations (Clements and DiTommaso, 2011; Maron *et al.*, 2004).

For an agricultural weed to be successful, it must adapt not only to a given climate but also to the constraints of the cropping system. Thus, the range shift of a weed species should not only be studied in the context of climate but should also include biotic parameters which have rarely been the case so far. The term "niche" is often used in the context of range shifts. According to Hutchinson (1957), the abiotic factors, and especially climate, determine the fundamental niche of a species while the interactions with the biotic factors together with the dispersal properties determine the realized niche of a species. Although the concepts of fundamental and realized niches are useful also for understanding range shifts of agricultural weeds, they do not take into account the economic impact of agricultural weeds. To overcome this shortcoming, McDonald *et al.* (2009) introduced the concept of the "damage niche." The latter is defined as "the environmental conditions that make specific weed abundant, competitive, and therefore damaging the production of particular crops." Determining the damage niche of a weed species requires information on the crop-weed interactions and is thus region and cropping system specific.

Despite its relevance to the impact of climate change on the distribution and impact of agricultural weeds, the "damage niche" concept has only been applied in a few other studies (Stratonovitch *et al.*, 2012; Bradley, 2013). Using the damage niche-based concept in conjunction with the HadCM3 projections for the periods 2046–2065 and 2080–2099, Stratonovitch *et al.* (2012) found a possible northward shift in the range of black grass (*Alopecurus myosuroides*) in winter wheat (Fig. 2.1). Moreover, they also observed a local-scale difference due to variations in soil type and water holding capacity. The competitive balance was predicted to shift in favour of the crop due to its deeper root system making the crop less prone than the competing weed to the more frequent drought stress events predicted. In short, this means, in this context, the damage niche was predicted to reduce, rather than increase, under climate change.

Fig. 2.1. Black grass (*Alopecurus myosuroides*) infestation in winter wheat crop that is subjected to a possible northward shift in Europe

2.4.2. Insect pests

Climate and weather patterns are of primary importance for the distribution, development, and population dynamics of insects. Although insects may regulate body temperature within a certain range, many species—especially smaller herbivores—are almost exclusively ectotherms (Willmer *et al.*, 2000). Insect physiology is therefore primarily driven by temperature which means that phenology, reproduction, and developmental rates significantly change when populations are exposed to different climatic regimes. Bioclimatic envelopes (the range and pattern of climatic parameters) and day length largely determine the potential distribution of insect species when host plants are available.

Insects are constantly adapting their life history patterns to the available climatic regimes or season lengths. Evolutionary adaptations in life cycles of insects occur so rapidly that ongoing climate changes are already reflected in according adaptations such as daylight signals that induce diapause (Bradshaw and Holzapfel, 2011). Increased pest outbreaks were judged as "virtually certain" in the previous IPCC Synthesis Report (IPCC, 2007), while according to the IPCC Synthesis Report 2014, increases in the frequency or intensity of ecosystem disturbances including pest outbreaks in many parts of the world are in some cases due to climate change (IPCC, 2014). Aside from population dynamics *per se*, current climatic projections predicts that the distribution of insect species will shift from lower latitudes to pole wards and from lower to higher altitudes (Hadley Centre, 2007). Such range shifts in insect distributions have been already observed in nature as a response to global warming (Battisti *et al.*, 2005; Gutierrez *et al.*, 2009; Parmesan, 2006; Walther *et al.*, 2002). For example, the stinkbug, *Acrosternum hilare* in England and Japan has shifted its range by more than 300 km northward with a temperature increase of only 2°C (Trumble and Butler, 2009). Likewise, the mountain pine beetle, a major forest pest in the USA and Canada, has extended its range northward by approximately 300 km when temperature rose by only 2°C (Logan and Powell, 2001).

Lower latitudes usually have more complex pest assemblages, i.e., more pest species and antagonists per crop than higher latitudes toward the poles. Climatic shifts are therefore expected to increase the number of harmful species in temperate regions, especially in those of the northern hemisphere (Logan and Powell, 2001; Parmesan, 2006). For example, higher temperatures may extend the distribution range of the European corn borer (*Ostrinia nubilalis*) to maize areas previously free of this species. This may not only lead to crop losses but also to more secondary infections by *Fusarium* moulds and consequent mycotoxin problems in this crop. Pests may also spread or invade into areas and find new host plants that were not infested before. Being the driving factor for establishment in new distribution ranges, global warming may also increase the traveling speed of invasive pests by decreasing the function of mountain ranges as cold barriers (Aluja *et al.*, 2011).

Climate change will lead also to phenological changes. The earlier onset and increased length of the growing season also means that indigenous multivoltine species may be able to complete more generations per year leading to higher population levels at the season's end (Ziter *et al.*, 2012). In this way, formerly univoltine species may also become bi- or multivoltine as reported for codling moth

(*Cydia pomonella*), a key pest in apple throughout several parts of Europe (Stoeckli *et al.*, 2012). Shifts in phenology due to changing temperatures may also change the synchrony between host plants, pest species, and their natural enemies. This may lead to unexpected interactions the outcome of which, in terms of more or less damage, is hard to predict. Furthermore, invasive species may have more opportunity to become established in areas that were formerly unsuitable as a habitat due to the climatic conditions (Trumble and Butler, 2009).

Besides higher population pressure, the potential for damage is expected to last longer within the growing season and crops will have to be protected accordingly. Established management strategies that are focused on low pesticide residues will have to be largely adapted to avoid the possible resistance build-up to plant protection chemicals thereby ensuring their sustainability in future use (Samietz *et al.*, 2014). Here, further knowledge about pest biology, especially the potential adaptation of insect pests to climatic changes, is necessary and needs to be combined in simulation studies in order to prepare management systems that can withstand the challenges of global warming.

Besides higher temperatures during the growing season, elevated winter temperatures are equally significant. The distribution of many species is limited by their low-temperature tolerance. Given milder winters under climate change, more species will be able to survive and colonize crops. Some species may skip their sexual stage and have new asexual generations throughout the year which may lead to early high population levels. There are a number of instances worldwide where aphids have shifted from holocyclic (primary and secondary hosts) to anholocyclic (only secondary hosts, i.e., only the crop plants) form (Radcliffe and Ragsdale, 2002). Such a shifting process in insect phenology is likely to have an impact on crops and increase pest management costs. An example of altered aphid phenology comes from Scotland whereby oilseed rape production was introduced with farmers growing vernalized oil seed rape (autumn planted with the crop remaining green during the winter). It was found that green peach aphids could overwinter as anholocyclic forms (asexual generations) on oilseed rape and colonize potato earlier than those aphids that overwintered on the primary host (Fenton *et al.,* 1998; Woodford, 1998). Increased temperatures allow aphids to remain on secondary hosts (either crop or non-crop), which can result in rapid colonization of crops early in the season. The phenology of pests will also be modified by increased temperatures. A 20-year study in the UK demonstrated that winter temperature was the dominant factor affecting aphid phenology and that a mere 1 °C rise in winter temperature advanced the migration phenology by 19 days (Zhou *et al.*, 1995). Overall, higher temperatures would allow insect populations to colonize crops earlier and develop faster. In this way, crop damage may occur more rapidly than what is currently observed. Finally, as some of these insect species are also vectors of viruses (such as whitefly, *Bemisia tabaci*), crops may also become more vulnerable due to earlier and more severe infections.

Besides temperature, elevated CO_2 levels were reported to favour insect population growth rates, such as those for pea aphids, by active elicitations of host responses that promote amino acid metabolism in both the host plant and its bacteriocytes (Guo *et al.*, 2013). In addition, rising CO_2 levels will increase the carbon-nitrogen balance in crop plants and hence their structure and palatability for leaf chewing insects. Each species may respond differently to all these changes, and this will affect concentrations of constitutive and induced defensive chemicals in plants, insect feeding behaviour, competition between pests, interactions among pests and natural enemies, and ultimately damage to crops (Trumble and Butler, 2009). However, our current knowledge on these aspects remains largely fragmentary.

Over recent decades, there has been constant growth in the number of reports of invasive alien species reaching new areas (Trumble and Butler, 2009). Invasions are primarily due to increasing anthropogenic impacts at global level, i.e., the increasing and rapid movement of people and agricultural commodities. However, warmer temperatures also mean that insects which could not previously survive in potential new areas are able to establish and colonize once introduced. A practical example is the potato psyllid (*Bactericera cockerelli*) which has migrated to California several times over the past century with the populations never lasting more than a year. Cool winter temperatures meant it could only survive in Mexico and southern Texas during this period of the year. However, in 2000, it became established in California inducing substantial losses in tomato, potato, and bell pepper crops (Liu and Trumble, 2007).

New emerging species often spread into completely new ecological settings where most of their natural enemies are missing. An example is the introduction of the B-biotype (*B. tabaci*) in Brazil which was responsible of vectoring of viruses present in native plants only onto cultured tomato crops, leading to new virus diseases in this crop (Fernandes *et al.*, 2008). Whether antagonists will also extend their ranges, thereby following the herbivores, is unknown for the moment. This is true especially in the case of introduction via globalization/international trade rather than the mere extension of an area of distribution via climate change. In-depth ecological knowledge about both pests and their natural enemies in the country of origin may help toward better preparedness and to develop more robust cropping systems. This may apply to both already introduced pests and vectored diseases such as *Tuta absoluta* (Desneux *et al.*, 2010; Zappalà *et al.*, 2013), *B. tabaci* (Tahiri *et al.*, 2006), tomato yellow leaf curl virus (TYLCV), *Plutella xylostella*, and its parasitoids (Sarfraz *et al.*, 2005), and to potential invasive species such as *Bactrocera* fruit flies (Stephens *et al.*, 2007). Robust cropping systems were already developed and tested in the tropics against a range of such pests (Licciardi *et al.*, 2008; Vayssieres *et al.*, 2009).

Insect pests in agricultural systems are the major cause of damage to yield quantity. Insect habitats and survival strategies are strongly dependent on patterns of climate. Insects are particularly sensitive to temperature because they are stenotherm's (cold-blooded). In general, insects respond to higher temperature with increased rates of development and with less time between generations. Warmer winters reduce winterkill and consequently induce increased insect populations in the subsequent growing season.

Precipitation – whether optimal, excessive, or insufficient – is a key variable that also affects crop-pest interactions. Drought stress sometimes brings increased insect pest outbreaks. It is well known that drought can change the physiology of host species, leading to changes in the insects that feed on them. Abnormally cool, wet conditions can also bring on severe insect infestations, although excessive soil moisture may drown out soil-residing insects.

If global warming raises the temperature by 2°C in the United States and slightly less in Africa, insects will multiply and prosper. During a growing season, some insects produce 500 off springs per female every two weeks. Rising temperatures will lengthen the breeding season and increase the reproductive rate. That, in turn, will raise the total number of insects attacking a crop and subsequently increase crop losses. In addition, some insects, such as the Southwestern corn borer, will be able to extend their range northward as a result of the warming trend (Chippendale, 1999).

Under the projected warming trend in the United States, farmers can expect a 25 to 100 per cent increase in losses due to insects, depending on the crop (Pimentel *et al.*, 1993). Because crop losses to insects in Africa are already high, the projected impacts on different crops range from minus 30 per cent for soybeans to plus 7 per cent for rice. West Africa's warm, moist conditions are ideal for insects.

Climate change is associated with warming, elevated CO_2 and regionally changed precipitation. Currano *et al.* (2008) conclude that global warming will in the long term increase insect herbivory. Global warming might therefore benefit many insect species in the temperate regions. A warmer climate in these regions may result in changes in geographical distribution, increased overwintering (i.e. more insects survive the winter), changes in population growth rates, increases in the number of generations, extension of the development season, changes in crop-pest synchrony, changes in interspecific interactions and increased risk of invasion by migrant pests (Bale *et al.*, 2002).

Under elevated CO_2, population densities of chewing insects are unaffected or decrease, but do not increase while sap sucker (phloem feeder) population densities might increase.

A meta-analysis of studies on elevated temperature and elevated CO_2 suggests that insect herbivore performance is adversely affected by elevated CO_2, favored by elevated temperature, and not modified when both parameters (temperature and CO_2 combined) were elevated.

2.4.3. Disease pathogens

Climate factors that influence the growth, spread, and survival of crop diseases include temperature, precipitation, humidity, dew, radiation, wind speed, circulation patterns, and the occurrence of extreme events. Higher temperature and humidity and greater precipitation results in the spread of plant diseases, as wet vegetation promotes the germination of spores and the proliferation of fungi and bacteria, and influences the life cycle of soil nematodes. In regions that suffer aridity, however, disease infestation lessens, although some diseases (such as the powdery mildews) thrive in hot, dry conditions, as long as there is dew formation at night.

Under the warmer-but-drier conditions projected for North America, crop losses due to plant diseases are expected to decline as much as 30 per cent. However, under the wetter conditions projected for Africa, losses from diseases will increase by more than 100 per cent for some crops.

2.4.3.1. Fungal pathogens: Plant disease outbreaks depend on complex interactions between many factors. Overall, the development of an aggressive strain within a diverse population, the presence of host plants missing appropriate resistance to this newly developed aggressive strain, plant architecture, uniform cropping system, weather conditions, and limited antagonistic activities of non-pathogenic populations play an important and inter-related role (Barbetti *et al.*, 2012; Burdon *et al.*, 2006; Garrett *et al.*, 2011; Jones and Barbetti, 2012; Pangga *et al.*, 2013). The mode of nutrition also has a non-negligible role. For example, biotrophic fungi deriving nutrition only from living host tissue are more successfully controlled than necrotrophs which derive nutrition from both live and dead plant materials (Beed *et al.*, 2011). Taking all these factors together, it is hard to speculate on the effects of climate change, particularly when long-term datasets from the past are missing to develop and test predictive models for the future. Nevertheless, our knowledge of the phenology of plant-disease interactions, population genetics of pathogens as well as crops, and examples of overwhelming establishment of new diseases in a region provides insights into how climate change may affect disease incidence and severity as summarized below.

Firstly, some features of climate change will influence disease phenology. Higher temperatures and/or elevated CO_2 will speed up the life cycle of some pathogenic fungi thereby increasing the development and availability of inoculum. The latter of which may lead to high levels of infection and accelerated evolution of new aggressive strains (Chakraborty and Dutta, 2003). Hence, crop protection practices must be timely, effective, precise, and complete to prevent a new wave of infection that leads to new disease epidemics due to overcoming crop resistance. Prolonged generations of pests will be able to infect crops at a later stage than at present. An example is *Phoma* which is estimated to cause a 10–50 % decrease in the total yield of oilseed rape in the UK depending on future climate conditions (Barnes *et al.*, 2010). Climate change was also reported to alter crop anthesis. For example, wheat in the UK will be developed earlier in the season being more favourable for *Fusarium* ear blight infection and consequent mycotoxins increases in cereal products (Madgwick *et al.*, 2011).

Secondly, climate change may affect the expression of plant resistance traits in a positive or negative way. Breeding for resistance is a lengthy process, and today's resistant varieties are specifically bred for present agricultural conditions. The expression of quantitative resistance against *Phoma* in oilseed rape dropped dramatically when there was an increase in temperature from 20 to 25 °C (Huang *et al.*, 2009). The per centage of leaf area infected increased from 5 to 50 %. Taken together, the expression of resistance genes in the host plant, and their efficacy, may decrease dramatically because of climate change. In addition, we could envisage that increased generation cycles of pests triggered by climate change might select more aggressive pathogen populations (Chakraborty, 2013). Such a selected population, combined with hampered resistance in the host, may lead to unpredicted, or unprecedented, epidemiological outbreaks.

Thirdly, when the genetic variation of a crop is low, particularly over a wide range, and continued series of cultivation, a new or adapted strain of a pathogen may arise and become dominant thereby leading to dramatic effects (Strange and Scott, 2005). A classic example is wheat production in the Central, West Asian, and North African regions which feeds more than one billion people. Although many wheat varieties are grown in this huge area, all have a similar genetic background. With a slightly increased temperature and decreased rainfall, as observed over recent decades, a new type of yellow rust caused by *Puccinia striiformis* (Hovmøller *et al.*, 2011) was able to extend from Africa to India within 15 years. This has led to widespread epidemics highlighting how vulnerable an area can be when slight changes in climate occur (Wellings, 2011). Indeed, it was reported that the new strains of the fungus have adapted to produce spores at warmer temperatures than usual for this pathogen. This may have probably increased the rate of disease expansion on a global scale (Milus *et al.*, 2009). The new strain of *P. striiformis* (Fig. 2.2) spread to at least three new continents within 3 years, faster than previously reported for any crop pathogen (Hovmøller *et al.*, 2008). This gave rise to severe epidemics in warmer wheat production areas (Hovmøller *et al.*, 2010) where yellow rust was previously absent or infrequent, implying that current cropping systems were not prepared for the new situation. At a temperature regime typical for these areas, isolates of the new strain produced three to four times more spores per day than strains found previously (Hovmøller *et al.*, 2011). The rapid spread of the new strain is probably the cumulative result of increased pathogen fitness, warmer environments, increase in the number of spores in the atmosphere, and long-distance dispersal of these by wind, increasing travel and commerce. Similar chains of events should be expected for other pathogens reinforcing the need for coordinated international surveillance (including long-range surveillance) and action to ensure sustainable crop disease management under climate change.

Fig. 2.2. New strains of yellow rust (*Puccinia striiformis*) on winter wheat adapted to produce spores at warmer temperatures than usual

Another example of increasing pathogen proliferation and disease outbreaks is the case of oomycete, *Phytopthora* sp. Root rot and canker disease caused by *Phytophthora cinnamomi* affects more than 1000 host species, especially in most temperate and subtropical areas of the globe (Sturrock *et al.*, 2011). Under climate change, this pathogen was reported to increase the instability and vulnerability of forest ecosystems by a shift toward central Europe (Jung, 2009). Increasing temperatures were predicted to boost potential range expansion of *P. cinnamomi* along the western coast of Europe up to a few hundred kilometers eastward from the Atlantic coast within one century (Bergot *et al.*, 2004). Likewise, *Phytophthora ramorum*, a pathogen of yet unknown origin, and not reported in Europe until the late 1990s, has rapidly established itself as a major threat to a range of plant species here. Within the last 15 years, over 20 different species of broad-leaved trees were found to be infected throughout southern England (Webber, 2008). Since autumn 2009, several dramatic outbreaks of the disease caused by this pathogen have been reported across the UK on several new host species (Brasier *et al.*, 2010; Webber *et al.*, 2010). Thanks to the introduction of new horticultural plant species in Europe (such as rhododendron), and altered weather conditions with humid summers and milder winters, this pathogen was not only able to cross the ocean (which it probably had done previously) but was also able to finally establish itself throughout most European regions. Today, this fungal pathogen is widely considered as endemic in Europe, infecting an increasing range of other host plants. Climate change scenarios based on CLIMEX model (that describes the response of a species to climate) projected that the area favourable, or very favourable, for *P. ramorum* will markedly decrease in the eastern USA but will increase in the west-coast states of Washington, Oregon, and California (Venette, 2009; Venette and Cohen, 2006). Numerous other examples of shift in fungal pathogens and disease outbreaks associated to changing climate were recently reviewed (Sturrock *et al.*, 2011).

Besides changes in host-pathogen interactions, climate change leads to other changes including fecundity, canopy size, and long- and short-distance disseminations (Chakraborty, 2013). Changes in the geographical distribution of hosts and pathogens, new pathogens, new remnant vegetation, and increased plant stress will all be important issues in the development of fungal diseases in changing climatic conditions. The complex interplay of changing resistance expression, new virulence of pathogens, and presence of other host plants and, overarching all these, more favourable infection conditions due to climate change warrant a careful inventory of the new knowledge needed to prevent unprecedented disease outbreaks in the future.

Several examples of fungal pathogen adaptation under climate change have been reported in the literature. Those include the substitution of species in the genus *Fusarium* and *Microdochium* sp. in Europe by *Fusarium pseudograminearum* with higher temperature optima and toxigenicity (Isebaert *et al.*, 2009). This replacement has led to the increased risk of mycotoxin contamination (Paterson and Lima, 2010). Similar example was reported also from Canada where adaptive evolution along environmental gradients has replaced existing *Fusarium graminearum* population with highly toxigenic forms (Ward *et al.*, 2008). In eastern USA, an aggressive high-temperature-tolerant strain has dominated stripe rust pathogen, *P. striiformis* f. sp. *tritici*, population since 2000 (Milus *et al.*, 2009).

As effects of climate change on fungal diseases are challenging to predict in general terms, a practical method might be detailed through modeling of each individual crop-pathogen situation in a projected climate change condition of a certain geographic area. Looking at north-west Europe with predicted warmer and more humid winters, and warmer and drier summers, polycyclic and monocyclic fungi can be classified in seven "ecotypes" (West *et al.*, 2012). This classification can be based on parameters such as dissemination method of spores, infection condition requirements, and latent period to weather conditions (West *et al.*, 2012). Being classified in a specific ecotype, one can predict if climate change in that region will support an increased or a decreased infection of a certain fungus-crop combination. This methodology might be useful to test in other regions of Europe.

2.4.3.2. Bacterial pathogens: In a globalized world and under climate change conditions, plant pathogenic bacteria are expected to become an increasing threat to crop health (Jones and Barbetti, 2012). Firstly, apparently healthy but latently infected woody plants (Lamichhane, 2014; Lamichhane *et al.*, 2014) highly favour the long-distance dissemination of plant pathogenic bacteria through global trade of nursery materials. This often leads to new disease outbreaks throughout the range of cultivation. Secondly, new strains emerge from warmer regions and easily adapt to more temperate regions. A telling example in Europe can be found in potato crops where diseases caused by *Pectobacterium* sp. were always predominant. However, over the past decade, *Dickeya* species have taken over (Czajkowski *et al.*, 2011). The latter, previously identified as pathogens from tropical and subtropical regions, have become established in Europe due to today's milder climatic conditions. *Dickeya* sp. are more aggressive than *Pectobacterium* sp. and are now responsible for 50 to 100 % of field infections thereby leading to significant damage in potato and other crops across Europe (Czajkowski *et al.*, 2011).

European horse chestnut bleeding canker caused by *Pseudomonas syringe* represents another pertinent example (Fig. 2.3). Although this pathogen was reported to cause mild disease symptoms on a local horse chestnut species from India during the 1980s (Durgapal and Singh, 1980), the pathogen was never reported in Europe. However, in 2002, an aggressive population of *P. syringae* was found on European horse chestnut in the Netherlands which, since, has rapidly established itself as a major threat to this tree throughout a large number of North-western European countries (Lamichhane *et al.*, 2014). Similarly, an aggressive population of *P. syringae* has jeopardized the entire kiwi fruit industry at global level causing severe economic losses (Bartoli *et al.*, 2015; Lamichhane *et al.*, 2014). In this last case, global trade of infected plant materials and the cultivation of genetically uniform plant species were considered as the major means that favoured pathogen dissemination.

Fig. 2.3. Horse chestnut bleeding canker caused by the bacterial pathogen *Pseudomonas syringae* which leads to rapid tree death

Another example of bacterial pathogen evolution under climate change includes race 9 of *Xanthomonas oryzae* pv. *oryzae* infecting rice with the Xa7 resistance gene increased virulence and aggressiveness over 11 years of rising temperature to dominate the pathogen population (Webb *et al.*, 2011). However, in this case, the increased effectiveness of Xa7 at high temperature has continued to offer effective pathogen control.

As climate change might enhance threats from plant pathogenic bacteria, it is important to realize that bacteria might behave as "kingdom hoppers" moving from human to animal to plant to environment and *vice versa.* These phenomena have raised several concerns. Recent evidence has confirmed that human pathogenic Gram-negative bacteria and viruses can colonize and infect plants as alternate hosts (Holden *et al.*, 2009). The causal agent of nosocomial infections in humans, *Pseudomonas aeruginosa* (Mesaros *et al.*, 2007), is reported to cause new diseases on naturally grown tobacco (Yu *et al.*, 2008) and causes fruit rot of Tinda (Mondal *et al.*, 2012). Similarly, *Salmonella enteric typhimurium* internally colonizes tomato (Gu *et al.*, 2013; 2011) as well as lettuce (Klerks *et al.*, 2007) plants. On tomato, *S. enterica* has been reported to move systemically reaching fruits and seeds from leaves (Gu *et al.*, 2011). Once within the fruits, the pathogen can multiply to high densities (Gu *et al.*, 2011; Noel *et al.*, 2010). A recent study demonstrated that *Escherichia coli* can internally colonize both leaves and roots of lettuce and spinach (Wright *et al.*, 2013). Lastly, human nor virus and nor virus surrogates (*Murine nor virus* and *Tulane virus*) are also reported to colonize spinach stem and leaves (Hirneisen and Kniel, 2013). Because virulence effector genes can be frequently exchanged or transferred between bacteria (Baltrus, 2013), such "kingdom hopping" events represent a real threat thereby raising several concerns also for human and animal, not least plant health. Recently, it had been reviewed that bacterial species, especially those belonging to the large family of *Enterobacteriaceae*, can easily be adapted in humans, animals, and plant environments. Horizontal gene transfer (Baltrus, 2013) between species can easily occur in a joint habitat such as the soil/manure/plant-rhizosphere interphase, which might confer an enhanced fitness to bacteria to live on plant as well as human hosts. In analogy to zoonosis (bacteria transmitted from animal to human), this new class of plant bacteria is termed "phytonosis" (Van Overbeek *et al.*, 2014). One of the most important threats of phytonosic bacteria are acquired genes for antibiotic resistance (Lipsitch *et al.*, 2002), known for being abundantly present in the bacterial community of the plant rhizosphere.

2.4.4. Nematode pathogens

Herbivore nematodes feeds on plant parts mostly on roots. It is estimated that nematodes causes crop losses worth US$ 125 billion annually in agriculture (Chitwood, 2003).

Global warming resulting in elevated carbon dioxide (CO_2) and temperature in the atmosphere may influence plant pathogenic nematodes directly by interfering with their developmental rate and survival strategies and indirectly by altering host plant physiology. Severe droughts resulting in a reduction of soil water will most likely negatively affect soil nematodes.

Nematode developmental rate is directly influenced by the temperature with slower development at cooler and faster growth rate at warmer soil temperatures.

Therefore, increase in atmospheric temperature due to global warming is expected to result in more number of generations per season and expansion of their geographical distribution range. Other potential effects of elevated temperature on parasitic nematodes include altered sex ratio, host defense responses and interference in their survival strategies like dauer juveniles or egg diapauses in extreme environments.

Herbivorous nematodes showed neutral or positive response to CO_2 enrichment effects with some species showing the potential to build up rapidly and interfere with plant's response to global warming. The number of herbivore, bacterivore and fungivore nematodes was significantly higher under winter wheat and sugar beets grown under elevated CO_2 (550 ppm), while the number of carnivore was not changed. The total numbers of herbivore, bacterivore and fungivore nematodes were higher under elevated CO_2 wheat than under elevated CO_2 sugar beet, most likely due to the very different root system of both plant species (Sticht *et al.*, 2009).

2.5. PREDICTION MODELLING

Plausible estimates of climate change impacts requires combined use of climate, crops, pests, and economic models. However, climate change models often ignore possible effects on dynamics and infectivity of pests and diseases. The most important reason is that we do not have long-term monitoring data or an empirical approach to feed modelling systems that might be used to predict impacts and mitigation scenarios with sufficient levels of certainty (Scherm, 2004; Shaw and Osborne, 2011). In addition, this pattern of climate change factors is not equally distributed over the globe. It is anticipated that a general shift of a milder climate toward the poles will improve the potential of crop production (Butterworth *et al.*, 2010; Evans *et al.*, 2008). By contrast, hotter and drier conditions in many already semi-arid areas of the world will limit the possibilities for agriculture (Luo *et al.*, 2009). Therefore, it is unlikely that general models can ever be developed. However, it is worth noting that all climate models predict a higher frequency of extreme and fluctuating weather condition which would influence the interactions between crops, pests, and diseases in an unpredictable way. As a consequence, there may be increasing risk of complete failure of current and possibly future crop protection strategies.

2.6. FOOD SECURITY

The greatest challenge for humanity in the coming century is to double the present levels of food production to meet the needs of ever increasing population by sustainable use of shrinking natural resource base (Deka *et al.*, 2012). Plant pests and diseases could potentially deprive humanity of up to 82% of the attainable yield in the case of cotton and over 50% for other major crops (Oerke, 2006) and, combined with postharvest spoilage and deterioration in quality, these losses become critical, especially for resource-poor regions. The aggravating pest problems under changing climate regimes are expected to intensify the yield losses; threatening the food security of the countries with high dependency on agriculture (Patterson *et al.*, 1999; Gutierrez *et al.*, 2008; Parry *et al.*, 2007; IPCC, 2007; Chahal *et al.*, 2008). The climate change is likely to affect the extent of insect pollination by disrupting the synchrony between plant-pollinator life cycles (Kudo *et al.*, 2004), with an estimated risk of reduction in world food production by one-third (Klein *et al.*, 2007). This has major implication for food and nutritional security (FAO, 2008). This may have direct bearing on the livelihood of the rural poor as their survival is directly linked to outcomes from food production systems. The increased food prices resulting from declining food production may also impact negatively the urban population (IPCC, 2007; Chahal *et al.*, 2008).

From a food-security perspective, emphasis must now shift from impact assessment to developing adaptation and mitigation strategies and options. Two broad areas of empirical investigation will be essential; firstly, to evaluate under climate change the efficacy of current physical, chemical and biological control tactics including disease-resistant cultivars, and secondly, to include future climate scenarios in all research aimed at developing new tools and tactics. Transgenic solutions (Huang *et al.*, 2002) must receive serious consideration in integrated disease management strategies to improve food security.

2.7. ADAPTATION AND MITIGATION

Adaptation refers to an adjustment in natural or human systems in response to actual or expected climatic stimuli or their effects that moderate, harm or exploit beneficial opportunities.

Climate change mitigation encompasses the actions being taken, and those that have been proposed, to limit the magnitude and/or rate of long-term global warming induced climate change.

The following adaptation and mitigation practices have been recommended to manage pests:

- Biosecurity, quarantine, monitoring, and control measures can be strengthened to manage the spread of pests and diseases under a warming climate.
- More resilient/adaptable crop genotypes needed, especially with durable resistance to pests.
- Adoption of environmental conserving pest control activities such as organic farming, bio-control and integrated (ecological) pest management are the need of the hour.
- Application of natural mulches helps in suppression of harmful pests and diseases.
- A diverse fauna of natural enemy species can successfully suppress pests.
- Mulching and reduced tillage for example increases spider abundance which predate on insect pests.
- Specific relationships between pests and host plants are interrupted by crop rotations.
- A 'healthy' soil, with optimal physical, chemical and biological properties increases plant resistance to insects and diseases.
- Avoidance of excess use of nitrogen which can increase the severity of certain diseases and make a crop more susceptible to pests.

- Organic agriculture system uses crop rotation, green manure, compost, biological pest control and mechanical cultivation to control pests. Organic systems avoid the use of synthetic pesticides and rely on cultural practices such as crop rotations which break up pest cycles and encourage beneficial insects.
- Polyculture techniques such as crop rotation, multi-cropping and intercropping are less susceptible to pests than monoculture crops.
- In addition to the prudent application of pesticides, increased use of nonchemical pest controls (crop rotations, biological controls, altering planting dates and fertilizer and irrigation applications, and soil management and tillage) would help minimize crop losses and thereby help maintain crop yields.
- Genetically engineered plants have been designed to resist pests, diseases, nematodes and tolerate weeds without the need for pesticides.
- The growers have to change pest management strategies by rescheduling the crop calendars in accordance with the projected changes in pest incidence and extent of crop losses in view of the changing climate.
- Geographic Information System (GIS) is an enabling technology for crop protection scientists, which help in relating pest outbreaks to biographic and physiographic features of the landscape, hence can best be utilized in area wide pest management programmes.
- Pesticides with novel mode of actions such as neonicotinoid insecticides for controlling sucking pests which induces salicylic acid associated plant defense responses. More such compounds need to be identified for use in future crop pest management.
- Integrated pest management (IPM) is an effective and environmentally sensitive approach to pest management that uses current, comprehensive information on the life cycles of pests and their interaction with the environment to manage pest damage by the most economical means, and with the least possible hazard to people, property, and the environment.

2.8. RESEARCH NEEDS

2.8.1. Research and development

- Providing site-specific weather data to help researchers run predictive pest models and for farmers to make informed decisions on pest management.
- Research on pest-crop-weather interactions for developing simple operational and predictive models that can be used in agro-advisory services.
- Integrate biotechnology with traditional agricultural practices, metabolomics and bioinformatics systems to design novel insecticide molecules for studying interactions with the DNA and protein models.
- Develop new bio-pesticides and technologies on pest management through sterile insect techniques, new botanicals, semiochemicals (repellents, pheromones, allomones etc.) and endophytic microbial metabolites for pest control, and disease forecasting.

2.8.2. Technologies and practices

- Develop effective surveillance systems for invasive species based on semiochemicals.
- Streamlining the flow of information of pest surveillance to reduce response time between detection and action to manage and prevent pests and diseases.
- Promotion of bio-intensive integrated pest management at large scale.
- Plant protection measures to be tailored to meets the threat to crops arising from outbreak of vector-borne diseases.

2.9. CONCLUSIONS

Climate-driven changes present challenges and opportunities for sustainable agriculture programmes based on IPM. As suggested by Anderson *et al.* (2004), research on the effect of climate change on pest evolution must shift from passive cataloging to analysis and interpretation of the factors favouring outbreaks. If agricultural production is to keep pace with a growing food demand while satisfying environmental goals, we will need new cultivars, major changes in IPM programmes, increased funding, and improved response time to new and evolving pests that lead to outbreaks.

Although climate change is expected to impact agriculture by influencing the stability of crop yield, information on the real consequences are lacking, and studies in this regard remain in their infancy. Our inability to make confident predictions highlights a need for new research on several fronts. Policies are needed to fill this uncertainty gap by considering a wide range of possible scenarios. Hence, research based on a broader collaborative approach should be made in order to develop anticipatory adaptive strategies resulting in more resilient cropping strategies and stabilized yields. Toward this ultimate aim, enhanced interregional, transnational, and global networking of researchers and stakeholders, at all levels across the plant health/crop protection spectrum, considering the scale and increasing pressures on crop production and yields, is an effective way to better use the limited resources in order to address this twenty-first century challenge in an ever increasing integrated global context.

In the future, the global warming trend is expected to continue, especially with increasing greenhouse gas concentrations. Changes in temperature, precipitation and carbon dioxide will affect plant growth, reduce the nutritive value of crops and influence the spread of pests and diseases. Global warming and climate change will have serious consequences on diversity and abundance of pests, and the extent of losses due to pests, which will impact both crop production and food security. Prediction of changes in geographical distribution and population dynamics of pests will be useful to adapt the pest management strategies to mitigate the adverse effects of climate change on crop production. Pest outbreaks might occur more frequently, particularly during extended periods of drought, followed by heavy rainfall. Some of the components of pest management such as host plant resistance, biopesticides, natural enemies, and synthetic chemicals will be rendered less effective as a result of increase in temperatures and UV radiation, and decrease in relative humidity. Climate change will also alter the interactions between the pests and their host plants. As a result, some of the cultivars that are resistant to pests, may exhibit susceptible reaction under global warming. Adverse effects of climate change on the activity and effectiveness of natural enemies will be a major concern in future pest management programmes. Rate of pest multiplication might increase with an increase in CO_2 and temperature. Therefore, there is a need to have a concerted look at the likely effects of climate change on crop protection, and devise appropriate measures to mitigate the effects of climate change on food security.

Section II

INSECT AND MITE PESTS

3

Climate Change and Insect Pest Scenarios

3.1. INTRODUCTION

Crop plants live in a very complex ecosystem. They are attacked by insects and mites. All of these species interact with each other. Globally, pests are estimated to destroy about one third of our crops and are increasingly serious constraints to crop production, in spite of the advances in pest control technology over the last half century. Many people believe that global warming as predicted would increase pressure from pests. Higher temperatures and longer growing seasons could result in increased pest population in temperate regions of Asia. In general, however, most pest species are favoured with warm and humid conditions. Pest infestations often coincide with changes in climatic conditions, such as early or late rains, drought, or increases in humidity, which can reduce yields. Climate change might have an influence on increased pesticide use due to presence of pests.

Agro-ecosystem environment is largely governed by interactions between abiotic (temperature, humidity, rainfall, soil factors, pollutants etc.) and biotic (crop plants, weeds, insect pests, disease pathogens, nematodes, etc.) components. The abiotic stress factors modulate the effects of biotic stresses and are most harmful when occur in combination (Mittler, 2006), greatly influencing crop growth and productivity to the extent of 80% (Oerke *et al*., 1994; Theilert, 2006). Climate change (CC) resultant abiotic environment especially changes in hydrological cycles (Rowntree, 1990) and temperature regimes may alter the composition of agro-ecosystems resulting in altitude wise shift in distribution and range of pest and crop species (Porter *et al.,* 1991; Sutherst *et al.*, 1991; NACCAP, 2008). Hence, in the context of global climate change, it is an utmost need of the hour to address multiple stresses threatening sustainability of agricultural production systems.

Climate change can induce elevated temperatures, droughts, rising sea levels, floods, changes in wind patterns and abnormal weather. These events can cause changes in ecosystem balance that can disrupt ecosystem services and favour crop pest development resulting in heavy crop losses, heavy/ injudicious pesticide use and in extreme cases, poverty, due to severe disruption of livelihoods.

At the moment, farmers' pest management practices and pesticide use are causing environmental and human health concerns in food crop agro-ecosystems. Climate change is adding new dimensions to the multiple challenges faced by agricultural policy makers and farmers. However, links between climate change and pest scenarios are not well known, and much less known are adaptation options that may be adopted by agricultural policy makers and farmers to reduce vulnerability of crops to abnormal pest infestations and to build resilience against climate change induced pests.

Pest menace under the influence of climatic factors, at various stages of crop growth is one of the factors limiting agricultural productivity (Oerke *et al.*, 1994). In India, pest damage varies considerably in different agro-climatic regions across the country mainly due to differential impacts of several abiotic factors such as temperature, humidity and rainfall (Reed and Pawar, 1982; Sharma *et al.*, 2005; 2010). This has major implication for the intensification of yield losses due to potential changes in crop diversity and increased incidence of insect-pests in the context of impending climate change.

Recent expansions and advances of insect infestations may be associated with regional increases in mean or minimum temperatures. Crop pests have made significant pole-ward shifts of 2.7 ± 0.8 km per year since 1960 threatening global food security by the CC-aided emergence and spread of crop pests. Spread is facilitated primarily by human transportation, but climate change allows establishment. For instance, *Nezara viridula*, a rice pest, shifted its distribution range about 70 km north from 1960 to 2000.

Insect-pests of crop plants are the real candidates most affected by global climate change. Complex physiological effects exerted by the increasing temperature and CO_2 may affect profoundly, the interactions between crop plants and insect-pests (Hare, 1992; Caulifield and Bunce, 1994; Roth and Lindroth, 1995). It has been reported that, global climate warming may lead to altitude wise expansion of the geographic range of insect-pests (Hill and Dymock, 1989; Parry and Carter, 1989; Elphinstone and Toth, 2008), increased abundance of tropical insect species (Cannon, 1998; Patterson *et al.*, 1999; Bale *et al.*, 2002; Diffenbaugh *et al.*, 2008), decrease in the relative proportion of temperature sensitive insect population (Petzoldt and Seaman, 2010; Sharma *et al.*, 2005; 2010), and more incidence of insect transmitted plant diseases through range expansion and rapid multiplication of insect vectors (Petzoldt and Seaman, 2010). Thus, with changing climate it is expected that the growers of crops have to face new and intense pest problems in the years to come.

Changes in insect-pest status due to climate change will perilously affect agricultural production and the livelihood of farmers in the country where larger portion of work force is directly dependent on climate sensitive sectors such as agriculture (Chahal *et al.*, 2008; Deka *et al.*, 2008). This envisages an urgent need to modify crop protection measures with changed climate in order to attain the goal of food security of the nation. The impacts of climate change on crop production mediated through changes in populations of serious insect-pests should be given careful attention for planning and devising adaptation and mitigation strategies for pest management.

Climate change also affects natural enemies and their interactions with pests in terms of predation behaviour, mobility, tolerances and adaptive responses to temperature changes and wind patterns. However, there is a dearth of comprehensive information about how climate change will actually affect predator-prey interactions and the biological control services provided.

Many pests exhibit considerable capacity for generation, recombination, and selection and thus, there is little doubt that any new opportunities resulting from climate change will be exploited by them. The extent to which crop pests, their natural enemies and pathogens adapt in response to global warming and the impact of this on food production is still largely unknown. Better understanding of such mechanistic effects on crops will be needed to develop realistic predictions on crop production deviations on a regional scale and thereby, assist in the development of more robust regional food security policies and adaptation strategies.

Changes in intensity and wind patterns may have contributed to China's recent pest outbreaks in rice production. Increase in drought frequencies and intensities can also cause outbreak of pests such as armyworms and locusts. Failure to take into account climate change induced pest scenarios in land use planning and farm management can exacerbate their impacts.

3.2. INSECT PESTS UNDER CHANGING CLIMATE

Losses due to insect damage are likely to increase as a result of changes in crop diversity and increased incidence of insect pests due to global warming. Current estimates of changes in climate indicate an increase in global mean annual temperatures of 1°C by 2025, and 3 °C by the end of the next century. The date at which an equivalent doubling of CO_2 will be attained is estimated to be between 2025 and 2070, depending on the level of emission of greenhouse gasses (IPCC, 1990a, b). Mean annual temperature changes between 3 and 6 °C are estimated to occur across Europe, with greatest increases occurring at high latitudes.

Pest menace under the influence of climatic factors, at various stages of crop growth is one of the factors limiting agricultural productivity (Oerke *et al.*, 1994). Climate is an important determinant of abundance and distribution of species. The rising concentrations of CO_2 will have a variety of direct effects on plants and may also have indirect effects on herbivores and their natural enemies. The climate has profound effects on the populations of invertebrate pests like insects, mites and others and affects their development, reproduction and dispersal systems. Extreme weather events such as intense rainstorms, high wind or elevated temperatures also affect their survival. The climate change impacts on pests may include shifts in species distributions with species shifting their ranges to higher latitudes and elevations, changes in phenology with life cycles beginning earlier in spring and continuing later in autumn, increase in population growth rates and number of generations, change in migratory behaviour, alterations in crop-pest synchrony and natural enemy-pest interaction, and changes in interspecific interactions (Root *et al.*, 2003). Changes in community structure and extinction of some species are also expected (Thomas *et al.*, 2004).

For pest species to survive in the changing climates, they must either adapt *in situ* to new conditions or shift their distributions in pursuit of more favourable ones. Many insects have large population sizes and short generation times, and their phenology, fecundity, survival, selection and habitat use can respond rapidly to the climate change. These changes to insect life-history may in turn produce rapid changes in their abundance and distribution. Due to recent climate changes, widespread, generalist species at their cool range margins have expanded their distribution ranges, whereas the ranges of localized, habitat-specialist species and those at their warm margins have narrowed (Konvicka *et al.*, 2003). An array of methods including surveys, experiments and modelling have been used to study the impact of climate change on pest abundance and distribution.

Insects being poikilotherms, temperature is probably the single most important environmental factor influencing their behaviour, distribution, development, survival, and reproduction (Bale *et al.*, 2002; Petzolet and Seaman, 2010). Therefore, it is highly expected that, the major drivers of climate change i.e. elevated CO_2, increased temperature and depleted soil moisture can impact population dynamics of insect pests (Fig. 3.1) and the extent of crop losses, significantly (Petzoldt and Seaman, 2010).

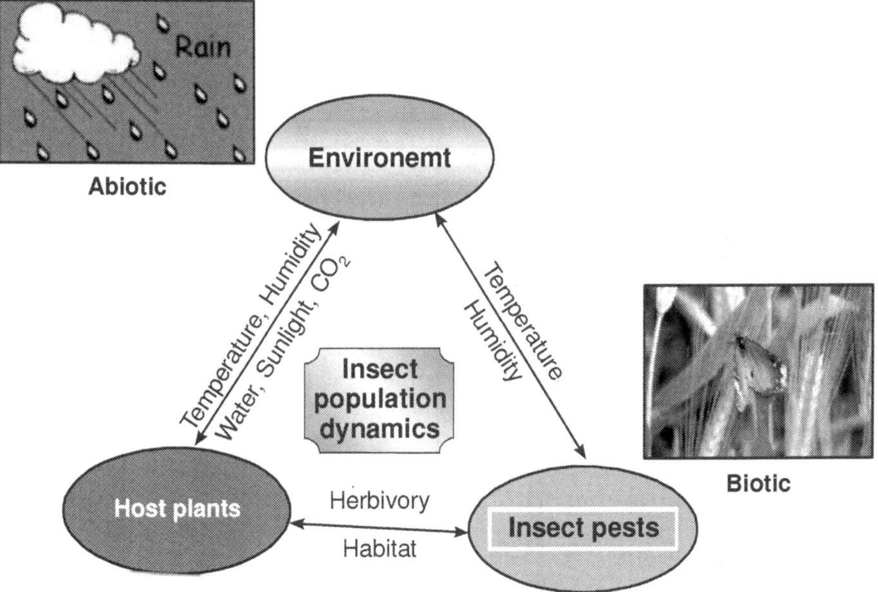

Fig. 3.1. Interactions between abiotic and biotic factors in crop-pest-environment triangle

High mobility and rapid population growth will increase the extent of losses due to insect pests. Many species may have their diapause strategies disrupted as the linkages between temperature and moisture regimes, and the day length will be altered. Genetic variation and multi-factor inheritance of innate recognition of environmental signals may mean that many insect species will have to adapt readily to such disruption.

Global warming and climate changes will result in:

- Extension of geographical range of pests.
- In cooler latitudes, global warming brings new species.
- Increased risk of invasion by migrant pests.
- Reduced effectiveness of crop protection technologies.
- A 2.4 to 2.7-fold increase in pesticide use by 2050 (Tilman *et al.*, 2001).
- Increased probability of pests developing faster resistance to pesticides.
- Warmer winter temperatures would reduce winter kill, favouring the increase of pest populations.

- Rising temperatures extend the growing season.
- Overall temperature increases may influence crop pest interactions by speeding up pest growth rates which increases reproductive generations per crop cycle.

Climate change will also result in increased problems with insect transmitted diseases. These changes will have major implications for crop protection and food security, particularly in the developing countries, where the need to increase and sustain food production is most urgent. Long-term monitoring of population levels and insect behaviour, particularly in identifiably sensitive regions, may provide some of the first indications of a biological response to climate change. In addition, it will also be important to keep ahead of undesirable pest adaptations, and therefore, to carefully consider global warming and climate change for planning research and development efforts for pest management and food security in future.

3.3. CLIMATE CHANGE AND EMERGING PEST PROBLEMS IN INDIA

Several insect pests, that were important in the past or the minor pests, are likely to become more devastating with global warming and climate change (Sharma, 2014) (Table 3.1). Many insect species, that will move to newer areas as invasive pests will also pose a major threat to crop production and food security, as they find more suitable climatic niches in the new areas. The invasive species are likely to cause more harm in the absence of natural enemies in the new habitats. Pest outbreaks due to climate change include pod borer, *Helicoverpa armigera* damage on pigeon pea following wet weather conditions in September – October; mealy bug, *Ceroplastodes cajaninae* infestation in pigeon pea under prolonged hot and dry conditions; beet armyworm, *Spodoptera exigua* damage in chickpea triggered by winter rains in October – November; and pink stem borer, *Sesamia inferens* damage in sorghum due to hot and dry conditions during the post-rainy season (Fig. 3.2).

Table 3.1. Insect pests that have become or are likely to become serious pests due to climate change and changes in cropping patterns.

Insect pest	Scientific name	Crop(s)
American bollworm	*Helicoverpa armigera*	Cotton, chickpea, pigeon pea, sunflower, tomato, etc.
Beet armyworm	*Spodoptera exigua*	Chickpea in southern India
Spotted pod borer	*Maruca vitrata*	Pigeon pea, cowpea, lab-lab beans
Diamondback moth	*Plutella xylostella*	Cabbage, cauliflower
Pink stem borer	*Sesamia inferens*	Maize, sorghum, wheat
Whitefly	*Bemisia tabaci*	Cotton, tobacco
Brown plant hopper	*Nilaparvata lugens*	Rice
Green leaf hopper	*Nephotettix* sp.	Rice

Insect pest	Scientific name	Crop(s)
Serpentine leaf miner	*Liriomyza trifolii*	Cotton, tomato, cucurbits, several other vegetables
Fruit fly	*Bactrocera* sp.	Fruits and vegetables
Mealy bugs	*Paracoccus marginatus* *Phenococcus solenopsis* *Cerolastodes gajaninae*	Several field and horticultural crops
Thrips	Several species	Groundnut, cotton, chillies, citrus, pomegranate
Wheat aphid	*Macrosiphum miscanthi*	Wheat, barley, oats
Pod sucking bugs	*Clavigralla* sp.	Pigeon pea
Gall midge	*Orseolia oryzae*	Rice
Termites, white grubs	Several species	Many crops
Sugarcane woolly aphid	*Ceratovacuna lanigera*	Sugarcane

Source: Prasad and Bambawale (2010), Fand *et al.* (2012) and Sharma (2014)

Fig. 3.2. Pink stem borer (*Sesamia inferens*) damage on sorghum due to hot and dry conditions during the post-rainy season.

3.4. EFFECTS OF CLIMATE CHANGE ON INSECT PESTS

Insects are cold-blooded organisms; the temperature of their bodies is approximately the same as that of the environment. Therefore, temperature is probably the single most important environmental factor influencing insect behaviour, distribution, development, survival, and reproduction. Insect life stage predictions are most often calculated using accumulated degree days from a base temperature and bio-fix point. Some researchers believe that the effect of temperature on insects largely overwhelms the effects of other environmental factors (Bale *et al.*, 2002). It has been estimated that with a 2°C temperature increase, insects might experience one to five additional life cycles per season (Yamamura and Kiritani, 1998).

In colder regions (higher latitudes) with distinctive seasons, insects have broader thermal tolerance and are living in climates that are currently cooler than their optima (Deutsch *et al.*, 2008). Global warming might therefore benefit many insect species in the temperate regions.

Reduced winter mortality of two insect pests was observed in Japan, when mean temperature in January was above 4°C. Species which already develop at low temperatures and need a smaller number of warm days will benefit from increasing average temperatures, and produce more generations and appear early when spring temperatures are high. Species which require higher temperature before they lay their eggs tend to have increased reproductive activity.

There is a general agreement between scientists that the reduced nutrient quality of C_3 plants might lead to a compensation by increased feeding of many, but not all, herbivorous species (DeLucia *et al.*, 2008). Under elevated CO_2, population densities of chewing insects are unaffected or decrease, but do not increase while sap sucker (phloem feeder) population densities might increase.

A meta-analysis of studies on elevated temperature and elevated CO_2 suggests that insect herbivore performance is adversely affected by elevated CO_2, favoured by elevated temperature, and not modified when both parameters (temperature and CO_2 combined) were elevated.

It seems that current knowledge does not allow a generalization regarding the impact of climate change in herbivorous insects, especially not for the tropics. Even the trend of a northward shift of insects must not coercively translate into a pest problem – ecosystems are not that simple and human influence is quite strong. Basically, it would be necessary to investigate at least over three trophic levels with several generations of plants, herbivores, predators/parasites, under elevated temperature and elevated CO_2.

There are fewer scientific studies on the effect of precipitation on insects than temperature. Some insects are sensitive to precipitation and are killed or removed from crops by heavy rains - in some north-eastern US states, this consideration is important when choosing management options for onion thrips (Reiners and Petzoldt, 2005). For some insects that overwinter in soil, such as the cranberry fruit worm and other cranberry insect pests, flooding the soil has been used as a control measure. One would expect the predicted more frequent and intense precipitation events forecasted with climate change to negatively impact these insects. Other insects such as pea aphids are not tolerant of drought (Macvean and Dixon, 2001). As with temperature, precipitation changes can impact insect pest predators, parasitoids, and pathogens resulting in a complex dynamic. Fungal pathogens of insects are favoured by high humidity and their incidence would be increased by climate changes that lengthen periods of high humidity and reduced by those that result in drier conditions.

Precipitation – whether optimal, excessive, or insufficient – is a key variable that also affects crop-pest interactions. Drought stress sometimes brings increased insect pest outbreaks. It is well known that drought can change the physiology of host species, leading to changes in the insects that feed on them (Mattson and Haack, 1987). Abnormally cool, wet conditions can also bring on severe insect infestations, although excessive soil moisture may drown out soil-residing insects.

Distribution and frequency of rainfall may also affect the incidence of pests directly as well as through changes in humidity levels. It is being predicted that under the climate change, frequency of rainfall would decline while its intensity would increase. This would lead to heavy showers and floods on one hand and drought spells on the other. Under such situations, incidence of small pests such as aphids, jassids, whiteflies, mites, etc. on crops may be reduced as these get washed away by the heavy rains.

Drought stress sometimes brings increased insect pest outbreaks. It is well known that drought can change the physiology of host species, leading to changes in the insects that feed on them (Mattson and Haack, 1987).

3.5. IMPACTS OF CLIMATE CHANGE ON INSECT PESTS

Climate change will have a major impact on geographic distribution of insect pests, and low temperatures are often more important than high temperatures in determining geographical distribution of insect pests (Hill, 1987). Increasing temperatures may result in a greater ability to overwinter in insect species limited by low temperatures at higher latitudes, extending their geographical range (EPA, 1989; Elphinstone and Toth, 2008), and sudden outbreaks of insect pests can wipe out certain crop species, and also encourage the invasion by exotic species (Kannan and James, 2009). Spatial shifts in distribution of crops under changing climatic conditions will also influence the distribution of insect pests in a geographical region (Parry and Carter, 1989). Some plant species may be unable to follow the climate change, resulting in extinction of species that are specific to particular hosts (Thomas *et al.*, 2004). However, whether or not an insect pest would move with a crop into a new habitat will depend on other environmental conditions such as the presence of overwintering sites, soil type, and moisture, e.g., populations of the corn earworm, *Heliothis zea* in the North America might move to higher latitudes/altitudes, leading to greater damage in maize and other crops (EPA, 1989).

A warmer climate in temperate regions may result in changes in geographical distribution, increased overwintering (i.e. more insects survive the winter), changes in population growth rates, increases in the number of generations, extension of the development season, changes in crop-pest synchrony, changes in interspecific interactions and increased risk of invasion by migrant pests (Bale *et al.*, 2002). In Japan, warmer climate led to the northward migration of the green stinkbug (*Nezara viridula*) a major agricultural pest damaging soybean, rice, cotton and many other crops (Musolin, 2007).

Theoretically, it is hypothesized that the winter survival of insects will be improved by an increase in winter temperature, but the evidence for this hypothesis is rather scarce. Effect of climate change on winter mortality/survival can be examined by collecting long-term data on winter survival at fixed over-wintering sites. Relationship can then be established between winter mortality and temperature.

Being poikilotherms, insects have limited ability of homeostasis with external temperature changes. Hence they have developed a range of strategies such as behavioural avoidance through migration and physiological adaptations like diapause to support life under thermally stressful environments (Bale and Hayward, 2010). Diapause is a period of suspended developmental activities, the manifestation of which is governed by environmental factors like temperature, humidity and photoperiod. As an adaptive trait, diapause plays vital role in seasonal regulation of insect life cycles because of which the insects have better advantage to survive great deal of environmental adversities. There are two main types of insect diapause; aestivation and hibernation to sustain life under high and low temperature extremes, respectively (Chapman, 1998).

Biological control of insect-pests is one of the important components of integrated pest management, safeguarding the ecosystem. Natural enemies of crop pest's viz., predators, parasitoids and pathogens are prompt density responsive in their action subjected to the action of abiotic components. Being tiny and delicate, natural enemies of the insect pests are more sensitive to the climatic extremes like heat, cold, wind and rains. Precipitation changes can also affect predators, parasitoids and pathogens of insect-pests resulting in a complex dynamics. With changing climate, incidence of entomopathogenic fungi might be favoured by prolonged humidity conditions and obstinately be reduced by drier conditions (Newton *et al.*, 2011). Natural enemy and host insect populations may respond differently to changes in climate. Hosts may pass though vulnerable life stages more quickly at higher temperatures, reducing the window of opportunity for parasitism which may give great set back to the survival and multiplication of parasitoids (Petzoldt and Seaman, 2010).

Effects of climate change on the efficiency of pathogens depend on the environment they live in. In general, fungi and bacteria benefit from warm and moist environments, therefore mild and wetter winters as predicted in temperate zones will benefit them, especially those living in the soil (e.g. *Beauveria bassiana*). Since many larvae or pupae of pests also overwinter (pass through or wait out the winter season) in soils, fungi and bacteria might affect them more strongly.

Parasitoids which live on crop pests belong to the third trophic level. Thus they are indirectly or directly affected by any changes of the first (plant) and second level (herbivore). It is not at all clear what happens to herbivores under climate change, therefore conclusions for parasitoids are speculative. However, there are some ecological 'laws' which imply certain scenarios. If an herbivore reproduces less, because of low nutritional value, less potential hosts are available for the parasitoid. If the host changes it seasonal appearance or behaviour due to climatic changes the parasitoid might not be able to locate the host. Finally parasitoids might be adversely affected, if the host dies too early due to additional environmental stress. However, in temperate zones milder winters might enhance survival of parasitoids. Legrand *et al.* (2004) have shown that parasitoids of cereal aphids are active in winter and this winter activity can considerably reduce spring aphid populations.

Expression of the host plant resistance is greatly influenced by environmental factors like temperature, sunlight, soil moisture, air pollution, etc. Under stressful environment, plant becomes more susceptible to attack by insect-pests because of weakening of their own defensive system resulting in pest outbreaks and more crop damage (Rhoades, 1985). Thermal and drought stress associated breakdown of plant resistance have been widely reported (Rhoades, 1985; Sharma *et al.*, 2005).

Insects play vital role in providing various ecosystem services, a foundation for human life on earth (Kannan and James, 2009). One of the important ecosystem service provided by insects is pollination as they are excellent pollinators for many of the economically important crops (Sidhu and Mehta, 2008). The majority of the flowering plants require insect pollinators like flies, butterflies, moths, beetles and especially bees for their reproduction and formation of fruits and seeds (Ricketts *et al.*, 2008). Honey bees are perhaps the best known pollinators because of their floral fidelity. Insect pollination, mostly by bees, is necessary for 75% of all crops that are used directly for human food worldwide. Thus the insect pollination is a fundamental process essential for the production of about one third of the world human food (Klein *et al.*, 2007).

Certain effective cultural pest management practices like crop rotation, early/late planting, etc. will be less or not effective with changed climate because of shrinking of crop growing seasons, colonization of crops by early insect arrival and/or increased winter survival (Harrington *et al.*, 2001; Sharma *et al.*, 2005; Petzoldt and Seaman, 2010). Disruption of synchrony between insect-pests and their natural enemies may upset the natural biological control (Petzoldt and Seaman, 2010). Certain pesticides like pyrethroids, organophosphates and especially the bio-pesticides being highly thermo-unstable degrade faster at higher temperatures. Altered temperature regimes may render many of these products to be less or not effective in pest control, necessitating frequent insecticide applications for effective control (Musser and Shelton, 2005). This may intensify the pest problems due to the increased chances of resistance development in insects. Ultimately it will add to increased cost of crop protection to the farmers and in turn environmental cost (Musser and Shelton, 2005; Petzoldt and Seaman, 2010). The forewarning models for predicting insect arrival/ infestations based on earlier climate profiles need to be revised in accordance with location specific changes in climate in order to provide precise and accurate forecast of the pest incidence.

It is likely that farmers will experience extensive impacts on insect management strategies with changes in climate. Entomologists expect that insects will expand their geographic ranges, and increase reproduction rates and overwintering success. This means that it is likely that farmers in the north-eastern US will have more types and higher numbers of insects to manage. Based on current comparisons of insecticide usage between more southern states and more northern states, this is likely to mean more insecticide use and expense for north-eastern farmers. New York conditions currently require 0-5 insecticide applications against lepidopteron insect pests to produce marketable sweet corn (Stivers, 1999a); Maryland and Delaware conditions require 4-8 insecticide applications (Whitney *et al.*, 2000); Florida conditions require 15-32 applications (Aerts *et al.*, 1999). It is apparent that for sweet corn pests, warmer temperatures translates to increased insecticide applications to produce a marketable crop. Insecticides and their applications have significant economic costs for growers and environmental costs for society. Additionally, some classes of pesticides (pyrethroids and spinosad) have been shown to be less effective in controlling insects at higher temperatures (Musser and Shelton, 2005).

3.6. PREDICTION MODELLING FOR INSECT PESTS

Models have become important analytical tools for the evaluation, understanding and prediction of the dynamics of insect populations in agro-ecosystems and in phytosanitary risk assessments. CIP developed a temperature-driven phenology model for the potato tuber moth (*Phthorimaea operculella*) that provides overall good predictions of population growth potential for the wide range of climatic conditions where the pest prevails today. Linked with geographic information systems (GIS) and atmospheric temperature the model allows simulating risk indices on a worldwide scale which are used to predict future changes of the species distribution due to global warming. The approach applied for the development and prediction of *P. operculella* can be utilized for other insect species. Hence, CIP developed Insect Life Cycle Modelling software (ILCYM) facilitates the development of insect phenology models.

ILCYM consists of three main modules: The "Model builder", which facilitates the development of insect phenology models based on life table data derived from constant temperature studies; the "Validation and simulation module", which provides tools for validating the developed phenology model using insect life tables established at fluctuating temperature conditions; and the "Potential population distribution and risk mapping module", in which the phenology models are implemented in a GIS environment that allows for spatial – global or regional – simulations of species activities and mapping.

In a collaborative effort between three International Agricultural Research Centres (CIP, IITA, and ICIPE), the University of Hohenheim, Germany and partners at NARI in Africa – funded by BMZ, Germany – innovative phenology modelling and risk mapping is used to understand the effects of rising temperatures caused by climate change on the distribution and severity of major insect pests on important food crops in Africa. The ILCYM software will be further improved and adapted to cover a wide range of insects of different orders and families. The results will contribute to fill the current knowledge gap about climate change effects on economically important insect herbivores and their natural enemies, especially in the tropics.

3.7. ADAPTATION AND MITIGATION STRATEGIES

3.7.1. Use of synthetic pesticides

In a broad sense, pesticide is any agent used to kill or control undesired insects, weeds, rodents, fungi, bacteria or other organisms. Schillhorn (1999), in his review of agricultural pest management noted that in Asia, pesticides are often sold through unlicensed dealers and shopkeepers. Some of these may be familiar with agrochemicals, but in many rural shops one may find pesticides sold next to food products such as milk or bread. Although the availability of pesticides in rural areas is important, the benefit of such availability has to be weighed against consumer safety, proper storage, and the need to educate farmers about proper use. For example, it has been claimed that in certain parts of India, the daily food borne intake of pesticide residues is approximately 0.51 mg (Alam, 1994). This is well above accepted levels, but the impact on public health and reproduction has not yet been fully investigated.

3.7.1.1. Screening of pesticides with novel modes of action: It has been reported by some researchers that the application of neonicotinoid insecticides for controlling sucking pests induces salicylic acid associated plant defence responses which enhance plant vigour and abiotic stress tolerance, independent of their insecticidal action (Ford *et al.*, 2010). This gives an insight into investigating role of insecticides in enhancing stress tolerance in plants. Such more compounds needs to be identified for use in future crop pest management.

3.7.2. Use of biopesticides (plant extracts)

Many farmers in Asia and Africa have been using plant extracts such as neem (*Azadiractha indica*), wild castor (*Calotropis procera*) wood ash and dried chillies among others for controlling and repelling some insect pests (Ahmed *et al.*, 2005). The neem extracts have been reported to be broad-spectrum in activity, degrade rapidly to harmless metabolites and therefore, leave no residues in the environment where they are applied (Schmutterer, 1990; Deka and Singh, 2001). The use of neem extracts and other medicinal plants can form an important component of pest management strategies, especially in developing countries (Obeng-Ofori and Ankrah, 2002). Many scientists and farmers have reported that the use of crude or formulated plant pesticides for pest management in Asian and African countries (Tsado and Tanko, 2002; Tang'an *et al.*, 2002; Anjorin *et al.*, 2004). In sub-Saharan Africa, the use of plant derivatives for pest control was said to have been common before the advent of synthetic pesticides, and the preparation and application of botanicals for crop protection for increased food production were linked to the folklores and tradition of the farmers (Saxena, 1987). The benefits of neem and other plants as source of pesticide include:

- It is relatively cheap and easily available.
- It possesses complex mixture of active ingredients which function differently on various parts of the insects' life cycle that makes it difficult for pests to develop resistance to it (Anjorin, 2008).
- It is systematic, thereby protecting the plant from within (Anjorin, 2008).
- Neem has also been shown to be effective in controlling pathogens, such as root-knot nematode, *Meloidogyne*; fungus, *Rhizoctonia*; and rice stunt virus (Anon, 1992).
- Khalid and Shad (2002) specifically reported that their toxic effect is normally of an ephemeral nature disappearing within 2 – 3 weeks.
- Eze *et al.* (2006) reported that alcohol extracts of *Azadiractha indica, Oscimum gratissimum, Xylopia ethiopica* and *Zingiber officinale* effectively inhibited population growth of weevils in stored yam chips and recommended that farmers should be educated on the economic and environmental benefits of botanical extracts.

Governments and development agencies in the developing world should therefore, encourage and promote the preparation and adoption of neem and other plant products at the farm level by resource poor farmers to protect their crops against insect pest infestation so as to ensure food security and safety.

3.7.3. Cultural practices

Because pests destroy potential crop production worldwide, use of appropriate cultural technologies to reduce pest losses would increase crop yields (Pimentel *et al.*, 1993). Common cultural practices that contribute to reduction in pest build-up include crop rotation, tillage, varying timing of planting or harvesting, planting trap crops and adjusting row width. Roller and Poehling (2004) reported that some insect pests and disease-causing organisms are host specific giving example of how rice stem borer was controlled by planting other crops that were not attacked by the stem borer. It has been reported that tillage destroys pest's shelters and disrupt their life cycles, exposes pests to predators and unfavourable conditions, distributes soil nutrients throughout the soil, controls weeds and makes other cultural practices easier to undertake (FAO, 2002). However, tillage disrupts the life cycle of other beneficial soil organisms and needs more labour cost for soil preparation among other disadvantages.

3.7.3.1. Intercropping: Intercropping is the cultivation of two or more crops simultaneously on the same field. It also means the growing of two or more crops on the same field with the planting of the second crop after the first one has completed its development. Facknath (2000) noted that the rationale behind intercropping was that the different crops planted were unlikely to share the same insect pests. The report further noted that intercropping reduces the insect/mite pest population because of the diversity of crops grown when other crops are present in the field, the insect/mite pests become confused and they needed more time to look for favourable plant hosts.

3.7.3.2. Trap cropping: Trap cropping is the planting of a crop which could attract and harbour a pest(s) to protect main crop from a certain pest or several pests. Grundy (2003) noted that trap cropping prevents pest attack that comes from all sides of the field and works best on pests that are found near the borderline of the farm. Row intercropping could be achieved by planting of the trap crop in alternating rows within the main crop. A planned trap cropping practice showing the trap crop and main crop, method of planting and pest control as demonstrated by Olkowski et al. (1995) is shown in Table 3.2.

Table 3.2. Examples of trap cropping practices

Trap crop	Main crop	Method of planting	Pest controlled	Reference
Basil and marigold	Garlic	Border crop	Thrips	MMSU, 2003
Castor	Cotton	Border crop	*Heliothis* sp.	Hasse, 1987
Chervil	Vegetables, Ornamentals	Among plants	Slugs	Ellis and Bradley, 1996
Chinese cabbage, mustard, and radish	Cabbage	Planted after every 15 rows of cabbage	Cabbage webworm, Flea hopper, Mustard aphid	Facknath, 2000
Chick pea (Grundy; Short)	Cotton	Block trap crop at 20 plants/ m^2	*Heliothis* sp.	Grundy, 2003

Trap crop	Main crop	Method of planting	Pest controlled	Reference
Corn	Cotton	Row intercrop, planted after every 20 rows of cotton or every 10-15 m	*Heliothis* sp.	Hasse, 1987
Cowpea	Cotton	Row intercrop after every 5 rows of cotton	*Heliothis* sp.	CIKS, 2000
Desmodium	Corn, Cowpea, Millet, Sorghum	Row intercrop	Stem borer, Striga	ICIPE, 2003
Onion and garlic	Carrot	Border crops or barrier crops in between plots	Carrot root fly, Thrips	—
Sickle pod	Soybean	Strip intercrop	Velvet bean caterpillar, Green stink bug	—
Soybean	Corn	Row intercrop	*Heliothis* sp.	—
Sudan grass	Corn	Intercrop or border crop	Stem borer	ICIPE, 2003

3.7.3.3. Pruning: Pruning is the selective removal of specific plant parts like shoots and branches. Roots, flower buds, fruits, and seed pods can also be pruned. French and Appleton (2001) reported that pruning done on regular basis as part of plant care makes the plant less dense, improves the air circulation and sunlight penetration that decreased the incidence of pests and the conditions that promoted fungal growth and gets rid of the pest infested parts among others.

3.7.3.4. Rescheduling of crop calendars: Global temperature increase and altered rainfall patterns may result in shrinking of crop growing seasons with intense problems of early insect infestations. As such certain effective cultural practices like crop rotation and planting dates will be less or not effective in controlling crop pests with changed climate. Hence there is a need to change the crop calendars according to the changing crop environment. The growers of the crops have to change insect management strategies in accordance with the projected changes in pest incidence and extent of crop losses in view of the changing climate.

3.7.4. Breeding climate-resilient varieties

In order to minimize the impacts of climate and other environmental changes, it will be crucial to breed new varieties for improved resistance to insect pests. Considering late onset and/or shorter duration of winter, there is chance of delaying and shortening the growing seasons for certain *Rabi*/cold season crops. Hence, there is a need to concentrate on breeding varieties suitable for late planting and those that can sustain adverse climatic conditions and pest incidences.

3.7.4.1. Transgenic crop cultivars for pest management: Another important issue regarding pest management in the future centers on the role of biotechnology in crop protection. The next 20 years will likely to see a substantial increase in the use of genetically engineered plants. Genetically engineered plants have been designed to resist pests such as stem borers and nematodes without the need for pesticides. Others are expected to combine both herbicide resistance and insect resistance in one seed.

3.7.5. GIS Based risk mapping of crop pests

Geographic Information System (GIS) is an enabling technology for entomologists, which help in relating insect pest outbreaks to biographic and physiographic features of the landscape, hence can best be utilized in area-wide pest management programmes. How climatic changes will affect development, incidence, and population dynamics of insect pests can be studied through GIS by predicting and mapping trends of potential changes in geographical distribution (Sharma *et al.*, 2010) and delineation of agro-ecological hot spots and future areas of pest risk (Yadav *et al.*, 2010).

3.7.6. Integrated pest management

Integrated pest management (IPM) is an effective and environmentally sensitive approach to pest management that uses current, comprehensive information on the life cycles of pests and their interaction with the environment to manage pest damage by the most economical means, and with the least possible hazard to people, property, and the environment.

Man in his efforts to destroy and or reduce his enemies (insect pests, rodents and pathogens) found that no one single method was perfect and even the very effective methods (chemicals) had negative impacts on the quality (food safety) of food crops. A combination of all the known methods of pest control (integrated pest management) was adopted. Hoyt (2001) described integrated pest management (IPM) as an effective and environmentally sensitive approach to pest management that relies on a combination of common sense practices. Sankoh (1999) noted that the quality of human life depends ultimately on the quality of the environment in which humans lives and ability of this environment to provide food (food security), shelter and natural resources needed to generate employment and well secured life. Since no single pest control method can guarantee food security and safety, IPM approach appears to bring hope but the impact in the rural farm communities is low. A research on integrated pest management adoption among farmers carried out in Nigeria by Ofuoku *et al.* (2008) revealed that majority of the farmers (84.37%) had not adopted IPM, while only 15.63% had adopted it. This according to the report could be because most of the farmers considered IPM as a computer technology that was not suitable for their own environment or not affordable at the moment. This is probably true because Agbamu (2006) had earlier reported that certain research findings, which were deemed to improve farm production, may be beyond the understanding of the rural farmers, even with the interpretation of the extension agents. It is also probable that the availability of local pest control technique provided cheaper alternatives to improved one like IPM.

3.8. FUTURE THREATS

Direct effects of climate change on herbivorous insects can be divided into the following two main categories:

- Geographical distribution range shift.
  - Natural expansion due to global warming.
  - Accidental introduction caused by the increase of world trade, resulting in the establishment of herbivorous insects outside of their natural distribution range.
- Increase of the number of generations per year.

Examples of plant pests whose geographical distribution and pest incidence and/or intensity will be substantially influenced by climate change (i.e. change in ecological conditions will substantially widen the area where the plant pest could be introduced and/or could substantially increase its potential economic importance) and of which economic (including environmental) consequences will be substantial. Some examples of insect and mite pests are given below (Table 3.3):

Table 3.3. Insect and mite pests that can assume serious proportions in future due to climate change

Insect/mite pests	Anticipated Effect of Climate Change
Geographical distribution range shift	
Colorado potato beetle, *Leptinotarsa decemlineata*	An average warming of 2.3°C, the pest could expand its potential range in the UK by 120% (Baker *et al.*, 1998). This would extend the potential northern limit of the species by 400 km, thus posing a potential risk to more than 99% of potato producing areas.
Western corn rootworm, *Diabrotica virgifera virgifera*	Pest was introduced into Europe near Belgrade in 1992, and has since been spreading east and northwards. This major pest has not yet spread to its potential limits under existing climatic conditions, but global warming has the potential for increase its range considerably northwards. Under current climate conditions, it appears to be at the edge of its range in the UK, but by 2050, under global warming, a large area of SE England will be much more suitable for this species (Baker *et al.*, 2003).
Leaf miner, *Liriomyza huidobrensis*	Warmer winter temperatures are likely to increase the probability of the non-indigenous South American leaf miner overwintering outdoors in the UK (Baker *et al.*, 1996).
Diamondback moth, *Plutella xylostella*	At the northern extremity of its range, this species generally occurs at relatively low population densities and it does not usually overwinter. If however, as a result of global warming, overwintering occurred more frequently, the pest status of this species would increase dramatically (Dosdall, 1994).
European corn borer, *Ostrinia nubilalis*	The pest may be able to produce additional generations, relative to current conditions, in a given locale, with a potentially greater impact on their host plants. Pest is predicted

Insect/mite pests	Anticipated Effect of Climate Change
	to become bivoltine – i.e. to produce two generations per season rather than one – in the Czech Republic as a result of predicted increases in temperatures during the period 2025-50 (Trnka *et al.*, 2007). This species is at present near the northern limit of its range in southern England. A 3° C increase in mean annual temperature would advance the limit for maize providing an opportunity for a substantial northward shift in the distribution of *O. nubilalis* (Porter, 1995).
Old world bollworm, *Helicoverpa armigera*	Migrant moths have shown a "phenomenal increase" over the period 1969-2004, penetrating inland more frequently (Parsons and Davey, 2007), although this has not (yet) led to an increase in its pest status in the UK. However, there have been outbreaks in areas at the northern edge of its range in Europe, including the first finding in 2003, of *H. armigera* in Austria; and a damaging outbreak, also in 2003, in Baden-Württemberg, Germany, on a great diversity of plants grown outdoors (EPPO, 2014).
Cottony cushion scale, *Icerya purchasi*	Outdoor populations appear to be naturally spreading northwards perhaps as a consequence of global warming. In 1999, a severe infestation was reported in the Jardin des Plantes, Paris and it has also recently been found breeding outdoors in London (Watson and Malumphy, 2004).
Cottony camellia scale, *Pulvinaria floccifera*	*P. floccifera* has become much more common in the UK, extending its range northwards in England and increased its host range in the last decade or so. This is almost certainly a response to climate change.
Laurel scale, *Aonidia lauri*	Pest has been found breeding outdoors sporadically in the London area since about 1996 (Malumphy, 1997). It is likely to respond to climate change and become more common.
Cabbage white butterfly, *Pieris brassicae*	Most butterflies are predicted to increase in diversity, range and abundance in the UK under a warmer climate, but according to Roy *et al.* (2001) *P. brassicae* would likely to decline.
Cabbage root fly, *Delia radicum*	A 3°C increase in mean daily temperatures would cause root fly populations to become active about a month earlier in the UK (Collier *et al.*, 1991), which may translate into increased pest severity.
Fruit flies	Invasion of the Japanese Archipelago.
Green stink bug, *Nezara viridula*; Rice bug, *Leptocorisa chinensis*	Northward range shift from tropical and subtropical countries to the temperate zones in the world.
Rice gall midge, *Orseolia oryzae*	Northward range shift to the subtropical Asia.

Insect/mite pests	Anticipated Effect of Climate Change
Sweet potato weevil, *Cylas formicarius*	Accidental introduction from tropical and subtropical countries to the temperate zones and establishment there under warmer conditions.
Citrus psyllid, *Diaphorina citri*	Northward range shift from tropical and subtropical countries to the te mperate zones.
Rice plant hoppers	Northward range shift from tropical and subtropical countries to the temperate zones (long distant migration).
Increase in the number of generations per year	
Aphis gossypii, Myzus persicae, Aulacorthum solani	Increase of the number of generations a year due to lower developmental zero point and low thermal totals required for one generation, resulted in more infestation.
Helicoverpa armigera, Mamestra brassicae, Spodoptera litura S. exigua	Increase of the number of generations a year, resulted in more infestation.
Rust Mites	Increase of the number of generations a year due to lower developmental zero point and low thermal totals required for one generation, resulted in more infestation.
Shift from greenhouse to field	
Whiteflies, aphids, mites, etc.	In Japan, previously pests in greenhouses were all domestic species, but in recent years about 50% of them are alien species. They came to Japan and can overwinter in greenhouses. Some of them will be able to go into the field from the greenhouses in future.
Camellia scale, *Pulvinaria floccifera*	In Sweden this species was previously only known as a greenhouse species, but is now established as an outdoor species (Gertsson, 2005).

3.9. FUTURE LINES OF WORK

- Emission of greenhouse gases can be minimized through adoption of conservation agriculture like direct seeded rice, aerobic rice, SRI (System of Rice Intensification), crop diversification, organic farming, bio-organic fertilizers and biological pest control, etc.
- Development of crop growth simulation models will be of greater relevance in new emerging research areas such as climate change impacts on pests and crop yield.

  For effective pest management there is need to:
- Predict and map trends of potential changes in geographical distribution and study how climatic changes will affect development, incidence and population dynamics of insect pests.

- Understand the influence of climate change on species diversity and cropping patterns and their influence on the abundance of insect pests and their natural enemies.
- Understand the changes in expression of resistance to insect pests and identify stable sources of resistance and pyramid the resistance genes in commercial cultivars.
- Exhaustive studies on the effect of climate change on the efficacy of transgenic crops in pest management.
- Assess the efficacy of various pest management tools under variable environmental conditions and develop viable pest management strategies to mitigate the effects of climate change on food security.

3.10. CONCLUSIONS

Global warming and climate change will have serious consequences on diversity and abundance of arthropods, and the extent of losses due to insect pests, which will impact both crop production and food security. Prediction of changes in geographical distribution and population dynamics of insect pests will be useful to adapt the pest management strategies to mitigate the adverse effects of climate change on crop production. Pest outbreaks might occur more frequently, particularly during extended periods of drought, followed by heavy rainfall. Some of the components of pest management such as host plant resistance, biopesticides, natural enemies, and synthetic chemicals will be rendered less effective as a result of increase in temperatures and UV radiation, and decrease in relative humidity. Climate change will also alter the interactions between the insect pests and their host plants. As a result, some of the cultivars that are resistant to insect pests, may exhibit susceptible reaction under global warming. Adverse effects of climate change on the activity and effectiveness of natural enemies will be a major concern in future pest management programmes. Rate of insect multiplication might increase with an increase in CO_2 and temperature. Therefore, there is a need to have a concerted look at the likely effects of climate change on crop protection, and devise appropriate measures to mitigate the effects of climate change on food security.

4

Effects of Climate Change Variables on Insect Pests

4.1. INTRODUCTION

Insect habitats and survival strategies are strongly dependent on patterns of climate. Therefore, it is highly expected that, the major drivers of climate change *i.e.* elevated CO_2, increased temperature and depleted soil moisture can impact population dynamics of insect-pests and the extent of crop losses, significantly. In addition, temperature and moisture may influence survival, development and reproduction of arthropods (Fig. 4.1). It has been reported that, global climate warming may lead to altitude wise expansion of the geographic range of insect pests (Elphinstone and Toth, 2008), increased abundance of tropical insect species (Diffenbaugh *et al.*, 2008), decrease in the relative proportion of temperature sensitive insect population (Sharma *et al.*, 2010), more incidence of insect transmitted plant diseases through range expansion and rapid multiplication of insect vectors (Petzoldt and Seaman, 2010). Thus, with changing climate it is expected that the growers of crops have to face new and intense pest problems in the years to come. The aggravating pest problems under changing climate regimes are expected to intensify the yield losses; threatening the food security of the countries with high dependency on agriculture. The climate change is likely to affect the extent of insect pollination by disrupting the synchrony between plant-pollinator life cycles, with an estimated risk of reduction in world food production by one-third (Klein *et al.*, 2007).

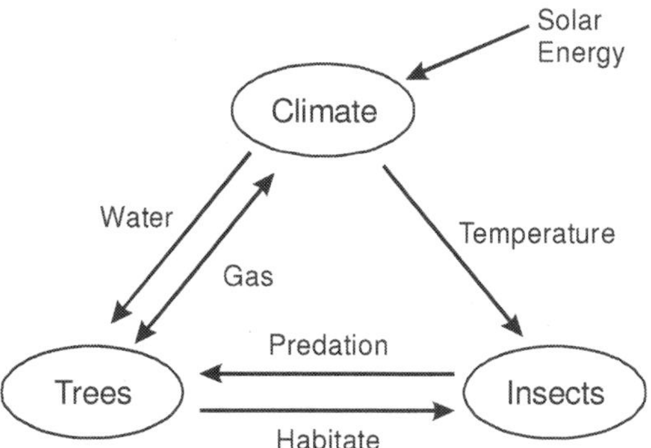

Fig. 4.1. Effects of climate change on insect pests

Effects of climate change on crop pests include:

- Increased levels of temperature and carbon dioxide due to climate change have a profound influence on crop pests (arthropods).
- Insects are poikilothermic (cold blooded organism); the temperature of their body is proximately the same as that of the environment.
- Temperature increases associated with climate change affect their reproduction, development and survival.
- Rising temperatures extend geographical range of insect pests.
- Rising winter temperatures reduce winter mortality.
- Rising temperatures extend the growing season.
- Rising temperatures accelerate insect life cycles- greater generation numbers. It has been estimated that with a 2 °C temperature increase, insects might experience one to five additional life cycles per season.
- UV radiation weakens the health of plants which leaves them more vulnerable to pests and diseases which are earlier minor.
- A rise in the carbon dioxide concentration increases the C: N ratio of plants reducing the nutritional quality for the protein which increases insect's food intake (compensatory feeding). There is dilution of N by 10-30% which causes 40% increase in food consumption by herbivores.
- Increased CO_2 may cause a slight decrease in N based defences (alkaloid) and slight increase in carbon based defences (tannins) in plants which breaks down resistance of plants to insects (e.g., sorghum midge, *Stenodiplosis sorghicola*).
- Pest outbreaks are more likely to occur with stressed plants as a result of weakening of plants defensive system, and thus increasing the level of susceptibility to insect pests.
- Altered wind patterns may change the spread of wind-borne pests.

4.2. CHANGE IN CROP PEST BEHAVIOUR DUE TO CLIMATE CHANGE

- Unusually severe draught increases the damage by spotted stem borer in sorghum.
- Boll worm incidence in cotton may increase by 4.17% with an increase of maximum temperature by 1°C.
- Aphid abundance increases with an increase of carbon dioxide and temperature.
- Increased temperatures will accelerate occurrence of cabbage maggot, onion maggot, European corn borer and Colorado potato beetle.
- High temperatures with humidity are subjected to severe cut worm damage. It will lead to earlier infestation by *Helicoverpa zea* in North America and *H. armigera* in North India.
- Shorter dry spell lead to higher mite and mealy bug infestation.
- Longer growing seasons with warmer climate will enable insect pests such as grasshoppers to complete a greater number of reproductive cycles during spring, summer and autumn.

- Increase in rainfall will largely affect the species particularly the beetle with poor dispersal capabilities, which will limit their ability to expand their home.
- Lower winter mortality of insects due to warmer winter temperatures could be important in increasing insect populations.
- Changes in climatic variables have led to increased frequency and intensity of outbreaks of insect-pests. Outbreak of sugarcane woolly aphid, *Ceratovacuna lanigera* in sugarcane belt of Karnataka and Maharashtra states during 2002-03 resulted in 30% yield losses.
- Changes in climatic variables have led to increased frequency and intensity of outbreaks of fruit fly in cucurbits particularly due to high temperature and high RH.
- Increasing infestation of rice crop by swarming caterpillar, hispa, stem borer and bacterial blight.

4.3. INCREASED TEMPERATURES

Climate change resulting in increased temperature could impact crop pest populations in several ways. Most researchers seem to agree that warmer temperatures in temperate climates will results in more types and higher populations of insects. The increased temperature can potentially affect insect survival, development, geographic range and population size. It can impact on physiology and development directly or indirectly through the physiology or existence of hosts. Some insects that take several years to complete one life-cycle (cicadas) will tend to moderate temperature variability over course of their life history. The potato tuber worm incidence under climate changes showed that the population of the pest at Ismalia gave the highest number of generations as compared with El Beheira location under current climate. However, the expected generation numbers of the tuber worm in 2050 and 2100 are expected to be 9 to 11 and 10 to 12 generations per year, respectively. The higher temperature in the future may thus increase the damage on crops, by increasing the number of generations of the pest (Abolmaaty *et al.*, 2011).

The influence of temperature on the developmental times and survival of argentine ant, *Linepithema humile* (Hymenoptera: Formicidae) can largely determine their distribution. Temperature affects both the complete brood development from egg to adult worker and each of the immature stages separately. Higher the temperature, shorter will be the developmental time. Brood survival from egg to adult was low, with the maximum survival rate being only 16% at 26 °C. Temperature also affected survival of each of the immature stages differently: eggs were negatively affected by high temperatures, while larvae were negatively affected by low temperatures, and the survival of pupae were apparently independent of environmental temperature. At 32 °C, no eggs survived, while at 18 °C less than 2% of the eggs hatched into larvae. The data of the study are essential for developing prediction models about the distribution range of this tramp species based on its physiological needs in relation to temperature (Abril *et al.*, 2008).

Elevated temperature decreased final plant biomass while leaf nitrogen concentration increased. Aphid, *Myzus persicae* abundance was enhanced by both CO_2 and temperature treatment. Parasitism rates remained unchanged in elevated

CO_2, but showed an increasing trend in conditions of elevated temperature. It shows that *M. persicae* might increase its abundance under changing climatic conditions (Bezemer *et al.*, 1998).

Many insects are contributors to global warming because of the CO_2 they emit. Bug and termites are major contributors of global warming. With every degree the global temperature rise, the life cycle of each bug will be shorter. The quicker the life cycle, the higher will be the population of pests (Deka *et al.*, 2012).

Insects are poikilothermic (cold-blooded) organisms; the temperature of their bodies is approximately the same as that of the environment. Therefore, temperature is probably the single most important environmental factor influencing insect behaviour, distribution, rates of development, survival, and reproduction in tropical and subtropical insects (Fig. 4.2) (Yamamura and Kiritani, 1998; Bale *et al.*, 2002; Petzolet and Seaman, 2010). Consequently, insects will be capable of completing more number of generations per year and ultimately it will result in more crop damage (Yamamura and Kiritani, 1998; Petzoldt and Seaman, 2010). Insect life stage predictions are most often calculated using accumulated degree days from a base temperature and bio-fix point. Some researchers believe that the effect of temperature on insects largely overwhelms the effects of other environmental factors (Bale *et al.*, 2002). It has been estimated that with a 2°C temperature increase insects might experience one to five additional life cycles per season (Yamamura and Kiritani, 1998).

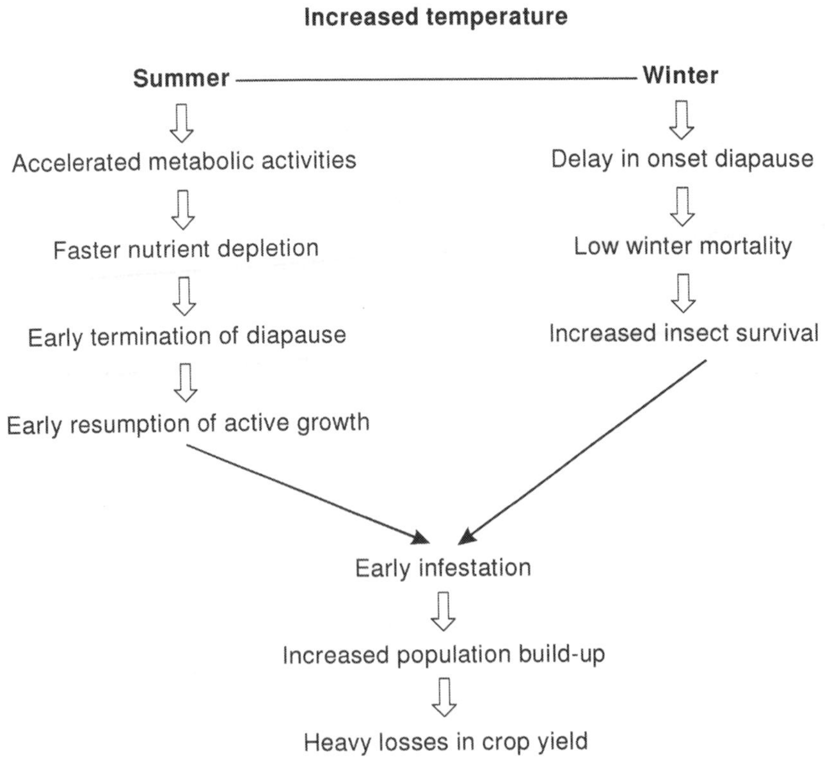

Fig. 4.2. Effect of increased temperature on insect population build-up

In cooler regions (higher latitudes) with distinctive seasons, insects have broader thermal tolerance and are living in climates that are currently cooler than their optima (Deutsch *et al.*, 2008). Global warming might therefore benefit many insect species in the temperate regions.

Reduced winter mortality of two insect pests were observed in Japan, when mean temperature in January was above 4° C. Species which already develop at low temperatures and need a smaller number of warm days will benefit from increasing average temperatures, and produce more generations and appear early when spring temperatures are high. Species which require higher temperature before they lay their eggs tend to have increased reproductive activity.

Musolin (2007) concludes that warming in temperate regions may have manifold effects on bugs. The responses of slight and substantial temperature increases on bugs (Heteroptera) are presented in Table 4.1 (Musolin, 2007).

Table 4.1. Expected responses of Heteroptera species and communities under two scenarios of further climate change

Category of responses	Slight temperature increase (< 2° C)	Substantial temperature increase (> 2° C)
Distribution range	Likely to shift in some species, especially those capable of long distance flights and associated with ornamental plants and/or urban habitats	Likely to change in many species
Abundance	Likely to increase in multi-voltine species with flexible life cycles	Likely to change, depending on the community response
Phenology	Slight to moderate advance of early season events	Substantial advance of early season and some delay of late season events
Voltinism	An additional generation in some multivoltine species with flexible life cycles	One or more additional generation(s) in some multivoltine and univoltine species (with facultative diapause)
Physiology and behaviour	Slight/undetectable changes	Evident/detectable changes (e.g. in parameters of photoperiodic responses)
Community structure	Similar to currently observed	Increased species richness; Substantial changes in structure

Studies on aphids and moths have shown that increasing temperatures can allow insects to reach their minimum flight temperature sooner, aiding in increased dispersal capabilities (Zhou *et al.*, 1995). Increased temperatures will accelerate the development of cabbage maggot, onion maggot, European corn borer, Colorado potato beetle–possibly resulting in more generations (and crop damage) per year.

Positive physiological responses to increasing temperatures will allow for faster insect growth and movement. Additionally, milder winters will allow for earlier insect growth and a reduction in overwinter deaths. The expansion or shift in ranges coupled with an increase in growth and numbers will likely result in an increase in insect invasions.

The effect of enhanced atmospheric temperature on crop pest dynamics is presented in Table 4.2.

Table 4.2. Effect of enhanced atmospheric temperature on crop pest dynamics

Increasing atmospheric temperature leads to	Reference/s
Increasing	
Northwards migration	Parmesan, 2006
Migration up elevation gradients	Epstein *et al.,* 1998
Insect development rates and oviposition	Regnier, 1983
Potential for insect outbreaks	Bale *et al.*, 2002
Invasive species introductions	Dukes and Mooney, 1999
Insect extinctions	Thomas *et al.*, 2004
Decreasing	
Effectiveness of insect biocontrol by fungi	Stacy and Fellowes, 2002
Reliability of economic threshold levels	Trumble *et al.*, 2009
Insect diversity in ecosystems	Erasmus *et al.*, 2002
Parasitism	Hance *et al.*, 2007; Fleming and Volney, 1995

The temperature increases associated with climatic changes could impact crop pest insect populations in several complex ways like,

- Extension of geographical range.
- Increased over wintering.
- Changes in population growth rate.
- Increased number of generations.
- Extension of development season.
- Changes in crop pest synchrony.
- Changes in interspecific interactions.
- Increased risks of invasions by migrant pests.
- Introduction of alternative hosts and over wintering hosts.

But all these effect of temperature on insects largely overwhelms the effects of other environmental factors (Bale *et al.*, 2002).

4.3.1. How rising temperatures affects insect pests

Climate change resulting in increased temperature could impact crop pest insect populations in several complex ways. Although some climate change temperature effects might tend to depress insect populations, most researchers seem to agree that warmer temperatures in temperate climates will result in more types and higher populations of insects.

4.3.1.1. Increase in pest insect populations: Researchers have shown that increased temperatures can potentially affect insect survival, development, geographic range, and population size. Temperature can impact insect physiology and development directly or indirectly through the physiology or existence of hosts. Depending on the development "strategy" of an insect species, temperature can exert different effects (Bale *et al.*, 2002). Some insects take several years to complete one life cycle – these insects (cicadas, arctic moths) will tend to moderate temperature variability over the course of their life history. Some crop pests are "stop and go" developers in relation to temperature – they develop more rapidly during periods of time with suitable temperatures. We often use degree-day or phenology based models to predict the emergence of these insects and their potential to damage crops (cabbage maggot, onion maggot, European corn borer, Colorado potato beetle). Increased temperatures will accelerate the development of these types of insects – possibly resulting in more generations (and crop damage) per year.

"Migratory" insects (corn earworm in northern parts of the northeast) may arrive in the Northeast earlier, or the area in which they are able to overwinter may be expanded. Natural enemy and host insect populations may respond differently to changes in temperature.

Parasitism could be reduced if host populations emerge and pass through vulnerable life stages before parasitoids emerge. Hosts may pass though vulnerable life stages more quickly at higher temperatures, reducing the window of opportunity for parasitism. Temperature may change gender ratios of some pest species such as thrips (Lewis, 1997) potentially affecting reproduction rates. Insects that spend important parts of their life histories in the soil may be more gradually affected by temperature changes than those that are above ground simply because soil provides an insulating medium that will tend to buffer temperature changes more than the air (Bale *et al.*, 2002).

Lower winter mortality of insects due to warmer winter temperatures could be important in increasing insect populations (Harrington *et al.*, 2001). Higher average temperature might result in some crops being able to be grown in regions further north – it is likely that at least some of the insect pests of those crops will follow the expanded crop areas. Insect species diversity per area tends to decrease with higher latitude and altitude (Gaston and Williams, 1996; Andrew and Hughes, 2005), meaning that rising temperatures could result in more insect species attacking more hosts in temperate climates (Bale *et al.*, 2002). Based on evidence developed by studying the fossil record, some researchers (Bale *et al.*, 2002) conclude that the diversity of insect species and the intensity of their feeding have increased historically with increasing temperature.

4.3.1.2. Decrease in pest insect populations: Some insects are closely tied to a specific set of host crops. Temperature increases that cause farmers not to grow the host crop any longer would decrease the populations of insect pests specific to those crops. The same environmental factors that impact pest insects can impact their insect predators and parasitoids as well as the pathogens that infect the pests, resulting in increased attack on insect populations. At higher temperatures, aphids have been shown to be less responsive to the aphid alarm pheromone they release when under attack by insect predators and parasitoids – resulting in the potential for greater predation (Awmack *et al.*, 1997).

4.3.2. Effect of temperature on natural enemies

Natural enemy and host insect populations may respond differently to changes in temperature. Parasitism could be reduced if host populations emerge and pass through vulnerable life stages before parasitoids emerge. Temperature may change gender ratios of some pest species such as thrips (Lewis, 1997) by potentially affecting reproduction rates. At higher temperatures, aphids have been shown to be less responsive to the aphid alarm pheromone they release when under attack by insect predators and parasitoids – resulting in the potential for greater predation.

Changes in temperature may have highly nonlinear effects on tri-trophic interactions of host, pathogen, and biocontrol agent. In wheat, a rise in temperature from 17 to 22° C resulted in an increase in aphid (*Sitobion avenae*) reproduction by 10%; at the same time, however, predatory activity by lady beetle (*Coccinella septempunctata*) adults increased by 250%. Aphid damage was reduced further because of earlier maturity of the crop.

4.3.3. Effect of temperature on specific insect pests

Global warming will lead to earlier infestation by *Heliothis zea* in North America (EPA, 1989), and *Helicoverpa armigera* in North India (Sharma *et al.*, 2010), resulting in increased crop loss. Rising temperatures are likely to result in availability of new niches for insect pests.

An increase of 3° C in mean daily temperature would cause the carrot fly, *Delia radicum* to become active a month earlier than at present (Collier *et al.*, 1991), and temperature increases of 5 to 10° C would result in completion of four generations each year, necessitating adoption of new pest control strategies.

The studies on impacts of climate change on seasonality of insects have shown declined survival rate of brown plant hopper, *Nilaparvatha lugens* and rice leaf folder, *Cnaphalocrocis medinalis* at higher temperature which indicates that the impacts of rising temperature could change the pest population dynamics of rice ecosystem (Bale *et al.*, 2002).

Mild winters have allowed Queensland fruit fly (Qfly), *Bactrocera tryoni* to survive, resulting in population explosions in early spring. This had major impacts on many fruit and fruiting vegetable producers, especially when combined with the withdrawal of some key insecticides previously used to control the pest. Much of Victoria is now considered endemic for Qfly. Conversely, hot conditions in summer may be reducing Qfly populations in northern Queensland and the Northern Territory. Qfly does not do well if conditions are hot and dry, even if irrigation is provided. According to CSIRO modelling, increases of 1-2° C will see Qfly disappear from some areas.

Groundnut leaf miner (GLM), *Aproaerema modicella* is favoured by the hot dry conditions of the post-rainy season (Srinivasa Rao *et al.*, 2010). The life cycle is completed in 21-28 days and this species completes 3-4 generations in a crop season in south India. Dry weather, bright sunshine and occasional rains favour pest build-up (Ghule *et al.*, 1989).

Several observations can be made from literature on influence of elevated temperature on insect growth and development.

- Increased temperature influenced the larval development and fecundity of Winter moth, *Operophtera brumata* (Dury *et al.,* 1998) and long term exposure to increased temperature 3.5° C shortened the insect development (Williams *et al.,* 2003). The temperature enhancement increased the relative growth rate of chrysomellid beetles (Veteli *et al.,* 2002).
- Increased temperatures can potentially affect the insect survival, development and population size. Insects proliferate more readily in warmer climate, since conditions for growth and multiplication is more favourable compared to cooler conditions. Thus, incidence of insect pests will be more in the areas with increased temperature conditions.
- Because of warm winter and spring, many aphid species started their spring migration much earlier than 'normal' and the peach-potato aphid (*Myzus persicae*), in particular, has been captured in unprecedented numbers in the traps.
- The rate at which most pests develop is dependent on temperature and every species has a particular 'threshold temperature' above which development can occur, and below which development ceases. As temperature rises, some pest species might be able to complete more generations in a year. This effect may be most noticeable in insects with short life-cycles such as aphids and the diamond-back moth. On the other hand, the temporary exposure of populations to extremely high temperatures may delay the development of surviving individuals and thus delay the subsequent generation.
- Warmer winter temperatures might allow larvae to survive the winter where they are limited by the cold. Thus, greater infestation is expected in those areas.
- Climate change may affect the ability to control pests. For example, high temperature is reported to reduce the effectiveness of some pesticides. If pests are able to complete more generations in a season then this may lead to greater pesticide use.
- More number of applications of pesticide use which in turn might lead to the more rapid development of pesticide resistance. Strategies to avoid the development of resistance needs to be planned in advance and rely on the availability of a range of pesticides and/or alternative control methods.

In addition to the above observations some more predictions and generalizations were made by several researchers and are given in the Table 4.3.

Table 4.3. Impact of increased temperature on insect pests

Anticipated /Expected effect	References
Under climate change scenario increased asynchrony between host plant and insect herbivore, with obvious adverse consequences.	Dewer and Watt 1992
Higher temperatures keeping all other variables equal allow faster development of insects and may allow for additional generations of insect pest within a year.	Pollard and Yates, 1993
Climatic warming will allow the majority of 'temperate' insect species to extend their ranges to higher latitudes and altitudes.	Gaston and Williams, 1996

Anticipated /Expected effect	References
Expand their geographical ranges to higher latitudes and altitudes, as has already been observed in a number of common butterfly species.	Parmesan et *al.*, 1999
Elevated temperature is known to alter the phyto-chemistry of the host plant and affect the insect growth and development directly or indirectly through effect on host plants.	Williams *et al.*, 2000
Diversity of insect herbivores and the intensity of herbivore increases with rising temperatures at constant latitude. Individuals may develop faster at higher temperature and survival may even be enhanced, but these insects may consequently have lower adult weight and fecundity.	Bale *et al.*, 2002

4.4. ELEVATED CARBON DIOXIDE (CO_2) LEVELS

Generally, CO_2 impacts on insects are thought to be indirect-impact on damage results from changes in the host crop. During the early season, soybean grown in elevated CO_2 atmosphere had 57% more damage from insects (primarily Japanese beetles, potato leafhopper, western corn rootworm and Mexican bean beetle) than those grown in today's atmosphere. It is thought that measured increase in the levels of simple sugars in the soybean leaves may have stimulated the additional insect feeding (Hamilton *et al.*, 2005). Larval life-span of *Helicoverpa armigera* increased by 5.49, 7.02 and 10.26% and larval survival rate decreased by 7.35, 9.52 and 11.48% in first, second and third generations, respectively under elevated CO_2 compared with ambient CO_2. Consumption and frass per larva of bollworm fed on cotton bolls showed significant increase for the first, second and third generations under elevated CO_2. Significantly, lower relative growth rate was observed in the first, second and third generations (Gang *et al.*, 2007).

Brachypterous females laid more eggs on rice plants exposed to elevated than ambient CO_2. Elevated CO_2 exhibited positive effect on BPH multiplication and resulted in more than doubling of its population at peak incidence compared to ambient CO_2 (Prasannakumar *et al.*, 2012).

The lifespan of *H. armigera* was delayed and larvae fed more artificial diet and produced more frass under elevated CO_2 compared with those under ambient CO_2. Furthermore, elevated CO_2 marginally influenced the artificial diet-utilization efficiency of *H. armigera* larvae that decreased in relative growth rate (RGR), relative consumption rate (RCR), and efficiency of conversion of ingested food (Wu *et al.*, 2006a).

The direct effects of elevated CO_2 on multiple generations of cotton aphid, *Aphis gossypii* were weak, even not existing. Moreover, the impact of elevated CO_2 level on the growth, development and fecundity of *A. gossypii* was mainly indirect, even though the host plants growing under elevated CO_2 levels were directly affected (Chen *et al.*, 2005). Longer larval life-span for the third generation and lower pupal weight for all generations were observed in *H. armigera* fed on milky grains of spring wheat grown in elevated CO_2. Moreover, the consumption, frass per *H. armigera* larva and RGR significantly increased under elevated CO_2 compared to ambient CO_2 (Wu *et al.*, 2006b).

Elevated atmospheric CO_2 expected in the near future as a consequence of increasing emissions will alter the quantity and quality of plant foliage, which in turn can influence the growth and development of insect herbivores. The impact of elevated CO_2 on the phytochemistry of the plants was well documented (Coviella and Trumble, 1999; Hunter, 2001) (Fig. 4.3). The atmospheric CO_2 concentrations have increased by above 20% and elevated CO_2 affects the plant growth and range of physical and chemical characteristics of the plant/crop. These include reduction in the leaf nitrogen content, changes in the defence compounds, water content, carbohydrates and leaf thickness. Indications are that exposure to elevated CO_2 levels will increase the plant photosynthesis, growth, above ground biomass, leaf area, yield, carbon, and C: N ratio. These changes can influence the food quality for herbivorous insects and was well reviewed (Hunter, 2001). These changes in the leaf quality are likely to have varied effect on the performance of insect herbivores. The information on effect of elevated CO_2 on insect pests was compiled and presented by Srinivasa Rao *et al.* (2008).

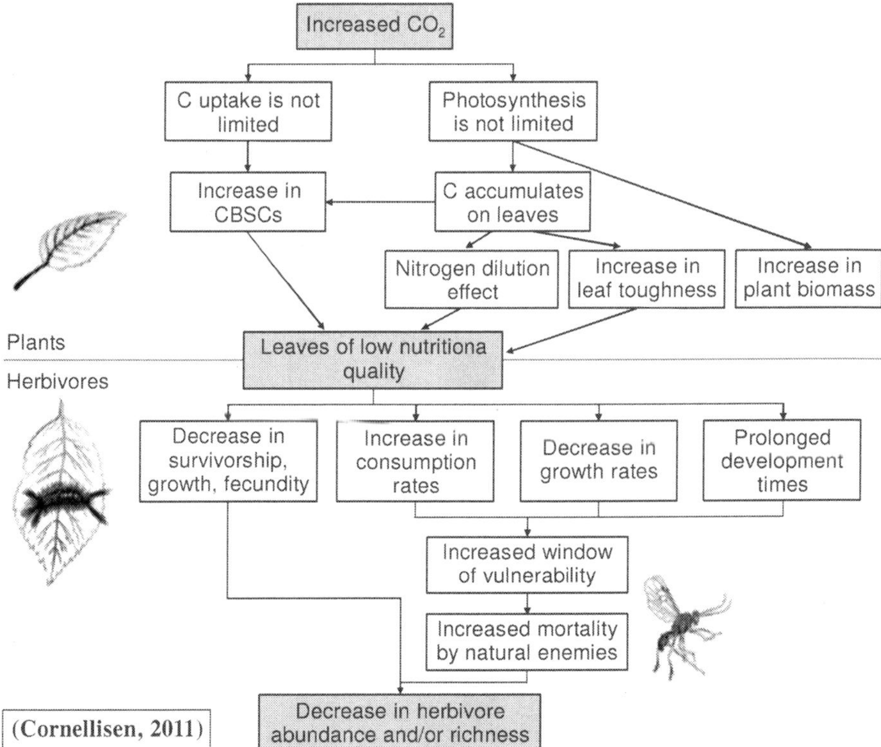

Fig. 4.3. Predicted effects of elevated CO_2 on plants and insect herbivores (Cornelliisen, 2011)

Succinctly, the information revealed that the performance of the same insect varied from host to host indicating host species specificity. Published data on impact of elevated CO_2 on insect pests indicated a general decrease in foliar nitrogen concentration and increase in carbohydrate and phenolic based secondary metabolites (Bezemer and Jones, 1998; Whittaker, 1999). The consumption by herbivores was related primarily to changes in nitrogen and carbohydrate levels. The leaf-mining insects could only partially compensate by increased consumption and pupal weights did decline. The phloem-feeding and whole-cell feeding insects responded positively to elevated CO_2, with increases in population size and decreases in development time.

One of the most studied aspects of climate change is the effect of increasing concentrations of carbon dioxide on crop and pest. In plants the carbon dioxide increases will increase the photosynthetic rates of most crop plants especially C_3 plants. In addition to growing more quickly, many crop plants would become more drought-tolerant. This is because the openings in the leaves (stomata) that let carbon dioxide in also let water vapour out, and if there is more carbon dioxide, then the stomata do not need to be open as much. A rise in CO_2 generally increases the carbon to nitrogen ratio of plant tissues (Ainsworth *et al.*, 2008), reducing the nutritional quality for protein thus diluting the nitrogen content of the tissues (Coviella *et al.*, 2000). The first reaction expected from herbivores to the increase of the C/N ratio is compensatory feeding. Insects may accelerate their food intake to compensate for reduced leaf nitrogen content (Fig. 4.4).

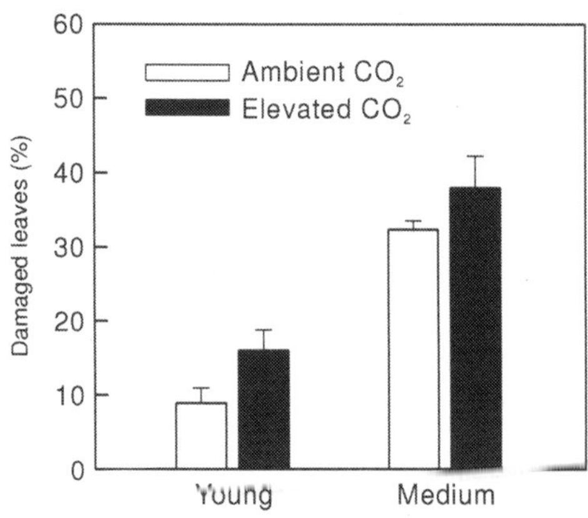

Fig. 4.4. Effect of CO_2 levels on crop damage

There is a general agreement between scientists that the reduced nutrient quality of C_3 plants might lead to a compensation by increased feeding of many, but not all, herbivorous species (DeLucia *et al.*, 2008). Under elevated CO_2, population densities of chewing insects are unaffected or decrease, but they do not increase, while sap sucker (phloem feeder) population densities might increase.

However, the results from experiments with aphids (phloem feeders) feeding on plants grown under elevated CO_2, and/or at elevated temperature have not shown consistent results. In some cases aphid performance was not (significantly) influenced by either elevated temperature and/or elevated CO_2 (Diaz *et al.*, 1998). In one research trial two species (e.g. *Brevicoryne brassicae* vs. *Myzus persicae* on *Brassica napus* s. sp. *oleifera*) responded very differently under same conditions (increased aphid infestation as well as reduced infestation in response to elevated CO_2). Experiments by Awmack and Harrington (2000) showed significant effects of elevated CO_2 on pea aphids (shoot, root weight, flower number), although the aphid density was unaffected.

Research with three generations of *H. armigera* reared on milky grains of spring wheat grown in ambient CO_2 concentrations and at 750 ppm showed again that bollworm fecundity was significantly decreased for the second and third generations under elevated CO_2 levels. While the consumption per larva and relative consumption rate significantly increased in elevated CO_2, the potential population consumption was significantly reduced by elevated CO_2 in the second and third generations. Therefore, the researchers suggests that net damage of cotton bollworm on wheat will be less under elevated atmospheric CO_2 levels, because increased consumption is offset by slower development and reduced fertility (Chen *et al.*, 2005).

In a similar experiment (larvae reared on milky wheat grain under 750 ppm CO_2) the researchers included a parasitoid wasp (*Microplitis mediator*) widely used as bio-control agent of *H. armigera*. The researchers found no significant changes in wheat consumption by *H. armigera* population under elevated CO_2 or in the parasitic rate of *M. mediator*. The results indicated that the population relationship between *H. armigera* and *M. mediator* is unlikely to vary due to future elevated atmospheric CO_2 concentrations.

A multiple generation experiment compared consumption, growth and performance of *H. armigera* feeding on transgenic *Bt* cotton versus conventional cotton grown under elevated CO_2 (750 ppm) versus ambient CO_2 (375 ppm). The results suggest that on the one hand, damage caused by the cotton bollworm might be higher under elevated CO_2 conditions, regardless of the cotton variety. On the other hand, population abundance might be lower under elevated CO_2 compared to that under ambient CO_2 (Chen *et al.*, 2007). The researchers explain both observations with nutritional changes under elevated CO_2 (compensatory feeding), but did not determine the nutrient content of the different experimental cotton groups.

An experiment by Coll and Hughes (2008) investigated the effects of elevated CO_2 on *H. armigera* and an omnivorous bug, which feeds on plants but also preys on the bollworm. Bollworm larvae feeding on elevated CO_2-grown pea plants (at 700 ppm) were significantly smaller than those grown on ambient-grown plants. The omnivorous bug required prey to complete its development, and performed best on a mixed plant-prey diet, regardless of CO_2 level. The bugs performed best when fed larvae from the elevated-CO_2 treatment apparently because these prey were smaller and thus easier to overcome. Taken together, results indicate that elevated CO_2 may benefit generalist predators through increased prey vulnerability, which would put pest species under higher risk of predation.

Recently, free air gas concentration enrichment (FACE) technology was used to create an atmosphere with CO_2 and O_2 concentrations similar to what climate change models predict for the middle of the 21st century. FACE allows for field testing of crop situations with fewer limitations than those conducted in enclosed spaces. During the early season, soybeans grown in elevated CO_2 atmosphere had 57% more damage from insects (primarily Japanese beetle, potato leaf hopper, western corn rootworm and Mexican bean beetle) than those grown in today's atmosphere, and required an insecticide treatment in order to continue the experiment. It is thought that measured increases in the levels of simple sugars in the soybean leaves may have stimulated the additional insect feeding (Hamilton *et al.*, 2005).

Researchers have observed that insects sometimes feed more on leaves that have lowered nitrogen content in order to obtain sufficient nitrogen for their metabolism (Hunter, 2001). Increased carbon to nitrogen ratios in plant tissue resulting from increased CO_2 levels may slow insect development and increase the length of life stages vulnerable to attack by parasitoids (Coviella and Trumble, 1999).

Elevated CO_2 can increase levels of simple sugars in leaves and lower their nitrogen content. These can increase the damage caused by many insects, who will consume more leaves to meet their metabolic requirements of nitrogen. Thus, any attack will be more severe. Higher temperatures from global warming, mainly due to elevated CO_2, will mean that more number of pests will survive the winter season.

Alteration in food quality of chickpea due to elevated CO_2 significantly affected the growth parameters of *H. armigera* in the form of increased food consumption, gain in larval weight and more faecal matter production. Elevated CO_2 concentration also affects the larval duration of *H. armigera* as compared to ambient CO_2 concentration (Table 4.4 and Fig. 4.5).

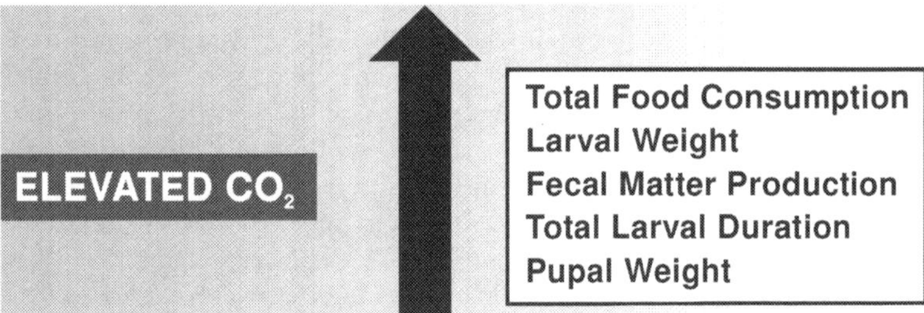

Fig. 4.5. Effect of elevated CO_2 on growth and development of *H. armigera* on chickpea (Khadar *et al.*, 2014)

Table 4.4. Effect of elevated CO_2 on growth and development of gram caterpillar, *Helicoverpa armigera* on leaf mediated artificial diet

CO_2 conc. (ppm)	Total diet consumed (mg)	Larval weight (mg)	Faecal matter produced (mg)	Larval duration (days)	Pupal wt. (g)
380 (Ambient)	1099.50 ± 52.24	367.91 ± 9.66	372.29 ± 9.94	14.06 ± 0.24	0.257 ± 0.004
550 (Elevated-I)	1887.50 ± 24.45	381.25 ± 7.61	602.50 ± 18.73	15.16 ± 0.37	0.258 ± 0.005
700 (Elevated-II)	1997.50 ± 50.71	384.16 ± 4.98	597.90 ± 12.92	15.27 ± 0.44	0.259 ± 0.006

Source: Khadar *et al.* (2014)

In one study two-spotted spider mites (*Tetranychus urticae*) were raised on common beans grown at 600 and 700 ppm CO_2. A significant decrease in the number of the offspring in the first and second generations (34% and 49%, respectively) was observed compared to ambient CO_2 (Joutei *et al.*, 2000). A similar experiment was conducted, where *T. urticae* were raised on clover (*Trifolium repens*) grown at different CO_2 levels (395-748 ppm). The results showed a quite opposite effect: under elevated CO_2 spider mite reproduction increased significantly compared to lower CO_2 (Heagle *et al.*, 2002).

In general elevated CO_2 widely assumed to increase N fixation in legumes (Soussana and Hartwig, 1996; Zanetti *et al.*, 1996; Hungate *et al.*, 1999). One mechanism is that root nodules can become more numerous and bigger resulting in more N fixed (Fig. 4.6).

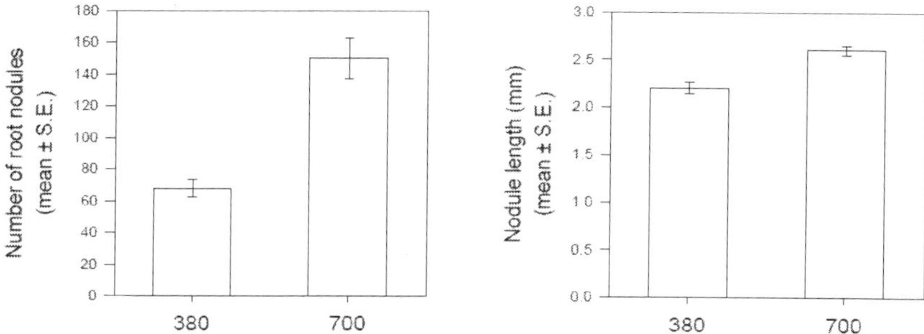

Fig. 4.6. Effect of elevated CO_2 on number and length of root nodules

However, more root nodules can cause a surge in root pests such as weevil, *Sitona* sp. Elevated CO_2 results in more weevils which develop faster. New born stages of *Sitona* sp. specifically target root nodules (Fig. 4.7).

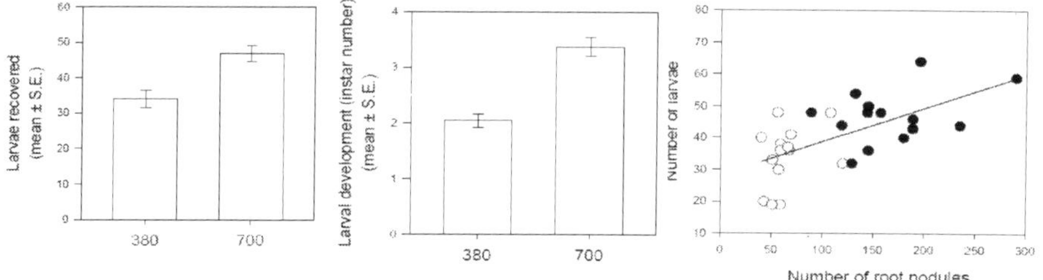

Fig. 4.7. Effect of elevated CO_2 on development of weevil, *Sitona* sp. on nodules

In Brassicas, elevated CO_2 increases shoot and root biomass. When either root or shoot insects are included, biomass levels return to ambient levels even when the insect is not feeding on that part of the plant (Fig. 4.8).

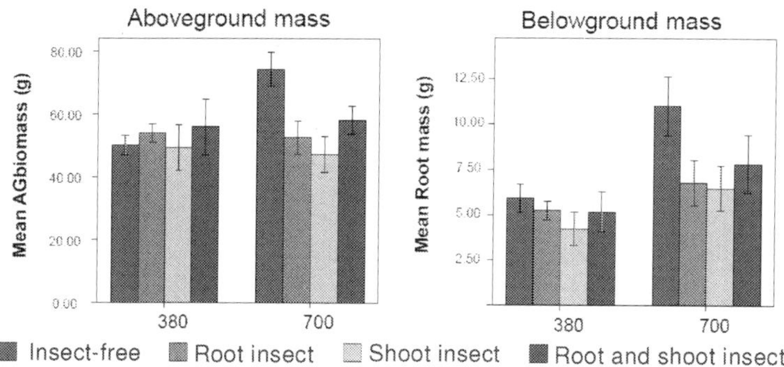

Fig. 4.8. Effect of elevated CO_2 on shoot and root biomass in Brassicas

4.4.1. Nutrient value

Fruits and vegetables contain significant levels of biologically active components that impart health benefits beyond basic nutrients. Elevated CO_2 has been found to increase concentrations of many biologically active components including sugars, ascorbic acid, phenols, starch, anthocyanins and flavonoids. However, increased CO_2 may reduce protein and mineral content in produce. Increased CO_2, temperature and water stress can affect photosynthesis, reproductive growth and mineral uptake, which may result in poor growth of vegetable crops. This could lead to lower nutritive values of horticultural produce.

Effect of increased CO_2 effects on insect-plant interaction has been presented in Tables 4.5 to 4.8.

Table 4.5. Effect of increased CO_2 effects on insect-plant interaction

Order	Herbivore species	Host species	Effects	Comments
Acarina	*Tetranychus urticae* (Red spider mite)	*Trifolium repens* (White clover)	–	
		Gossypium hirsutum (upland cotton)	–	
		Phaseolus vulgaris (kidney bean)	+	
Coleoptera	*Papillio japonica* (Japanese beetle)	*Glycine max* (Soybean)	–	Beetles and aphids generally perform better to the detriment of the plants
	Diabrotica virgifera (Western corn rootworm)	*Glycine max* (Soybean)	–	
	Sitona lepidus (Clover root weevil)	*Trifolium repens* (White clover)	–	
Diptera	*Pegomya nigritarsis* (leaf-mining fly)	*Rumex crispus* (Invasive dock)	–	
		R. obtusifolius (Invasive dock)	–	
	Chromatomyla syngenesiae (leaf-mining fly)	*Sonchus oleraceus* (Invasive sow thistle)	+	
	Bemesia tabaci (Sweet potato whitefly)	*Gossypium* (cotton)	ǿ	
Hemiptera	*Aulocorthum solani* (glasshouse potato aphid)	*Vicia faba* (broad bean)	–	Beetles and aphids generally perform better to the
	Stiobion avenae (grain aphid)	*Triticum aestivum* (spring wheat)	–	

Order	Herbivore species	Host species	Effects	Comments
	Myzus persicae (green peach aphid)	*Poa annua* (grass)	–	detriment of the plants
		Brassica napus (oilseed rape)	+	
	Brevicoryne brassicae (cabbage aphid)	*Brassica napus* (oilseed rape)	ǿ	
	Aphis glycines (soybean aphid)	*Glycine max* (soybean)	–	
Hymenoptera	*Aphidius matricariae* (green peach aphid parasitoid)	*Poa annua* (grass)	ǿ	
Lepidoptera	*Pseudoplusia includens* (soybean looper)	*Glycine max* (soybean)	–	Caterpillars generally eat more to compensate, but enhanced plant growth results in little net effect
	Trichoplusia ni (cabbage looper)	*Phaseolus lunata* (lima bean)	ǿ	
	Spodoptera eridania (southern armyworm)	*Menthe piperita* (peppermint)	ǿ	
	Spodoptera frugiferda (fall armyworm)	*Festuca arundinanceae* (tall fescue)	ǿ	
	Pectinophora gossypiella (pink bollworm)	*Gossypium hirsutum* (upland cotton)	ǿ	
	Helicoverpa armigera (cotton bollworm)	*Gossypium* (cotton)	+	

Table 4.6. Effect of increasing atmospheric CO_2 on plant-insect interaction

Increasing atmospheric CO_2 leads to	Reference
Increasing	
Food consumption by caterpillars	Osbrink *et al.*, 1987
Reproduction of aphids	Bezemer *et al.*, 1998
Predation by lady beetle	Chen *et al.*, 2005
Carbon based plant defences	Coviella and Trumble, 1999
Effect of foliar application and *Bacillus thuringiensis*	Coviella and Trumble, 2000

Increasing atmospheric CO_2 leads to	Reference
Decreasing	
Insect development rates	Osbrink *et al.*, 1987
Response to alarm pheromones by aphids	Awmack *et al.*, 1997
Parasitism	Roth and Lindroth, 1995
Effect of transgenic *B. thuringiensis*	Coviella *et al.*, 2000
Nitrogen-based plant defence	Coviella and Trumble, 1999

Source: Das et al. (2011)

Table 4.7. Effect of elevated CO_2 on plant-insect interaction

Common name	Scientific name	Effect	Reference
Elevated CO_2 and insect pests			
Colorado beetle	*Leptinotarsa decemlineata*	23.8% larger dry mass	Vaccari *et al.*, 2000
Aphids	*Myzus persicae*	Not negatively affected by increased CO_2	Hughes and Bazzaz, 2001
Bollworm	*H. armigera*	Significant differences in growth and development	Wu *et al.*, 2006a
Elevated CO_2 on transgenic plants and insect pests			
Tobacco bud worm	*Spodoptera exigua*	Limited N produced less toxin	Coviella *et al.*, 2000
	Spodoptera exigua	N based toxin was affected	Coviella *et al.*, 2000
Boll worm	*H. armigera*	Decrease in Bt toxin level	Ge Feng *et al.*, 2005

Table 4.8. Pest complex under elevated CO_2

Insects	Elevated CO_2 (570 ppm)	References
Japanese beetle Potato leaf hopper Western corn root worm Mexican bean beetle	57% more damage	Trumble and Butler, 2009
Thrips	90% more feeding	Heagle, 2003
Cereal aphid	Higher population	Newman, 2004

4.4.2. Influence on plant defences

First of all, plants have two groups of chemical defences: carbon-based and nitrogen-based. *Carbon-based defences* include tannins, lignins and phenolics that are designed to slow insect growth and development. These types of defences are present at higher concentrations under elevated levels of carbon dioxide. *Nitrogen-based defences* include alkaloids and cyanogenic glycosides that are directly toxic or repellent to the insect pests. These types of defences are present at lower concentrations under higher levels of CO_2. As such, chewing insect pests such as caterpillars and beetles that consume more plant tissues when plants are exposed to increased levels of CO_2 may actually ingest more toxins and thus may be killed faster and in greater numbers.

Whittaker (1999) concluded from his review of studies on insects and elevated CO_2 that so far, population density of chewing insects are unaffected or decrease, but do not increase; while sap sucker (phloem feeder) population density might increase.

4.4.3. Impact on natural enemies

Few studies explored the impact of elevated CO_2 on natural enemies (parasitoids and predators) at third trophic level (Bezemer *et al.*, 1999; Stacey and Fellows, 2002). The growth and development of insect herbivores varied with the nutrition quality of their diet (host plant) and the dietary differences showed varied effects on parasitoids. The population size of the insects significantly differed under elevated CO_2 and in turn influencing the insect fecundity. Thus, any dietary differences that prolong developmental time, increase food consumption, and reduce growth by herbivores serve to increase the susceptibility of herbivores to natural enemies (Roth and Lindroth, 1995). On the other hand, poor host nutrition could also decrease parasitoid fitness.

Chen *et al.* (2005) showed that increasing CO_2 concentrations could alter the preference of lady beetle to aphid prey and enhance the biological control of aphids by lady beetle in cotton crop. This study provided the first empirical evidence that changes in prey reared on host plant grown at different levels of CO_2 altered the feeding preferences of the predator.

4.5. VARYING PRECIPITATION PATTERNS

There are fewer scientific studies on the effect of precipitation on insects than temperature. Precipitation – whether optimal, excessive, or insufficient – is a key variable that also affects crop-pest interactions. Large scale changes in rainfall will have a major effect on the abundance and diversity of arthropods. Analysis of precipitation data over the past 100 years showed that the total precipitation did not change, but the frequency of light rain decreased and the frequency of heavy rainfall increased (Das *et al.*, 2011). Some insects e.g. onion thrips are sensitive to precipitation and are killed or removed from crops by heavy rains in some north-eastern US states (Reiners and Petzoldt, 2005). For some insects that overwinter in soil, such as the cranberry fruit worm and other cranberry insect pests, flooding the soil has been used as a control measure (Vincent *et al.*, 2003). Decreasing snowfall promotes the expansion of pine moth, *Thaumetopoea pityocampa* into high elevation stands of mountain pine. More than 50 species of butterflies showed northward range expansions and 10 species of previously migrant butterflies have been established on Nansei Islands during 1966 to 1987.

One would expect that the predicted, more frequent and intense precipitation events forecasted with climate change, to negatively impact these insects. Other insects such as pea aphids are not tolerant of drought (Macvean and Dixon, 2001). As with temperature, precipitation changes can impact insect pest predators, parasitoids, and pathogens resulting in a complex dynamics. Fungal pathogens of insects are favoured by high humidity and their incidence would be increased by climate changes that lengthen periods of high humidity and reduced by those that result in drier conditions. Guiterrez *et al.* (2008) found that during the normally wet Northern California winter, the fungal pathogen (*Pandora neoaphidis*) causes catastrophic mortality to pea aphid (*Acyrthosiphon pisum*)

Drought stress sometimes brings increased insect pest outbreaks. It is well known that drought can change the physiology of host species, leading to changes in the insects that feed on them (Mattson and Haack, 1987). Abnormally cool, wet conditions can also bring on severe insect infestations, although excessive soil moisture may drown out soil-residing insects.

Chen and McCarl (2001) investigated the relationship of precipitation and pesticide costs for several crops in the USA and concluded that increases in rainfall leads to increases in average pesticide costs for corn, cotton, potatoes, soybeans and wheat.

Distribution and frequency of rainfall may also affect the incidence of pests directly as well as through changes in humidity levels. It is being predicted that under the climate change, frequency of rainfall would decline while its intensity would increase. This would lead to heavy showers and floods on one hand and drought spells on the other. Under such situations, incidence of small pests such as aphids, jassids, whiteflies, mites, etc. on crops may be reduced as these get washed away by the heavy rains.

Armyworm, *Mythimna separata* reaches outbreak proportions after heavy rains and floods. Lever (1969) had analysed the relationship between outbreaks of armyworm and to a lesser extent *Spodoptera mauritia* and rainfall from 1938 to 1965 and observed that all but three outbreaks occurred when rainfall exceeded the average 89 cm. The effect of rainfall on pests can be studied by simulating various rainfall intensities through sprinklers. Aphid population on wheat and other crops was adversely affected by rainfall and sprinkler irrigation (Chander, 1998).

Masters *et al.* (1998) have carried out novel manipulations of local climate to investigate how warmer winters with either wetter or drier summers would affect the Homopteran insects - a major component of the insect fauna of grasslands. Direct and indirect effects of climate manipulation were observed. It was observed that the supplemented summer rainfall resulted in an increase in the vegetation cover, leading to an increase in the abundance of the insects. Summer drought, however, caused a decrease in the vegetation cover, but this did not lead to a corresponding decrease in the abundance of the insects. Egg hatch and the termination of nymphal hibernation occurred earlier in the winter-warmed plots, but the rate of nymphal development was unaffected.

High humidity may favour some insects e.g. aphids. Thrips and whiteflies are sensitive to precipitation and are killed or removed from crops by heavy rains. Outbreaks of *Amsacta moorei* are directly related to heavy and frequent rains (Saini and Ram, 2000). Reductions in summer rainfall can alter the diversity and abundance of plant communities and their associated invertebrates (Morecroft *et al.*, 2004)

It is likely that rainfall may indirectly influence groundnut leaf miner (GLM), *Aproaerema modicella* populations by increasing humidity which will favour GLM pathogens. Temperature was also positively correlated with GLM incidence and accounted for more of the variation than rainfall (Lewin *et al.*, 1979). Logiswaran and Mohanasundaram (1986) noted negative associations of light-trap catches with maximum and minimum temperatures and wind speed, and positive relationships with total rainfall and morning evening relative humidity. Tobacco caterpillar, *Spodoptera litura* is widely distributed throughout Asia and Oceania and is highly polyphagous in nature. There is pronounced peak in the flight activity of moths (adults) during late January.

Water stress diminishes plant vigour and alters C/N ratio lowering plant resistance to insects.

4.6. DROUGHT

Droughts are likely to decrease multi-trophic diversity and change the composition of arthropod communities which in turn might affect the other associated taxa. Enhanced summer rainfall and drought conditions promote rapid increase in the population of wireworm (*Agriotes lineatus*) in the upper soil. Drought conditions severely affect egg viability of *Scopelosaurus lepidus* eggs and did not hatch at all under dry conditions (Karuppaiah and Sujiyanad, 2012).

Drought stress sometimes brings increased insect pest outbreaks. It is well known that drought can change the physiology of host species, leading to changes in the insects that feed on them (Mattson and Haack, 1987) (Fig. 4.9).

4.6.1. Effect of drought on population dynamics

Drought may result in various physiological changes in plants that can increase their attractiveness and susceptibility to herbivorous insects. Droughts increase pathogen and insect survival and growth due to change in nutrient level particularly nitrogen, decrease in plant defence system and creates more favourable conditions for pest attack.

In some cases plant temperature conditions influence the oviposition or feeding activity of the insects, which generates an insect population selection pressure on the crops under extreme drought conditions. Some studies have shown that under drought conditions the most favoured pests are those known as invasives (Hamada, 2011). It has been confirmed that insects belonging to the order Hemiptera and Thysanoptera, such as bugs and thrips, are the most benefited under these conditions, since the increase in temperature favours the reproductive rate of these insects (Vasquez, 2011).

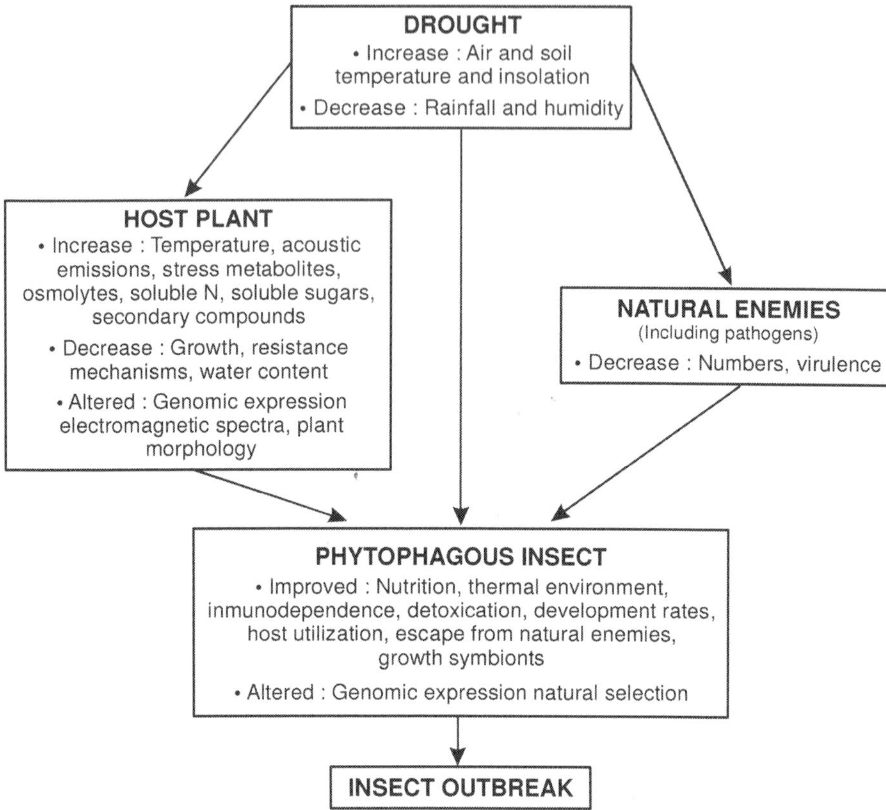

Fig. 4.9. Drought influences on host plants, phytophagous insects and their natural enemies leading to insect outbreaks (Mattson and Haack, 1987)

4.7. COMBINED EFFECT OF MULTIPLE FACTORS

A meta-analysis of studies on elevated temperature and elevated CO_2 suggests that insect herbivore performance is adversely affected by elevated CO_2, favoured by elevated temperature, and not modified when both parameters (temperature and CO_2 combined) were elevated.

Increased CO_2, temperature and water stress can affect photosynthesis, reproductive growth and mineral uptake, which may result in poor growth of vegetable crops. This could lead to lower nutritive values of horticultural produce.

Rate of insect multiplication might increase with an increase in CO_2 and temperature. Therefore, there is a need to look into the effects of climate change on crop protection and devise appropriate measures to mitigate the effects of climate change on food security.

Increased temperature would reduce crop duration and causes the change in the growth and development of the insect pests. The duration of the insect life cycle gets reduced under increased temperature resulting in more number of generations per year. Under elevated CO_2 conditions, higher consumption of foliage by leaf chewing insects with extended larval duration is predicted in majority cases. The studies that combine the effects of elevated CO_2 (enriched plants) and increased temperatures on insect performance are rare. The documented information reveals that some pests become more serious while others may decline. The impact of increased temperature and elevated CO_2 on crop and insect herbivore interactions is still unknown as both these variables are counteracting.

It seems that current knowledge does not allow a generalization regarding the impact of climate change in herbivorous insects, especially not for the tropics. Even the trend of a northward shift of insects must not coercively translate into a pest problem – ecosystems are not that simple and human influence is quite strong. Basically, it would be necessary to investigate at least over three trophic levels with several generations of plants, herbivores, predators/parasites, under elevated temperature and elevated CO_2.

The greatest challenge facing humanity in the coming century will be the necessity to double our global food production to meet a booming increase in population by using less land area, less water, less soil nutrients, droughts from global warming. Various insect responds differently to atmospheric temperature and carbon dioxide rise and it is obvious of having varied impact depending on insect and regions.

However, complex interactions among temperature, host plant quality, and insect performance make pest predictions more complicated. For example, the reduced rates of insect growth that have been observed under elevated CO_2 may be masked by increased temperatures. There is a need to have long-term, multifactorial experiments under field conditions to study the interactive and confounding effects of elevated CO_2, temperature and other ecological variables on the insects.

4.8. CONCLUSIONS

Climatic variability, together with increase in atmospheric temperature and carbon dioxide, do have a lot of implication in agriculture sector. Latest assessment report from the Intergovernmental Panel on Climate Change (IPCC) predicts an increment in mean atmospheric temperature from 1.1 to 6.4°C toward the year 2100 with equally increasing atmospheric carbon dioxide (CO_2). Such climatic changes could profoundly affect insects both directly i.e. seasonal shifts of insects and indirectly i.e. changing plant productivity and quality, and changing predator populations. Studies carried out around the globe on impact of increasing atmospheric temperature and carbon dioxide on crop pest population and crop-pest interaction reported both harmful and beneficial impacts, however benefits are least reported. A substantial expansion in high latitude and elevation of various crop insect pest distributions is critical in temperate part of the world. Increased frequent insect outbreaks have been observed where changing climate may interfere with the induction of extended diapauses. Increase in atmospheric temperature resulted in reduction in survival and increase in developmental rate, resulting in more generations (and more crop damage) per year. New studies show that insect species living in warmer areas are more likely to undergo rapid population growth because they have higher metabolic rates and reproduce more frequently, thus climate change would create favourable conditions for growth in insect populations. Other environmental parameters like relative humidity and CO_2 are potentially important. More CO_2 in the atmosphere makes plants to take up more carbon and leaves become less nutritious and results in voluminous feeding. Further studies on crop-pest interaction focusing on spatio-temporal variations are needed to have a better understanding and to quantify its future impacts.

Understanding how climate change will impact on various pests especially crop pests, helps agricultural scientist to orient their research on various futuristic possibilities that can help in mitigating and adapting to menace of anticipated climate change.

5

Impacts of Climate Change Induced Consequences on Insect Pests

5.1. INTRODUCTION

The occurrence of climate change is evident from increase in global average temperature, changes in the rainfall pattern and extreme climatic events. Climate and weather can substantially influence the fauna, flora, population dynamics development and distribution of insects. Anthropogenically induced climatic change arising from increasing levels of atmospheric greenhouse gases would, therefore, be likely to have a significant effect on agricultural insect pests. Current best estimates of changes in climate indicate an increase in global mean annual temperatures of 1°C by 2025 and 3°C by the end of the next century. Such increases in temperature have a number of implications for temperature-dependent insect pests in mid-latitude regions. Changes in climate may result in changes in geographical distribution, increased overwintering, changes in population growth rates, increases in the number of generations, extension of the development season, changes in crop-pest synchrony, changes in interspecific interactions and increased risk of invasion by migrant pests (Fig. 5.1) (Menendez, 2007).

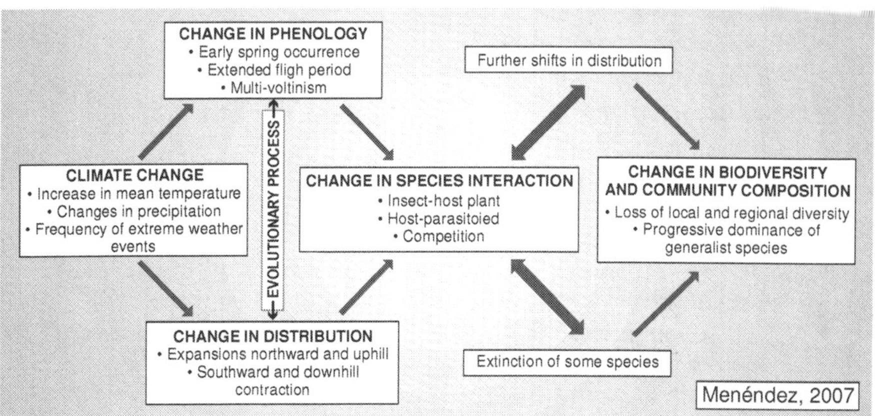

Fig. 5.1. Impact of climate change on insect pests (Menendez, 2007)

Losses due to insect damage are likely to increase as a result of changes in crop diversity and increased incidence of insect pests due to global warming. High mobility and rapid population growth will increase the extent of losses due to insect pests. Geographical distribution of insect pests confined to tropical and subtropical regions will extend to temperate regions along with a shift in the areas of production of their host plants, while distribution and relative abundance of some insect species vulnerable to high temperatures in the temperate regions may decrease as a result of global warming. These species may find suitable alternative habitats at greater latitudes. Many species may have their diapause strategies disrupted as the linkages between temperature and moisture regimes, and the day length will be altered. Genetic variation and multi-factor inheritance of innate recognition of environmental signals may mean that many insect species will have to adapt readily to such disruption.

Overwintering of insects will increase as a result of climate change, producing larger spring populations in the following season. These may be vulnerable to parasitoids and predators if the latter also overwinter more readily. There may also be increased dispersal of air-borne insects in response to atmospheric disturbances. Many insects such as *Helicoverpa* sp. are migratory and therefore may be well adapted to exploit new opportunities by moving rapidly into new areas as a result of climate change (Sharma, 2005).

Global warming and climate changes will result in:

- Extension of geographical range of insect pests
- Prolonged over-wintering and rapid population growth
- Changes in insect – host plant interactions
- Increased risk of invasion by migrant pests
- Impact on arthropod diversity and extinction of species
- Changes in synchrony between insect pests and their crop hosts
- Introduction of alternative hosts as green bridges
- Reduced effectiveness of crop protection technologies
- Extending the growing season
- Altering the timing of pest emergence
- Rapid growth and development rates
- Shortening generation times and increases in the number of generations
- Shorten predation window
- Changes in inter-specific interactions

5.1.1. Projected impacts of climate change on insect pests

- Increased insect outbreaks (virtually certain).
- Many tropical insect species may become extinct as they are already living at environmental temperatures close to their optimum.
- Reduced nutrient quality of C_3 plants might lead to compensation by increased feeding of many insects (Lincoln *et al.*, 1986, Whittaker, 1999, Emmerson *et al.*, 2004, DeLucia *et al.*, 2008, Barbehenn *et al.*, 2004).
- Population densities of chewing insects would be unaffected or decrease, but do not increase; while sap sucker population densities might increase under increased CO_2 concentrations (Whittaker, 1999)

5.2. EXPANSION OF GEOGRAPHICAL DISTRIBUTION

Climate change will have a major effect on geographic distribution of insect pests, and low temperatures are often more important than high temperatures in determining geographical distribution of insect pests (Hill, 1987). Increasing temperatures may result in a greater ability to overwinter in insect species limited by low temperatures at higher latitudes, extending their geographical range (Fig. 5.2) (EPA, 1989; Elphinstone and Toth, 2008; Fand *et al.*, 2012), and sudden outbreaks of insect pests can wipe out certain crop species, and also encourage the invasion by exotic species (Kannan and James, 2009). Spatial shifts in distribution of crops under changing climatic conditions will also influence the distribution of insect pests in a geographical region (Parry and Carter, 1989). Some plant species may be unable to follow the climate change, resulting in extinction of species that are specific to particular hosts (Thomas *et al.*, 2004). There are several examples of change in the geographic distribution of several insect species as a result of climate change. The cotton bollworm/ legume pod borers, *H. armigera* and *M. vitrata* will move to temperate regions in northern Europe. *Helicoverpa armigera* has already reached Brazil as an invasive pest, and is likely to move to North America (Czepak *et al.*, 2013). For all the insect species, higher temperatures, below the species' upper threshold limit, will result in faster development, resulting in rapid increase of pest populations as the time to reproductive maturity is reduced. However, whether or not an insect pest would move with a crop into a new habitat will depend on other environmental conditions such as the presence of overwintering sites, soil type, and moisture, e.g., populations of the corn earworm, *Heliothis zea* in the North America might move to higher latitudes/altitudes, leading to greater damage in maize and other crops (EPA, 1989).

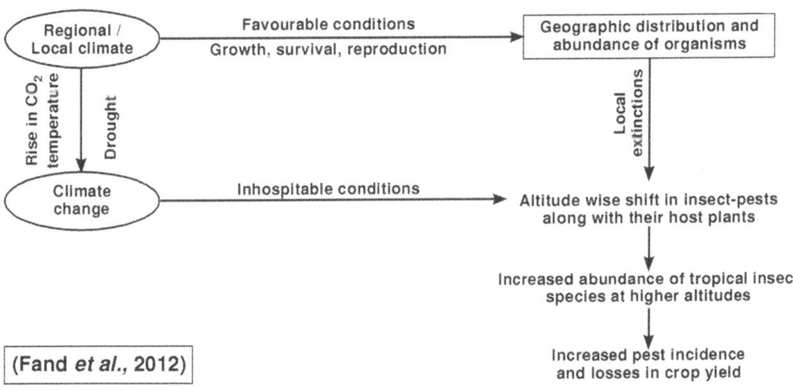

Fig. 5.2. Impact of climate change on range expansion of insect pests

A warmer climate in temperate regions may result in changes in geographical distribution, increased overwintering (i.e. more insects survive the winter), changes in population growth rates, increases in the number of generations, extension of the development season, changes in crop-pest synchrony, changes in interspecific interactions and increased risk of invasion by migrant pests (Bale *et al.*, 2002). In Japan, warmer climate led to the northward migration of the green stinkbug (*Nezara viridula*) a major agricultural pest damaging soybean, rice, cotton and many other crops (Musolin, 2007).

Geographical distribution of insect pests confined to tropical and subtropical regions will extend to temperate regions, which leads to increased abundance of tropical insect species and sudden outbreaks of insect-pests (Kannan and James, 2009). Along with a shift in the areas of production of their host plants, while distribution and relative abundance of some insect species vulnerable to high temperatures in the temperate regions may decrease as a result of global warming. These species may find suitable alternative habitats at greater latitudes.

An increase of 1 and 3°C in temperature will cause northward shifts in the potential distribution of the European corn borer, *Ostrinia nubilalis* up to 1,220 km, with an additional generation in nearly all regions where it is currently known to occur (Porter *et al.*, 1991).

Diamondback moth, *Plutella xylostella* overwintered in Alberta in 1994 (Dosdall, 1994), and if overwintering becomes common, the status of this insect as a pest in North America will increase dramatically. Many insects such as *Helicoverpa* sp. are migratory, and therefore, may be well adapted to exploit new opportunities by moving rapidly into new areas as a result of climate change (Sharma *et al.*, 2005).

The traditional range of Queensland fruit fly, *Bactrocera tryoni* was focused on coastal areas of northern NSW and Queensland, extending south just past Sydney. While incursions occurred in more southern and western areas, generally during mid to late summer, the pest was unable to establish permanent populations.

There is clear evidence that climate change is altering the distribution, incidence and intensity of plant pests and diseases whose distribution is shifting north in the United Kingdom and other parts of Europe. For example, migrant moths of the Old World bollworm (*Helicoverpa armigera)* had a phenomenal increase in the United Kingdom from 1969-2004 and there have been outbreaks at the northern edge of its range in Europe. Similarly, the cottony cushion scale (*Icerya purchasi)* populations appear to be spreading northwards perhaps as a consequence of global warming. Likewise, the cottony camellia scale (*Pulvinaria – Chloropulvinaria – floccifera)* has become much more common in the United Kingdom, extending its range northwards in England and increasing its host range in the last decade or so, which is almost certainly in response to climate change. In Sweden, this species was previously only known as a greenhouse species, but is now established as an outdoor species. *Nezara viridula*, a tropical and subtropical crop pest, is gradually moving northward in south-western Japan, possibly due to global warming, replacing the more temperate species *Nezara antennata.*

Tea mosquito bug, *Helopeltis antonii* is a serious constraint in cashew (west coast-Kerala, Karnataka, and east coast-Tamil Nadu). Cashew tracts of Andhra Pradesh, West Bengal and Orissa are free from this pest. The hot spot areas of the bug across the cashew tract of the whole country have been demarcated, taking into consideration the optimum temperature during flushing and following stages. The pest may spread to new areas under current scenario of climate change and states like Andhra Pradesh, West Bengal and Orissa may come under this pest attack in the changed situation (Gupta *et al.*, 2009).

New England hemlock trees are under assault from the woolly adelgid, an aphid-like insect that has migrated northward with warmer winters.

Published reports of empirical studies on impact of climate change on future geographic range and distribution of insect pests are presented in Table 5.1.

Table 5.1. Impact of climate change on future geographic range and distribution of insect pests

Insect pest	Order/ Family	Host plant/s	Impact on insects/ behavioural response	Reference
Corn earworms, *Heliothis zea* and *Helicoverpa armigera*	Lepidoptera: Noctuidae	Maize	Altitudes wise range expansion and increased overwintering survival in USA.	Diffenbaugh *et al.*, 2008
European corn borer, *Ostrinia nubilalis*	Lepidoptera: Crambidae	Maize	Northward shifts in the potential distribution up to 1220 km are estimated to occur. An additional generation per season.	Porter *et al.*, 1991
104 common microlepidoptera species inhabitant in Netherlands	Lepidoptera	Many crops of agricultural importance	Changing patterns in phenology and distribution of microlepidoptera in the Netherlands. Advancement of flight peak dates almost by 12 days since 1975-1994. Changes in the species composition of the local fauna.	Kuchlein and Ellis, 1997
Old world Bollworm, *Helicoverpa armigera*	Lepidoptera: Noctuidae	Many crops of agricultural importance	Phenomenal increase in the United Kingdom from 1969-2004 and outbreaks at the northern edge of its range in Europe.	Cannon, 1998
Cottony cushion scale, *Icerya purchasi*	Hemiptera: Monophlebidae	Cotton	Populations appear to be spreading northwards	Cannon, 1998
Oak processionary moth, *Thaumetopoea Processionea*.	Lepidoptera: Thaumetopoeidae	Oak	Northward range extension from central and southern Europe into Belgium, Netherlands and Denmark.	Cannon, 1998
Cottony camellia scale, *Chloropulvinaria floccifera*	Hemiptera: Coccidae	Wide host range	More abundant in the United Kingdom. Extending its range northwards in England and increasing its host range in the last decade.	Cannon, 1998

Insect pest	Order/ Family	Host plant/s	Impact on insects/ behavioural response	Reference
35 species of non-migratory European butterflies	Papilionidae, Lycaenidae, Nymphalidae, Satyrinae	Many crops of agricultural importance	Pole ward shift of the geographic range and distribution.	Parmesan and Yohe, 2003
Cotton bollworm/ Pulse pod borer, *Helicoverpa armigera*	Lepidoptera: Noctuidae	Cotton, Pulses, vegetables	Expansion of geographic range in Northern India. Adult flights/ migratory behaviour.	Sharma *et al.*, 2010

Projected trends for several pests are summarized in Table 5.2. Examples of the mid-point projections of WCR infestation areas in maize are shown in Table 5.3.

Table 5.2. Outline trends in selected pests by 2020.

Species/crop	General trend	Europe	North America
Western corn rootworm in maize	Spreads with temperature rise	Extension from current outbreaks	Spreads to 50% of crop area
European corn borer (*Ostrinia nubilalis*) in maize	More generations; spreads north	Moves with crop	More frequent outbreaks
Colorado beetle in potato	Adaptable; moves north	Could become established in the UK and Scandinavia	More frequent in Canada

Table 5.3. Estimated impact of climate change on western corn rootworm infestations in maize.

Region	Crop area (million ha)		Crop area infested (million ha)	
	2004	2020	2004	2020
Europe	15.1	18.1	< 0.1	2.1
North America	30.8	32.3	12.0	19.0

Projected trends to 2020 are expected to continue up to 2050 and beyond, and are useful strategic indicators for the plant breeding and crop protection industries.

If crop pests continue to march pole wards as the earth warms, the combined effects of a growing world population and the increased loss of crops to pests will pose a serious threat to global food security.

5.3. INCREASE IN NUMBER OF GENERATIONS

As stated earlier the temperature being the single most important regulating factor for insects (Yamamura and Kiritani, 1998; Bale *et al.,* 2002; Petzolet and Seaman, 2010), global increase in temperature within certain favourable range may accelerate the rates of development, reproduction and survival in tropical and subtropical insects. Consequently, insects will be capable of completing more number of generations per year and ultimately it will result in more crop damage (Yamamura and Kiritani, 1998; Petzoldt and Seaman, 2010). The impacts of climate change on seasonality of insects have been studied by many workers (Porter *et al.,* 1991; Bale *et al.,* 2002; Walther *et al.,* 2002).

With every 2° C rise in temperature, multivoltine insects may have 1-5 additional generations (Yamamura and Kiritani, 1998). Rice stem borer, *Chilo suppressalis* would extend its distribution range in Japan and produce two generations a year after 2° C warming (Morimoto *et al.*, 1998).

In Alaska, spruce bark beetles have denuded 4 million acres of conifers on the Kenai Peninsula, as warming allows the beetles time for an extra generation each year (Kerlin, 2002). Dead stands are then vulnerable to fire.

5.4. INCREASED OVERWINTERING SURVIVAL

Being poikilotherms, insects have limited ability of homeostasis with external temperature changes. Hence they have developed a range of strategies such as behavioural avoidance through migration and physiological adaptations like diapause to support life under thermally stressful environments (Bale and Hayward, 2010). Diapause is a period of suspended developmental activities, the manifestation of which is governed by environmental factors like temperature, humidity and photoperiod. As an adaptive trait, diapause plays vital role in seasonal regulation of insect life cycles because of which the insects have an advantage to survive great deal of environmental adversities. There are two main types of insect diapause; aestivation and hibernation to sustain life under high and low temperature extremes, respectively (Chapman, 1998).

The studies have shown that, global warming is occurring notably in winter than in summer and is greatest at high latitudes (IPCC, 2007; IMD, 2010). Looking at the past 100 year's climate profile of India, warming was more pronounced during winter season and it was the minimum and not the maximum temperature where significant increase was observed (IMD, 2010). The temperature in India is expected to increase by 1-5^{0}C within next 100 years (De and Mukhopadhyay, 1998; Lal, 2003; IPCC, 2007; IMD, 2010). Thus, insects undergoing a winter diapause are likely to experience the most significant changes in their thermal environment (Bale and Hayward, 2010). Increase in temperature in the range of 1-5^{0}C would increase insect survival (Sharma *et al.*, 2010).

Theoretically, it is hypothesized that the winter survival of insects will be improved by an increase in winter temperature, but the evidence for this hypothesis is rather scarce. Effect of climate change on winter mortality/survival can be examined by collecting long-term data on winter survival at fixed over-wintering sites. Relationship can then be established between winter mortality and temperature.

Accelerated metabolic rates at higher temperatures shorten the duration of insect diapause due to faster depletion of stored nutrient resources (Hahn and Denlinger, 2007). Warming in winter may cause delay in onset and early summer may lead to faster termination of diapause in insects, which can then resume their active growth and development (Fig. 5.3). This gives an important implication that increase in temperature in the range of 1^{0}C to 5^{0}C would increase insect survival due to low winter mortality, increased population build-up, early infestations and resultant crop damage by insect-pests under global warming scenario (Harrington *et al.*, 2001; Sharma *et al.*, 2005; 2010). Very few studies have concentrated on the direct effects of higher winter temperatures on rates of development and reproduction in insects (Coulson *et al.*, 2000; Bale and Hayward, 2010).

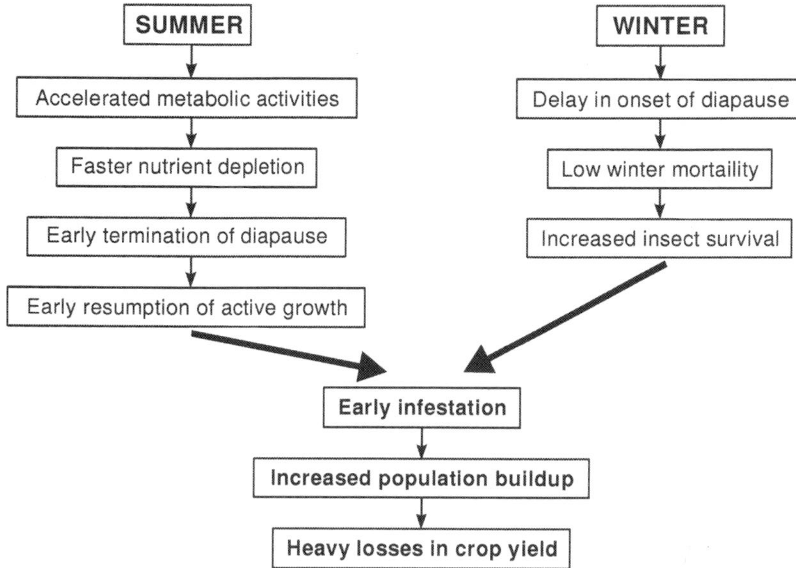

Fig. 5.3. Thematic presentation of impact of climate change on insect survival and population build-up

Kiritani (1971) had examined the winter mortality of adults of *Nezara viridula* in the late March at fixed over-wintering sites from 1962 to 1967 in Wakayama. A regression between winter mortality (Y) and the mean temperature in January (X) [Y= –16.45X+147.08 (R2 = 0.6127, P<0.0001)], suggested that every 1°C rise around X = 4°C would result in a decrease in winter mortality by about 16.5%.

In New York a network of pheromone traps in sweet corn fields has been used to monitor corn earworm (*Helicoverpa zea*) throughout the central and western part of the state for over 10 years. Corn earworm (CEW) is thought not to overwinter in upstate New York and is generally considered to be a late season, migratory pest of sweet corn, so trapping was initiated in mid-July. The graphs in Fig. 5.4 compare the trap catches in 1995 with those in 2003 in Eden Valley, New York.

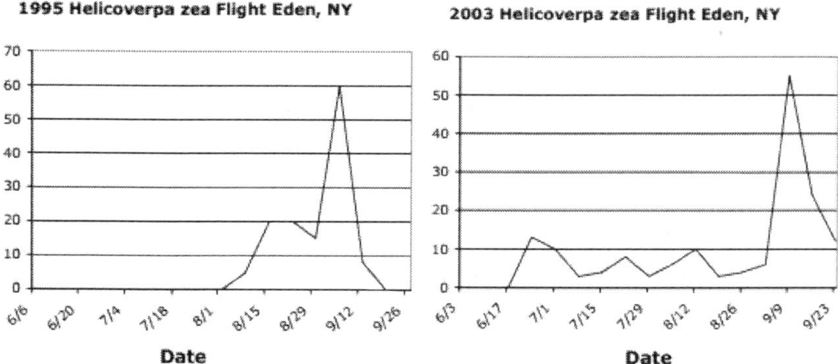

Fig. 5.4. Trap catches data indicating possible over wintering of corn earworm in western NY

During the early years of the trap network, CEW traps remained empty until mid-late August. After an unexpected early season infestation in Eden in 1999 trapping was initiated in early June, and typically, low levels of moths are caught through the early season, increasing when the migratory flight arrives. It is yet to be determined if the earlier arrival of corn earworm indicates it is overwintering in Eden, but since CEW management recommendations are based on trap catches, it is clear that control of this pest is already costing farmers more than it did nine years ago.

5.5. PEST POPULATION DYNAMICS AND OUTBREAKS

The population dynamics is the aspect of population ecology dealing with factors affecting changes in population densities. The seasonal effects of weather and on-going changes in climatic conditions will directly lead to modifications in dispersal and development of insect species. The changes in surrounding temperature regimes certainly involve alterations in development rates, voltinism and survival of insects and subsequently act upon size, density and genetic composition of populations, as well as on the extent of host plant exploitation (Bale *et al.*, 2002). The change in the environment affects the pest population dynamics in two ways either directly or indirectly by altering the host physiology. Host availability and the probability of pest outbreaks are further determined by the incidence and character of abiotic disturbances. The developmental success of insect herbivores also indirectly depends on climate, as environmental parameters impact on plant physiology. Insects and plants are exposed to complex interactions among changes in temperature, precipitation, and increased levels of CO_2 and variations in nutrient availability.

Climate change resultant abiotic environment (increased temperature, elevated CO_2 and depleted soil moisture) will affect significantly the diversity and abundance of insect-pests through geographic range expansion, increased overwintering survival and more number of generations per year, thereby increasing the extent of crop losses. It may result in upsetting ecological balance because of unpredictable changes in the population of insect-pests along with their existing and potential natural enemies.

Changes in climate and weather could profoundly affect the population dynamics and status of insect pests of crops. These may arise not only as a result of direct effects on the distribution and abundance of pest populations but also indirect effects on the pests' host plants, competitors and natural enemies (Porter *et al.*, 1991). Some pests which are already present but only occur in small areas, or at low densities may be able to exploit the changing conditions by spreading more widely and reaching damaging population densities (Porter *et al.*, 1991; Bale *et al.*, 2002).

The seasonal and long term climate changes would affect the fauna, flora and population dynamics of insect pests. The abiotic parameters are known to have direct impact on insect population dynamics through modulation of developmental rates, survival, fecundity, voltinism and dispersal. Among the climatic factors, temperature is an important factor. The studies showed that, declined survival rate of brown plant hopper, *Nilaparvatha lugens* and rice leaf folder, *Cnaphalocrocis medinalis* at higher temperature indicates the impacts of rising temperature could do the changes in the pest population dynamics of rice ecosystem. The alteration in the voltinism also could be the results of warming and it is of more profit to multivoltine species and voltinism could be reflected in changes in the geographical distribution. Beside these, elevated CO_2 also showed some impact on pest's population abundance, the crop grown under the elevated CO_2 could alter the nutritional value of plants; it may alter the insect abundance and increase the consumption rate of herbivores. Therefore, climate change would result in changes in the population dynamics of insect pests. Thus, temperature rise plays a pivotal role in insect population dynamics.

Most threatening are outbreaks of migratory pests because of their sudden emergence, often without or with limited warning, triggered by changing ecological conditions or agricultural practices which favour an explosive increase in the pest population and its rapid spreading into neighbouring areas. The unexpected appearances of such massive pest outbreaks hit the countries which in most cases are ill prepared, with no or only poor capacities to effectively cope with rapidly evolving and fast moving threats.

Various species of mealy bugs have started appearing in serious proportions on field crops, vegetables, fruits and ornamentals (Tanwar *et al.*, 2007). In fact, mealy bugs have become indicator insects for the current ecosystem alterations due to slow changes in climate during the period from 2002 to 2005. Among these, *Phenacoccus solenopsis* on cotton and *Paracoccus marginatus* on papaya have become quite serious. The papaya mealy bug, *P. marginatus,* has become quite alarming in Tamil Nadu, challenging the pesticides or other IPM measures. *Maconellicoccus hirsutus* causes extensive damage to ornamentals, though its host range extends to 76 families and over 200 genera, including beans, chrysanthemum, citrus, coconut, coffee, cotton, croton, cucumber, grape, groundnut, guava, *Hibiscus*, maize, mulberry, pumpkin and rose (Tanwar *et al.*, 2007; Rajendran, 2009).

Some of the insect pests of vegetable crops have become major and are gradually attaining the major pest status in different regions of the country due to changes in the ecosystem and habitats. Fruit borer, *Helicoverpa armigera* on tomato; whitefly, *Bemisia tabaci* and serpentine leaf miner *Liriomyza trifolii* on tomato and cucurbitaceous crops; fruit fly on fruits and vegetables; mealy bugs on several horticultural crops; gall midge on brinjal; okra stem fly; bitter gourd leafhopper; red spider mite on okra, brinjal, cowpea, Indian bean; and nematodes on several vegetable crops are some of the examples. In recent times, there were outbreaks of gall midge (*Asphondylia capparis*), which was known to be a minor pest and gradually becoming a regular problem in chilli, bell pepper and brinjal in the states of Andhra Pradesh and Karnataka, and on brinjal in Chhattisgarh. Various species of mealy bugs in cotton, vegetables and papaya have intensified their severity of occurrence in different parts of the country and have become indicator insects for the current ecosystems, due to slow changes in climate in the last one decade. Several national and international polyphagous pests like termite, white grub, hairy caterpillar, gram pod borer and *Spodoptera litura* are also becoming severe and expanding their host-horizon. This envisages that there is a need to relook into the cropping systems for reorientation and vegetables should be given importance with adequate insect pest management measures. In this context, integrated pest management (IPM) system was developed out of the need for sustainable crop protection strategy against the backdrop of increasing pesticide use and deleterious effect of residues in the environment.

Changes in climatic variables have led to increased frequency and intensity of outbreaks of insect-pests. Outbreak of sugarcane woolly aphid, *Ceratovacuna lanigera* in sugarcane belt of Karnataka and Maharashtra states during 2002-03 resulted in 30% yield losses. These situations of increased and frequent pest damage to the crops have made another big hole in the pockets of already distressed farmers by increasing the cost of plant protection and reducing the margin of profit (Table 5.4).

5.6. RISK OF INTRODUCING INVASIVE ALIEN SPECIES

Even though the causes of biological invasions are manifold and multifaceted, changes in abiotic and/or biotic components of the environment (climate change, biological control) are recognized as primary drivers of species invasion (Dukes and Mooney, 1999; IPCC, 2007). Globalization and liberalization of world agricultural trade coupled with the rapid transport and communication means nowadays, have substantially and plausibly increased the chances of exotic introductions.

According to the Convention on Biological Diversity (CBD), invasive alien species are the greatest threat to loss of biodiversity in the world and impose high costs to agriculture, forestry and aquatic ecosystems by altering their regional structure, diversity and functioning (Mooney and Hobbs, 2000; Sutherst, 2000; Timoney, 2003).

It is expected that global warming may exacerbate ecological consequences like introduction of new pests by altering phenological events like flowering times especially in temperate plant species (Fitter and Fitter, 2002; Parmesan and Yohe, 2003; Willis *et al.,* 2008) as several tropical plants can withstand the phenological changes (Corlett and LaFrankie, 1998). Invasion of new insect-pests will be the major problem with changing climate favouring the introduction of insect susceptible cultivars or crops (Gregory *et al.,* 2009).

5.7. CROP-PEST INTERACTIONS

The increasing temperature and CO_2 have been found to exert both bottom-up and top-down effects on the tri-tropic interactions between crops, insects and natural enemies by means of certain physiological changes especially related to host suitability and nutritional status (Table 5.5) (Roth and Lindroth, 1995; Coviella and Trumble, 1999; Gutierrez *et al.*, 2008). The CO_2 enriched environment reduces the nitrogen content of the plant tissue due to widening of Carbon: Nitrogen (C: N) ratio, thus cause a slight decrease in nitrogen-based defences like alkaloids and in turn may increase in carbon-based defences such as tannins (Roth and Lindroth, 1995; Coviella and Trumble, 1999; Gutierrez *et al.*, 2008). This enhances the feeding by insect herbivores in order to obtain sufficient nitrogen for their metabolism (Coviella and Trumble, 1999). Ultimately, it slows down the insect development and increases the length of life stages resulting in more foliage feeding than the normal (Coviella and Trumble, 1999).

Table 5.4. Recorded instances of recent insect pest outbreaks in relation to changing climate scenario in India

Insect pest	Order/ Family	Host plant/s	Region/ location	Probable reason/s	Impact of pest outbreak	Reference
Sugarcane woolly aphid, *Ceratovacuna lanigera*	Hemiptera: Aphididae	Sugarcane	Sugarcane belt of Karnataka & Maharashtra States during 2002-03	Recent abnormal weather patterns. Insecticide misuse.	30% yield losses. Reduced cane recovery.	Joshi & Viraktamath, 2004; Srikanth, 2007
Rice plant hoppers, *Nilparvata lugens* & *Sogatella furcifera*	Hemiptera: Fulgoridae	Rice	North India	Recent abnormal weather patterns. Insecticide misuse.	Crop failure over more than 33,000 ha paddy area	IARI News, 2008
Mealy bug, *Phenacoccus solenopsis*	Hemiptera: Pseudococcidae	Cotton, Vegetables, ornamentals	Cotton growing belt of the country	Recent abnormal weather patterns. Insecticide misuse. Changed cropping environment (introduction of *Bt* cotton).	Heavy yield loss (30-40 %) to cotton. Increased cost of crop protection due to overuse of pesticides.	Dhawan *et al.*, 2007
Papaya mealy bug, *Paracoccus marginatus*	Hemiptera: Pseudococcidae	Papaya	Tamil Nadu, Karnataka, Maharashtra	Recent abnormal weather patterns. Insecticide misuse.	Significant yield loss to the papaya growers.	Tanwar *et al.*, 2010

Table 5.5. Crop-insect pest interactions in the context of climate change

Insect pest	Order/ Family	Host plant/s	Climatic factors studied	Impact on host plant	Impact on insect pest	Reference
Many foliage feeding Lepidopterons	Lepidoptera: Lymantridae, Noctuidae, Pyralidae	Economically important agricultural and forest species	Increased CO_2	Reduced nitrogen content of the plant tissue due to widening of C: N ratio. Decrease in nitrogen-based plant defences like alkaloids. Increase in carbon-based defences such as tannins.	Enhanced feeding by insects in order to obtain sufficient nitrogen for their metabolism. Slower development. Increased length of life stages. More foliage feeding than the normal.	Coviella & Trumble, 1999
				Less stomatal opening. Increased water use efficiency. Reduced water loss through stomata.	Extension of plant life spans. Ease in nutrient assimilation and digestion especially the nitrogen. Longer periods of habitat suitability for insects.	Reitz *et al.*, 1997
Gypsy moth, *Lymantria dispar*	Lepidoptera: Noctuidae	Red maple, *Acer rubrum;* Sugar maple, *Acer saccharum*	Temperature × CO_2 combination (Ambient and elevated)	Reduced leaf water content. Increased concentration of soluble sugars. Declined nutritional quality of foliage.	Reduced larval weight gain. Increased larval feeding. Prolonged development.	Williams *et al.*, 2000

Insect pest	Order/ Family	Host plant/s	Climatic factors studied	Impact on host plant	Impact on insect pest	Reference
Midge, *Stenodiplosis sorghicola* and Spotted stem borer, *Chilo partellus*	Diptera: Cecidomyidae Lepidoptera: Pyralidae	Jowar, *Sorghum bicolor*	High temperature, drought/ water stress	Breakdown of resistance against target insect pests. Heavy loss in yield due to increased pest damage.	—	Sharma *et al.*, 2005; 2010
Bollworms, *Heliothis virescens, Helicoverpa armigera* and *H. punctigera*	Lepidoptera: Noctuidae	Cotton	High temperature, drought/ water stress, photoperiod	Negative impacts on transgene expression in *Bt* Cotton. Reduced production of *Bt* toxins. Enhanced susceptibility of crops to insect pests.	—	Kaiser, 1996; Hilder and Boulter, 1999

In CO_2 enriched atmosphere water use efficiency of plants increases owing to the reduced water loss through less stomatal opening (Groninger *et al.*, 1996). Increased water content in plants is beneficial for most of the herbivorous insects as it helps in nutrient assimilation and digestion especially nitrogen (Reitz *et al.*, 1997). Increased water use efficiency enables the plants to extend their life spans providing longer periods of habitat suitability for transient insects. Contrarily, under elevated temperature, the concentrations of certain allellochemicals like terpenes and phenolic compounds increases in plants that act as defensives against attacking insect-pests (Roth and Lindroth, 1995; Coviella and Trumble, 1999; Gutierrez *et al.*, 2008).

Temperature and photoperiod have been found to affect profoundly the critical events such as stem elongation, flowering and fruiting in the life cycle of plants (Cleland *et al.*, 2007). Global warming lead increased temperatures may accelerate the life cycles in some of the plant species (Parmesan and Yohe, 2003: Willis *et al.*, 2008) which may affect significantly, feeding and reproduction patterns in associated insect-pests like aphids, jassids, mealy bugs, etc. Such increases can greatly exacerbate the negative ecological and economic consequences (Timoney, 2003).

5.8. LOSS OF ECOLOGICAL BIODIVERSITY

Arthropods (insects, spiders, and mites) are the most abundant and diverse group of organisms (Kannan and James, 2009; Gregory *et al.*, 2009). They are the most important and diverse component of terrestrial ecosystems and occupy a wide variety of functional niches and microhabitats (Kremen *et al.*, 1993). We can take advantage of the terrestrial arthropod diversity as a resource for conservation and management of different eco-systems. Monitoring of terrestrial arthropods can provide early warnings of ecological changes due to climate change. Arthropods can be used as indicators of environmental change more rapidly than the vertebrates (Scherm *et al.*, 2000; Gregory *et al.*, 2009). For monitoring purposes, indicator assemblages should exhibit varying sensitivities to environmental changes, and exhibit diversity in life-history and ecological interactions. Realistic information on arthropod diversity must be integrated into policy planning and management practices, if ecosystems are to be managed for use by future generations.

The biodiversity signifies the biological wealth of habitat by means of species richness in an ecosystem. For sustainable agriculture development in any given country, biodiversity is of paramount importance (UN-HABITAT, 2004; Murugan, 2006). South Asia in general and India in particular is blessed with ecologically rich natural and crop-related biodiversity due to its unique geographic location and diversified climatic conditions. India is one of the 12 mega-biodiversity centres with three out of 34 biodiversity hotspots in the world (Myers *et al.* 2000; UN-HABITAT, 2004; Murugan, 2006). Due to change in the climate pattern in recent decades owing to increasing industrialization and over-exploitation of natural resources for various anthropogenic developmental activities, many species of plants, animals and insects are decreasing at an alarming rate (Costanza *et al.*, 1987; Murugan, 2006; Sachs, 2008). The loss of biological diversity is still accelerating which may reduce the ecosystem's resilience to the climatic changes.

Insects comprise the largest group of animal kingdom and play vital role in providing various ecosystem services (Kremen *et al.*, 1993; Kannan and James,

2009). The insect diversity in a habitat indicates the health status of an ecosystem as they are very good indicators of environmental change (Gregory *et al.*, 2009), play an important role in food chains, are excellent pollinators for many of the economically important crops (Ingram *et al.* 1996; Klein *et al.* 2007; Ricketts *et al.*, 2008) and contribute directly to the human economies through valuable products like silk, lac, honey and wax (Myers *et al.*, 2000; Murugan, 2006; Sidhu and Mehta, 2008). About 6.83% of world insect species are inhabitant in India (Alfred, 1998). The climate change may affect the relative abundance of different insect species and the species unable to adapt the changes may be lost in the due course of time (Thomas *et al.*, 2004). The Western Ghats in India is the only habitat to many rare, endemic and exotic species of colourful butterflies in the world (Hampson, 1908; Anand and Pereira, 2008). In the present day scenario, many butterfly species are under a real threat due to depletion of the natural vegetation for various anthropogenic developmental activities (Costanza *et al.*, 1987; Sachs, 2008; Sidhu and Mehta, 2008).

The negative effects of climate change are accelerating the rate of biodiversity loss, worldwide. According to the Millennium Ecosystem Assessment (2005), more than one-third of species in the world are at the risk of extinction and an estimated 60% of the Earth's ecosystems have been degraded in the last 50 years, with negative consequences for the ecosystem flow. Nearly 99.9% of all species that ever existed have become extinct. Up to 50% of the Asia's total biodiversity is at risk due to climate change. Many other species could also be extricated as a result of the climate change and habitat fragmentation (Ishigami *et al.*, 2005). According to the estimates of IUCN (1994), around 22 species of invertebrates (insects, earthworms, nematodes, crustaceans, spiders, etc.) are at the risk of extinction (Fig. 5.5). The relative abundance of different insect species may change rapidly due to climate change, and the species unable to withstand the stresses may be lost in the near future (Jump and Penuelas, 2005; Thomas *et al.*, 2004). The loss of biodiversity may impact negatively the structure, composition and functioning of ecosystems and wildlife habitat leading to outbreaks of destructive insect-pests and diseases (Timoney, 2003; UN-HABITAT, 2004; IPCC, 2007).

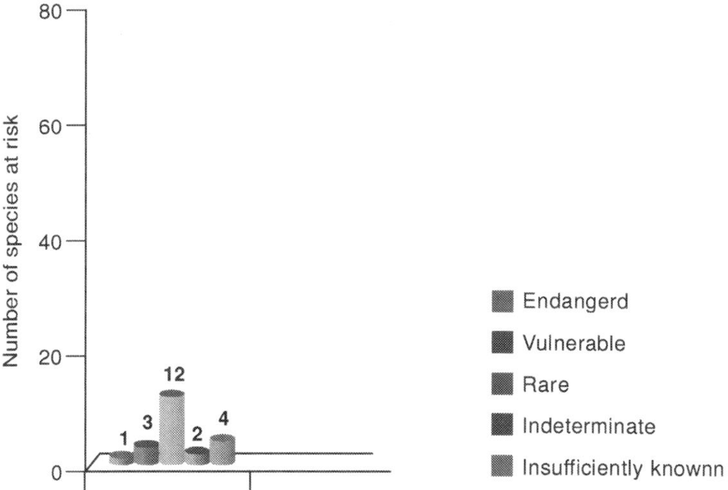

Fig. 5.5. Number of threatened invertebrate species (*Source*: IUCN threat categories, 1994)

Ecosystem baselines that document arthropod species assemblages in a manner comparable in space and time are key to interpretation and implementation of strategies designed to mitigate the effects of global warming and climate change on bio-diversity. Main effects of climate change and pollution on arthropod communities results in decreased abundance of decomposers and predators, and increased herbivores, which may have negative consequences for structure and services of the entire ecosystems. Responses of arthropods to pollution depend on both temperature and precipitation, and ecosystem-wide adverse effects are likely to increase under predicted climate change (Zvereva and Kozlov, 2009). Consequences of temperature increases of 1 to 2° C will be comparable in magnitude to the currently seen climate change in the Antarctic region (Bokhorst *et al.*, 2008).

Increase in rainfall in the Pampas region of Argentina will largely affect the species with poor dispersal capabilities, which will limit their ability to expand their home range. The most affected among the beetle species are the habitat specialists (Xannepuccia *et al.*, 2009). At higher trophic levels, an indirect effect in terms of habitat loss and a reduction in prey availability has also been observed. Large scale changes in rainfall due to climate change will have a major effect on the abundance and diversity of arthropods. Extreme climatic events such as drought are likely to decrease multi-trophic diversity and change the composition of arthropod communities, which in turn might affect the other associated taxa. In the forest eco-systems, chronic stress significantly altered community composition, and the trees growing under high stress supported 1/10th the number of arthropods compared to trees growing under more favourable conditions (Talbot Trotter *et al.*, 2008). Increasing tree stress was also correlated with an eight to 10-fold decline in arthropod species richness and abundance. Arthropod richness and abundance on individual trees were positively correlated with the tree's radial growth during drought, suggesting that tree ring analysis could be used as a predictor of arthropod diversity (Stone *et al.*, 2010).

5.9. CHANGES IN PHENOLOGY

Recent climate change has led to an ecological shift in time, with changes in species' phenology. Changes in insect phenology can be studied through long-term experiments with variable sowing dates for observing the appearance of pests on crops. Likewise, the timing of arrival of insect species can also be recorded through light traps, suction traps or pheromone traps. Analysis of long-term data on phenology would reveal changes in the timings of pest appearance under the climate change.

Analysis of suction trap data at the Rothamsted Insect Survey since 1964, has revealed that spring flights of the peach potato aphid (*Myzus persicae*) started two weeks earlier for every 1°C rise in combined mean temperature of January and February.

Likewise, long-term data from several insect-recording schemes in Europe and North America have provided evidence for species becoming active, migrating or reproducing earlier in the year due to increases in temperatures that lead directly to increased growth rates or earlier emergence from winter inactivity (Roy and Sparks, 2000). Increasing temperatures have also allowed a number of species to remain active for a longer period during the year or to increase their annual number of generations.

Under the All India Coordinated Rice Improvement Project (AICRIP) of ICAR, there is a widespread network of the Coordinating Centres all over India that collect light trap insect data round the year. Analysis of historical light trap data vis-à-vis current data can provide important information on the impacts of climate change on rice pests.

Some of the species that are able to adapt to the warmer climates may become major pests. Differences in the pattern of response to temperature changes would disrupt synchronization in phenology between insects and host plants or natural enemies (Kiritani, 2006).

5.10. INCREASED INCIDENCE OF INSECT VECTORED PLANT DISEASES

Interaction between insect vector, disease and climate change is more complicated as compared to interaction of disease or insect alone with climate change because of disease dependency on the vector. Climate change is anticipated to alter the abundance and distribution of insect vectors. Both vectors and vector-borne diseases are likely to follow the migrating hosts and may infect remenant vegetation of natural plant communities which are not previously exposed to the often more aggressive strains from agricultural crops (Coakley *et al.,* 1999). The impact of climate change on insect vectors is complex as it could alter patterns of disturbance from vector insects through direct effects on their development and survival, adaptation capability, availability of host plants and physiochemical changes in host defences and indirect effects from changes in abundance of natural enemies, mutualists, and competitors (Porter *et al.,* 1991; Patterson *et al.,* 1999; Ayres and Lombardero, 2000). The changes in insect vectors ecology especially is occurring evidently, which are consistent with predictions of climate change impacts (Harrington *et al.,* 2001).

The changes in vector over-wintering and over-summering are certain when a new vector species enters the crop fields and other effects of climate change on insects may have important influence on pathogen survival, movement and reproduction (Fabre *et al.,* 2005; Newman, 2004; 2005). Introduction of glassy-winged sharp shooter has led to increased patterns of infection of grape plants in winter; greatly altering infection rates is an excellent example of such impact. Temperature is the dominant climate factor, which directly affects over-wintering and potentially important in combination with photoperiod. There are little evidences of direct effects of CO_2, UV-B and precipitation on disease and insert vector interactions to draw a general conclusion (Bale *et al.,* 2002).

Vector diversity is the key to understand plant virus epidemiology. The plant disease dynamics may be more constrained by the host range of vectors than by the host range of virus. The importance of vector diversity for pathogen epidemiology and disease dynamics is not well-understood. Many viruses are strongly dependent on the vectors that transmit them among hosts. Analysis of a large database of 1673 total plant viruses, including 910 unique virus species, indicated that plant viruses have much more specialized relations with their vectors than with their hosts. In general, plant viruses have a low diversity of vectors and a high diversity of hosts, and the largest proportion of viruses fall into the category of host generalist/vector specialist (Reddy, 2010).

Climate change may lead to more incidence of insect transmitted plant diseases through range expansion and rapid multiplication of insect vectors (Petzoldt and Seaman, 2010; Sharma et al., 2005; 2010). Increased temperatures, particularly in early season, have been reported to increase the incidence of viral diseases in potato due to early colonization of virus-bearing aphids, the major vectors for potato viruses in Northern Europe (Robert *et al.,* 2000).

A majority of vector-transmitted viruses (58.4%) have a single known vector species, whereas only 9.9% of such viruses have a single known host plant.

5.11. DISRUPTION OF PLANT-POLLINATOR INTERACTIONS

Insects play vital role in providing various ecosystem services, a foundation for human life on earth (Kannan and James, 2009). One of the important ecosystem service provided by insects is pollination as they are excellent pollinators for many of the economically important crops (Sidhu and Mehta, 2008). Majority of the flowering plants require insect pollinators like flies, butterflies, moths, beetles and especially bees for their reproduction and formation of fruits and seeds (Ricketts *et al.*, 2008). Honey bees are perhaps the best known pollinators because of their floral fidelity. Insect pollination, mostly by bees, is necessary for 75% of all crops that are used directly for human food worldwide. Thus, the insect pollination is a fundamental process essential for the production of about one third of the world human food (Klein *et al.*, 2007).

According to Millennium Ecosystem Assessment Report 2005, pollination is one of the 15 major ecosystem services currently under threat from mounting pressures exerted by growing population, depleting natural resource base and global climate change (Sachs, 2008). Earlier studies have clearly shown that the population abundance, geographic range and pollination activities of important pollinator species like bees, moths and butterflies are declining considerably with changing climate (FAO, 2008). The climatic factors like temperature and water availability have been found to affect profoundly the critical events like flowering, pollination and fruiting in the life cycle of plants (Cleland *et al.*, 2007). Many pollinators have synchronized their life cycles with plant phenological events. Impending climate change is expected to disrupt the synchrony between plant-pollinator relationships by changing the phenological events in their life cycles and may thus affect the extent of pollination (Ricketts *et al.*, 2008). The quality and the quantity of pollination have multiple implications for food security, species diversity, ecosystem stability and resilience to climate change (FAO, 2008).

Although pollination is a critical issue, it appears to be neglected and overlooked for other ecosystem services such as water and air quality, climate regulation and food availability. The pollination services and associated risks are not addressed properly in determining the actions needed for conserving pollinators. The high degree of uncertainty regarding the risks related to pollination services implies the need for well-focused research to understand scientifically the pollination processes.

5.12. REDUCED EFFECTIVENESS OF PEST MANAGEMENT STRATEGIES

Certain effective cultural pest management practices like crop rotation, early/late planting, etc. will be less or not effective with changed climate because of shrinking crop growing seasons, colonization of crops by early insect arrival and/or increased winter survival (Harrington *et al.*, 2001; Sharma *et al.*, 2005; Petzoldt and Seaman, 2010). Disruption of synchrony between insect-pests and their natural enemies may upset the natural biological control (Petzoldt and Seaman, 2010). Certain pesticides like pyrethroids, organophosphates and especially the biopesticides being highly thermo-unstable degrade faster at higher temperatures. Altered temperature regimes may render many of these products to be less or not effective in pest control, necessitating frequent insecticide applications for effective control (Musser and Shelton, 2005). This may intensify the pest problems due to the increased chances of resistance development in insects. Ultimately, it will add to increased cost of crop protection to the farmers and in turn environmental cost. (Musser and Shelton, 2005; Petzoldt and Seaman, 2010). The forewarning models for predicting insect arrival/ infestations based on earlier climate profiles need to be revised in accordance with location specific changes in climate in order to provide precise and accurate forecast of the pest incidence.

A number of cultural practices that can be used by farmers could be affected by changes in climate – although it is not clear whether these practices would be helped, hindered, or not affected by the anticipated changes. Using crop rotation as an insect management strategy could be less effective with earlier insect arrival or increased overwintering of insects. However, this could be balanced by changes in the earliness of crop planting times, development, and harvest. Row covers used for insect exclusion might have to be removed earlier to prevent crop damage by excessive temperatures under the covers – would the targeted early insects also complete their damaging periods earlier or be ready to attack when the row covers were removed?

There will be an increase in variability in insect damage as a result of climate change. Higher temperatures will make dry seasons drier, and conversely, may increase the amount and intensity of rainfall, making wet seasons wetter than at present. Natural plant products, entomopathogenic viruses, fungi, bacteria, and nematodes, and synthetic pesticides are highly sensitive to the environment. Increase in temperatures and UV radiation, and a decrease in relative humidity may render many of these control tactics to be less effective, and such an effect will be more pronounced on natural plant products and the biopesticides (Isman, 1997). Rapid dissipation of insecticide residues due to increases in temperature and precipitation will require more frequent application of insecticides. Therefore, there is a need to develop appropriate strategies for pest management that will be effective under situations of global warming in future.

5.12.1. Reduced effectiveness of transgenic crops for pest management

Environmental factors such as soil moisture, soil fertility, and temperature have strong influence on the expression of *Bacillus thuringiensis* (*Bt*) toxin proteins deployed in transgenic plants (Sachs *et al.,* 1998). Cotton bollworm, *Heliothis virescens* destroyed *Bt*-transgenic cottons due to high temperatures in Texas, USA (Kaiser, 1996). Similarly, *Helicoverpa armigera* and *H. punctigera* destroyed the *Bt*-transgenic cotton in the second half of the growing season in Australia because of the reduced production of *Bt* toxins (Hilder and Boulter, 1999). *Cry1Ac* levels in transgenic plants decreases with the plant age, resulting in greater susceptibility of the crop to insect pests during the later stages of crop growth (Sachs *et al.*, 1998; Greenplate *et al.*, 2000; Adamczyk *et al.*, 2001; Kranti *et al.*, 2005). Possible causes for the failure of insect control in transgenic crops may be due to inadequate production of the toxin protein, effect of environment on transgene expression, *Bt*-resistant insect populations, and development of resistance due to inadequate management (Sharma and Ortiz, 2000). It is therefore important to understand the effects of climate change on the efficacy of transgenic plants for pest management.

In summary, the effect of climate change on different crop protection technologies include:

- Increase in temperature and precipitation lead to rapid dissipation of insecticide residues for which more frequent application is needed.
- Increase in temperature, UV radiation and decrease in RH may render natural plant products, entomopathogenic virus, fungus, bacteria, nematode to be less pathogenic and less effective.
- Due to high temperature, the synthetic pyrethroids and biopesticides show lesser efficacy due to degradation.
- Breakdown of natural resistance in crop plants to insect pests due to increased temperature.
- Activity and effectiveness of natural enemies (parasitoids and predators) is also affected.
- Soil moisture, soil fertility and temperature have strong influence on expression of *Bacillus thuringiensis* toxin deployed in transgenic plants. Example - Cotton bollworm (*Heliothis virescens*) destroys the Bt transgenic cotton due to high temperature in Texas. Possible reason is inadequate production of toxin protein and effects of environment on transgenic crop.

5.12.2. Breakdown of host plant resistance

Global warming may result in breakdown of resistance to certain insect pests. Sorghum varieties exhibiting resistance to sorghum midge, *Stenodiplosis sorghicola* in India, became susceptible to the pest under high humidity and moderate temperatures near the Equator in Kenya (Sharma *et al.*, 1999). Lower foliar nitrogen content due to CO_2 causes an increase in food consumption by the herbivores up to 40%, while unusually severe drought increases the damage by insect species such as stem borer, *Chilo partellus* in sorghum (Sharma *et al.*, 2005). Global warming may also change the flowering times in temperate regions, leading to ecological consequences such as introduction of new insect pests and attaining of a pest status by non- insect pests (Willis *et al.*, 2008).

Host plant resistance to insects is one of the most environmental friendly components of pest management. It is one of the eco-friendly options for managing harmful insect-pests of crops wherein the plant can lessen the damage caused by insect pests through various mechanisms like antixenosis, antibiosis and tolerance

(Dhaliwal and Dilawari, 1993). However, expression of the host plant resistance is greatly influenced by environmental factors like temperature, sunlight, soil moisture, air pollution, etc. Global warming may result in breakdown of resistance to certain insect pests. Under stressful environment, plant becomes more susceptible to attack by insect-pests because of weakening of their own defensive system, resulting in pest outbreaks and more crop damage (Rhoades, 1985). Thermal and drought stress associated breakdown of plant resistance have been widely reported (Rhoades, 1985; Sharma *et al.*, 2005). Increased CO_2 may also cause a slight decrease in nitrogen-based defences (e.g., alkaloids) and a slight increase in carbon-based defences (e.g., tannins). With global temperature rise and increased water stress, tropical countries like India may face the problem of severe yield loss in sorghum due to breakdown of resistance against midge *Stenodiplosis sorghicola* and spotted stem borer *Chilo partellus* (Sharma *et al.*, 2005).

There will be increased impact on insect pests which benefits from reduced host defences as a result of the stress caused by the lack of adaptation to sub-optimal climatic conditions. Chemical composition of some plant species changes in direct response to biotic and abiotic stresses; as a result, their tissues are less suitable for growth and survival of insect pests (Sharma, 2002). However, problems with new insect pests will occur, if climatic changes favour the introduction of insect susceptible cultivars or crops. The introduction of new crops and cultivars to take advantage of the new environmental conditions is one of the adaptive methods suggested as a possible response to climate change (Parry and Carter, 1989).

Down-regulation of genes for defence signalling (Iox7, Iox8 and acc-s) reduced the production of cysteine proteinase inhibitors (CystPI), which are specific deterrents of Coleopteran herbivores resulting in increased colonization by the pests.

Development of insect resistant transgenic crops opened new avenues for exploiting host plant resistance in integrated pest management. A gene encoding delta-endotoxin proteins from entomopathogenic soil bacterium *Bacillus thuringiensis* is deployed in transgenic plants (Kranthi *et al.*, 2005). However, expression of *Bt* toxins in transgenic plants is greatly influenced by environmental factors like temperature, soil moisture and plant age (Dhaliwal and Dilawari, 1993; Kranti *et al.*, 2005). The environmental factors like high temperature have been found affecting transgene expression in *Bt* cotton resulting in reduced production of *Bt* toxins. This lead to enhanced susceptibility of the crops to insect-pests like bollworms viz., *Heliothis virescens* (Kaiser, 1996), *Helicoverpa armigera* and *H. punctigera* (Hilder and Boulter, 1999).

European large raspberry aphid (*Amphorophora idaei*) is the most significant insect pest of raspberry production. Aphids are vectors of at least four plant viruses that reduce plant vigour and can cause death. Introduction of two resistance genes in particular, which have become at least partially overcome (A1 resistance – weakened in the UK, A10 resistance – weakened in England).

5.12.3. Reduced activity and abundance of natural enemies

Relationships between insect pests and their natural enemies will change as a result of global warming, resulting in both increases and decreases in the status of individual pest species. Changes in temperature will also alter the timing of diurnal activity patterns of different groups of insects (Young, 1982), and changes in interspecific interactions could also alter the effectiveness of natural enemies for pest management (Hill and Dymock, 1989). Quantifying the effect of climate change on the activity and effectiveness of natural enemies for pest management will be a major concern in future pest management programmes. The majority of insects are benign to agro-ecosystems, and there is considerable evidence to suggest that this is due to population control through interspecific interactions among insect pests and their natural enemies – pathogens, parasitoids, and predators (Price, 1987). Oriental armyworm, *Mythimna separata* populations increases during extended periods of drought (which is detrimental to the natural enemies) followed by heavy rainfall because of the adverse effects of drought on the activity and abundance of the natural enemies of this pest (Sharma, 2002). Aphid abundance increases with an increase in CO_2 and temperature, however, the parasitism rates remain unchanged in elevated CO_2. Temperatures up to 25°C will enhances the control of aphids by Coccinellids (Freier and Triltsch, 1996). Temperature not only affects the rate of insect development, but also has a profound effect on fecundity and sex ratio of parasitoids (Dhillon and Sharma, 2008; 2009). The interactions between insect pests and their natural enemies need to be studied carefully to devise appropriate methods for using natural enemies for pest management.

Biological control of insect-pests is one of the most important components of integrated pest management, safeguarding the ecosystem. Natural enemies of crop pests viz., predators, parasitoids and pathogens are prompt density responsive in their action subjected to the action of abiotic components. Being tiny and delicate, natural enemies of the insect pests are more sensitive to the climatic extremes like heat, cold, wind and rains. Precipitation changes can also affect predators, parasitoids and pathogens of insect-pests resulting in a complex dynamics. With changing climate, incidence of entomopathogenic fungi might be favoured by prolonged humidity conditions and obstinately be reduced by drier conditions (Newton *et al.*, 2011). Natural enemy and host insect populations may respond differently to changes in climate. Hosts may pass though vulnerable life stages more quickly at higher temperatures, reducing the window of opportunity for parasitism which may give great set back to the survival and multiplication of parasitoids (Petzoldt and Seaman, 2010).

Ecologists argue that the tritrophic interactions between plants, herbivorous insects, and their natural enemies (predators, parasitoids and pathogens) result from a long co-evolutionary process specific to a particular environment and relatively stable climatic conditions (Hance *et al.*, 2007). Abrupt environmental changes as induced by current climatic change and elevated CO_2 may influence the biology of each component of a system differently, provoking a destabilization in their population dynamics that may lead to the extinction of part of the system. Specialists, for example many host specific parasitoids, which evolved under rather stable conditions might be especially endangered.

Atmospheric CO_2 levels may affect the performance of natural enemies and/or susceptibility of prey via a variety of indirect effects. Some of these impacts, which potentially make prey more susceptible to their enemies, include:

- Herbivores that feed on poor host plants under elevated CO_2 conditions often spend more time in the more vulnerable, early stages of development, and thus may suffer greater mortality from natural enemies.

- Herbivores may be physically weakened while feeding on poor hosts under elevated CO_2 conditions, and are thus less able to defend themselves against predators and parasitoids; and enriched CO_2 may alter enemy-avoidance behaviour. Some aphids, for example, show reduced responses to alarm pheromones under elevated CO_2, potentially making them more susceptible to enemy attack (Awmack *et al.*, 1997). Such effects would increase the susceptibility of herbivores to natural enemies, reducing herbivore population size under elevated CO_2 conditions (Coll and Hughes, 2008).

Elevated temperature basically favours adult hunting insects and spiders, and it seems that the lethal temperature of many spiders is much above the temperature expected by climate change. Skirvin *et al.* (1997) modelled the interaction of ladybird (*Coccinella septempunctata*) with aphid populations (*Sitobion avenae*) and predicted that in hot summer Coccinellids reduces aphids more strongly than in moderate summers.

5.12.3.1. Pathogens: Fungi, bacteria, microsporidia and viruses can successfully affect rodents, insect pests and mites. They are widely used in biological control, with the bacteria *Bacillus thuringiensis* and the fungi *Beauveria bassiana* being prominent examples.

Effects of climate change on the efficiency of pathogens depend on the environment they live in. In general, fungi and bacteria benefit from warm and moist environments, therefore mild and wetter winters as predicted in temperate zones will benefit them, especially those living in the soil (e.g. *Beauveria bassiana*). Since many larvae or pupae of pests also overwinter (pass through or wait out the winter season) in soils, fungi and bacteria might affect them more strongly.

Guiterrez *et al.* (2008) found that during the normally wet Northern California winter, the fungal pathogen (*Pandora neoaphidis*) causes catastrophic mortality to pea aphid (*Acyrthosiphon pisum*), but during hot dry periods, the impact of the pathogen declines.

Most entomopathogenic fungi have optimal growth temperatures between 25 and 35° C. *Beauveria bassiana* grows at a wide temperature range (from 8 to 35° C) with a maximum thermal threshold for growth at 37° C. Higher temperatures, low humidity, as well as direct exposure to UV radiation reduces efficiency of pathogens.

However, each pathogen responds to temperatures differently and behaviour of the host in response to temperature is important as well (Blanford and Thomas, 1999). Manning and von Tiedemann (1995) showed that direct exposure of bacteria and fungi to high levels of CO_2 concentrations often inhibits their growth.

Some pathogens, which always live in the host body might not be affected directly by climatic changes, they basically follow the development of their hosts. The effects of higher temperature on the impact of the microsporidia, *Nosema lymantriae* on the gypsy moth (*Lymatria dispar*) clearly showed a much higher and earlier mortality of gypsy moth larvae at higher temperatures (Pollan, 2009).

Pathogens, especially viruses, become more deadly if the vector/host is weakened, therefore environmental stress such as high or low temperature might lead to higher mortality. Considering that herbivorous pests are potentially weakened by the lower nutritional quality of C_3 plants grown under elevated CO_2, it could be assumed that mortality of pests feeding on C_3 crops increases when infected with pathogens (with potentially serious consequences also for some natural ecosystems). However, it seems that no one has investigated this kind of interactions so far.

5.12.3.2. Parasitoids: Parasitoids, which live on crop pests belong to the third trophic level. Thus, they are indirectly or directly affected by any changes of the first (plant) and second level (herbivore). It is not at all clear what happens to herbivores under climate change, therefore conclusions for parasitoids are speculative. However, there are some ecological 'laws' which imply certain scenarios. If an herbivore reproduces less, because of low nutritional value, less potential hosts are available for the parasitoid. If the host changes it seasonal appearance or behaviour due to climatic changes, the parasitoid might not be able to locate the host. Finally parasitoids might be adversely affected, if the host dies too early due to additional environmental stress. However, in temperate zones milder winters might enhance survival of parasitoids. Legrand *et al.* (2004) have shown that parasitoids of cereal aphids are active in winter and this winter activity can considerably reduce spring aphid populations.

No experiments have been conducted to investigate changes of all three trophic levels together (plant-herbivore-parasitoid) under climate change (elevated CO_2 and temperature). Bezemer *et al.* (1998) conducted an experiment involving several plant species, aphids and parasitoids under elevated CO_2 (+200 ppm of ambient concentrations), and showed that elevated CO_2 did not influence parasitism. Elevated temperature (+2° C of ambient temperature) increased parasitism by about 300% on average, but due to high variation between the replicates no significance could be detected.

A mathematical model has been developed that predicts responses of grasses, cereal aphids and parasitoids to combined effects of elevated CO_2 and elevated temperature. Their results suggest that aphid and parasitoid populations will develop more similar to current ambient conditions than expected from the individual effects of CO_2 or temperature increases.

In one experiment with cotton bollworm, larvae reared on milky wheat grain under 750 ppm CO_2, researchers included a parasitoid wasp (*Microplitis mediator*) widely used as bio-control agent of the cotton bollworm (*Helicoverpa armigera*). The researchers found no significant changes in wheat consumption by *H. armigera* population under elevated CO_2 or in the parasitic rate of *M. mediator*. The researchers concluded that the population relationship between *H. armigera* and *M. mediator* is unlikely to vary due to further elevated atmospheric CO_2 concentrations.

The development of a parasitoid wasp (*Glyptapanteles liparidis*) of gypsy moth (*Lymantria dispar*) feeding on three different tree species fumigated with 540 ± 20 ppm CO_2 was not adversely affected by changes in food quality when compared to ambient CO_2. However, it must be taken into account that the effects of elevated CO_2 on mature trees might not be comparable to annual plants or tree seedlings.

In some cases, extreme climatic events not only have an effect on the incidence of pests and diseases, there have conditions also been observed where beneficial insects are present in the crops, for example the eggs of the parasitoid, *Trichogramma carverae* used for oat pest control at temperatures above 35° C decreases fertility by 50% compared to a temperature of 30° C (Thompson *et al.*, 2010).

The precipitation variability seems to be a key factor influencing parasitism in 15 Lepidopteron (butterfly) rearing programmes from a broad spectrum of climatic regimes and locations, from the region between southern Canada and central Brazil. A higher variability led to a decrease in parasitism. These findings basically support the theory that interaction, which evolved due to stable conditions, are weakened when frequent changes occur.

In general, host-specific parasitoids should be more sensitive to variations in host emergence time or developmental rate when compared to generalists. Specialist parasitoids may miss narrow windows of vulnerability of their particular hosts. In contrast, because generalists exploits a variety of hosts that might individually respond to climatic cues in different ways, they should be less susceptible to the host population's lags and asynchronies associated with climatic unpredictability.

5.12.3.3. Predators: Like parasitoids, predators which prey on crop pests belong to the third trophic level. Thus, they are indirectly or directly affected by any changes of the first (plant) and second level (herbivore).

In some cases extreme climatic events not only have an effect on the incidence of pests and diseases, there have also been observed conditions in beneficial insects present in the crops, for example the eggs of the predator *Cyrtorhinus lividipennis* used for the control of skipped rice leaves caused by *Nilaparvata lugens* decreases if there is an increase in temperature of 32-35°C, indicating that the activity of the predator will be reduced above 35°C (Karuppaiah and Sujayanad, 2012).

In summary, the impact of climate change on natural enemies includes:

- Climate change can have diverse effects on natural enemies of pests species, like enhanced mortality with increasing temperatures, their decreased effectiveness in controlling pests when pest distribution shift into regions outside the distribution of their natural enemies, affected by climate.
- Changes in temperature will also alter the timing of diurnal activity patterns of different groups of insects and change in interspecific interactions could also alter the effectiveness of natural enemies for pest management.
- Temperature not only affects the rate of insect development, but also has a profound effect on fecundity and sex ratio of parasitoids.
- Natural enemies of the insect-pests are more sensitive to the climatic extremes like heat, cold, wind and rains.
- Precipitation changes can also affect predators, parasitoids and pathogens of insect-pests resulting in a complex dynamics.
- With changing climate, incidence of entomopathogenic fungi might be favoured by prolonged humidity conditions and obstinately be reduced by drier conditions.

5.12.4. Increased pesticide usage

Biopesticides and synthetic insecticides are highly sensitive to environment. Increase in temperature and UV radiation and decrease in relative humidity may render many of these control tactics to be less effective and such an effect will be more pronounced on natural plant products and biopesticides (Isman, 1997).

It is likely that farmers will experience extensive impacts on insect management strategies with changes in climate. Entomologists expect that insects will expand their geographic ranges, and increase reproduction rates and overwintering success. This means that it is likely that farmers in the north-eastern US will have more types and higher numbers of insects to manage. Based on current comparisons of insecticide usage between more southern states and more northern states, this is likely to mean more insecticide use and expense for north-eastern farmers. New York conditions currently require 0-5 insecticide applications against lepidopteron insect pests to produce marketable sweet corn (Stivers, 1999); Maryland and Delaware conditions require 4-8 insecticide applications (Whitney *et al.*, 2000); Florida conditions require 15-32 applications (Aerts *et al.*, 1999). It is apparent that for sweet corn pests, warmer temperatures translate to increased insecticide applications to produce a marketable crop. Insecticides and their applications have significant economic costs for growers and environmental costs for society. Additionally, some classes of pesticides (pyrethroids and spinosad) have been shown to be less effective in controlling insects at higher temperatures (Musser and Shelton, 2005).

Entomologists predict additional generations of important pest insects in temperate climates as a result of increased temperatures, probably necessitating more insecticide applications to maintain populations below economic damage thresholds. A basic rule of thumb for avoiding the development of insecticide resistance is to apply insecticides with a particular mode of action less frequently (Shelton *et al.*, 2001). With more insecticide applications required, the probability of applying a given mode of action insecticide more times in a season will increase, thus increasing the probability of insects developing resistance to insecticides.

Tilman *et al.* (2001) foresee a 2.4 to 2.7-fold increase in pesticide use by 2050 related to population growth and conversion of natural ecosystems to agriculture, but the effect of climate change is not considered by the authors. Chen and McCarl (2001) investigated the relationship of temperature, precipitation and pesticide costs for several crops in the USA and concluded, that increases in rainfall leads to increase in average pesticide costs for corn, cotton, potatoes, soybeans, and wheat; while hotter weather increases pesticide costs for corn, cotton, potatoes, and soybeans but decreases the cost for wheat. A simulation by the same authors applying different climate change scenarios showed uniform increases in average pesticide costs for corn, soybeans, cotton, and potatoes and mixed results for wheat.

The questions to address at this point are: Will we have more problems with certain arthropod pests, and what will be the potential impact on pest management? In general, this could lead to increased pesticide use (in this case, insecticides and acaricides) throughout the growing season with more frequent applications, resulting in higher incidences of resistance occurring in arthropod pest populations because of the increased selection pressure placed on these populations. In addition, this may directly and indirectly impact natural enemies thus influencing any natural mortality.

Additional factors to consider in regard to pest management include long-term harmful effects to plants (phytotoxicity) due to the amount of pesticide applications, and associated infestations of arthropod pests for extended periods of time. This may lead to natural infestations of multiple arthropod pest guilds such as chewers (caterpillars and beetles), suckers (aphids, scales and leafhoppers) and/or wood-boring insects occurring simultaneously on the same plant.

Finally, what about the potential effects of climate change on pesticides? This may include decreased residues or persistence, reduced toxic action on arthropod pests, increased treatment rates or dosages required, increased number of pesticide applications, and fewer times suitable for pesticide application (influenced by temperature and wind, which may prevent or delay application).

5.13. CONCLUSIONS

Climate changes have adverse effects on host plant resistance, transgenic plants, natural enemies, biopesticides and synthetic chemicals used in pest management. Climate change alters the interactions between the insect pests and their host plants. As a result, some of the cultivars that are resistant to specific insect may become susceptible. Adverse effects of climate change on the activity and effectiveness of natural enemies will be a major concern in future pest management programmes.

The principal components of climate change are increased temperature and atmospheric carbon dioxide (CO_2) concentrations. Since climate change is believed to be more certain now, it is time for researchers to develop management strategies to cope with the increased incidence of insect pests as a result of climate change. Of the several environmental factors temperature extremes play critical role in influencing the population of insect pests. Past experience indicated that there is a 'shift' in the pest status of several key species in recent periods, though these shifts may not be solely attributable to climate change. In the present situation, one can expect significant changes in the growth, development and population dynamics of various insect pests. The duration of the insect life cycle is altered under increased temperature and elevated CO_2 concentrations resulting in variable number of generations per year. Under elevated CO_2, higher consumption of foliage by leaf chewing insects with extended larval duration is observed in many studies. Published data reveals that some pests become more serious while others may decline. Evaluating the impact of climate change on insect pests is a very complex exercise and requires greater understanding of interactive factors. A more critical database on biotic stresses and their relationship with drivers of climate change is required for evolving adaptation strategies by farmers.

6

Development of Prediction Models for Insect Pests

6.1. INTRODUCTION

Impact of climate change would depend upon complex interactions of climatic and biological factors with technological and socio-economic changes that are difficult to predict. Therefore, these interactions are not amenable to qualitative analyses. Hence, quantitative (modelling) approaches, which allow investigating multiple scenarios and interactions simultaneously, will become more important for the impact assessment (Coakley and Scherm. 1996). Sutherst *et al.* (1996) have given a framework for such model-based assessment of impacts of climate change. Some of these approaches are discussed here.

6.2. INFOCROP MODELS

InfoCrop-maize, coupled with holometabolous population dynamics model, was used to simulate population dynamics of maize stem borer, *Chilo partellus* as well as crop-pest interactions. Maize stem borer acts as a stand reducer and causes loss of leaf area, leaf weight, stem weight and panicle weight to the crop. Larva being the damaging stage of the pest, larval population from population dynamics model was linked to the processes of leaf area, leaf weight, stem weight and ear weight in the crop model. Depending upon larval population and feeding rate of a larva, these crop growth processes were affected. The coupled model was calibrated and validated with field experimental data on larval population and the corresponding maize yield. Validated model was used to simulate effect of 0.5-3.0° C rise in both maximum and minimum temperatures compared to the ambient conditions on pest dynamics as well as crop-pest interactions. Simulation of pest dynamics showed a decline in the pest severity thereby reducing the pest-induced yield losses under global warming. However, maize productivity also depicted reduction even without pest stress under climate change, indicating that despite reduction in pest stress, crop productivity as such may be adversely affected by the global warming.

Climate change impact assessment through coupled rice brown plant hopper (BPH) InfoCrop model, in the light of the projected climate change scenario for Indian subcontinent, showed a decline in BPH population of 3.5% by 2020 and of 9.8-14.0% by 2050, during the rainy season at New Delhi, while the pest population exhibited only a small decline of 2.1-3.5% during winter at Aduthurai, Tamil Nadu even by 2050. Simulation attributed the decline in BPH population to reduction in fecundity and survival. Concomitant to its population decline, the BPH-induced yield loss also indicated a declining trend with temperature rise. However, the study considered the effect of only CO_2 and temperature rise on the BPH population and crop yield, and not that of probable changes in the feeding rate and adaptive capacity of the pest.

The impact of climate change on pink borer, *Sesamia inferens*, population and crop-pest interaction was analyzed through coupled InfoCrop model. A rise of 0.5-1.0°C temperature showed a small effect on various pest developmental stages, but a further increase had a significant adverse effect on them. In accordance with climate change projections for Indian subcontinent during *kharif* season, the study indicated that population of *S. inferens* developmental stages might decline to the extent of 5.82-22.8% by 2020 and 19.01-42.74% by 2050. Following decline in pest population, yield loss due to *S. inferens* also revealed a declining trend with temperature rise.

6.3. CLIMATE MATCHING

Climate matching involves the calculation of a "match index" to quantify similarity in the climate between two or more locations. The match index is based on variables such as monthly minimum and maximum temperatures, precipitation, and evaporation rates. Software packages for climate matching include BIOCLIM122 Climate Change Impact, Adaptation and Mitigation in Agriculture (Busby. 1991), CLIMEX (Sutherst and Maywald, 1985), HABITAT (Walker and Cocks, 1991) and WORLD. Climate matching may be used for climate change impact assessment by identifying those locations on the globe with a current climate that is similar to the predicted future climate at the location of interest. An analysis of the plant disease problems at the matching locations based on disease distribution maps made it possible to predict the future disease risk at the location of interest.

The use of climate matching to improve the success rate of introductions of biological control agents into new environments is well-established (DeBach, 1964). Similarly, there have been robust examples where the risk of establishment of invasive species has been successfully defined *a priori* using climatic modelling. These include: *Leptinotarsa decemlineata* in Europe (Sutherst et al., 1991), *Amblyomma variegatum* and *A. hebraeum* in Zimbabwe (Bruce and Wilson, 1998); *Chrysomya bezziana* in Ethiopia (Hall and Wall, 1995) and *Boophilus microplus* in east and southern Africa (Sutherst, 2001).

CLIMEX modelling software has been used to predict the future distribution ranges of two central European serious forest pest species, the nun moth (*Lymantria monacha*) and the gypsy moth (*L. dispar*).

CLIMEX software can be used to generate distribution maps of insect species and to assess the possible distributions of these insects in changing climate (Sutherst and Maywald, 1985). The software has also been used extensively in the fields of biological control, climate change and pest risk assessment with positive results in many countries. It holds the weather data for monthly long-term average maximum and minimum temperatures, rainfall and relative humidity from 2031 meteorological stations worldwide from 1931 to 1960. Additional weather data can be added into CLIMEX meteorological database from different meteorological stations.

6.3.1. CLIMEX model

Sutherst and Maywald (1985) gave the original description of the model. Since then many other features, insights and caveats have been described by Maywald and Sutherst (1991), Sutherst *et al.* (1995, 1996, 1999, 2000), Sutherst (1998, 2000, 2001) and Yonow and Sutherst (1998).

CLIMEX can predict species potential distribution through weather parameters of its current habitat range or directly by the species biological parameters such as minimum, maximum and optimum temperatures for development. On the basis of biological parameters of the species, CLIMEX generates a map for the potential geographical distribution of the species by counting an eco-climatic index (EI). EI is a numerical value for climatic suitability and relative abundance of the species. CLIMEX calculates EI from an annual growth index, describing conditions favourable for population growth together with stress factors that limit population growth during unfavourable season in the following manner:

$$EI = [100/52\,(TIw \times MIw \times DIw)] \times [(1-CS/100)\,(1-HS/100)\,(1-DS/100)\,(1-WS/100)]\ \Sigma\ 52\ W = 1$$

Where, TIw is the growth index counts for weekly temperature index, MIw is the moisture index, DIw is the diapause index and w is the week of the year. Each of the stress indices are calculated on weekly basis and expressed as a sum over the year as annual heat (HS), cold (CS), wet (WS) and dry stress (DS) indices, all indicative of the climatic requirements of the species.

The temperature index consists of the lower temperature threshold (DV0), the lower and upper optimum temperatures (DV1 and DV2), and the upper temperature threshold (DV3). Further, the number of degree-days (PDD) required to complete a generation cycle is also used. Four parameters are used in the calculations of the Moisture Index (MI); these are lower and upper soil moisture thresholds (SM0 and SM3) and the lower (SM1) and upper (SM2) bounds of optimum range. Diapause index is composed of diapause induction day length (DPDO), diapause induction temperature (DPT0), diapause termination temperature (DPT1) and diapause development days (DPD). The stress indices used are heat stress (HS), dry stress (DR) and wet stress (WS).

The eco-climatic index (EI) values range from 0 to 100, describing climatic suitability of the location for the species. At the EI value of 0, the species cannot establish a viable population at the location. Values over 20 indicates a very favourable climate for the species.

The CLIMEX model is designed to extract maximum information on the response of a species to climate out of minimal field data. It derives weekly and annual indices that describes the responses of a nominated species to temperature and moisture, and light in the case of plants. The model is based on a conceptual model that there are two types of seasons each year – one with population growth and the other with population decline (Fig. 6.1). These are referred to as *growth* and *survival* or *stress* seasons, respectively. A population has to be able to survive during the stressful season, with its negative effects on population density, in order to exist at the start of the favourable season (unless it is a migratory species). The overall climatic suitability of a nominated location for a given organism is provided by an 'Eco-climatic Index' (EI, scaled from 0 to 100), which combines the annual potential for population growth with the annual stress. As the EI is an aggregate measure, it alone is inadequate to explain an organism's response to climate. The model parameter values for growth and stress constitutes the description of the organism's response to climate and, as such, they represent the hypotheses to be tested. Testing demands that, like any mathematical model, the parameter values are published, so that the results can be repeated and the values are challenged and refined as further observations become available.

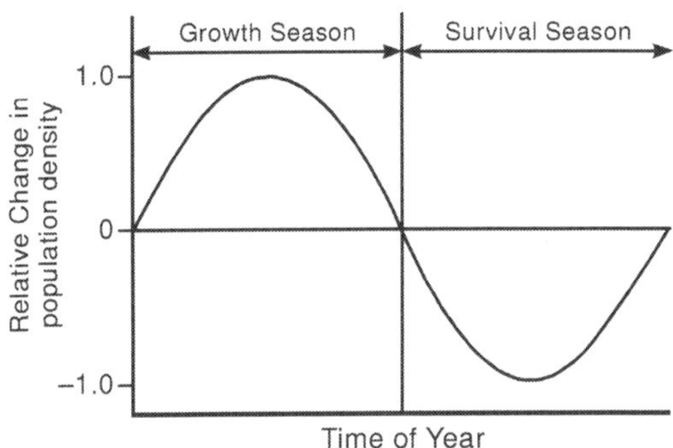

Fig. 6.1. Schematic diagram showing the year conceptually divided into two seasons, one for population growth and one for survival of stressful climatic conditions, when using the CLIMEX model.

CLIMEX incorporates a hydrological model to integrate the effects of rainfall and evaporation into a 'moisture index', which describes the availability of soil moisture to plants or animals. The index is then combined with a temperature response curve to produce weekly and annual 'Growth Indices' (GI_W and GI_A, respectively, with the former scaled between 0 and 1 and the latter between 0 and 100). Any suboptimal weekly values of either temperature or moisture reduces the values of the growth index from its potential.

Extreme values and duration of temperature and moisture limits the survival of populations, and so set the ultimate bounds of the potential geographical range as determined by the climate. In CLIMEX, their effects are described by four annual 'Stress Indices': wet, dry, cold and hot, that estimate the threat to that species posed by prolonged or intensely extreme periods of adverse conditions, and their interactions where appropriate. The weekly values of each stress index are accumulated using a nonlinear function to give an annual value. If that value exceeds 100, the species is deemed not to be able to persist in that environment.

The EI gives an overall measure of the suitability of a given location for permanent occupation by a species. It is therefore, a measure that is both species-specific and location-specific, and implies that establishment is only possible when the value exceeds zero. In practice, with variable seasons, values of EI < 10 indicate that the location is marginal for the species and large annual fluctuations in numbers are likely. A location can only have a perfect suitability for a species (EI = 100) if it provides ideal conditions throughout the year for that species. We saw above, how the year can be divided into favourable and stressful seasons. One consequence of this assumption is that in temperate regions or other regions with distinct wet and dry seasons, the *maximum* achievable value of the EI is likely to be around 50. In practice, values in excess of 20 have been found to support substantial population densities and values in excess of 50 are rare and usually confined to the wet tropics.

CLIMEX analyses usually require that a species' climatic requirements are inferred from information on its known geographical distribution, relative abundance and seasonal phenology. This 'top-down' procedure is called 'inverse' or 'inferential' modelling and it is the reverse of the reductionist approach usually used to build mechanistic models. Where laboratory data are available, they can be used to bolster the model fitting process or to explain the processes. The aim is to capture the core features of the species climatic requirements from minimal observations, not to describe the population dynamics in detail. CLIMEX analyses have proved very illuminating even with species that have been studied intensively. This is partly because, CLIMEX starts with the geographical distribution, which is the result of the integration of the influences of all the factors affecting the life cycle. The model is designed to mimic the integration of all the climatic influences on populations at each location. Given the huge range of possible, seasonal permutations of temperature and moisture associated with locations spread across geographical space (Sutherst, 1998), the fitting process is, in fact, very demanding of the model. In order to function on a global scale, the parameter values of a CLIMEX model for a given species needs to be valid over a very wide range of values of temperature and moisture. These demands partly explain why the results can be quite robust. In contrast, with a reductionist approach, it is difficult to define limiting climatic factors over a useful range of parameter values. Thus both reductionist and inferential modelling approaches have strengths and weaknesses, and they complement each other in ecological investigations. Indeed, one should never start a fresh ecological study (or overseas travel) without doing a CLIMEX analysis first, because it provides such a useful geographical and climatic context, within which site-specific observations can more readily interpreted (Sutherst, 2001).

6.3.2. Match climates

The second function in the CLIMEX software package is 'Match Climates'. This function is designed simply to compare meteorological data from different locations. That is often the only option available, for example when targeting collection sites for searches of new, as yet unidentified biological control agents that will be adapted to the climate at the proposed destination. We refer to the Match Climates function as the *poor man's* option, because it addresses the situation where no knowledge is available on the distribution of the species and it therefore provides much less intelligent information than the CLIMEX model. The function generates a climate *Match* Index. The index is analogous to a correlation coefficient but it is scaled to make it more sensitive to small deviations in the range of interest, i.e. 0.6–1.0. Based on past experiences, values of MI < 0.60 indicates poor matches between locations, with the potential for significantly different limiting effects of some variables in some of the locations. Hence, as a rule-of-thumb only, similarities of this order or less are not considered to be promising in terms of finding well-adapted species to introduce into the matching location. A number of caveats are necessary when using the function. Sutherst et al. (2000) emphasized the need to support the index with visual assessment of the data to see whether deviations between locations were positive or negative. Deviations from a given location in the direction of a species range boundary will give opposite conclusions to deviations in the direction of the core of the species climatic envelope, as illustrated schematically in Fig. 6.2. While the climates at the extremities of the circle at points B and C have the same similarity (Match Index) to the location marked with a star, B is more favourable for the species while C is unfavourable. Thus, there is an asymmetry in the index when compared with the geographical distribution A, which is indefinable under these circumstances where the geographical distribution is unknown.

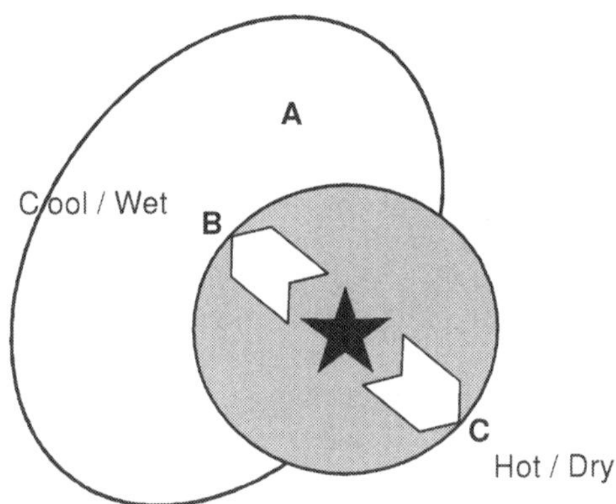

Fig. 6.2. Schematic illustration of the overlap between the range of a CLIMEX Match Climates index (shaded circle) and a hypothetical geographical distribution of a species in the unshaded oval 'A'. Point 'B' lies inside the distribution of the species while point 'C' lies outside it, despite both locations having the same CLIMEX Match Index.

6.4. EMPIRICAL MODELS

Empirical models based on long-term data on pest incidence and weather variables can be used to assess the likely impact of climate change on pest status in a region.

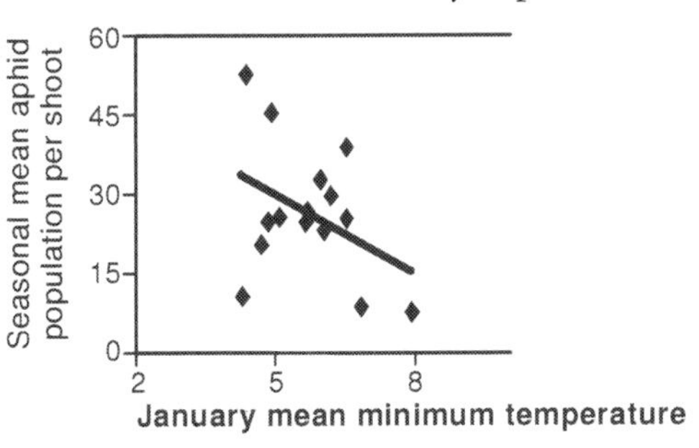

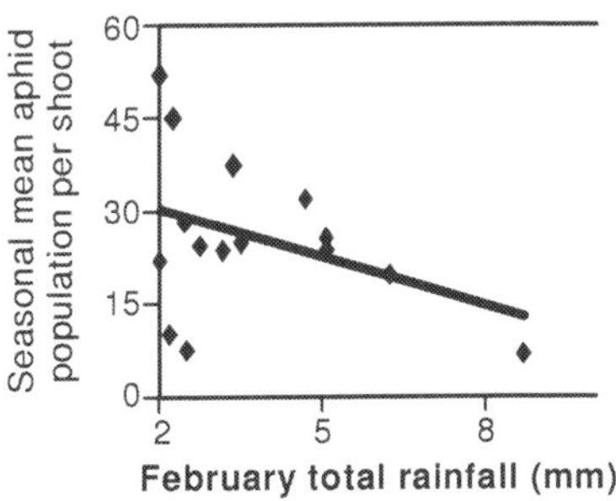

Fig. 6.3. Effect of minimum temperature and rainfall on incidence of barley aphids

Chander *et al.* (2003) have related the aphid incidence on barley crop variety 'DL-70' during *rabi* season from 1985-86 to 1999-2000 to weather parameters. There was appreciable inter-annual variation in the aphid incidence on barley, perhaps due to inter-annual climatic variability. The aphid population on barley exhibited a declining trend with time. The aphid population showed a negative relationship with the January mean minimum temperature (r = -0.37, Fig. 6.3), while it was not related to the February mean minimum temperature. The February total rainfall and aphid population were also found to be negatively related (r = -0.27, Fig. 6.3). Therefore, the rise in minimum temperature and more intense rains in future as speculated, might reduce aphid incidence on barley.

6.5. SIMULATION MODELS

Simulation models have been used widely to assess the impact of climate change on yields of various crops in different agro-ecological zones.

6.5.1. Simulation of population dynamics

Insect population dynamics model can be devised based on various bio-ecological factors, viz. fecundity, sex ratio, migration, abiotic and biotic mortality factors, threshold of development and thermal constant. Chander *et al.* (2009) have developed such a model for rice gundhi bug, *Leptocorisa acuta*, which is comprised of state, rate, driving and auxiliary variables (Fig. 6.4). State and rate variables approach was followed in developing the model. Insect population converted from one stage to another in proportion to the ratio of effective temperature to thermal constant of that stage and stage-specific mortality. Depending upon the rate of population change, total insect number in each developmental stage was updated daily.

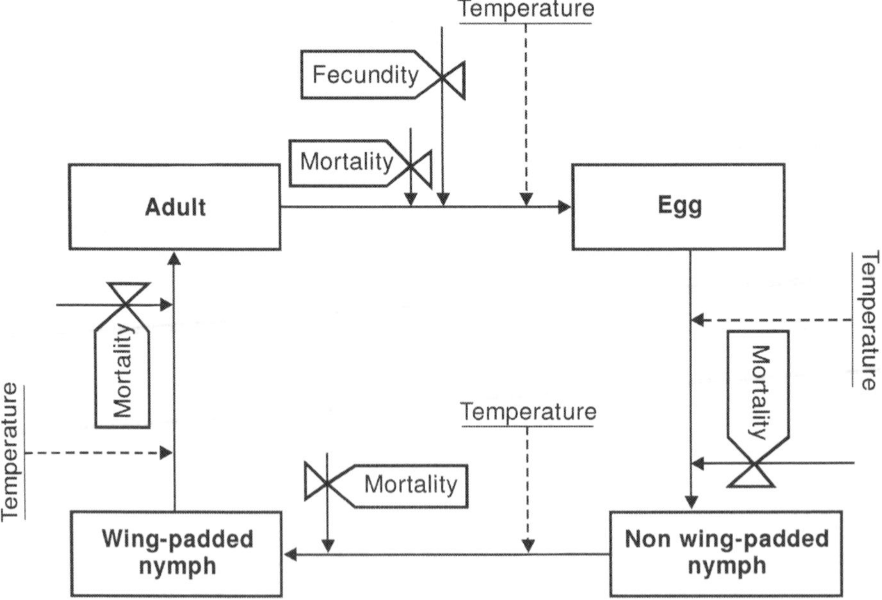

Fig. 6.4. Relational diagram showing the variables of population dynamics simulation model of rice gundhi bug

The model was used to simulate the impact of global warming on the rice bug by altering daily minimum and maximum temperatures. It could be concluded that up to 1°C rise in daily average temperature over the present temperature of Delhi, would not affect the gundhi bug population much, but further increase would cause appreciable decline in its population.

6.5.2. Coupling of population dynamics model to crop growth model

The population dynamics model can be coupled to crop growth model at the appropriate level of plant processes, depending on the pest damage mechanisms. The damaging pest stage affects plant growth and yield based on its number and feeding rate of individuals. Crop-pest model can then be used to analyse the impact of climate change on crop productivity, insect dynamics as well as crop pest interactions.

6.6. INSECT PHENOLOGY BASED MODELLING

Models have become important analytical tools for the evaluation, understanding and prediction of the dynamics of insect populations in agro-ecosystems and in phytosanitary risk assessments. International Potato Centre (CIP) has developed a temperature-driven phenology model for the potato tuber moth (*Phthorimaea operculella*) that provides overall good predictions of population growth potential for the wide range of climatic conditions where the pest prevails today. Linked with geographic information systems (GIS) and atmospheric temperature the model allows simulating risk indices on a worldwide scale which are used to predict future changes of the species distribution due to global warming. The approach applied for the development and prediction of *P. operculella* can be utilized for other insect species. Hence, CIP developed Insect Life Cycle Modelling software (ILCYM) to facilitate the development of insect phenology models (Fig. 6.5).

ILCYM consists of three main modules:

- The "Model builder", which facilitates the development of insect phenology models based on life table data derived from constant temperature studies.
- The "Validation and simulation module", which provides tools for validating the developed phenology model using insect life tables established at fluctuating temperature conditions.
- The "Potential population distribution and risk mapping module", in which the phenology models are implemented in a GIS environment that allows for spatial – global or regional – simulations of species activities and mapping.

Innovative phenology modelling and risk mapping is used to understand the effects of rising temperatures caused by climate change on the distribution and severity of major insect pests on important food crops in Africa. The ILCYM software will be further improved and adapted to cover a wide range of insects of different orders and families. The results will contribute to fill the current knowledge gap about climate change effects on economically important insect herbivores and their natural enemies, especially in the tropics.

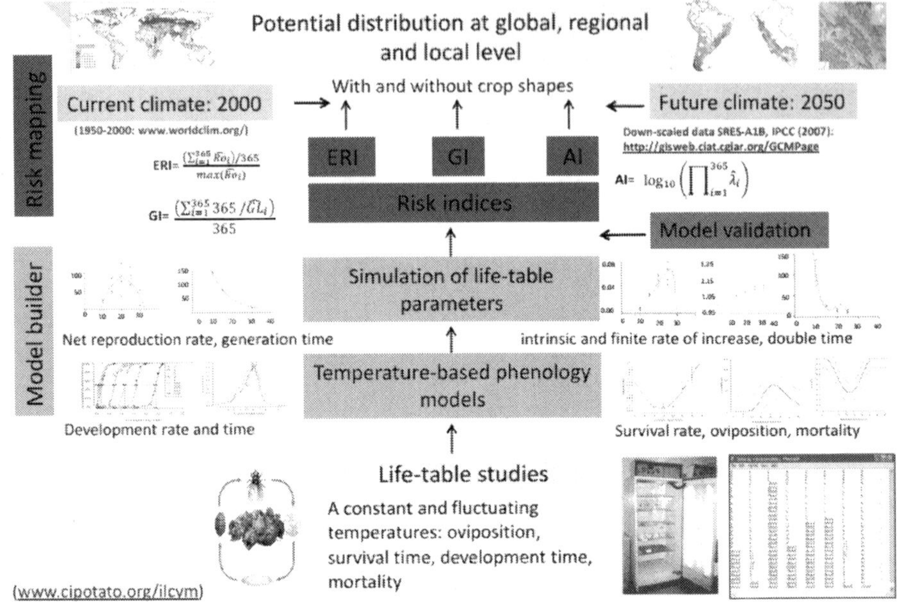

Fig. 6.5. Pest risk mapping using ILCYM

6.6.1. Insect life cycle modelling (ILCYM)

- Required input: life-table data collected under constant and fluctuating temperatures.
- Model builder:
  - o Fitting functions for describing temperature driven development processes under constant temperatures, i.e. development rate, mortality and reproduction.
- Validation and simulation module
  - o Uses life-table data established under fluctuating temperature; can be used for deterministic and stochastic simulations.
- Potential population and risk mapping module
  - o Establishment index
  - o Generation index
  - o Activity index

6.6.2. ILCYM applications

- Pest Risk Analysis (PRA) on different scales
- Climate change and adaptation planning
- Regional assessments
- Decision support tool in pest management

6.6.3. Integrated pest management (examples)

- Simulation of pest life-table parameters under prevailing weather conditions as decision support for pest management.

- Identification of potential release sites for parasitoids (classical biocontrol).
- Simulation of field performance of biopesticides and application rates.

6.6.4. Main advantages of ILCYM modelling

- Complex simulation of an entire insect life system.
- It uses process-based functions that provide a mechanistically detailed representation of the insects life history, and hence population growth.
- Model functions are built based on statistical analysis of experimental data that allows determining errors in particular processes.
- Full life history (development, survival, reproduction) is modelled and the software compiles the overall model automatically (reducing programming skills and time of users).
- Models have parameters with straight forward interpretation of the biological system.
- Individual functions of a model can be validated and improved.
- The information can be used in different modelling types; i.e. spatially (SDM through the use of specific risk indices), or in time (in a given area) for different inquiry (from basic ecological understanding to prediction, management and policy-making).
- Since the model works with probability distributions for specific processes, it is possible to conduct both deterministic and stochastic simulations.
- The simulation uses a cohort-updating algorithm and quantifies the population age-stage specific.
- Input data can be real or simulated temperature at variable intervals.
- The simulation is implemented in R-language, which is open source software.

6.7. CONCLUSIONS

Scientists now face the challenge of explaining and predicting the consequences of global climate change. The use of simple theoretically based models including exogenous perturbations may improve our predictive capacity. Moreover, prediction models approach provides an opportunity to examine the link between invasive species and climate change (Ward and Masters, 2007) and how new suitable habitat may become available for species whose niche space is limited in some degree by climatic conditions. Finally, the use of different scenarios is absolutely necessary when the reliability of the predictions is not clear, which is the case of the predicted future climate change (Cammell and Knight, 1992). The use of different scenarios may allow us to examine the sensitivity of the predictions and to improve the communication of our researches to the general public and decision-makers, a key topic in pest management.

7

Climate Change Adaptation and Mitigation Strategies for Insect Pests

7.1. INTRODUCTION

The changes in insect-pest status due to climate change will perilously affect agricultural production and the livelihood of farmers in the country, where larger portion of work force is directly dependent on climate sensitive sectors such as agriculture (Chahal *et al.*, 2008; Deka *et al.*, 2012). This envisages an urgent need to modify crop protection measures with changed climate in order to attain the goal of food security of the nation. At the moment, farmers' pest management practices and pesticide use are causing environmental and human health concerns in food crop agro-ecosystems. Climate change is adding new dimensions to the multiple challenges agricultural policy makers and farmers are facing. However, links between climate change and pest scenarios are not well known, and much less known are adaptation and mitigation options that may be adopted by agricultural policy makers and farmers to reduce vulnerability of crops to abnormal pest infestations and to build resilience against climate change induced pests.

Pest management has potential to provide eco-friendly sustainable solution to obnoxious pest problems. However, the relative efficacy of pest management components such as, host-plant resistance, bio-pesticides, natural enemies, and synthetic chemicals are liable to change as a result of global warming (Sharma, 2010). It, therefore, becomes important to assess climate change impact on insect populations and adopt suitable pest management adaptations for their effective management.

With the increasing resistance of many insect pest species to chemical insecticides and an increasing organic food market, pest control strategies are slowly shifting towards more sustainable, ecologically sound and economically viable options. Biologically based pest management strategies present such opportunities through predation or parasitism of pests and plant direct or indirect defense mechanisms, that can all be important components of sustainable integrated pest management programmes.

7.2. ADAPTATION

Adaptation is defined by IPCC as an adjustment in natural or human systems in response to actual or expected climatic stimuli or their effects, which moderates harm or exploits beneficial opportunities. Various types of adaptation can be distinguished, including anticipatory, autonomous and planned adaptation.

- **Anticipatory adaptation:** Adaptation that takes place before impacts of climate changes is observed. Anticipatory adaptation is also referred to as proactive adaptation.
- **Autonomous adaptation:** Adaptation that does not constitute a conscious response to climatic stimuli, but is triggered by ecological changes in natural systems and by market or welfare changes in human systems. Autonomous adaptation is also referred to as spontaneous adaptation.
- **Planned adaptation:** Adaptation that is the result of a deliberate policy decision, based on an awareness that conditions have changed or are about to change and that action is required to return to, maintain, or achieve a desired state.

Historically, people whose livelihoods depend on agriculture have developed ways to cope with climate variability autonomously. Today, the current speed of climate change will modify known variability patterns to the extent that people will be confronted with situations they are not equipped to handle. Thus, anticipatory and planned adaptation is an immediate concern. However, vulnerabilities are mostly local and, thus, adaptation should be highly location specific (Fig. 7.1).

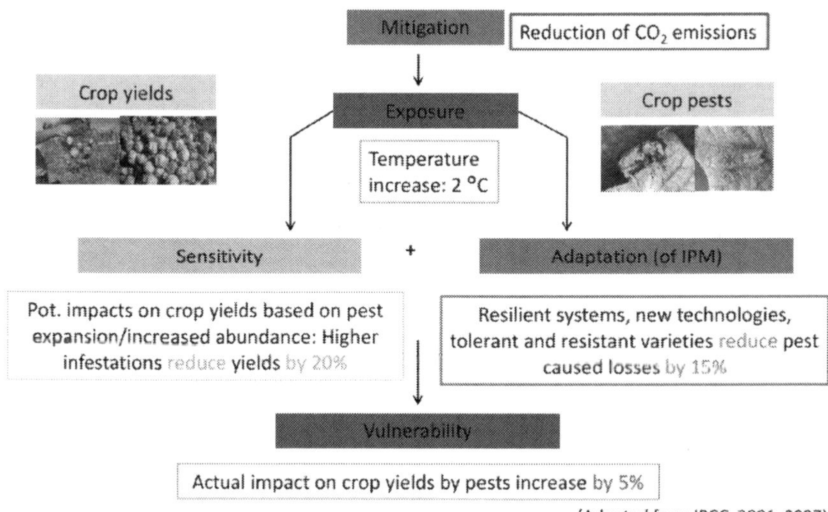

Fig. 7.1. Reducing crop vulnerability by adaptation to new and emerging pests

The following adaptation strategies have been recommended for the management insect pests under climate change.

7.2.1. Promotion of resource conservation technologies

Shrinking resource base due to anthropogenic developmental activities is a major challenge ahead for humanity. Conservation of natural resources can be promoted by giving incentives to the farmers those who are adopting environmental conserving pest management activities such as organic farming, biological control, integrated pest management, habitat conservation for important insect pollinators, etc. Strategies for adaptation and coping could benefit from combining scientific and indigenous technical knowledge (ITK), especially in developing countries where technology is least developed. ITK is helpful to adapt the adverse effects of changing climate e.g. application of natural mulches helps in suppression of harmful pests besides moderating soil temperatures and conservation of soil moisture. Furthermore, study towards integrating indigenous adaptation measures in global adaptation strategies and scientific research is required.

7.2.2. Increased biodiversity

Natural pest control by the natural enemies comes for free. A diverse fauna of natural enemy species can successfully suppress pests (Cardinale *et al.*, 2003). Intercropping can attract natural enemies (pull) and repel pests (push) (Cook *et al.*, 2007), partial weediness (as long as weeds are not host to disease pathogens or problematic pests), mulching and reduced tillage for example, increases spider abundance. Spraying pesticides to control weeds and pests usually kills natural enemies and/or destroy their homes. As a consequence, an increase in the pest population may occur (resurgence) and need to spray more frequently, and resistant pests might emerge.

A return to crop rotations would substantially reduce soil erosion and water runoff and improve the control of insect pests. They are sound agricultural practices that should be widely used in agriculture. Noxious pests establish slower because specific relationships between pests and host plants are interrupted (Dhawan and Peshin, 2009).

7.2.3. Avoidance of one 'high input variety' or one breed of crop variety

Breeds should be mixed and changed. A broad genetic variability serves as a foundation for robust crops. In addition, it seems more recent traditional breeding has not selected for CO_2 responsiveness, which simply means newer breeds do not benefit from elevated CO_2 as much as older breeds (Ainsworth *et al.*, 2008).

7.2.4. Ecologically based pest management

Ecologically based pest management (EBPM) considers below-ground and above-ground habitat management equally important. A 'healthy' soil, with optimal physical, chemical, and biological properties increases plant resistance to insects and diseases (Altieri *et al.*, 2005). Excess of nitrogen can increase the severity of certain diseases and make a crop more susceptible to pests (Altieri *et al.*, 2005).

7.2.5. Facilitating adaptation to climate change

Variability in seasonal conditions has long driven changes in farming practices as has the development of new tools and crop varieties to optimise yield (Olmstead and Rhode, 2011). For example, decreased frost frequency and the availability of new wheat varieties have allowed wheat crops in central Queensland, Australia, to be planted 3–4 weeks earlier (Howden *et al.,* 2010). In this context, adaptation can be defined as adjustments to farming systems in order to moderate the pest risk associated with likely changes.

Examples of several management options that a farmer might consider if climate change models suggest their risk from certain pests is likely to increase, decrease or not change (Table 7.1). Some of these options are associated with more sustainable pest management under today's climate, regardless of future change in risk.

Table 7.1. Pest management options that could be recommended based on the change in potential projected from models and other information sources

Management option	Modelled change in pest risk*			Degree of difficulty**
	Risk reduced	Risk increased	No change (or not sure about direction)	
Change planting time to avoid period at which crop is susceptible.	×	✓	×	2
Change planting density or row spacing to make micro-climate less favourable.	✓	✓	✓	1
Change seeding density to compensate for early plant loss.	×	✓	✓	1
Change crop variety, consider them less susceptible to damage (and better adapted to local climate conditions).	✓	✓	✓	1
Change crop choice, consider those not a host for pest (and better adapted to local climate conditions).	×	✓	×	3
Change in-field weed control to reduce alternate plant hosts.	×	✓	×	2
Gradually reduce insecticide-use, and monitor for frequency of outbreaks.	✓	×	✓	1
Use forecasting tools /monitoring networks to help decide when and where to scout, and then make spray decision.	×	✓	×	2
Use border sprays only.	✓	×	×	1
Manage weed growth surrounding the field.	×	✓	×	2
Use strategic tillage to reduce stubble that provides an ideal micro-climate for some pests.	✓	×	✓	2
Use conventional tillage.	×	✓	×	1
Improve pest and natural enemy identification skills. Understand the non-target impact of pesticides.	✓	✓	✓	1
Employ a crop protection specialist to help with pest management decisions.	×	✓	×	2

Management option	Modelled change in pest risk*			Degree of difficulty**
	Risk reduced	Risk increased	No change (or not sure about direction)	
Establish a support network of neighbours, crop protection specialists and researchers to help when new pest issues arise.	✓	✓	✓	1
Increase crop diversity at the whole-farm level.	×	✓	×	3
Increase plant diversity at the whole-farm level.	×	✓	×	3

* A tick (✓) indicates that the management option is suitable given the change in risk profile, *and* the degree of difficulty associated with adoption of the management option. A cross (X) indicates that the management option is unsuitable given both the modelled change in pest risk, *and* the degree of difficulty associated with adoption.

** The management change may be costly or complex to implement. 1, not difficult to implement; 2, moderately difficult (i.e. will require small changes in other areas of farm business, has small costs); 3, very difficult to implement (i.e. will require large changes in other areas of farm business, is costly and difficult to go back).

7.3. MITIGATION

Climate change mitigation encompasses the actions being taken, and those that have been proposed, to limit the magnitude and/or rate of long-term global warming induced climate change.

Strategies to mitigate the effects of climate change includes cultural practices, natural enemies, host plant resistance, bio-pesticides, and synthetic pesticides are now being widely used for pest management. However, many of these methods of pest control are highly sensitive to the environment. Therefore, there is a need to develop appropriate strategies for pest management that will be effective under situations of global warming in future.

- Host plant resistance to insects is one of the most environmentally friendly components of pest management. It is important to identify and develop cultivars that are stable in expression of resistance to the target pests under variable climate.
- Transgenic crops have been developed for controlling some of the most intractable pest problems. There is need to combine host plant resistance from the germplasm with the transgene expression that is stable across environments, and locations.
- Crop diversification is one of the most effective methods of increasing the activity and abundance of natural enemies. There is a need to develop crop cultivars that are hospitable to the natural enemies, in addition to identifying cropping systems that will enhance the diversity of natural enemies for pest management.

- There is a need for a greater understanding of the effect of climate change on the efficacy of synthetic insecticides, their persistence in the environment. Therefore, there is need to develop pesticide formulations and the application schedules that will be least affected by climate change. Finally, we need to use an integrated pest management system that takes into consideration the change in pest spectrum, cropping patterns and effectiveness of different components of pest management for sustainable crop production.

7.3.1. Pest monitoring

Strategies that identify the vulnerabilities of cropping systems prior to pest introduction and establishment are worthwhile preventative approaches.

International and/or global coordination in the immediate and long-range surveillance and monitoring has also been identified as an invaluable tool to monitor both invasive alien species and the appearance of new and more aggressive strains of existing pest species. The creation of a truly global database, or integrated information exchange platform, on harmful organisms accessible by all countries could be another useful step, for example, integrating the expertise and intelligence gathering and surveillance of regional plant protection organizations toward an overarching view.

Early detection of the appearance of pests followed by rapid and accurate identification is essential if correct control measures are to be deployed. Nucleic acid sequencing and advances in DNA bar-coding, microarray technologies, and lateral flow devices promise to revolutionize plant diagnostics in the near future (Boonham *et al.*, 2008). DNA microarrays printed on the bottom of an Eppendorf, which can be read with a regular document scanner, have the capability to detect many pests simultaneously. Little training is required and thus the technology can be implemented in any laboratory without the need of specialized or expensive equipment. DNA bar-coding, on the other hand, relies on the generic amplification and sequencing of a nucleic acid sequence that provides a 'bar-code' unique to any specific pest, enabling its rapid and precise identification.

Farmer Field Schools (FFS) should be started, where farmers can learn more about pests and their enemies and their management. Thorough knowledge of the pest life cycle, the ecological and behavioural interactions with the environment and natural management factors are the basic foundation for successful management strategies (Conlong and Rutherford, 2009).

7.3.2. Rescheduling of crop calendars

Global temperature increase and altered rainfall patterns may result in shrinking of crop growing seasons with intense problems of early insect infestations. As such certain effective cultural practices like crop rotation and planting dates will be less or not effective in controlling crop pests with changed climate. Hence, there is need to change the crop calendars according to the changing crop environment. The growers of the crops have to change insect management strategies in accordance with the projected changes in pest incidence and extent of crop losses in the view of the changing climate.

7.3.3. GIS based risk mapping of crop pests

Geographic Information System (GIS) is an enabling technology for entomologists, which helps in relating insect pest outbreaks to biographic and physiographic features of the landscape, hence can best be utilized in area-wide pest management programmes. How climatic changes will affect development, incidence, and population dynamics of insect pests can be studied through GIS by predicting and mapping trends of potential changes in geographical distribution (Sharma *et al.*, 2010) and delineation of agro-ecological hot spots and future areas of pest risk (Yadav *et al.*, 2010).

The ability to accurately assess the health of agro-ecosystems is crucial in determining the need for and effects of any pest management strategy. Remote sensing is one emerging tool available to IPM for ecosystem surveillance. Remote sensing is a very flexible tool with the potential to offer new insights into pest management at temporal and spatial scales that would have required intensive human efforts in the past. Remote sensing has the potential to be an excellent tool for large-scale assessment and management of crop pests. Different forms of spectral assessment are being used in predicting the development of pest prior to symptom expression, in crop breeding for the early detection of resistance, and for pest assessment in ecosystems.

7.3.4. Screening of pesticides with novel modes of action

It has been reported by some researchers that the application of neonicotinoid insecticides for controlling sucking pests induces salicylic acid associated plant defense responses which enhance plant vigour and abiotic stress tolerance, independent of their insecticidal action (Ford *et al.*, 2010). This gives an insight into investigating role of insecticides in enhancing stress tolerance in plants. Such more compounds needs to be identified for use in future crop pest management.

7.3.5. Improved pest control

Because insects destroy potential crop production worldwide, use of appropriate technologies to reduce pest losses would increase crop yields. In addition to the prudent application of pesticides, increased use of non-chemical pest controls would help minimize crop losses (Pimentel *et al.,* 1993). Nonchemical controls include crop rotations, biological controls, altering planting dates and fertilizer and irrigation applications, and soil management and tillage. These technologies could help minimize projected pest losses and thereby help maintain crop yields.

7.3.6. Behaviour modification

Semio-chemicals control the communication of insects both interspecific (allelochemicals) and intraspecific (pheromones). They are used in pest management either alone for pest monitoring and decision-making, for mass trapping or mating disruption, or in combination with insecticides, sterilants, or insect pathogens, the so-called 'attract-and-kill' strategy (El-Sayed *et al.*, 2009). Additionally, semio-chemicals released by plants can repel insect pests from the crop ('push') and attract them into trap crops ('pull').

The scientists at the International Center of Insect Physiology and Ecology (ICIPE), Kenya; and Rothamsted Experiment Station, the United Kingdom; have developed a push–pull strategy for the management of stem borer (*Chilo partellus*) and *Striga* weed on maize and sorghum. They used repellent intercrops such as molasses grass (*Melinis minutiflora*), leguminous silver leaf Desmodium (*Desmodium uncinatum*) and green leaf Desmodium (*Desmodium intortum*) for driving the stem borers away ('push'), and attractive trap crops like Napier grass (*Pennisetum purpureum*) and Sudan grass (*Sorghum vulgare sudanense*) in the crop border to attract female moths ('pull') to lay eggs (Fig. 7.2) (Khan *et al.*, 2011). When the eggs hatch, 80% die as the Napier grass also produces a sticky substance that traps the larvae (restricts larval development, causing few to survive) (Khan *et al.,* 2006). Besides controlling stem borers, Molasses grass also increased stem borer parasitism by a natural enemy, *Cotesia sesamiae*, when intercropped with maize (Khan *et al.,* 1997).

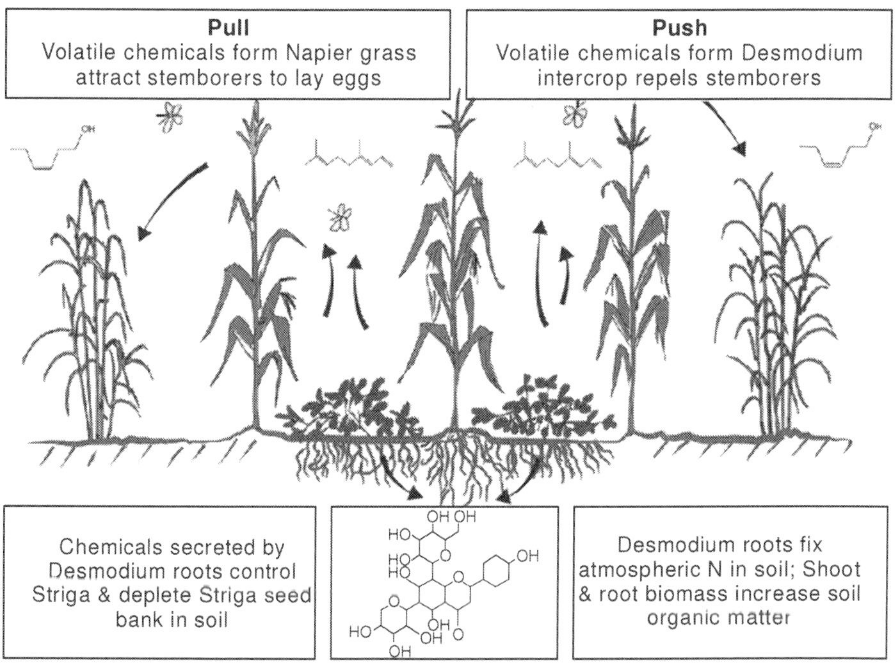

Fig.7.2. Push-pull strategies for the management of cereal stem borers

In this way, the push-pull approach has been developed for controlling insect pests for subsistence farming systems in Africa, and has been adopted by over 25,000 maize smallholder farmers in East Africa. There, maize yields have subsequently increased from about 1 t/ha to 3.5 t/ha with minimal inputs (Khan *et al.*, 2008). The potential use of semio-chemicals for pest management on small-scale farms in developed countries remains underexploited. Similarly, a clearer understanding of the behaviour of insects, including their migration capacities and spatial dispersal, could enable simple systems of pest management to be developed (Kroschel *et al.*, 2009).

7.3.7. Precision pest management

Although this is often considered expensive and primarily useful for large farm operations, it is adaptable in developing countries with the necessary infrastructure and larger farm units. Precision technology can be used to accurately determine the presence and impact of pests on a crop in a particular field or part of a field. Pests can then be mapped to GPS coordinates to target the delivery of specific resistant varieties, bio-control agents, or pesticides. Precision technology can have an impact in extensive production systems (i.e., for rice, maize, and wheat) where it allows the optimum use of inputs for pest management and crop productivity, while reducing the need for large-scale sampling and extension input (Oerke *et al.*, 2010). Many small farmers already practice a form of precision agriculture without any technological aides. They know the variability in their fields and try to use certain inputs on a refined scale. They are well-positioned to improve if the information is made available on what to do, where, and when, in response to the needs of small but variable land plots.

7.3.8. Breeding for pest resistance

In order to minimize the impacts of climate and other environmental changes, it will be crucial to breed new varieties for improved resistance to abiotic and biotic stresses. Considering late onset and/ or shorter duration of winter, there is chance of delaying and shortening the growing seasons for certain *rabi*/ cold season crops. Hence, we should concentrate on breeding varieties suitable for late planting and those can sustain adverse climatic conditions and pest incidences.

7.3.8.1. Transgenic crops for pest management: Another important issue regarding pest management in the future centers on the role of biotechnology in crop protection. The next 20 years will likely to see a substantial increase in the use of genetically engineered plants that have been designed to resist pests such as stem borers and nematodes without the need for pesticides. Others are expected to combine both herbicide resistance and insect resistance in one seed.

Environmental factors such as soil moisture, soil fertility, and temperature have strong influence on the expression of *Bacillus thuringiensis* (*Bt*) toxin proteins deployed in transgenic plants (Sachs *et al.*, 1998). Cotton bollworm, *Heliothis virescens* destroyed Bt-transgenic cottons due to high temperatures in Texas, USA (Kaiser, 1996). Similarly, *Helicoverpa armigera* and *H. punctigera* destroyed the Bt-transgenic cotton in the second half of the growing season in Australia because of reduced production of Bt toxins (Hilder and Boulter, 1999). Cry1Ac levels in transgenic plants decreases with the plant age, resulting in greater susceptibility of the crop to insect pests during the later stages of crop growth (Sachs *et al.*, 1998; Kranthi *et al.*, 2005). Possible causes for the failure of insect control in transgenic crops may be due to inadequate production of the toxin protein, effect of environment on transgene expression, Bt-resistant insect populations, and development of resistance due to inadequate management (Sharma and Ortiz, 2000). It is therefore, important to understand the effects of climate change on the efficacy of transgenic plants for pest management.

Colorado potato beetle (CPB) resistance in potato has been achieved through the incorporation of a gene from the *B. thuringiensis* (Bt) protein into potatoes (Russet Burbank, Atlantic).

Several developing-country potato varieties have been transformed with the Bt gene to express resistance to the potato tuber moth.

- In Central Africa (Rwanda, Burundi, Uganda, Congo) resistant varieties include Mabondo, Sangema, Murca and Cruza.

- For the Andean region (Peru, Bolivia, Ecuador), tuber moth resistance is now in Tomasa Condemayta, Costanera, Achirana INTA, María Tambeña and Revolución.
- In Colombia, Pardo Pastusa has been transformed.
- Costa Rica now has resistant Atzimba.
- For both North Africa (Egypt, Tunisia, Morocco) and the Southern cone of South America (Argentina, Chile), Desiree has been transformed with the Bt gene.

Desiree potato plants were genetically modified to resist attack by insect species belonging to the orders Coleoptera (Colorado potato beetle - CPB) and Lepidoptera (potato tuber moth-PTM and European corn borer-ECB), through the insertion of such a hybrid gene, SN19 under the control of a chrysanthemum ribulose-1, 5-bisphosphate carboxylase/oxygenase small subunit (Rubisco SSU) promoter and terminator (Naimov *et al.*, 2003). Transgenic plants were shown to be resistant against CPB larvae and adults, PTM larvae, and ECB larvae. These are the first transgenic plants resistant to pests belonging to two different insect orders.

Bt Brinjal has been developed by inserting a gene cry1Ac from a soil bacterium, *B. thuringiensis* through an *Agrobacterium*-mediated gene transfer. The Event EE1 was introgressed by plant breeding into various local varieties such as Malpur local, Manjari gota, Kudachi local, Udupi local, 112 GO, and Pabkavi local (Fig. 7.3).

Fig. 7.3. Damage caused by brinjal fruit and shoot borer (left) and damage-free Bt brinjal developed by MAHYCO (right)

Bt Soybean (*Glycine max*) MON 87701 was developed by *Agrobacterium*-mediated transformation of soybean using the 2T-DNA plasmid vector PV- GMIR9, which produces one *B. thuringiensis* subsp. *kurstaki* protein, Cry1Ac. This protein is intended to provide protection from feeding damage caused by a number of Lepidopteron pests, such as the velvet bean caterpillar (*Anticarsia gemmatalis*) (Fig. 7.4), soybean looper (*Pseudoplusia includens*), soybean axil borer (*Epinotia aporema*), and sunflower looper (*Rachiplusia nu*).

Fig. 7.4. Velvet bean caterpillar infestation-free transgenic Bt soybean (*left*) as compared to non-transgenic control on the right showing extensive defoliation (*right*)

Many rice varieties have been transformed with genes encoding various Bt crystal (Cry) proteins and have been shown to be resistant to one or more Lepidopteron pests of rice, the most important of which are the yellow stem borer (*Scirpophaga incertulas*), the striped stem borer (*Chilo suppressalis*) and several species of leaf-folders (*Marasmia* sp. and *Cnaphalocrocis medinalis*) (Fig. 7.5) (Pathak and Khan, 1994).

Fig. 7.5. Transgenic Bt rice (MH63) showing resistance to yellow stem borer (*left*) and leaf folder (*right*). T – Transgenic, C - Control.

Commercially available genetically modified crops resistant to infect pests are presented in Table 7.2.

Table 7.2. Commercially available genetically modified crops resistant to infect pests

Crop	Trait Phenotype*	Target trait gene(s)	Trait designation	Originating company	Year of first commercial sale	Trade name
Cotton	Resistance to Lepidopteron insects	cry1Ac	MoN531	Monsanto	1996	Bollgard, Ingard
		cry1Ac, cry2Ab2	MoN15985	Monsanto	2003	Bollgard II
		cry1Fa, cry1Ac, pat	281–24–236 × 3006–210–23	Dow AgroSciences	2005	WideStrike
Corn	Resistance to European corn borer and other lepidopteron insects	cry1Ab, pat	Bt11	Northrup King (now Syngenta)	1996	YieldGard Attribute
		cry1Ab	MoN810	Monsanto	1997	YieldGard (Corn Borer)
		cry1F, pat	TC1507	Dow AgroSciences; Pioneer Hi-Bred Intl.	2003	Herculex I
	Resistance to corn rootworm	cry3Bb1	MoN863	Monsanto	2003	YieldGard (Rootworm)
		cry1Ab, cry3Bb1	MoN863 x MoN810	Monsanto	2005	YieldGard Plus

Source: Castle *et al.* (2006)

* Many insect resistance and herbicide tolerance traits are also available in combinations as stack traits.

7.3.9. Biological control

Adaptation and mitigation practices to combat climate change and its variability such as increasing crop diversity on farms, reduced tillage, use of mulches, increasing vegetation diversity both on farms and landscapes, are likely to influence the abundance of insect pests and their natural enemies. Positive effects on parasitoid and predator abundance and diversity are likely with resultant benefits on increased natural regulation of pests (Table 7.3), sustainable crop production and mitigation of climate change (Thomson *et al.*, 2010).

Table 7.3. Effect of climate change adaptation and mitigation practices on natural enemies

Adaptation and mitigation practices	Effect on pests and natural enemies	Reference
Increased crop diversity on farms through intercropping / mixed cropping / strip cropping		
Pigeon pea + sorghum	Increased egg parasitisation of *H. armigera*; transfer of parasites from sorghum to short- and medium duration pigeon pea	Romies *et al.*, 1999 Duffield and Reddy, 1999
Pigeon pea + green gram/ sorghum / groundnut	Increased abundance of coccinellid predators and spiders	Srinivasa Rao *et al.*, 2003
Cotton + groundnut/ cowpea /soybean/ green gram	Low incidence of *H. armigera*, positive effect on natural enemies	Venugopal Rao *et al.*, 1995
Cotton + cluster bean/ green gram	Low incidence of leaf hoppers, aphids, thrips and whiteflies; higher populations of predators	Balasubramanian *et al.*, 1998
Increased field border diversity and remnant vegetation	Increased abundance and diversity of natural enemies on many crops	Gurr *et al.*, 2003; Olson and Wackers, 2007
Conservation agricultural practices		
Zero till or low tillage plantings and mulches with crop residues	Increased abundance of Hymenopteran parasitoids; Dipteran parasitoids and predators; and Hemipteran predators in vineyards	Thomson and Hoffmann, 2007
	Increased parasitisation and predation of cereal aphids	Schmidt *et al.*, 2004
Re-vegetation of agricultural landscapes	Increased abundance of carabid predator; long-term conservation of natural enemies in agro-ecosystems	Nash *et al.*, 2008 Tscharntke *et al.*, 2007

7.3.10. Chemical control

An increase in probability of intense rainfall could result in increased pesticide wash-off and reduced pest control. In contrast, increased metabolic rate at higher temperature could result in faster uptake by plants and higher toxicity to pests. Likewise, increased thickness of epicuticular wax layer under high CO_2 could result in slower or reduced uptake by host, while increased canopy size may hinder proper spray coverage and lead to a dilution of the active ingredient in the host tissue. The rates of pesticide application thus have to be modified according to new situations. Granular formulations may prove more effective as these are less liable to be washed by rainfall. Reliable medium-range weather forecast can help in predicting rainfall and avoid pesticide application under imminent rainy conditions. Likewise, regular monitoring can help in undertaking control interventions at right time, thereby, helping in managing small population levels easily and effectively. Properly timed pesticide application and application coinciding with egg hatching prove more effective than many ill-timed applications. Likewise, proper placement of pesticide is also equally important to ensure its efficacy against pests. In case of rice plant hoppers, pesticide application needs to be targeted at plant stems, because foliar spray on canopy proves of little or of no use.

7.3.11. Integrated pest management

Integrated pest management lies at the center of insect control. The combination of farming strategies, biological control agents, and necessary pesticide use has helped California farmers address pest problems using a variety of methods. Climate change may make IPM solutions less effective and harder to implement. Additionally, the spread of pests and disease through human vectors will continue to become a problem, especially as they become more tolerant to environmental conditions.

IPM using Indian mustard as a trap crop involves planting of paired rows of mustard after every 25 rows of cabbage/cauliflower (Fig. 7.6) and spraying of 4% NSKE at primordial formation. Two more sprays of 4% NSKE may be given at 10-15 day interval after the first spray. The IPM plots gave 152% more returns than pure cabbage crop. IPM controls diamondback moth (*Plutella xylostella*); leaf webber (*Crocidolomia binotalis*); stem borer (*Hellula undalis*); aphids (*Brevicorne brassicae, Hyadaphis erysimi*) and bug (*Bagrada cruciferarum*) (Srinivasan and Krishna Moorthy, 1991). The IPM gave 60% more yield and 152% more returns than pure cabbage crop (Khaderkhan *et al.*, 1998).

Fig. 7.6. IPM in cabbage using Indian mustard as a trap crop.

IPM using of African marigold (*Tagetes erecta*) as a trap crop for the management of fruit borer, *Helicoverpa armigera* on tomato involves planting one row of 45 day-old marigold seedlings after every 16 rows of 25 day-old tomato seedlings and spraying of *Ha* NPV at 250 LE/ha or 4% NSKE or 4% pulverized NSPE, 28 and 45 DAP coinciding with peak flowering (Srinivasan *et al.*, 1994) (Fig. 7.7). In IPM validation studies conducted at three locations (Bangluru, Varanasi and Ranchi), IPM fields recorded higher tomato fruit yields of 74.038, 14.250 and 22.293 MT/ha as compared to 45.056, 13.000 and 18.772 MT/ha in non-IPM fields, respectively (Amerika Singh *et al.*, 2004)

Fig. 7.7. *Left* - IPM of tomato fruit borer using African marigold as a trap crop. *Right* – Fruit borer larva trapped in marigold flower

7.4. CONCLUSIONS

Climate change is imminent, and it has started showing its effect on insects, which are of no exception for it. Climate change will also impinge upon efficacy of pest management components. It thus, becomes very important to assess climate change impact on insects and pest management components and adopt appropriate adaptation and mitigation measures to sustain agricultural productivity. Simulation models have been used for several applications in the area of pest management, which helped to increase the efficiency of field research greatly. These will be of even greater relevance in new emerging research areas such as climate change impacts on pests and crop yield, impacts of transgenics on environment, pest risk analysis for sanitary and phytosanitary requirements, and pest forecasting.

Now, emphasis must shift from impact assessment to developing adaptation and mitigation strategies as also options thereof. We have to systematically evaluate the efficacy of current physical, chemical and biological control tactics, including pest-resistant cultivars under climate change, and to include future climate scenarios in all research aimed at developing new tools and tactics. Pest risk analysis based on host–pathogen/pests interactions should be done, and research on host response and adaptation should be conducted to understand how an imminent change in the climate could affect crop pests. India is fortunate enough to have such a diverse climate suitable to grow various types of crop plants with diverse pest population, which can help to counter the pest problems in the changing climate scenario.

Understanding abiotic stress responses in crop plants, insect-pests and their natural enemies is an important and challenging topic ahead in agricultural research. Impacts of climate change on crop production mediated through changes in populations of serious insect-pests need to be given careful attention for planning and devising adaptation and mitigation strategies for future pest management programmes.

Section III
DISEASE PATHOGENS

8

Climate Change and Disease Pathogen Scenarios

8.1. INTRODUCTION

Crops can be damaged by diseases caused by fungi (rust, blight, mildew, and rot), bacteria/phytoplasma (wilt) and viruses. The occurrence of plant fungal and bacterial pathogens depends on climate and weather. Evaluation of the limited literature in this area suggests, that the most likely impact of climate change will be felt in three areas: in losses from plant diseases, in the efficacy of disease management strategies and in the geographical distribution of plant diseases.

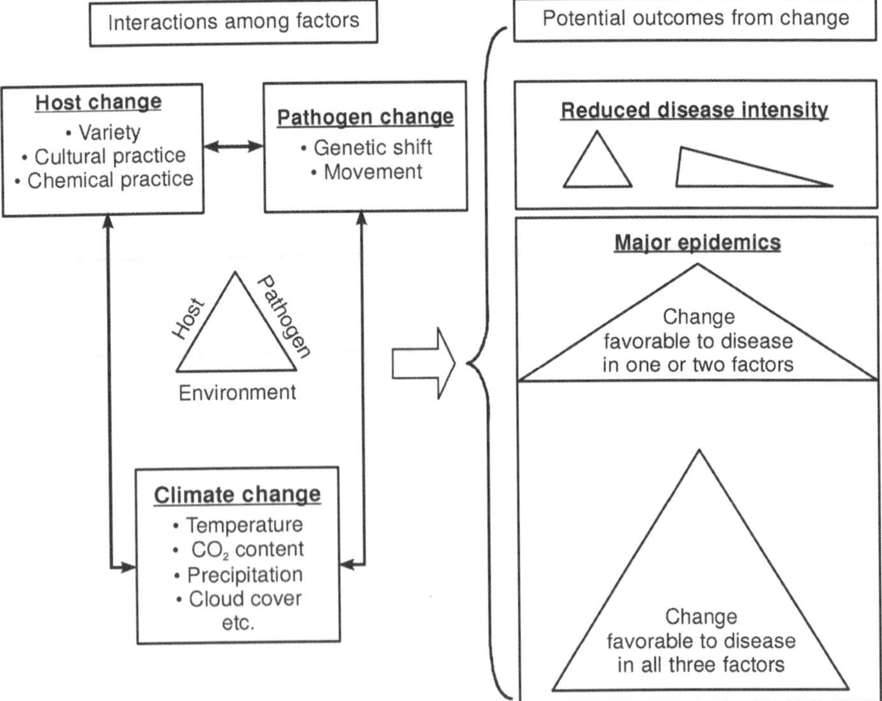

Fig. 8.1. Interactions among host, pathogen and environmental factors in disease triangle

They are also strongly influenced by agricultural practices. Higher temperature and humidity and greater precipitation result in the spread of plant diseases, as wet vegetation promotes the germination of spores and the proliferation of fungi and bacteria, and influences the life cycle of soil nematodes. In regions that suffer aridity, however, disease infestation lessens, although some diseases (such as the powdery mildews) thrive in hot, dry conditions, as long as there is dew formation at night. They constitute a fundamental group of biological indicators that needs to be evaluated regarding climate change impacts, besides being responsible for production losses, and potential threats to agro-ecosystem sustainability. Plant diseases require varying environmental conditions to develop; thus, it is vital to understand the environmental requirements of individual plant pathogens before predicting responses to climate change (Fig. 8.1).

Weather influences all stages of host and pathogen life cycles as well as the development of disease. Relationships between weather and disease are routinely used for forecasting and managing epidemics, and disease severity over a number of years can fluctuate according to climatic variation (Coakley, 1988; Scherm and Yang, 1995).

Climate factors that influence the growth, spread, and survival of crop diseases include temperature, levels of CO_2, precipitation, humidity, dew, drought, radiation, wind speed, circulation patterns, and the occurrence of extreme events. Typically, the two most important environmental factors in the development of plant disease epidemics are temperature and moisture. Other important factors influencing plant diseases are air pollution, particularly ozone and UV-B radiation (Manning and von Tiedemann, 1995) as well as nutrient (especially nitrogen) availability (Thompson *et al.*, 1993). The global climate change, especially increased levels of CO_2 and temperature (Pachauri and Reisinger, 2007), is thought to influence or change all the elements of a disease triangle, viz. host, pathogen and weather factors and their interactions (Fig. 8.2) (Legreve and Duveiller, 2010).

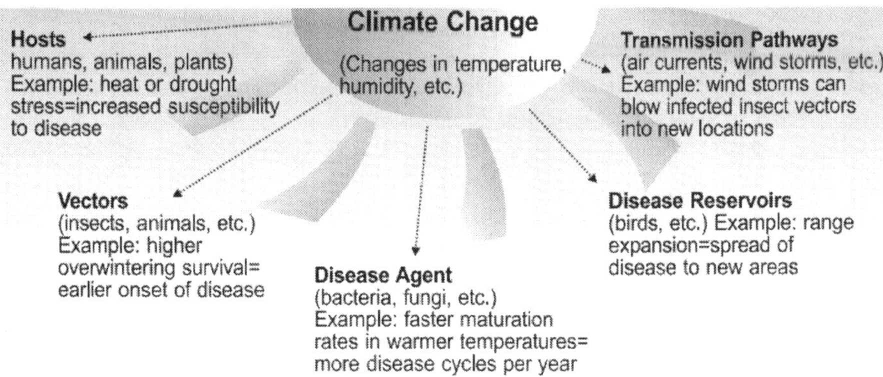

Fig. 8.2. The effects of climate change on disease pathogens

It is likely that climate change will have positive, negative or neutral impacts on specific host–pathogen systems (Coakley *et al.*, 1999; Chakraborty *et al.*, 2000). In general, climate change has the potential to modify host physiology and resistance, and to alter stages and rates of development of the pathogen (Coakley *et al.*, 1999).

Climate change could alter stages and rates of development of the pathogen, modify host resistance, and result in changes in the physiology of host-pathogen interactions. The most likely consequences are shifts in the geographical distribution of host and pathogen and altered crop losses, caused in part by changes in the efficacy of control strategies. Recent developments in experimental and modelling techniques offer considerable promise for developing an improved capability for climate change impact assessment and mitigation. Changes in levels of CO_2, ozone and UV-B will influence disease by modifying host physiology and resistance. In addition, changes in temperature, precipitation and the frequency of extreme events will influence disease epidemiology. Changes in geographical distribution will potentially alter the relative importance and spectrum of diseases and new disease complexes may arise. Evolution of pathogen populations may accelerate from enhanced UV-B radiation and/or increased fecundity in elevated CO_2. As a result, host resistances may overcome more rapidly. Disease management will be influenced due to altered efficacy of biological and chemical control options. Given the multitude of atmospheric and climatic factors, possible change scenarios and the number of disease systems, modelling approaches to impact assessment need to be strengthened. Changes in both mean temperature and its variability are equally important in predicting the potential impact of climate change.

Changes in disease situations have already been experienced as some minor diseases have become major diseases in the Indian subcontinent due to climate change, global change or shifts in seasonality. Current shift in the disease scenario in India, especially in rice and wheat, is a case in point. In rice, bacterial leaf blight, *Xanthomonas oryzae* pv. *oryzae* has become a global biotic threat despite constant efforts to improve resistance through exploitation of host R-gene. Sheath blight, *Rhizoctonia solani* and tungro virus that were of minor importance, have emerged as major problems in most of the rice-growing areas. Spot blotch, *Bipolaris sorokiniana*, once unknown or of minor importance, has become a serious problem in wheat. Increasing trend in winter temperature probably provides a favourable situation to the spot blotch in the north-western India (Duveiller *et al.*, 2007). Dry root rot, *Rhizoctonia bataticola* in chickpea is becoming more severe in the rain-fed environments due to moisture stress and higher temperatures (Pande *et al.*, 2010). Excess moisture on the other hand is favouring some of the dreaded soil-borne diseases caused by *Phytophthora, Pythium, Rhizoctonia solani* and *Sclerotium rolfsii*, especially in pulses (Sharma *et al.*, 2010).

In temperate regions, most plant pathogens are not active in late fall, winter, and early spring because of low temperatures. Some diseases are favoured by cool temperatures, while others are favoured by moderate or hot conditions. Disease often occurs when temperatures are more stressful for the plant than for the pathogen. Moisture, in the form of free water or high humidity, is necessary for many pathogens to infect, reproduce, and spread, although some can cause diseases in dry conditions.

Both climatic variability and climate change are the relevant drivers of plant disease epidemics and are expected to alter the synchrony between crop phenology and disease patterns. This change in climatic patterns also affects the spatial distribution of agro-ecological zones, habitats, and distribution patterns of plant diseases which can have significant impacts on food production.

Plant diseases are one of the important factors which have a direct impact on global agricultural productivity and climate change will further aggravate the situation (IPCC, 2007). Further, plant diseases are estimated to cause yield reduction of almost 20% in the principal food and cash crops worldwide (Thind, 2012). The plant pathologists' should provide long-term as well as short-term climate change adaptation measures to reduce the risk of crop loss due to sudden emergence of diseases so far unknown or economically negligible. Therefore, regular disease monitoring and surveillance is important since disease management does not work like fire-fighting. Disease risk in advance requires for short-term or tactical disease management and disease scenario based on climate forecasts are to be the basis of future strategies. Identifying and quantifying the impacts of climate changes on plant diseases is a complex phenomenon (Coakley and Scherm, 1996) as there is a great deal of uncertainty about the accurate climate forecast.

The analysis of the potential impacts of climate change on plant diseases is essential for the adoption of adaptation measures, as well as for the development of resistant cultivars, new control methods or adapted techniques, in order to avoid more serious losses (Chakraborty and Pangga, 2004; Ghini, 2005). For sustainable food production, disease management strategies should be reoriented in changing climate. Although, plant diseases play an important role in agriculture (Agrios, 2005), a limited amount of information on the potential impacts of climate change on plant diseases is available (Harvell *et al.*, 2002; Garrett *et al.*, 2006).

8.2. EFFECTS OF CLIMATE CHANGE ON DISEASE PATHOGENS

8.2.1. Direct effects

Several aspects of the biology of a pathogen can be directly influenced by environmental factors. More severe epidemics occur due to prolonged periods of environmental conditions [temperature, precipitation and relative humidity (RH)] which are close to the optimal for the development of the pathogen. Thus, as temperatures increase, many pathogens will spread into new geographic areas, where they will come into contact with new potential hosts. Pathogen survival in the absence of a host (e.g., over-wintering and over-summering) can also be affected by temperature and RH. The strongest consequences of global warming are expected to be observed in the tropics, because tropical species have a narrow temperature growth range and are, therefore, relatively sensitive to changes in temperature. These species are also currently living very close to their optimal temperature conditions (Ghini *et al.*, 2011). Pathogens in higher latitudes usually live in climates cooler than their physiological optima; therefore, warming is expected to enhance their fitness and the risk of epidemics of the diseases with which they are associated. Epidemics involving polycyclic pathogens are strongly influenced by the number of generations of the pathogen within a particular time period. Temperature and moisture govern the rate of reproduction of many pathogens (Caffarra *et al.*, 2012).

The longer growing seasons that will result from global warming will extend the amount of time available for pathogen reproduction and dissemination. Climate change may influence the sexual reproduction of pathogens thereby increasing the evolutionary potential of individual populations. Environmental factors that cause or accelerate tissue death, such as high temperatures or ozone levels, may favour infection by necrotrophic pathogens.

The most important environmental factor affecting fungal and bacterial plant diseases is availability of free water for infection and spread of the pathogenic organisms. Most fungal pathogens require free water to infect their host plants. Rainfall is often required for dispersal of fungal spores, although some fungal spore types are wind dispersed under dry conditions. For plant pathogenic bacteria, free water is required for infection, multiplication and dispersal. Although areas with higher rainfall generally have higher disease risk, the weather variable that best represents availability of water for plant disease development is the duration of surface wetness. Wetness is the key variable in most weather-based plant disease prediction models.

The powdery mildews are an important group of fungal plant pathogens with less dependence on free water for infection than other fungi, and they tend to be inhibited by heavy rainfall. Very few powdery mildew disease risk models have been developed and there is none that utilizes a negative association between rainfall and disease development. Therefore, examination of powdery mildew risk in relation to climate change rainfall predictions would not have been useful.

The rainfall variable that is often highly correlated with plant disease development is rainfall frequency (e.g. number of days per month with >1 mm rainfall), rather than rainfall amount (mm). This is important in relation to climate change studies, because published climate change data (MfE, 2008) usually indicates amount rather than frequency of rainfall. For some of the patho-systems, predicted changes in rainfall amount were interpreted as changes in rainfall frequency. Although this was an approximation, it did allow useful inferences to be made about the likely effects of rainfall changes on the selected patho-systems.

Temperature affects the rate of disease development. Many fungal and bacterial pathogens have temperature optima between 15 and 20°C. Below the optimum, increasing temperature increases development rate, whereas above the optimum, increasing temperature decreases development rate. In temperate climates, field temperatures during spring, when many crop disease epidemics begin, are most often sub-optimal. Therefore, increased average temperature would tend to increase disease risk.

8.2.2. Indirect effects

Diseases can also result from indirect effects, where plants have their defences weakened by an abiotic factor and are predisposed to infection by plant pathogens. Several important plant diseases are initiated by abiotic stresses, including forest decline diseases, which are an example of a disease complex caused by a combination of plant predisposition and a repetitive sequence of plant stresses that weaken a plant to become susceptible to weak pathogens that then can often infect and kill the plant. These weak pathogens, called saprogens, are often ubiquitous inhabitants of soil and decaying plant material and, normally, they do not cause disease in healthy, unstressed plants. However, under conditions of environmental stress, plants can become susceptible to these saprogens and their opportunistic infections. One of the more prevalent examples of these saprogens is the girdling fungi of the genus *Armillaria*. As climate changes, new combinations of host–stress–

saprogens will be encountered and might give rise to new types of decline diseases, particularly in tree species. In temperate climates, plants that are stressed by biotic or abiotic factors during a growing season are often predisposed to freezing damage during the subsequent winter. Plant diseases associated with interactions of biotic and abiotic stresses, or disease complexes, are a unique and important area of consideration for assessing the influence of climate change on plant diseases. In particular, forest declines are an example of plant diseases that result from a combination of interacting biotic and abiotic factors. Such diseases are characterized by a variety of disease symptoms and signs, are typically scattered in a random pattern throughout a population within a region, and are often host-specific, although more than one tree species in a region may have its own specific decline symptoms. Decline diseases are one example where a strong association between climate change and disease incidence and (or) severity has already been established in several forest species. Facts of extensive forest declines have been documented in Europe and North America in ash, birch, balsam fir and maple, and a strong relationship was evident between climate warming in the Northern Hemisphere and the onset of crown dieback in 1925, 1937, and 1981 on selected species of northern hardwoods in eastern Canada.

8.3. IMPACTS OF CLIMATE CHANGE ON DISEASE PATHOGENS

Climate change is predicted to have a direct impact on the occurrence and severity of diseases in crops, which will have a serious impact on food security. Temperature increases are predicted to lead to the geographic expansion of pathogen and vector distributions, bringing pathogens into contact with more potential hosts and providing new opportunities for pathogen hybridization. Pathogen evolution rates are determined by the number of generations of pathogen reproduction per time interval. Temperature governs the rate of reproduction for many pathogens. Longer seasons that result from higher temperatures will allow more time for pathogen evolution. Pathogen evolution may also be more rapid when large pathogen populations are present, so increased overwintering and over-summering rates will contribute as well. Climate change may also influence whether pathogen populations reproduce sexually or asexually; in some cases, altered temperatures may favour overwintering of sexual propagules, thus increasing the evolutionary potential of a population.

In a majority of the examples reviewed by Chakraborty *et al.* (2000) disease severity increased with elevated CO_2 concentration, but in some examples it decreased. CO_2 can also affect pathogen fecundity which was shown to increase under elevated CO_2 levels leading to enhanced rates of pathogen evolution (Chakraborty and Datta, 2003). Elevated CO_2 also reduced expression of induced resistance (Pangga *et al.*, 2004) or affected expression of resistance more directly (Plazek *et al.*, 2001; Plessl *et al.*, 2005), but in resistant cultivars, it also enhanced resistance by boosting phytoalexin production (Braga *et al.*, 2006).

The climate change could alter stages and rates of development of the pathogen, modify host resistance, and result in changes in the physiology of host-pathogen interactions. The most likely consequences are shifts in the geographical distribution of host and pathogen and altered crop losses, caused in part by changes in the efficacy of control strategies. The review of available literature suggests that potential impacts will be felt in three areas:

- Changes in crop loss due to diseases;
- Effectiveness of management strategies; and
- Geographical distribution of diseases.

In conclusion, the climate change has the following impacts on disease pathogens:

- Breakdown of host plant resistance
- Emergence of disease pathogens
- Expansion of geographic ranges
- Impact on crop-pathogen interactions
- Impact on pest population dynamics and disease outbreaks
- Increase in number of generations
- Increased overwintering survival
- Increased pesticide usage
- Emergence of new and intense disease problems
- Reduced effectiveness of biological control agents
- Reduced effectiveness of disease management strategies
- Risk of introducing invasive alien disease pathogens
- Introduction of alternative hosts as green bridges
- Extending the growing season
- Altering the timing of pest emergence

8.4. PREDICTIVE MODELS IN PLANT PATHOSYSTEMS

Predictive models have been developed for a few plant pathogen systems, including Phoma stem canker, *Leptosphaeria maculans* of oilseed rape in the United Kingdom (Evans *et al.*, 2007); downy mildew, *Plasmopara viticola* (Salinari *et al.*, 2006) and powdery mildew, *Erysiphe necator* (Caffarra *et al.*, 2012) of grapevine in northern Italy; coffee rust, *Hemileia vastatrix* in Brazil (Ghini *et al.*, 2011b); *Cercospora leaf* spot, *Cercospora beticola* of sugar beet in Germany (Richerzhagen *et al.*, 2011); and banana black sigatoka, *Mycosphaerella fijiensis* in Brazil (Ghini *et al.*, 2007).

Powdery mildew, *Erysiphe necator* is one of the most important diseases of grapevine. The European grapevine moth, *Lobesia botrana* is one of the most noxious vineyard pests in the European and Mediterranean regions. Caffarra *et al.* (2012) combined detailed phenological models of grapevine with phenological models of grape powdery mildew and the European grapevine moth, applied the models to climate change scenarios for the eastern Italian Alps and considered potential changes in the interactions between these species.

The current and future impact of downy mildew on vineyards was studied in Piedmont, Italy. Two global circulation models (GCMs) predicted increased temperatures and decreased precipitation in this region. The simulations obtained by combining the disease model with the two GCM outputs predicted increased levels of disease pressure in each of the three tested future decades following more favourable temperature conditions during the months of May and June. The increased temperatures are predicted to promote the epidemics in ways that will be only partially counteracted by the effects of reduced precipitation, which alone would have diminished disease pressure (Salinari *et al.*, 2006).

8.5. ADAPTATION AND MITIGATION STRATEGIES

Climate change can affect disease management by altering efficacy of biological and chemical control options (Chakraborty *et al.*, 1998; Coakley *et al.*, 1999). For example, heavy rains reduces fungicide residue. Crop plants growing under elevated CO_2 could be altered morphologically or physiologically, affecting uptake, translocation, and metabolism of systemic fungicides. For example, increased thickness of the epicuticular wax layer on leaves could result in slower and/or reduced uptake by the host, while increased canopy size could negatively affect spray coverage. Conversely, if higher temperatures increase plants' metabolic rates, they may take up chemicals more quickly, which may result in greater toxicity (Coakley *et al.*, 1999).

New dimensions of climate change may add extra uncertainty in management strategies for diseases caused by different pathogens. Impacts of climate change on plant patho-systems would occur chiefly through influences on host resistance or chemical and biological control agents. Delayed planting to avoid a pathogen may become less reliable. Particular attention is needed to identify cases where the efficacy of disease management may be reduced under climate change.

In conclusion, the following management practices have been recommended:

- Crops may require more fungicide spray treatments or higher application rates, thus increasing costs for farmers, prices for consumers, and the likelihood of development of fungicide resistance.
- Agronomic practices, such as crop rotation, multiple cropping, tillage, fertilization, irrigation, selection of the production site, use of resistant/tolerant varieties, and sanitation to reduce the amount of overwintering inoculum.
- Fungicides may continue to serve as common disease suppression agents.
- Alternative measures, such as cultural methods and biological control, should be developed.
- Host resistance should be deployed.
- Addition of biochar to soil alters microbial populations in the rhizosphere.

8.6. FUTURE THREATS

Examples of disease pathogens whose incidence and/or intensity will be substantially influenced by climate change (i.e. change in ecological conditions will substantially widen the area where the disease pathogen could be introduced and/or could substantially increase its potential economic importance) and of which economic (including environmental) consequences will be substantial. Some examples of disease pathogens are given below (Table 8.1):

Table 8.1. List of disease pathogens likely to become serious in future due to climate change

Disease pathogens	Anticipated effect of climate change
Fungal diseases	
Fusarium head blight, *Fusarium graminearum, F. culmorum*	With increasing rainfall at critical time of the growing season, more wheat areas will be subjected to mycotoxin infections and this can affect trade since international regulations on authorized mycotoxin contents are becoming more stringent. Climate change expected to affect cropping systems with more maize being grown in some areas and pathogen surviving on a new crop in the rotation. In North Africa, unusual high rainfall is already showing more head blight outbreaks. If 'sporadic' rains occur more frequently in India with climate changes, Fusarium head blight may turn-out as a huge problem for public health (food) and industry (feed). More fungicides will also be used.
Wheat blast, *Magnaporthe grisea*	It emerged in Brazil in the mid 80's. Heavy losses may occur due to absence of resistance. Adaptation of the pathogen in warm environments (i.e. South Asia) would be a disaster. Increased yield losses, infestation in new areas due to ability/possibility of pathogen to survive in unknown alternate hosts.
Wheat spot blotch, *Cohliobolus sativus*	Heavy losses may occur due to insufficiency of satisfactory resistance. Adaptation of the pathogen is likely to occur in warm environments (i.e. South Asia). More areas will be affected in South Asia as 50% of areas considered presently under favourable wheat growing conditions will become more subject to heat and water stress. Observations over 6 years in India, Nepal and Bangladesh have already shown that climate change is taking its toll as a result of the increase of average night temperature penalizing grain filling and increasing the spot blotch severity. This effect is likely to increase in more optimum growing areas.
Wheat yellow rust, *Puccinia striiformis*	New severe epidemics are likely to occur due to change of races. Indications exist that the pathogen has become more tolerant to heat. Average heat increase triggering races adapted to new thresholds of temperature requirements. More areas are affected by epidemics or epidemics have become more frequent.
Wheat leaf rust, *Puccinia tritici*	New epidemics may occur due to change of races. In south Asia, the disease tends to appear late in the growing season when temperature is rising. Climate change can lead to disease spread in areas where its importance is minor at present, e.g. facultative and winter wheat growing areas of China, parts of Europe, Pacific Northwest of US, winter and facultative wheat areas of Central Asia.
Wheat stem rust, *Puccinia graminis*	New epidemics may appear due to change of races and lack of genetic resistance. Change in heat average and relative humidity may trigger races adapted to new thresholds of environmental requirements. There is a study done in the UK indicating where in Europe Ug99 can become important with global warming.

Disease pathogens	Anticipated effect of climate change
Viral diseases	
Indian peanut clump virus (IPCV)	Yield losses are likely to occur. Transmitted by seed and soil fungi (*Polymixa*). No resistance in wheat or largely unknown. Climate change is likely to affect cropping systems patterns. Changes in rotations or cropping sequences due to high cereal prices may induce new infections of field or spread of the disease.
Barley yellow dwarf virus (BYDV)	Yield losses may occur. Not enough resistance in wheat. Resistance on virus diseases in developing countries is presently largely under-funded, delaying to capacity of reaction against the problem. Climate change (drought) will increase the survival and spread of the aphid vectors. In Pakistan, it has been observed that some years, dry air favours aphid flights.

8.7. RESEARCH NEEDS

Research needs to be undertaken on the following lines:

- Better understanding of gene expression in plants and pathogens in response to climatic factors.
- Integrated omic studies of host and pathogen responses, as well as communities of soil and plant associated microbes.
- Multifactor studies of climate change effects.
- Better models of adaptation rates.
- Better data and models related to dispersal, current levels of intraspecific diversity, strength of selection under different climate change scenarios, and heritability of traits.
- Long-term large-scale records of pathogen and host distributions.
- Models of regional processes that incorporate disease.
- Data and models describing dispersal of propagules and vectors.
- Integrated multi-disciplinary international networks for data collection and synthesis.

Research gaps in this rapidly developing area include effects of climate change on the following plant pathosystems:

- Mountain and boreal ecosystems (Roy *et al.*, 2004; Rohrs-Richey *et al.*, 2011; Witzell *et al.*, 2011);
- Mediterranean and tropical regions (Garbelotto, 2008; Zocca *et al.*, 2008; Thompson *et al.*, 2010; Savary *et al.*, 2011a, b);
- With climate change aspects other than temperature (e.g. precipitation: Hawkes *et al.*, 2011);
- With multiple climate change factors (Paajanen *et al.*, 2011);
- In relation to endophytes and viruses, two key factors for plant health (Jones, 2009; Brosi *et al.*, 2010);
- In relation to mycorrhiza diversity and productivity (Deslippe *et al.*, 2010; Egli, 2011);

- With long-term datasets (Hannukkala *et al.*, 2007; Fitt *et al.*, 2011);
- Under various climate change scenarios (Watt *et al.*, 2011);
- Following various climate change management options (e.g. assisted migration of plant species; McDonald-Madden *et al.*, 2011; Garbelotto and Pautasso, 2012); and considering various plant disease control methods (Ghini *et al.*, 2011a).

8.8. FUTURE PROSPECTS

As a discipline, plant pathology is dedicated to ensuring sustainable production in agricultural systems through the management and improvement of plant health. In the foreseeable future, global change will impact on plant health and its management to influence productivity. The role of plant pathology must be to provide an assessment of how plant diseases will impact on agricultural systems under climate change, in order to minimize loss in production and quality. Outcomes from this research will have important implications for decisions on amelioration and mitigation strategies. Due to uncertainties in climate change predictions (IPPC, 1996) a `no regrets approach', where the proposed actions have definite and quantifiable benefits, with or without the effects of climate change factored in, should provide the rationale for this research. An example would be an improved understanding of a disease cycle; this will enhance our capacity to predict and manage the disease under current climatic conditions in addition to improving our capacity to respond to climate change.

Climate change may have positive, negative or neutral impact on diseases. Research in this area is as much about identifying new opportunities as preparing to minimize negative impacts. Success will require an improved understanding of the causes, impacts and consequences of climate change from which emerges amelioration and mitigation strategies. The shortage of critical epidemiological data on individual plant diseases needs to be addressed using experimental approaches. In the first instance, studies in a controlled environment may be used to formulate hypothesis and to determine critical relationships to help develop process-based approaches. Field-based research examining the influence of a combination of interacting factors (Senft, 1995; Norby *et al.*, 1997) would be needed to provide a more realistic appraisal of impacts.

Methodology to model climate change impact has not been fully developed. Some of the impact assessments (Podger *et al.*, 1990; Brasier and Scott, 1994) are `first pass analysis' using climate-matching softwares such as BIOCLIM (Busby, 1991), HABITAT (Walker and Cocks, 1991) or CLIMEX (Sutherst and Maywald, 1985). Some have used and advocated the use of simulation models (Goudriaan and Zadoks, 1995; Luo *et al.*, 1995; Teng *et al.*, 1996). However, their use is currently limited due to lack of hard data on impacts. Empirical procedures for assessing long-term climate and disease interactions are only just beginning to emerge (Calvero *et al.*, 1994; Scherm and Yang, 1995; Coakley and Scherm, 1996). There is a need to look beyond the science of plant pathology to seek and invite concepts and ideas from other relevant disciplines for a reappraisal of priorities. Developments in information technology can help in the quest for knowledge and its dissemination (Bridge *et al.*, 1998).

Knowledge needs to be acquired, synthesized and generalized at a scale relevant to an environmental unit. Impact on an agricultural system must include on- and off-farm effects determined at a landscape scale of spatial resolution. Historically, plant pathology research using site-specific knowledge of individual patho-systems has served well in understanding, predicting and managing diseases. Environmental variables at the microclimate level have been utilized at this spatial scale. In contrast, climate systems operate at a global scale and General Circulation Models (GCMs) are better at explaining climate at this coarse level of resolution. This difference in the level of understanding between plant pathology and biometeorology/ climatology at the various spatial and temporal scales has hampered interdisciplinary interaction. The need to bridge this gap in knowledge has been recognized (Kennedy, 1997). In recent years, a number of attempts have been made to downscale GCM outputs to a biologically relevant mesoscale (Bardossy, 1997). Lack of epidemiologically relevant weather variables has been an impediment to the application of GCMs and other climate models to plant disease modelling. Duration of surface wetness and relative humidity, which critically influence infection and disease development by many plant pathogens, have not been easily obtained from GCM output until recently. Usefulness of remotely gathered site-specific wetness and other data for plant pathology research has been variable (Gleason *et al.*, 1997) and Seem *et al.* (2000) provides a more detailed discussion on this topic.

8.9. CONCLUSIONS

If changes in atmospheric composition and global climate continues in the future as predicted, there will be relocation of crops and their diseases and impacts will be felt in economic terms from crop loss. Changes in levels of CO_2, ozone and UV-B will influence disease by modifying host physiology and resistance. In addition, changes in temperature, precipitation and the frequency of extreme events will influence disease epidemiology. Changes in geographical distribution will potentially alter the relative importance and spectrum of diseases and new disease complexes may arise. Evolution of pathogen populations may accelerate from enhanced UV-B radiation and/or increased fecundity in elevated CO_2. As a result, host resistance may be overcome more rapidly. Disease management will be influenced due to altered efficacy of biological and chemical control options. Information gathered so far has been fragmented and a comprehensive analysis of climate change impacts on diseases is not possible with present knowledge. Experimental research on a diverse range of disease systems is necessary to improve comprehension of climate change impacts. Given the multitude of atmospheric and climatic factors, possible change scenarios and the number of disease systems, modelling approaches to impact assessment need to be strengthened. For instance, changes in both mean temperature and its variability are equally important in predicting the potential impact of climate change (Scherm and van Bruggen, 1994). Given that climate change is a global issue, the focus needs to shift from paddock-based assessment on specific diseases to a more ecologically relevant spatial unit (Scherm *et al.*, 2000) to consider climate with other associated changes in land use and vegetation cover (Luo *et al.*, 1995), among others.

There has been only a limited research on impact of climate change on plant diseases under field conditions or disease management under climate change.

However, some assessments are now available for few countries, regions, crops and particular pathogens which concern with food security. Now, emphasis must shift from impact assessment to developing adaptation and mitigation strategies and options. First, there is a need to evaluate under climate change the efficacy of current physical, chemical and biological control tactics, including disease-resistant cultivars, and secondly, to include future climate scenarios in all research aimed at developing new tools and tactics. Disease risk analysis based on host–pathogen interactions should be performed, and research on host response and adaptation should be conducted to understand how an imminent change in the climate could affect plant diseases.

9

Effects of Climate Change Variables on Disease Pathogens

9.1. INTRODUCTION

Plant disease pathogens are one of the important factors which have a direct impact on global agricultural productivity and climate change will further aggravate the situation (IPCC, 2007). A plant disease is the result of interaction between a susceptible host plant, a virulent pathogen(s) and the environment. Changes in environmental conditions are strongly associated with differences in the level of losses caused by a disease because the environment significantly (directly or indirectly) influences plants, pathogens and their antagonists. The changes associated with global warming (i.e., increased temperatures, changes in the quantity and pattern of precipitation, increased CO_2 and ozone levels, drought, etc.) thus, may affect the incidence and severity of plant disease and influence the further co-evolution of plants and their pathogens (Chakraborty, 2005; Burdon *et al.*, 2006; Garrett *et al.*, 2006; Crowl *et al.*, 2008; Eastburn *et al.*, 2011). There are already some examples of the effect of past climate changes on some patho-systems. Changes in the environmental conditions under which plants are grown may affect not only the growth and productivity of these plants but also the populations of microorganisms living on plant leaf and root surfaces. Plant infecting viruses operate in intimate association with their hosts and vectors. Climate change may affect both host plant and insect vector populations, thereby affecting the spread of plant viruses. Climate change adds an extra layer of complexity to plant protection. It is predicted that because of changes in the relative importance and distribution of different pathogens, the fungicide market will certainly change.

The analysis of the potential impacts of climate change on plant diseases is essential for the adoption of adaptation measures, as well as for the development of resistant cultivars, new control methods or adapted techniques, in order to avoid more serious losses (Chakraborty and Pangga, 2004; Ghini, 2005).

The potential effect of climatic variations on plant diseases of agricultural crops includes discussion on the influence of temperature, precipitation, CO_2, ozone, ultraviolet light as well as insect vectors of plant disease (Pautasso *et al.*, 2012).

Climate change influences plant diseases in the following ways:

- Increased frequency of heat and drought
    - May contribute to disease susceptibility/resistance.

  - Drought can aggravate the effects of soil borne diseases, like *Macophomina*, *Fusarium* and others.
- Temperature governs the rate of reproduction for many pathogens.
- Elevated CO_2 levels
  - Changes plant structure- increased leaf thickness, higher leaf area, higher plant biomass- all these would influence infection by pathogens.
  - CO_2 increases pathogen load on C_3/C_4 plants/grasses.
- Elevated ozone (O_3)
  - Can change the leaf surface structure.
  - Affecting physical topography and chemical composition, structure of epicuticular wax.
  - May influence pathogen infection.
  - Likely enhanced infection by necrotrophic pathogens and root-rot fungi.

9.2. CO_2 ENRICHMENT

Increased CO_2 levels can impact both the host and the pathogen in multiple ways. Some of the observed CO_2 effects on disease may counteract others. Researchers have shown that higher growth rates of leaves and stems observed for plants grown under high CO_2 concentrations may result in denser canopies with higher humidity that favours pathogens. Lower plant decomposition rates observed in high CO_2 situations could increase the crop residue on which disease organisms can overwinter, resulting in higher inoculum levels at the beginning of the growing season, and earlier and faster disease epidemics. Pathogen growth can be affected by higher CO_2 concentrations resulting in greater fungal spore production. However, increased CO_2 can result in physiological changes to the host plant that can increase host resistance to pathogens (Coakley *et al.*, 1999).

Photosynthesis, leaf area, plant height, total biomass (shoot and root) and crop yield, sugar and starch content, water-use efficiency, growth and yield are increased in the presence of higher levels of CO_2. These effects often result in changed plant architecture and the development of larger plant organs. Because many foliar pathogens can take advantage of the more humid microclimate caused by denser plant growth and the higher availability of host tissue, pathogen infection rates of these pathogens usually increase at higher CO_2 levels. However, the final effect of increased CO_2 concentrations on the disease depends on the interaction between the effects on the pathogen and the effects on the plant under the specific environmental conditions.

Elevated CO_2 may increase C_3 plant canopy size and density, resulting in a greater biomass with a much higher microclimate relative humidity. This is likely to promote plant diseases such as rusts, powdery mildews, leaf spots and blights (Manning and von Tiedemann 1995; Chakraborty *et al.*, 2003). However, Kobayashi *et al.* (2006) concluded from literature reviews that it is not clear whether the disease severity is enhanced or diminished by a higher CO_2 level. Research on rice leaf blast and rice sheath blight in the temperate climates of Japan showed, that elevated CO_2 increased the potential risks for infection from leaf blast and epidemics of sheath blight (Kobayashi *et al.*, 2006).

In soybean, elevated concentration of CO_2 and O_3 (ozone) altered the expression of three soybean diseases, namely downy mildew, *Peronospora manshurica*; brown spots, *Septoria glycines*; sudden death syndrome, *Fusarium virguliforme* and plant response to the diseases varied considerably. Elevated CO_2 alone or in combination with ozone (O_3) significantly reduced downy mildew, *Peronospora manshurica* disease severity by 39–66% across a 3 year study. In contrast, elevated CO_2 alone or in combination with O_3 significantly increased brown spot, *Septoria glycines* severity, but the increase was small in magnitude (Eastburn *et al.*, 2010).

In wheat, grown at elevated atmospheric CO_2 (700 ppm) and under different fertilization and water regimes, the host water content, the plant N content and the infection rate with powdery mildew were investigated. In all fertilization regimes, the mean per cent leaf area infected with mildew was significantly reduced under elevated atmospheric CO_2, compared to ambient CO_2.

In a moderate water supply treatment (3.6 mm/day), the plants grown in elevated atmospheric CO_2 concentrations had significantly reduced N contents (9.9%) and significantly increased water content (4%), the amount of mildew infection was unchanged. At higher water supply (5.4 mm/day) host water content at elevated CO_2 was similar to that of ambient CO_2, but N content was significantly reduced. As a consequence severity of powdery mildew caused by *Erysiphe graminis* infection was significantly reduced as compared to ambient CO_2. At lower water supply (1.8 mm/day) the results were quite different. Host water content at elevated CO_2 was higher compared to that of ambient CO_2, but N content showed no difference. As a consequence, severity of powdery mildew infection was significantly increased compared to ambient CO_2. It seems that severity of mildew infection is more sensitive to host water content than to host nitrogen content (Thompson *et al.*, 1993).

Astonishing results were gained in an experiment on oat, *Avena sativa* grown under elevated CO_2 (700 ppm) and infected by barley yellow dwarf virus (BYDV). Root mass of virus infected plants increased by 37–60% with CO_2 enrichment, but was largely unaffected in healthy plants. CO_2 enrichment increased photosynthesis and water use efficiency by 34 and 93% in healthy plants and by 48 and 174% in infected plants – basically the infected plant performed better under elevated CO_2 than at ambient CO_2 (Malmström and Field, 1997).

Chakraborty *et al.* (2000a) studied dispersal of and infection by *Colletotrichum gloeosporioides* under ambient weather conditions in the field on *Stylosanthes scabra* plants that had been raised under 1× or 2× CO_2 in controlled environment chambers. Plants from the two CO_2 environments were exposed to naturally occurring inoculum in the field on different dates, and conidial dispersal and infection were monitored. The enlarged canopy of plants grown under elevated CO_2 trapped more conidia that, together with increased humidity in the denser canopy, led to more severe anthracnose than on plants grown under 1× CO_2 (Fig. 9.1).

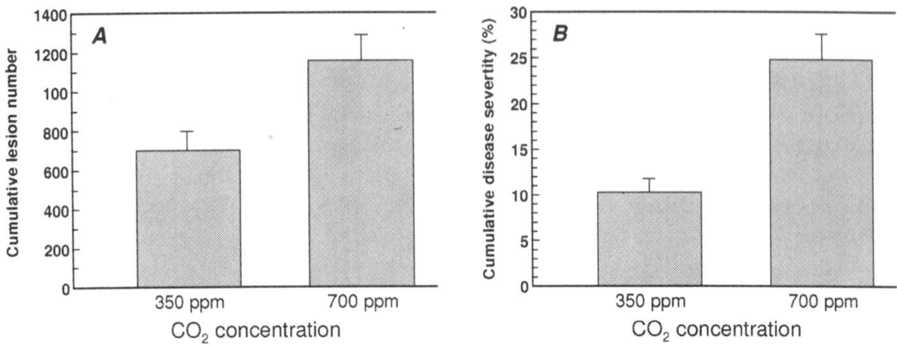

Fig. 9.1. Cumulative number of lesions
(*A*) disease severity expressed as per cent leaf area affected
(*B*) caused by *Colletotrichum gloeosporioides* on susceptible *Stylosanthes scabra* plants

Decomposition of plant litter is important for nutrient cycling and in the saprophytic survival of many pathogens. Because of high C: N ratio of litter as a consequence of plant growth under elevated CO_2, decomposition will be slower. Increased plant biomass, slower decomposition of litter, and higher winter temperature could increase pathogen survival on over-wintering crop residues and increase the amount of initial inoculum available for subsequent infection, and earlier and faster disease epidemics.

Some fungal patho-systems under elevated CO_2 revealed two important trends. First, delay in the initial establishment of the pathogen because of modifications in pathogen aggressiveness and/or host susceptibility. For example, reduction in the rate of primary penetration of *Erysiphe graminis* on barley and a lengthening of latent period in *Maravalia cryptostegiae* (rubervine rust) has been observed under elevated CO_2. Here, host resistance may have increased because of the change in host morphology, physiology, nutrients and water balance. A decrease in stomatal density increases resistance to pathogens that penetrate through stomata. Under elevated CO_2 barley plants were able to mobilize assimilates into defence structures including the formation of papillae and accumulation of silicon at sites of appressorial penetration of *E. graminis.*

At elevated CO_2, increased partitioning of assimilates to roots occurs consistently in crops such as carrot, sugar beet, and radish. If more carbon is stored in roots, losses from soil-borne diseases of root crops may be reduced under climate change. In contrast, for foliage diseases favoured by high temperature and humidity, increases in temperature and precipitation under climate change may result in increased crop loss. The effects of enlarged plant canopies from elevated CO_2 could further increase crop losses from foliar pathogens.

The second important effect is an increase in the fecundity of pathogens under elevated CO_2. Following penetration, established colonies of *Erysiphe graminis* grew faster and sporulation per unit area of infected tissue was increased several-fold under elevated CO_2. It has been also observed that under elevated CO_2 out of the 10 biotrophic pathogens studied, disease severity was enhanced in six and reduced in four and out of 15 necrotrophic pathogens, disease severity increased in nine, reduced in four and remained unchanged in the other two (Chakraborty *et al.*, 1998).

It has been observed that oats infected with Barley yellow dwarf virus (BYDV) showed greater biomass accumulation to CO_2 enrichment than the healthy plant. Tobacco plants grown at increased CO_2 concentrations showed a markedly decreased spread of virus. It appears that CO_2 rise in the air may have some positive effects, which may likely offset the negative effects of virus infection.

Some diseases can cause more severe reduction in plant growth under twice ambient compared to ambient CO_2 at least in controlled environments. For example, in barley powdery mildew, an acclimation of photosynthesis at elevated CO_2 and an infection-induced reduction in net photosynthesis caused larger reductions in plant growth at elevated CO_2. Changes brought by high CO_2 concentration like reduced stomatal density, production of papillae and accumulation of silicon at the sites of appressorial penetration and changed leaf chemistry increased resistance to powdery mildew, *Blumeria graminis* in barley (Hibberd *et al.*, 1996b).

Thompson *et al.* (1993) reported a significant reduction in wheat powdery mildew at twice-ambient CO_2, but final severity was dependent on nitrogen and water status of plants.

Despite initial delays and reduction in host penetration, established colonies grow faster inside host tissues at elevated CO_2 (Hibberd *et al.*, 1996a; Chakraborty *et al.*, 2000b). Fecundity of both biotrophs and necrotrophs (Chakraborty *et al.*, 2000b) studied so far has increased under elevated CO_2. A combination of increased fecundity and a favourable microclimate within enlarged canopies will provide more opportunities for infection. There is evidence of adaptation for increased aggressiveness in some pathogens (Kolmer and Leonard, 1986) within three sexual generations and controlled crossing has shown that aggressiveness is heritable and may be polygenically controlled (Caten *et al.*, 1984). For sexually reproducing pathogen populations with broad genetic diversity, increased population size and the number of generations in favourable microclimates would increase the probability of more damaging pathotypes evolving more rapidly (Sutherst *et al.*, 1996).

9.3. ELEVATED TEMPERATURES

Temperature has potential impacts on plant disease through both the host crop plant and the pathogen. Research has shown that host plants such as wheat and oats become more susceptible to rust diseases with increased temperatures; but some forage species become more resistant to fungi with increased temperature (Coakley *et al.*, 1999). Many mathematical models that have been useful for forecasting plant disease epidemics are based on increases in pathogen growth and infection within specified temperature ranges. Generally, fungi that cause plant disease grow best in moderate temperature ranges. Temperate climate zones that include seasons with cold average temperatures are likely to experience longer periods of temperatures suitable for pathogen growth and reproduction if climates warm. For example, predictive models for potato and tomato late blight caused by *Phytophthora infestans* shows that the fungus infects and reproduces most successfully during periods of high moisture that occur when temperatures are between 7.2°C and 26.8°C (Wallin and Waggoner, 1950). Earlier onset of warm temperatures could result in an earlier threat from late blight with the potential for more severe epidemics and increase in the number of fungicide applications needed for control.

Harvell *et al.* (2002) considered the consequences of warmer temperatures on host-pathogen interactions and concluded that there will be three main effects:

- Increase in pathogen development rate, transmission, and generations per year.
- Increase in overwintering of pathogens.
- Changes in host susceptibility to infection.

Furthermore, they suggested that the most severe and unpredictable consequences would occur if populations of pathogen and host, which were formerly geographically separated due to climate constraints, converged.

Long and cold winters have reduced survival of organisms that cause disease, restricted the number of disease generations per year, and limited disease activity during the growing season. If the climate warms, these issues of disease survival, growth, and activity will change. For disease survival, climate change effects will depend on the way in which the pathogen presently survives adverse conditions (Table 9.1).

Table 9.1. Possible effects of climate change on plant pathogen survival

Survival mode of pathogen	Common types of pathogens	Effects of climate change
Disease organisms that survive in soil	Introduced diseases	Thick-walled spores and other survival structures produced by these pathogens should not be greatly affected by climate change.
Disease organisms that survive in plants or plant debris	Introduced diseases	Milder winters could increase survival of pathogens that overwinter on living and dead plants.
Disease organisms that are transferred by vectors	Viruses and fungi	Milder winters could increase survival of insect vectors; milder summers could result in increased development and reproductive rates.
Introduced diseases	Introduced diseases	Pathogens that do not presently survive winters may do so if winters become milder.

Increase in temperature can modify host physiology and resistance. Both temperature and the length of exposure are important in determining the effect of climate change on disease severity. Even if the temperature change may be well within the limits of current climatic variability, a modest warming can cause a significant increase in cumulative-temperature above a critical temperature threshold to affect crop physiology and resistance to a disease. Temperature can also affect disease resistance in plants, thus affecting the incidence and severity of the diseases. Temperature sensitivity to resistance has been reported for leaf rust, *Puccinia recondita* in wheat, black shank, *Phytophthora nicotianae* in tobacco and bacterial blight, *Xanthomonas oryzae* pv. *oryzae* in rice (Gregory *et al.*, 2009). Temperature change might lead to appearance of different races of the pathogens hitherto not active but might cause sudden epidemic. Change in temperature will directly influence infection, reproduction, dispersal, and survival between seasons and other critical stages in the life cycle of a pathogen.

At higher temperature, lignification of cell walls increased in forage species and enhanced resistance to fungal pathogens. Impact would, therefore, depend on the nature of the host-pathogen interactions and mechanism of resistance. A rise in temperature above 20° C can inactivate temperature sensitive resistance to stem rust in oat cultivars. Increase in temperature with sufficient soil moisture may increase evapotranspiration resulting in humid microclimate in crop canopy and may lead to incidence of diseases favoured under warm and humid conditions. Some of the soil-borne diseases may increase at the rise of soil temperature. If climate change causes a gradual shift of cropping regions, pathogens will follow their host. Analysis of long-term data of wheat and rice diseases in China has shown trends of an increase in minimum temperatures in association with the abundance of rice blast or wheat scab. In most locations, temperature changes had significant effects on disease development. However, these effects varied between different agro-ecological zones. In cool sub-tropical zones such as Japan and northern China, elevation of ambient temperature resulted in greater risk of blast epidemics. Situations in the humid tropics and warm humid subtropics were opposite to those in cool areas. A lower temperature resulted in greater risk of blast epidemics.

A simulation model was developed for rice leaf blast epidemics in Japan, China, Thailand, South Korea and the Philippines under increasing temperature and ultraviolet B (UV-B) radiation. Elevated CO_2 was not considered. The simulation showed that in the cooler regions of Japan and northern China a temperature increase might lead to more severe blast epidemics, while in humid tropics and warm humid tropics this risk might decrease (Luo *et al.*, 1995). The authors concluded that in these regions blast development is inhibited by high temperatures. UV-B radiation will enhance the severity of blast, but more in cooler, than in warmer regions.

Kaukoranta (1996) simulated yield loss in potato based on a 3-year long controlled environment study of late blight at ambient temperature and at 3°C higher than ambient. This study suggested that increases in yield loss of unprotected potato crop at the high temperature would wipe out any benefits from yield increases of around 2 t/ha dry matter per degree of warming.

A change in temperature could directly affect the spread of infectious diseases and their survival between seasons. There are indications of increased aggressiveness at higher temperatures of wheat stripe rust, *Puccinia striiformis* isolates, suggesting that rust fungi can adapt to and benefit from higher temperatures (Mboup *et al.*, 2012).

In India, in the last decade the disease scenario of chickpea has changed drastically; dry root rot (*Rhizoctonia bataticola*) of chickpea has emerged as a potential threat to the production of chickpea (Pande and Shanna, 2010). Higher risk of dry root rot has been reported in *Fusarium* wilt chickpea-resistant varieties in those years when the temperature exceeds 33°C (Dixon, 2012).

Agricultural crops and plants in natural communities may harbour pathogens as symptomless carriers (Dinoor, 1974), and disease may develop if plants are stressed in a warmer climate. Host stress is an especially important factor in decline of various forest species. Climate extremes such as drought may increase invasion by

Armillaria sp. that are not normally very pathogenic (Lonsdale and Gibbs, 1996). High temperatures may increase the damage caused by diseases such as Scleroderris canker on lodge pole pine (Lonsdale and Gibbs, 1996). Such projections, however, do not consider other factors that can enhance the resilience of forest ecosystems to climate change, which led Loehle (1996) to conclude that there is "a systematic bias toward alarmist predictions" in projections of tree health response to climate change.

Temperature is one of the most important factors affecting the occurrence of bacterial diseases such as *Ralstonia solanacearum*, *Acidovorax avenae* and *Burkholderia glumea*. Thus, bacteria could proliferate in areas where temperature-dependent diseases have not been previously observed (Kudela, 2009).

Most plant viruses are transmitted by vectors and majority by insects. Particularly aphids are expected to react strongly to environmental changes because of their short generation time, low developmental threshold temperatures and ability to survive mild winters without winter storms. An increase in the number of insect vectors will inevitably lead to a higher risk for viral infection of plants. The aphid transmissible complex of barley yellow dwarf viruses in cereals and potato virus Y in potato are amenable to show potential effects on the prevalence of infection because of climate change. In mild winters, high intensity of aphid movement during spring and a high frequency of PVY-infected potatoes have been reported. The severity of viral diseases is determined in large part by the amount of inoculum and the time of infection. The amount of virus inoculum is influenced by winter survival of its hosts. For some viruses, higher temperatures also cause more severe symptoms development. Aphids are expected to have increased survival with milder winter temperatures, and higher spring and summer temperatures will increase their development and reproductive rates and lead to more severe diseases. Milder winters are also expected to increase survival of alternate weed hosts of viruses. Increases in frequency and intensity of summer storms with high winds, rain, and hail will increase wounding of plants and result in increased transmission of viruses by mechanical means. Therefore, with predicted changes in climate, viral diseases of plants are expected to increase in importance. Potentially of greater importance will be the effects of diseases caused by newly introduced viruses that will be able to persist because of the changed climate. A warmer climate might also allow viruses that are present in greenhouses, such as Pepino mosaic virus (PepMV), to establish infection in the field. The main effect of temperature in temperate regions is to influence winter survival of vectors. Natural spread of vectors, pests and diseases is accelerated towards the North, as former climate barriers are no longer effective. This results in more severe outbreaks of plant-disease vectors like aphids, whiteflies, thrips or beetles, an extension of the period of disease infection further into the growing season and also introduction and establishment of new vector species. The described effects on vectors can have severe negative effects on food production or result in an increased use of plant protection products to control the vectors.

9.4. VARYING PRICIPITATION PATTERNS

Moisture can impact both host plants and pathogen organisms in various ways. Some pathogens such as apple scab, late blight, and several vegetable root pathogens are more likely to infect plants with increased moisture. The forecast models for these diseases are based on leaf wetness, relative humidity and precipitation measurements. Other pathogens like the powdery mildew species tend to thrive in conditions with lower (but not low) moisture. More frequent and extreme precipitation events that are predicted by some climate change models could result in more and longer periods with favourable pathogen environments. Host crops with canopy size limited by lack of moisture, might no longer be so limited and may produce canopies that hold moisture in the form of leaf wetness or high canopy relative humidity for longer periods, thus increasing the risk from pathogen infection (Coakley *et al.*, 1999). Some climate change models predict higher atmospheric water vapour concentrations with increased temperature also would favour pathogen and disease development.

With increased temperature, various models on climate change predict frequent and extreme rainfall events and higher atmospheric water vapour concentrations. These encourage the crops to produce healthier and larger canopies that retain moisture as leaf wetness and RH for longer periods and results in condition conducive for pathogens and diseases such as late blights and vegetable root diseases including powdery mildews (Coakley *et al.*, 1999). High moisture favours foliar diseases and some soil-borne pathogens such *Phytophthora, Pythium, Rhizoctonia solani* and *Sclerotium rolfsii.*

Bacteria are spread to their host plants mainly by water, usually in the form of rain splash and insects. In humid, wet conditions, infected plant tissues can exude masses of bacteria that are spread from host to host by rain splash and insects. Therefore, the warmer drier summers expected with climate change should limit bacterial diseases. However, bacteria often enter hosts through wounds, and the expected increase in frequency and intensity of summer storms with high winds, rain, and hail will increase wounding of plants and provide moisture for the spread of bacteria.

9.5. FREQUENCY AND MAGNITUDE OF EXTREME EVENTS

The increase in frequency and severity of extreme climate events (heat waves, cold snaps, droughts, floods, etc.) has become an important topic for the global change agenda and ecological research. The frequency and magnitude of extreme events are predicted to increase under future climate change.

Drought conditions enhance damage by disease pathogens. Water stress diminishes plant vigour and alters C/N ratio lowering plant resistance to disease pathogens. Attack by fungal pathogens of stems and roots are favoured by weakened plant conditions. Dry and warm conditions promote growth of insect vector populations, increasing viral epidemics.

- Outbreak of soybean cyst nematode correlated to drought conditions in north central US (1990).
- Increased incidence of *Aspergillus flavus* (producer of aflatoxin) in southern US (1997 and 1983).

Examples of extreme weather events such as hurricanes causing the spread of plant pathogens to new areas are common (Rosenzweig *et al.*, 2005) and are expected to increase with the projected increase in the frequency of extreme weather events under climate change. In addition to lost production, this can restrict market access, limiting valuable export earnings for some developing countries.

Floods and heavy rains caused by extreme weather events are responsible for the following crop damage (Rosenzweig *et al.*, 2000):

- Increased moisture benefits epidemics and prevalence of leaf fungal pathogens.
  - Rice leaf blight caused great famine in Bengal in 1942, 2 million people died.
  - Wheat stripe rust outbreak in major production regions of China contributed to the 1960's famine.
  - Fungal epidemics in corn, soybean, alfalfa and wheat in US Midwest in 1993.
  - Mycotoxin (produced by *Fusarium* sp.) reached a record high in US Great Plains in 1993. Mycotoxin increases are related to high humidity during harvest (East Africa and South America in 1990's).
  - Humid summers drive epidemics of grey leaf spot of maize (Iowa and Illinois in 1996).
- Water induced soil transport increases dissemination of soil-borne pathogens to non-infected areas.
  - Outbreaks of soybean sudden death syndrome in the north central US in 1993.
- Continuous soil saturation causes long-term problems related to rot development and increased damage by pathogens.
  - Crazy top and common smut in maize.

9.6. ELEVATED LEVELS OF ATMOSPHERIC POLLUTANTS

9.6.1. Ozone

Ozone (O_3) is considered to be the most phytotoxic of the common air pollutants. It can cause chlorotic and necrotic lesions on sensitive plant species, and even in the absence of visible symptoms, photosynthesis and growth can be inhibited. Ozone damage can lead to reduced competitive fitness of plants, and reduced vitality makes plants more susceptible to plant pathogens (Sandermann, 2000). Direct effects of ozone on fungal pathogens are not significant (Manning and Tiedemann, 1995), although interactions between ozone damage and infection by *Alternaria solani*, the causal agent of early blight of potato, have been reported (Holley *et al.*, 1985a, b). Researchers have reported both increased (Sandermann, 2000) and decreased (Coleman *et al.*, 1988) disease susceptibility in plants after ozone exposure. According to von Tiedemann and Firsching (2000), ozone effects on plant disease susceptibility may be strongly altered by interfering factors such as plant developmental stage, nutrient supply, and other atmospheric trace gases.

Most air pollutants indirectly influence diseases through their effect on host. Ozone induces reactions similar to those normally elicited by viral and other pathogens. Of the 49 bacteria and fungal pathogens examined, exposure to elevated ozone concentration enhanced disease in 25, did not affect in 10 and reduced in 14. Pollutant concentrations, which inhibit pathogen development, also injure the host. Similarly, infection by plant pathogens can alter ozone sensitivity of plants. Exposure to 5-10 ppm ozone for a few hours can cause visible injury to sensitive crops like barley, tomato, onion, potato, soybean, tobacco and wheat.

Elevated ozone concentrations may change the structure and properties of leaf surfaces in ways that may affect the inoculation and infection process. Ozone enhances senescence processes, may encourage necrosis and seems to promote attacks on plants by necrotrophic fungi.

Current climate change scenarios predict a further increase of tropospheric ozone, which is well known to inhibit plant photosynthesis and growth process. Ozone can also predispose plants to enhanced biotic attack, as proposed in particular for necrotrophic fungi, root rot fungi and black beetles. However, at present it does not seem possible to predict whether increased ambient ozone will lead to higher or lower disease likelihood in particular plant-pathogen system. Several root pathogens show a preference for stressed trees, although the direct role of ozone is not always evident. Onions injured by ozone exposure were more susceptible to *Botrytis cinerea,* but not to *B. squamosa.* Increased onion yields and reduced dieback when filters removed ambient ozone has been also observed in some experimental studies.

Apart from the losses of marketable yield in crops and the reduction in growth and productivity of tree species caused by exposure to high levels of O_3, pollutants can influence plant pathogens, and foliar pathogens can affect the leaf responses to O_3. Generally, ozone exposure tends to decrease the incidence of disease caused by obligate parasites, increasing the disease incidence caused by facultative parasites (Table 9.2).

Table 9.2. Effect of ozone (O_3) on some fungal plant diseases

Fungi	Host plant	Effect on disease*
Obligate biotrophs		
Puccinia graminis f. sp. *tritici*	Wheat	Decreased hyphal growth and urediospore production on O_3- injured leaves
Erysiphe graminis f. sp. *hordei*	Barley	Reduce rate of infection if exposed to sufficient O_3 during incubation
Necrotrophs		
Botrytis cinerea	Potato	Infection only in O_3 injured leaves
Lophodermium sp.	Pine	Increased severity of needle blight

Source: Manning and von Tiedemann (1995) (modified)

* Observed effect was dependent on the exposure O_3 dose before or after inoculation with pathogen.

9.6.2. Acid rain

Most studies on the effect of acid rain were done with simulated acid rain, since it is not easy to establish experiments under field conditions. In first year of experiment no effect of acid rain has been observed on any of four patho-systems: alfalfa leaf spot, peanut leaf spot (PLS), potato late blight (PLB), and soybean brown spot. In the second year, PLS severity decreased with increasing acidity and the dose response was linear; PLB severity showed a curvilinear response to acid rain.

9.6.3. Elevated ultraviolet B

Ultraviolet (UV) light has long been known to influence plant pathogenic fungi. This light may stimulate spore production in a wide range of fungi, but may also reduce spore survival during dispersal or early stages of infection (Paul *et al.*, 1998). Although an increase in solar UV-B radiation due to ozone depletion could promote sporulation of pathogenic fungi in a way that could greatly increase the frequency and intensity of epidemics, normal daylight already contains enough UV light to stimulate sporulation of light-dependent fungi.

There is considerable information on the effects of increased UV-B on crops and natural vegetation and on the growth and life cycle of pathogenic organisms such as fungi. Studies indicate that the UV-B component of solar radiation plays an important role in natural regulation on plant diseases. Stimulatory effect of near-UV light on reproduction of many fungi, and spore production in *Leptosphaerulina trifoli* peaks at 287 nm are reported. Fungi differ in their sensitivity to UV-B. Some strains of *Septoria tritici* are more sensitive to UV-B than others and *S. nodorum*, as a species, is more sensitive than *S. tritici*. UV-B radiation can modify the relative composition of phylloplane organisms, such as pink and white yeast. Continued exposure to enhanced UV-B radiation lowers the level of antifungal compounds in foliar parts. UV-B has been shown to reduce tolerance of rice to blast, (*Magnaporthe grisea*) and although higher UV-B reduced plant biomass and leaf area; there was no increase in blast severity. There are some evidences that sunlight can influence pathogen by causing accumulation of phytoalexins or protective pigments in plant tissues. Therefore, UV-B may affect plant diseases directly via the pathogen or indirectly via the host.

9.7. CONCLUSIONS

The impact of climate change on agriculture could result in problems with food security and may threaten the livelihood activities upon which much of the population depends. Climate change is expected to have significant impact on distribution, phenology and abundance of many plant pathogenic species. Temperature is the major factor in global climate change that directly affects the development of different pathogens, their reproduction and survival. Changes in rainfall pattern also has implications for difference in pathogen survival. Abiotic stress such as temperature and drought may contribute plant susceptibility to pathogens. An increase in atmospheric carbon dioxide level will have positively affected the growth and productivity of crops by increasing the photosynthesis. An intermittent rain from flowering to fruiting stage of the crop favours the infection and development of foliar diseases. Protected cultivation technology can be adopted to combat climate change. The application of biotechnological approaches in crop improvement can contribute significantly in minimizing the problem of plant pathogens.

10

Impacts of Climate Change Induced Consequences on Disease Pathogens

10.1. INTRODUCTION

Speed of climate change and the unpredictability of its characteristics are of great concern with respect to the disease pathogens that reduce crop yield. The classic disease triangle recognizes the role of climate in plant diseases as no virulent pathogen can induce disease on a highly susceptible host, if climatic conditions are not favourable. Climate influences all stages of host and pathogen life cycles, as well as development of disease. Disease severity over a period, can fluctuate according to climatic variation.

Climate change is predicted to have a direct impact on the occurrence and severity of diseases in crops, which will have a serious impact on food security. Climate change has the potential to modify host physiology and resistance and to alter the stages and rates of development of the pathogen. The most likely impacts would be shift in the geographical distribution of the host and pathogen, change in the physiology of host-pathogen interactions and change in crop losses. In many cases, temperature increases are predicted to lead to the geographic expansion of pathogen and vector distributions, bringing pathogens into contact with more potential hosts and providing new opportunities for pathogen hybridization (Baker *et al.*, 2000; Brasier, 2001). Pathogen evolution rates are determined by the number of generations of pathogen reproduction per time interval, along with other characteristics such as, heritability of traits (Garrett *et al.*, 2006). Temperature governs the rate of reproduction for many pathogens. Longer seasons that result from higher temperatures will allow more time for pathogen evolution. Pathogen evolution may also be more rapid when large pathogen populations are present, so increased overwintering and over-summering rates will contribute as well. Climate change may also influence whether pathogen populations reproduce sexually or asexually; in some cases, altered temperatures may favour overwintering of sexual propagules, thus increasing the evolutionary potential of a population (Pfender and Vollmer, 1999). Under climate change, due to increased biomass of crops and alternative host plants, necrotrophic pathogen will produce large quantities of inoculum that can infect subsequent crops, thereby often losing the advantage of using a partially resistant variety to reduce inoculums (Melloy *et al.*, 2010). Change in temperature regimes will provide wider opportunities for overwintering of sexual stages, thereby accelerating gene recombination and opportunities for the development of more aggressive pathogen strains (McElrone *et al.*, 2005). Some mechanisms of effects of elevated CO_2 on plants are fairly well understood, such as reduced stomatal opening and changes in leaf chemistry. In such situations, diseases caused by the pathogens that infect through stomata such as *Phyllosticta minima* (leaf spot of maple) may be reduced (McElrone *et al.*, 2005). In soil-borne pathogens, increase in disease development for autumn and winter-infecting root and stem pathogens has been predicted due to increased thermal time (Evans *et al.*, 2007; Butterworth *et al.*, 2010). Pathogens such as *Sclerotinia sclerotiorum*, which causes stem or white rot of oilseed rape and a wide range of vegetable crop diseases, are likely to release spores in synchrony with earlier flowering of crops like oilseed rape. Excess moisture on the other hand, favours some dreaded soil-borne diseases caused by *Phytophthora*, *Pythium*, *Rhizoctonia solani* and *Sclerotium rolfsii*, especially in pulses (Sharma *et al.*, 2010). Different climate change variables will have different effects on different soil microorganisms and associated biological processes as the soil is a highly complex ecosystem. Further, such changes are highly dependent on the particular soil conditions and few generalizations attributable to climate change can be made (Pritchard, 2011).

The climate change has the following impacts on disease pathogens:

- Expansion of geographical distribution
- Impact on host plant-pathogen interactions
- Impact on changes in crop loss
- Impact on vector-borne diseases
- Impact on disease pathogens
- Impact on disease complexes
- Impact on crop yield
- Breakdown of host plant resistance
- Impact on disease outbreaks
- Emergence of new and intense disease problems
- Increased pesticide usage
- Impact on food security
- Impact on efficacy of disease management strategies

10.2. EXPANSION OF GEOGRAPHICAL DISTRIBUTION

With changes in climate, plants will migrate to new areas, and their pathogens will follow. According to Coakley *et al.* (1999), pathogens tend to follow the host plant in its geographical distribution, but the rate at which pathogens become established in the new environment is a function of the mechanism of pathogen dispersal, suitability of the environment for dispersal, survival between seasons, and physiological and ecological changes in the host plant. If a host is chronically stressed due to less than optimum conditions, its health would deteriorate and its susceptibility to disease would increase, particularly in perennials (Chakraborty *et al.*, 1998). New diseases may establish in a region, while some established diseases may cease to be economically important. Although, climate changes may reduce the suitability of a crop for a region, it may continue to be grown for agro-ecological or economic reasons.

The northward expansion of the soybean sudden death syndrome (a soil-borne fungal disease caused by *Fusarium solani* f. sp. *glycines*) is an example (Roy *et al.*, 1997).

The disease was first reported in Arkansas in 1971; in the early 1980s it was found in southern Missouri, Illinois, and Indiana; by the early 1990s it was also found in southern Iowa, northern Illinois and northern Indiana; and in 1998 it was found in Ontario, Wisconsin and Ohio.

The grey leaf blight of corn caused by the fungus *Cercospora zeae-maydis* is also a disease whose range expansion was first noticed in the 1970s; in the last two decades, the disease has gradually developed into a major production problem in the Corn Belt. Although the increase in the abundance and epidemics of this disease may be due, in part, to the increase in the use of conservation tillage, the observed trends in minimum temperatures and precipitation in the region may also have contributed.

In the US, the frequency of the reported expansions of geographical range seems to have increased since 1970 (Table 10.1).

Table 10.1. Major crop diseases that appeared after 1970 in the United States

Crop	Disease/Causal agent	Expansion
Soybean	Sudden death syndrome, *Fusarium solani* f. sp. *glycines*	Appeared in Arkansas in 1971 and has spread to the northern soybean region as far as Ontario.
	Southern stem canker, *Diaporthe phaseolorum*	First observed in 1973, has developed into a devastating disease in the southern production region.
	Sclerotinia stem rot, *Sclerotinia sclerotiorum*	Re-emerged as a leading disease in 1990 in the north central soybean regions.
	Soybean cyst nematode, *Heterodera glycines*	Expanded to northern soybean regions.
Corn	Grey leaf spot, *Cercospora zeae-maydis*	First reported in the 1940s, became a concern in the 1970s in the eastern states and now is a major concern in the Corn Belt.
Potato	Late blight, *Phytophthora infestans*	Re-emerged in 1990 as a new threat to potato production after a new mating type was found in Mexico.
	Powdery scab, *Spongospora subterranea*	Increased damage in Washington and Oregon
Rice	Sheath blight, *Rhizoctonia solani*	Major rice disease worldwide since the 1970s.
Wheat	Wheat scab, *Fusarium* sp.	Re-emerged after 1990 as a leading wheat disease in the central and north regions.
	Barley yellow dwarf, Barley yellow dwarf virus	Listed by 14 wheat production states as a recently emerging disease.

Source: Plant Pathology Department, North Carolina State University.

Warming will generally cause a pole-ward shift of the agro-climatic zones and crops that grow in these zones. Carter *et al.* (1996) predicted that maize could be cultivated reliably in southern Finland by 2050 and warming will extend the northern limit for cereal cultivation. Warming will generally cause a pole-ward shift of the risk of damage from late blight of potato which would increase in all regions. Similar predictions of northward migration and increased severity in areas of current distribution have been made for the oak decline pathogen, *Phytophthora cinnamomi* (Brasier and Scott, 1994; Brasier, 1996).

Pathogens, in particular unspecialized necrotrophs, may extend their host range to cause new disease problems in migrating crops. There are atleast two well-known examples of an indigenous pathogen attacking an introduced plant when grown in close proximity. In its native habitat in the USA, the fire blight bacterium, *Erwinia amylovora*, attacks indigenous plants of the family Rosaceae without causing significant damage. When European settlers grew apples and pears in some regions, the bacterium caused serious losses. Similarly, the coffee rust epidemic in Asia during the late 1800's was facilitated by growing an introduced susceptible host, *Coffea arabica*, in a region where the native pathogen, *Hemileia vastatrix*, was already present on alternative hosts in the forests outlining the coffee plantations (Carefoot and Sprott, 1967). Expansion of host range may even occur in specialized biotrophs, as geographical proximity is as important as phylogenetic relatedness in influencing the host range of some rusts (Savile and Urban, 1982). As plants in both natural and agricultural communities can be symptomless carriers of pathogens, any early predictions of impending damage will be difficult (Dinoor, 1974).

10.3. HOST PLANT–PATHOGEN INTERACTIONS

Plants react to environmental changes by regulating their patterns of gene expression. These changes in the transcriptome allow the plants to respond to changing conditions (Garrett *et al.*, 2006). Climate change may directly affect several aspects of the biology of host plants, including their phenology (including senescence), sugar and starch contents, nitrogen and phenolic contents, root and shoot biomass, number and size of leaves, amount and composition of wax on leaves, changes in stomatal densities, and conductance and root exudation. Any change in any of these areas may influence infection and colonization by pathogens. For example, after infection, plant water potentials and nutritional content may affect the rate of colonization of host tissues, the production of new inoculum, and the expression of symptoms in the host (Colhoun, 1973).

A stronger, indirect effect of climate change that will act through changes induced in a plant can be assumed for those pathogens that can only infect a narrow range of closely related host biotypes (Gessler *et al.*, 2011) as opposed to pathogens that are able to cause disease in a wide range of host plants belonging to different families and genera (Elad *et al.*, 2004). Situations in which host defence reactions (i.e., nonspecific stress-induced responses) to very specific pathogens based on the presence of individual genes present both in the host and the pathogen [as in the gene for gene interaction (Gessler and Pertot, 2012)], are affected by climate change. Abiotic stress may induce the activation of general defence pathways, which not only increases resistance but also increase the susceptibility of the plant to certain pathogens.

Moisture and temperature can affect disease development by affecting the susceptibility of the host to infection and/or increasing the level of symptom expression (Boyer, 1995; McElrone *et al.*, 2001). Water stress causes stomatal closure and reduces photosynthesis. Leaf growth is inhibited, and changes in shoot architecture and the root/shoot ratio are commonly observed. In this manner, drought may impair production of plant defence substances or growth, favouring the progress of the pathogen. Conversely, several diseases are less severe when the availability of moisture is limited. Drought can reduce root growth, which reduces the chance that roots will come into contact with propagules of soil-borne pathogens, lowering the incidence of infection (Huisman, 1982). Drought and temperature stress have been shown to negatively affect a plant's ability to respond to biotic stresses via changes in endogenous abscisic acid levels that affect defence responses involving salicylic acid, jasmonic acid, or ethylene (Asselbergh *et al.*, 2008). The effect of increased temperature on plants depends on the season. Elevated temperatures reduces plant stress during the winter but increases the stress on the plant during the summer. Although any generalization is impossible, heat stress certainly affects the plant's susceptibility to pathogens. Heat stress causes symptoms similar to those caused by drought stress, including wilting, leaf burn, and phylloptosis (Wang *et al.*, 2003).

In some patho-systems, host–pathogen gene pairs, that are related to resistance have been found to interact differently at different temperature ranges (Browder and Eversmeyer, 1986). However, it is difficult to discriminate between the effects of temperature on host resistance genes and the effects of temperature on pathogen virulence. Similarly, efforts to identify the effects of water stress on plant resistance to infection are complicated by the fact that several foliar pathogens have lower infection rates under conditions of low RH (Huber and Gillespie, 1992). In some patho-systems, plants exhibit fewer symptoms when they are subjected to drought stress (Pennypacker *et al.*, 1991). However, in some others, plants are more susceptible under drought conditions (Schoeneweiss, 1975), above and beyond any additive effect of drought and disease stress.

Similar to temperature, CO_2 and ozone concentrations have increased since the Industrial Revolution. CO_2 levels are expected to double preindustrial levels by 2050. Both increased CO_2 and ozone concentrations may alter plant function (Adaros *et al.*, 1991; Amthor, 1995), but usually in opposite ways. In general, photosynthesis, leaf area, plant height, total biomass (shoot and root) and crop yield, sugar and starch content, water-use efficiency, growth, and yield are increased in the presence of higher levels of CO_2 (Eastburn *et al.*, 2011). Elevated CO_2 levels tend to reduce damage from ozone (von Tiedemann and Firsching, 2000). These effects often result in changed plant architecture and development of larger plant organs (Pritchard *et al.*, 1999). Because many foliar pathogens can take advantage of the more humid microclimate caused by denser plant growth and the higher availability of host tissue, pathogen infection rates usually increase at higher CO_2 levels. Carbon dioxide and ozone may also influence plant resistance (Plazek *et al.*, 2001). Fecundity of some pathogens may increase at elevated air CO_2 concentrations (Chakraborty and Datta, 2003), thereby accelerating the potential for accelerated pathogen evolution.

A recent review concerning the effects of CO_2 on plant disease reported increased levels of disease severity for 6 out of 10 pathosystems involving biotrophic fungi that were studied and decreased levels of disease severity in the other 4. Of the 15 necrotrophic fungi studied, an increase in disease severity was observed for 9, decreased disease severity was observed for 4 species, and severity levels were unchanged for 2 pathogens (Chakraborty *et al.* 2000b). These results indicate that the effect of increased CO_2 concentrations on a particular pathogen depends on the interaction between the effects of elevated CO_2 on the pathogen and the effects of this change on the specific plant under specific environmental conditions.

Similarly, it is difficult to generalize the effects of elevated ozone levels on disease (Sandermann, 2000). Elevated ozone concentrations may change the structure and properties of leaf surfaces (Karnosky *et al.*, 2002) in ways that may affect the inoculation and infection process (Karnosky *et al.*, 2002). Ozone enhances senescence processes, may encourage necrosis, and seems to promote attacks on plants by necrotrophic fungi (Sandermann, 2000). Ozone is an effective abiotic elicitor that influences secondary plant metabolism, as well as hypersensitive response and systemic acquired resistance (Heagle and Key 1973; von Tiedemann and Firsching, 1993; Sandermann, 2000).

10.4. CHANGES IN CROP LOSS

Diseases are responsible for losses of at least 10% of global food production, thus representing a threat to food security (Strange and Scott, 2005). The annual losses by disease are estimated to be US$ 220 billion (Agrios, 2005). Besides direct losses, the methods for disease control, especially the chemical ones, could lead to serious environmental contamination apart from the chemicals residues invading food chain and the social and economic tribulations. The close relationship between the environment and plant diseases suggests, that climate change would cause modifications in the current phytosanitary scenario. The impacts could be positive, negative or neutral, since there might be a decrease, an increase or no effect on the different patho-systems, in each region. The analysis of the potential impacts of climate change on plant diseases is essential for the approval of adaptation measures, as well as for the development of resistant cultivars, new control methods or adapted techniques are urgently needed in order to avoid more serious losses (Ghini, 2005).

As under current climate, crop loss from diseases in a changed climate condition would be determined by a large number of interacting factors that directly and indirectly influence plant diseases. Among direct effects, altered physiology and morphology of the host under elevated CO_2 would change the interception of light and precipitation and modify canopy structure and disease epidemiology. Some diseases could cause more severe reduction in plant growth under twice ambient compared to ambient CO_2, at least in controlled environments. For example, in barley powdery mildew, an acclimation of photosynthesis at elevated CO_2 and an infection-induced reduction in net photosynthesis caused larger reductions in plant growth at elevated CO_2 (Hibberd *et al.*, 1996). To protect potato crop from late blight, fungicide application would need to be extended by 10-20 days for each degree of warming. While the significance of such growth reductions on yield cannot be fully determined in the absence of field studies, their results suggest that predictions of bumper harvest due to CO_2 fertilization and increased water use efficiency, might be unrealistic. Among indirect effects, ozone could result in losses greater than those induced by sole pathogen, while the effect of UV-B is inconsistent (Manning and Tiedemann, 1995). However, increased severity under climate change does not always lead to increased losses (Luo *et al.*, 1995).

A recent modelling approach had been used to determine the potential impacts of climate change on the most important diseases of coffee, sugarcane, eucalyptus, cassava, citrus, banana, pineapple, cashew, coconut, mango and papaya, employing detailed knowledge of environmental conditions favourable for disease development and predicted climate change projected for the next several decades.

The potential impacts of climatic change on black Sigatoka (*Mycosphaerella fijiensis*), which is considered the most damaging and costly disease of banana in the world (Ploetz *et al.*, 2003), were measured using IPCC scenarios A2 and B2 to project distribution maps of the disease (Ghini *et al.*, 2007). These maps projected a reduction of the favourable area to the disease in Brazil, resulting from a reduction in relative humidity to levels below 70%. Such reduction is expected to be gradual for the 2020s, 2050s and 2080s and would be greater for scenario A2 than for B2. Despite that, these extensive areas would remain favourable to this disease, especially from November to April, which is currently the most preferable period.

10.5. VECTOR-BORNE DISEASES

Interaction between insect vector, disease and climate change is more complicated as compared to interaction of disease or insect alone with climate change because of disease dependency on the vector. Climate change is anticipated to alter the abundance and distribution of insect vectors. Both vectors and vector-borne diseases are likely to follow the migrating hosts and may infect remnant vegetation of natural plant communities which are not previously exposed to the often more aggressive strains from agricultural crops (Coakley *et al.*, 1999). The impact of climate change on insect vectors is complex as it could alter patterns of disturbance from vector insects through direct effects on their development and survival, adaptation capability, availability of host plants and physiochemical changes in host defences and indirect effects from changes in abundance of natural enemies, mutualists, and competitors (Porter *et al.*, 1991; Patterson *et al.*, 1999; Ayres and Lombardero, 2000).

The changes in insect vectors ecology is occurring evidently, which are consistent with predictions of climate change impacts (Harrington *et al.*, 2001). The changes in vector overwintering and over-summering are certain when a new vector species enters the crop fields and other effects of climate change on insects may have important influence on pathogen survival, movement and reproduction (Fabre *et al.*, 2005; Mondor *et al.*, 2005; Newman 2004, 2005). Introduction of glassy-winged sharp shooter has led to increased patterns of infection of grape plants in winter; greatly altering infection rates, is an excellent example of such impact (Almeida *et al.*, 2005). Temperature is the dominant climate factor, which directly effects overwintering and is potentially important in combination with photoperiod. There are little evidences of direct effects of CO_2, UV-B and precipitation on disease and insert vector interactions to draw a general conclusion (Bale *et al.*, 2002).

The vector diversity is the key to understand plant virus epidemiology. The plant disease dynamics may be more constrained by the host range of vectors than by the host range of viruses. The importance of vector diversity for pathogen epidemiology and disease dynamics is not well-understood. Many viruses are strongly dependent on the vectors that transmit them among hosts. Analysis of a large database of 1673 total plant viruses, including 910 unique virus species, indicated that plant viruses have much more specialized relations with their vectors than with their hosts. In general, plant viruses have a low diversity of vectors and a high diversity of hosts, and the largest proportion of viruses fall into the category of host generalist/vector specialist (Krishna Reddy, 2010). A majority of vector-transmitted viruses (58.4%) have a single known vector species, whereas only 9.9% of such viruses have a single known host plant.

Plant-infecting viruses operate in intimate association with their hosts and vectors. The risk of vector-borne disease at the local and regional level is limited by the climatic requirements of disease vectors (Malmstrom *et al.*, 2011). Climate change may affect both host plant and insect-vector populations, thereby affecting the spread of plant viruses (Jones, 2009). Global warming may influence the primary infection of the host, the spread of the infection within the host and/or the horizontal transmission of the virus to new hosts by the vector. Climate change may also affect the phenology and physiology of the host, which may affect its susceptibility to the virus and the ability of the virus to infect, as well as the geographic range and densities of alternative hosts/reservoirs. In their turn, the effects on host physiology may affect the attractiveness of the host to vectors and/or viral transmissibility. Modification of the geographic range of the potential vectors and/or vector phenology, as well as the vector's over-wintering; density, migration and activity can follow. Virus stability, replication and movement rates, as well as synergism and complementation between viruses can be affected. Although elevated CO_2 levels have been shown to have little direct effect on natural enemies of insect herbivores, they can influence the third trophic level indirectly, by altering the size and composition of populations of prey insects available to predators and/or by disrupting developmental synchrony for parasitoids (Sun *et al.*, 2011). Aphids are quite often efficient vectors of virus diseases.

The limited experimental evidence for the effects of elevated CO_2 levels indicates that higher concentrations of atmospheric CO_2 may increase plant resistance to certain viruses. In fact, the titre of viral coat protein in leaves of tobacco (*Nicotiana tabacum*) was markedly reduced under high CO_2 conditions (Matros *et al.*, 2006). A virus-resistant tobacco cultivar was found to be more sensitive to elevated ozone levels than a virus-sensitive cultivar (Ye *et al.*, 2012). Temperature increases may affect two types of antiviral resistance mechanisms in plants:

- Gene-silencing based on ribonucleic acid interference (RNAi).
- Some types of resistance based on protein–protein recognition.

RNAi defends plants against viruses by using small interfering RNA (siRNA) to target and destroy viral RNA. RNA silencing increases with increasing temperature, and it is manifested in the appearance of less symptomatic newly developed leaves (Singh *et al.*, 2004). The siRNA silencing of virus-generated double-stranded RNA (dsRNA) seems to be strongly affected by temperature. However, it is also possible that other mechanisms besides gene silencing reduces virus accumulation at high temperatures (Chellappan *et al.*, 2005). In protein-induced dominant resistance systems, a viral determinant (coded by the *AVR* gene) is recognized by the host, leading to the activation of resistance responses. This recognition usually induces a localized hypersensitive response that limits the virus to the initial site of infection. Some types of resistance are specifically temperature dependent. In such cases, increased temperatures negatively affects the inhibition of viral replication, so it actually breaks down the host's resistance (Fraser and Loughlin, 1982).

A number of studies have been conducted to investigate the effects of increased CO_2 and ozone levels and increased temperatures on viral vectors. The most common vectors of plant viruses are aphids, whiteflies, and leafhoppers. Some nematodes and mites can also transmit viruses. Only limited data are available concerning the effects of climatic factors on nematode vectors of viruses, except for the effects of temperature on their biology and numbers (Feil *et al.*, 1997). There is almost no information available concerning mites as virus vectors. Aphids are widespread throughout all temperate areas of the world, whereas whiteflies are confined to warmer regions. Compared with aphids, which are generally good nonspecific virus vectors, only a few whiteflies (e.g., *Trialeurodes vaporariorum* and *Bemisia tabaci*) have been shown to transmit plant viruses. Aphids and whiteflies will probably do well in a warming climate, thanks to the shorter developmental times and higher number of generations per year that they exhibit following a rise in temperature (Yamamura and Kiritani, 1998). Increases in temperature greatly affect the life cycles of both aphids and whiteflies (Muñiz and Nombela, 2001). Warmer autumns and winters will increase the risk for insect transmission of viruses into winter cereal crops. In northern countries, these crops are now sown when the number of active insect vectors is relatively low. Under increased temperatures, there may be a higher risk of virus infection because the vector populations will be large by the time the crop is planted (Roos *et al.*, 2011). Higher temperatures may also increase the size of overwintering populations and induce vector populations to migrate northward (Cannon, 1998). Some bacterial endosymbionts of aphids increases the tolerance of aphids to high temperatures (Dunbar *et al.*, 2007). Insects possessing these specific endosymbionts may be at an advantage in a warming climate.

Increased CO_2 concentrations have different effects on aphids and whiteflies, suggesting that we cannot generalize the effects of these conditions across different pest–plant systems (Bezemer *et al.*, 1998; Cannon, 1998; Sun *et al.*, 2011). These discrepancies can probably be attributed to the fact that the effects of CO_2 on vectors are mostly indirect (i.e., changes in carbon fixation rates in plants at high CO_2 levels affect C/N ratio or increased levels of phenolics and terpenes). Interactions between increased CO_2 levels and higher temperatures have also been noted in some insect-plant systems (Zvereva and Kozlov, 2006).

Modelling the complex interactions of viruses, plants, vectors, and their natural enemies in the face of climate change is quite a challenging task. However, projections have been carried out for some case studies. A decline in populations of the cereal aphid *Rhopalosiphum padi* is predicted in a warmer southern Britain by the end of this century (Newman *et al.*, 2003; Newman, 2005), as well as in parts of Canada (Newman, 2006). Effects of high temperatures on the growth and reproduction of corn aphids are predicted to affect the populations of this insect in the north-eastern Iberian Peninsula (Asin and Pons, 2001). In northern countries, the changing climate is projected to lead to higher temperatures, implying extended growing seasons and opportunities for the introduction of new crops, as well as new opportunities for insect vectors (Roos *et al.*, 2011).

The reliability of climate change projections strongly affects the prediction of changes in the distribution and abundance of vectors and the subsequent risk of virus transmission. Different climatic models may yield different predictions, with all of the consequential implications (Newman, 2006).

10.6. DISEASE PATHOGENS

10.6.1. Viral pathogens

Most of the climate change studies have been conducted on diseases caused by fungal pathogens. Viral diseases mostly have been ignored. Only few studies have reported the response of plants infected with viral diseases to various climate change components. It has been observed that oats infected with Barley yellow dwarf virus (BYDV) showed greater biomass accumulation to CO_2 enrichment than the healthy plant. Tobacco plants grown at increased CO_2 concentrations showed a markedly decreased spread of virus. It appears that CO_2 rise in the air may have some positive effects, which may likely offset the negative effects of virus infection. However, generalization is difficult without much information from different virus-host infections.

Average temperature increase of 0.7°C over the last hundred years and average of 6.3 °C predicted by the year 2100 will dramatically affect both mobile (fauna) and immobile (flora) organisms, resulting in both altered and novel forms of interactions between plants, pathogens and their vectors. Most plant viruses are transmitted by vectors and majority by insects. Particularly aphids are expected to react strongly to environmental changes because of their short generation time, low developmental threshold temperatures and ability to survive mild winters without winter storms. An increase in the number of insect vectors will inevitably lead to a higher risk for viral infection of plants. The aphid transmissible complex of barley yellow dwarf viruses in cereals and potato virus Y in potato are amenable to show potential effects on the prevalence of infection because of climate change. In mild winters, high intensity of aphid movement during spring and a high frequency of PVY-infected potatoes have been reported. The severity of viral diseases is determined in large part by the amount of inoculums and the time of infection. The amount of virus inoculum is influenced by winter survival of its hosts. For some viruses, higher temperatures also cause more severe symptoms development. Aphids are expected to have increased survival with milder winter temperatures, and higher spring and summer temperatures will increase their development and reproductive rates and lead to more severe diseases. Milder winters are also expected to increase survival of alternate weed hosts of viruses. Increases in frequency and intensity of summer storms with high winds, rain, and hail will increase wounding of plants and result in increased transmission of viruses by mechanical means. Therefore, with predicted changes in climate, viral diseases of plants are expected to increase in importance. Potentially of greater importance will be the effects of diseases caused by newly, introduced viruses that, because of the changed climate, will be able to persist. A warmer climate might also allow viruses that are present in greenhouses, such as Pepino mosaic virus (PepMV), to establish infection in the field. The main effect of temperature in temperate regions is to influence winter survival of vectors. Natural spread of vectors, pests and diseases is accelerated towards the North, as former climate barriers are no longer effective. This results in more severe outbreaks of plant-disease vectors like aphids, whiteflies, thrips or beetles, an extension of the period of disease infection further into the growing season and also introduction and establishment of new vector species. The described effects on vectors can have severe negative effects on food production or result in an increased use of plant protection products to control the vectors.

10.6.2. Bacterial pathogens

Milder and shorter winters are expected to have little effect on soil-borne bacterial pathogens; however, survival of host or debris-borne and vector-borne primary inoculum is expected to increase. Soil-borne, bacterial plant pathogens, such as *Agrobacterium tumefaciens*, may build up their populations in host plants and could be released into the soil where they can survive as primary inoculum in the next season. Host or debris-borne bacteria survive on and in host tissues. On perennial hosts, bacteria, such as *Erwinia amylovora* on apple, overwinter on infected host tissue, and primary inoculum is spread from host to host in the next season. On annual hosts, bacteria such as *Pseudomonas syringae* pv. *phaseolicola* may survive in host debris in soil or on the soil surface. Vector-borne bacterial pathogens, such as *Erwinia stewartii*, survive in insect vectors, and these vectors act as the source of primary inoculum in the next season.

Bacterial pathogens, such as *Pseudomonas syringae* pv. *tomato* and *Xanthomonas campestris* pv. *vesicatoria*, arise from infected seed and possibly also survive in debris, soil, and weeds. Bacteria are spread to their host plants mainly by water, usually in the form of rain splash, and insects. In humid, wet conditions, infected plant tissues can exude masses of bacteria that are spread from host to host by rain splash and insects. Therefore, the warmer drier summers expected with climate change should limit bacterial diseases. However, bacteria often enter hosts through wounds and the expected increase in frequency and intensity of summer storms with high winds, rain, and hail will increase wounding of plants and provide moisture for the spread of bacteria.

10.7. DISEASE COMPLEXES

Taking a traditional pathologist's approach, we might focus on pathogens, the 'causal agents' of disease and the effect of climate change on them. However, crop loss is attributable to direct and indirect effects of pathogens together with the effects of other biotic and abiotic agents. Direct effects of pathogens or other organisms can be the induction of resistance or susceptibility and its associated cost or benefit to the host plant (Newton *et al.*, 2010a). Disease symptoms such as development of necrotic lesions can be due to a combination of a pathogen or pathogenic complex and in some circumstances non-pathogens of that crop such as bacterial species, exploiting an advantageous trophic environment (Newton and Toth, 1999; Newton *et al.*, 2004). Each component of the host/pathogen/ microbe interaction and the pathogenic complex will have a different interaction with the environment, and thus will change under different climates (Stonard *et al.*, 2008; Xu and Nicholson, 2009). However, such associations of organisms in the disease complex are very variable due to their opportunistic nature and therefore represent another area of uncertainty in predicting the effects of climate change.

Many climate change experiments involve the manipulation of single parameters such as temperature or CO_2 and therefore do not simulate likely change scenarios where several parameters change simultaneously. These parameters will probably affect the host, disease and pathogen/pest complexes, in different ways through different mechanisms. Nevertheless, some trends can be shown that highlight where more resilience needs to be built into the system. Furthermore, the effects of many stresses are interactive, so it is frequently not possible to generalize.

For the major arable crops, particularly under intensive monoculture, a few pathogens often dominate. Epidemics can be both fast and severe where epidemiologically successful races of pathogens have adapted to the prevailing resistance mechanisms deployed. These are likely to have the most damaging economic impact, particularly rusts on cereals that are likely to have a large impact in terms of food security under climate change scenarios. The fact that the above examples are from monocultures under high input situations underlines the vulnerability of such systems to change.

10.8. IMPACT ON CROP YIELD

The major emphasis of climate change/food security research over recent years has been concerned with the impacts of climate change on crop yield. The grey leaf blight of corn caused by the fungus *Cercospora zeae-maydis* ranks number one in causing yield losses of corn in recent years. Gregory *et al.* (1999) summarized experimental findings on wheat and rice crops that indicated decreased crop duration (and hence yield) of wheat as a consequence of warming, and reductions in yields of rice of about 5% per °C rise above 32° C. More recently, the modification of crops and management systems to cope with changed temperatures demonstrated what adaptations might be possible. For example, simulation of production for cropping systems in northern and central Italy showed that the combined effects of increased CO_2 and climate change would depress crop yields by 10-40% if current management practices were unamended, largely because the warmer air temperatures would accelerate the phenology of current cultivars (Tubiello *et al.*, 2000). Through a combination of early planting of spring and summer crops and the use of slower-maturing winter cereal cultivars, though, the model indicated that it should be possible to maintain present yields. However, a major caveat to this conclusion was that 60-90% more irrigation water was required to maintain grain yields under conditions of climate change; this water was assumed to be available (Tubiello *et al.*, 2000). More recently, elements of crop adaptation to extreme weather events have been explored with genotypic variation and adaptation able to compensate for several of the negative impacts on unadapted productivity (Fuhrer, 2006; Challinor *et al.*, 2007).

Extreme events may also become more frequent in the future increasing the risks to crop yields, although this is not always so and is often dependent on the precipitation scenario that is considered (Porter and Semenov, 2005). The effect of extremes is illustrated by events in the summer of 2003 in parts of Europe, where temperatures were 6° C warmer than long-term means and precipitation deficits of up to 300 mm were recorded (Tubiello *et al.*, 2007). As a consequence, parts of the European Union such as the Po valley in Italy, reported a record reduction in maize yield of 36% (Ciais *et al.*, 2005).

In one of the most comprehensive studies using three different climate change scenarios and three general circulation models, Rosenzweig and Parry (1994) modelled the impact on wheat, maize, rice and soybean. The crop models were run for changes in +2 and +4° C, 20% precipitation and for a doubling of CO_2. Average crop yields at +2° C warming increased by 10-15% in wheat and soybeans, and by 8% in rice and maize. Yields of all four crops were reduced at +4° C warming indicating a threshold of compensation for the direct effect of CO_2.

Similarly, in sorghum, yield increases due to CO_2 are masked by effects of elevated temperature resulting in an overall reduction in yield in drier regions of India (Rao *et al.*, 1995).

A doubling of CO_2 has been conclusively shown to increase C_3 crop yield by about 33% and C_4 crop yield by about 10% in over 1000 studies (Kimball, 1985; Cure and Acock, 1986). Whether these benefits would still be realized in the presence of pests and diseases and other limiting factors is largely unknown.

10.9. DISEASE OUTBREAKS

Changes in rainfall patterns, temperature and wind direction could result in the establishment and emergence of new pests and diseases threatening agriculture and food security.

- Samoa's experience with outbreak of taro leaf blight (Fig. 10.1) in 1993 - 95% reduction in taro production
- Taken 16 years for breeding resistant varieties and resume exports.

Fig. 10.1. Taro leaf blight

Among the earliest attempts to relate historic records of meteorological conditions and crop pest damage were the studies of potato leaf roll outbreaks in North America and Europe. Analysis of the historic records from 1930 to 1991 suggests that the outbreaks of this aphid-borne viral disease were related to drought and sunspot cycles. In the US, the frequency of the reported outbreaks seems to have increased since 1970.

Another example of the linkage between meteorological variables and pests is the wheat stem rust disease in the US Great Plains. The epidemics of the disease from 1921 to 1962 seems to be related to the conditions during El Niño episodes (Yang and Scherm, 1997). In contrast, wheat stripe rust epidemics in the US Northwest may be more severe during La Niña years (Scherm and Yang, 1995).

10.10. BREAKDOWN OF HOST PLANT RESISTANCE

While physiological changes in host plants may result in higher disease resistance under climate change scenarios, host resistance to disease may be overcome more quickly by more rapid disease cycles, resulting in a greater chance of pathogens evolving to overcome host plant resistance.

Elevated temperatures may cause the breakdown of temperature-sensitive resistance in oat cultivars with Pg3 and 4 genes (Martens *et al.*, 1967). Even if temperature change is well within the limits of current climatic variability, a modest warming can cause significant increase in heat that sums above a critical temperature threshold to affect crop physiology and resistance to disease.

For sexually reproducing pathogen populations with broad genetic diversity, increased population size and the number of generations in favourable microclimates would increase the probability of more damaging pathotypes evolving more rapidly (Sutherst *et al.*, 1996b).

10.11. EMERGENCE OF NEW AND INTENSE DISEASE PROBLEMS

In recent decades, wheat Fusarium Head Blight (FHB) caused by *Fusarium graminearum* and *Fusarium culmorum* has re-emerged as a disease of global significance, causing yield loss and price discounts as a result of reduced grain quality, costing an estimated US$2.7 billion in the northern Great Plains and central USA from 1998 to 2000 (Goswami and Kistler, 2004). The production of trichothecene mycotoxin and oestrogenic zearalenone in infected host tissue, responsible for a loss of grain quality, is harmful to humans and animals, although fungal infection alone can reduce grain quality too. Trichothecenes have been associated with chronic and fatal toxicoses of humans and animals (Desjardins, 2006).

Weather is the most significant factor in determining incidence, severity and the relative importance of Fusarium Head Blight (FHB) and crown rot (CR) diseases. The greatest risk from FHB came from the predicted increase in the number of rainy days coinciding with critical wheat growth stages during the September–November period. Yield loss from CR is severe when there is a post-anthesis drought, when the restriction of the flow of water to the spike tissue by the pathogen causes 'white heads' with shrivelled or no grains (Chakraborty *et al.*, 2006; Luck *et al.*, 2011). FHB, on the other hand, is favoured by warm and wet weather at anthesis and causes partial or complete blighting of the head, reduced yield and quality (shrivelled kernels), reduced testa weight and bread-making quality and the production of one or more mycotoxins. The quantitative relationships between weather, cropping practices and mycotoxin concentrations (Champeil *et al.*, 2004) forms the basis of mycotoxin forecasts such as DONCAST (Schaafsma and Hooker, 2007). In general, the potential for high concentrations of mycotoxin in grains generally increases with the number of rainy days and days with relative humidity >75%, but decreases with temperature < 12 or 32° C (Schaafsma and Hooker, 2007).

The greatest risk from FHB came from the predicted increase in the number of rainy days coinciding with critical wheat growth stages during the September–November period.

10.12. INCREASED PESTICIDE USAGE

As world agriculture responds to challenges of securing sufficient, safe and nutritious food for the ever-expanding human population under changing climate, no doubt pesticide usage will increase even more.

The more frequent rainfall events predicted by climate change models could result in farmers finding it difficult to keep residues of contact fungicides on plants, triggering more frequent applications.

Systemic fungicides could be affected negatively by physiological changes that slow uptake rates, such as smaller stomatal opening or thicker epicuticular waxes in crop plants grown under higher temperatures.

Effectiveness of most plant protection chemicals depends on prevailing climatic conditions. Changed duration, intensity and frequency of rainfall events would impact on the effectiveness of chemical control measures. Data from field experiments and modelling studies suggests precipitation during the post-application period (Wagenet *et al.*, 1994; Schepers, 1996) is critical. Precipitation following fungicide application may improve its distribution (Schepers, 1996) but an increase in rainfall intensity can deplete fungicide residue on the foliage (Neuhaus *et al.*, 1974).

10.13. FOOD SECURITY

Food security can be defined as "when all people, at all times, have physical and economic access to sufficient, safe, and nutritious food to meet their dietary needs and food preferences for an active and healthy life" (FAO, 2003) or "fair prices, choice, access through open and competitive markets, continuous improvements in food safety, transition to healthier diets, and a more environmentally sustainable food chain" (Anonymous, 2008a), although a simpler definition could be 'the risk of adequate food not being available'. It is a combination of multiple food availability, food access and food utilization issues. Each of these is influenced by climate change. This has resulted in many food security 'hotspots' around the world.

The impacts of diseases on yield in current conditions are well known, but the consequences of climate change on diseases are complex and, as the preceding descriptions attest, are still only imperfectly understood. Scherm *et al.* (2000) highlighted the importance of diseases as important yield-reducing factor and as early indicators of environmental changes because of their short generation times, high reproductive rates and efficient dispersal mechanisms. It is already clear that some diseases will be able to invade new areas and become increasingly problematic for the maintenance of biodiversity, the functioning of ecosystems and the profitability of crop production. Some diseases which are already present, but only occur in small areas, or at low densities, may be able to exploit the changing conditions by spreading more widely and reaching damaging population densities.

To understand how best to control plant diseases, to improve food security in the context of climate change, crop protection professionals must work with societal change, defining its key processes and influencers to effect change. More specifically, there is a key role to play in improving food security. Crop pests and diseases could potentially deprive humanity of up to 82% of the attainable yield in the case of cotton and over 50% for other major crops (Oerke, 2006) and, combined with postharvest spoilage and deterioration in quality, these losses become critical, especially for resource-poor regions. Actual average losses for rice in the period 2001–2003 totalled 37.4% including 10.8% to pathogens and 1.4% to viruses (Oerke, 2006). Each year an estimated 10–16% of global harvest (Strange and Scott, 2005; Oerke, 2006) is lost to plant diseases. In financial terms, disease losses cost US$ 220 billion. There are additional post-harvest losses of 6–12%; these are particularly high in developing tropical countries lacking infrastructure (Agrios, 2005) and consequently are difficult to estimate. As attested by the infamous 19th century Irish potato famine (Fry, 2008) or the Bengal famine (Padmanabhan, 1973), devastations from plant diseases can be far reaching and alter the course of society and political history.

The 'fertilization effect' of increasing CO_2 increases crop biomass and grain yield (Ainsworth and Long, 2005) raising the possibility of increased food production. However, emerging evidence of reduced grain yield from high temperature and water limitations (Anwar *et al.*, 2007; Torriani *et al.*, 2007) makes a wholesale increase in crop productivity unlikely. Also, the impacts of plant diseases, mostly ignored in assessments of global food security under climate change, minimizes or reverses any benefit from the CO_2 fertilization effect (Fernandes *et al.*, 2004; Butterworth *et al.*, 2010). Nevertheless, grain production has doubled over the last 40 years as a consequence of changes in plant protection and other agricultural technology, including a 15–20-fold increase in pesticide use worldwide. Despite this, the overall proportion of crop losses has increased during this period and excessive use of insecticides has increased pest outbreaks and losses in some crops and areas (Oerke, 2006). As world agriculture responds to challenges of securing sufficient, safe and nutritious food for the ever-expanding human population under changing climate, no doubt pesticide usage will increase even more. Identifying key constraints to food security, primarily from a production perspective, there is a need to highlight how improving plant disease management can enhance global food security. The key disease management issues have to be addressed in improving food security in a changing climate.

Plant diseases are a major problem not only for food production but also for the quality and safety of important foodstuffs. In Europe, mycotoxins and pesticide residues are among the top food safety concerns associated with a changing climate. For example, the concentration of mycotoxin produced by *Fusarium* head blight in grain generally increases with the number of rainy days and days with high RH but decreases with low and high temperatures. Changes in both temperature conditions and atmospheric composition may influence the severity of outbreaks of *Fusarium* head blight and the production of mycotoxins. The most significant effects on mycotoxin presence generally occur during the production phase, but the entire wheat value chain can be affected (Chakraborty and Newton, 2011).

Many assessments of climate change effects on crops have focused on potential yields (e.g. Fischer *et al.*, 2001) but factors such as pathogens have major effects in determining actual yields (Gregory *et al.*, 1999). The importance of including pathogens can be seen, for example, in the work of Butterworth *et al.* (2010). They combined a crop yield simulation model with a weather-based epidemiological model and showed that under 2020 and 2050 climate change scenarios fungicide protected oilseed rape will increase in yield in Scotland by 15%, whereas in England yields could decrease by 50%.

Ingram *et al.* (2008) pointed out that more mechanistic inclusion of disease effects on crops would lead to more realistic predictions of crop production on a regional scale and thereby assist in the development of more robust regional food security policies. Economic and social issues are often dominant determinants of household food security, but changed disease incidence under climate change is an essential influence on future regional food security.

Another aspect of disease control is the CO_2 cost. Yield has a cost in terms of inputs that can be calculated in CO_2 equivalents per hectare. It therefore, stands to reason that preventing yield loss increases efficiency that can be given a value. If it is achieved through use of crop protectants, their costs of production and application must be deducted and similarly any additional cost of elite germplasm, if it is achieved through plant resistance. It is calculated that control of disease in oilseed rape in the U.K. by fungicides reduces emissions by 100 kg CO_2 eq. t^{-1} of seed (Mahmuti *et al.*, 2009).

For wheat, it is estimated that in the U.K. fungicides save 59 kg CO_2 eq. t^{-1} of grain and that a further 14 kg CO_2 eq. t^{-1} could be achieved if all diseases were eliminated by fungicides or resistance that carries no yield penalty (Berry *et al.*, 2008). Therefore, efficient disease control is important for both adaptation to and mitigation of climate change.

10.14. REDUCED EFFICACY OF DISEASE MANAGEMENT STRATEGIES

While physiological changes in host plants may result in higher disease resistance under climate change scenarios, host resistance to disease might be overcome more quickly by more rapid disease cycles, resulting in a greater chance of pathogens evolving to overcome host plant resistance. Fungicide and bactericide efficacy may change with increased CO_2, moisture, and temperature. The more frequent rainfall events predicted by climate change models could result in farmers finding it difficult to keep residues of contact fungicides on plants, triggering more frequent applications. Systemic fungicides could be affected negatively by physiological changes that slow uptake rates, such as smaller stomatal opening or thicker epicuticular waxes in crop plants grown under higher temperatures. These same fungicides could be affected positively by increased plant metabolic rates that could increase fungicide uptake. It is not well understood how naturally-occurring biological control of pathogens by other microbial organisms could change as populations of microorganisms shift under changed temperature and moisture regimes – in some cases antagonistic organisms may out-compete pathogens while in others pathogens may be favoured. Exclusion of pathogens and quarantines through regulatory means may become more difficult for authorities as unexpected pathogens might appear more frequently on imported crops.

New dimensions of climate change may add extra uncertainty in management strategies for diseases caused by different pathogens. Delayed planting to avoid a pathogen may become less reliable. There may be problems with applications of bio-control agents in the field because of the vulnerability of bio-control agent populations to environmental variations and environmental extremes. If appropriate temperature and moisture are not consistently available, bio-control agent populations may reach densities that are too small to have important effects, and may not recover as rapidly as pathogen populations when congenial conditions reoccur.

Effectiveness of most plant protection chemicals depends on prevailing climatic conditions. Changed duration, intensity and frequency of rainfall events would impact on the effectiveness of chemical control measures. Data from field experiments and modelling studies suggests precipitation during the post-application period (Wagenet *et al.*, 1994; Schepers, 1996) is critical. Precipitation following fungicide application may improve its distribution (Schepers, 1996) but an increase in rainfall intensity can deplete fungicide residue on the foliage (Neuhaus *et al.*, 1974). In addition to influences of the physical climate on biological control agents, changes in atmospheric composition will modify the community structure of phyllosphere and rhizosphere microflora (Manning and Tiedemann, 1995; Ayres *et al.*, 1996). In the rhizosphere, elevated CO_2 would interact with nitrogen and other soil factors to modify the number and type of mycorrhizal fungi to influence root health and nutrient uptake. Some short-term studies under controlled conditions have shown that elevated CO_2 can stimulate mycorrhizal colonization (Staddon and Fitter, 1998) due to faster plant growth. Colonization of roots by arbuscular mycorrhizal fungi is favoured in soils of poor nutritional status (Klironomos *et al.*, 1997). It is not clear if increases in soil carbon storage due to greater root and mycorrhizal growth under high CO_2 will influence mycorrhizal colonization. Climate change will modify host-pathogen interaction to determine the outcome of many diseases. Consequently, the greatest impact of climate change will be on management strategies that utilizes host resistance.

Modification in Rubisco biochemistry, stomatal physiology, anatomy, morphology and phenology is well known under elevated CO_2 (Bowes, 1993; Wolfe, 1995). Many of these changes in host physiology can potentially enhance host resistance. Significant increase in rates of net photosynthesis allows increased mobilization of resources into host resistance at elevated CO_2 (Hibberd *et al.*, 1996a). Other changes, including production of papillae and accumulation of silicon at sites of appressorial penetration (Hibberd *et al.*, 1996a); greater accumulation of carbohydrates in leaves; more waxes, extra layers of epidermal cells and increased fibre content (Owensby, 1994); lowered nutrient concentration, leading to partitioning of nitrogen from photosynthetic proteins to metabolism that is limiting to plant growth (Baxter *et al.*, 1994); and greater number of mesophyll cells (Bowes, 1993) can all influence host resistance. Reduced pathogen penetration results from a reduction in stomatal density (Chakraborty *et al.*, 2000) and stomatal conductance at high CO_2 (Hibberd *et al.*, 1996b). Thompson *et al.* (1993) reported a significant reduction in wheat powdery mildew at twice-ambient CO_2, but final severity was dependent on nitrogen and water status of plants. Increased lignification in some forage species at high temperatures (Wilson *et al.*, 1991) can potentially enhance the level of resistance (Strange, 1993).

Whether changes in host physiology would equally influence susceptible and resistant cultivars or resistance in both traditionally bred and transgenic cultivars is not well known. In the tropical legume, *Stylosanthes scabra*, reduction in plant growth due to *Colletotrichum gloeosporioides*-induced anthracnose was more than compensated by growth enhancements at elevated CO_2 in a susceptible but not in a resistant cultivar. Anthracnose severity in the susceptible cultivar was reduced at elevated CO_2 but not in the resistant cultivar (Chakraborty *et al.*, 2000a). This suggests that some forms of resistance may be less effective under climate change.

Stress induced by extremes such as drought can exacerbate disease by species of *Armillaria* (Rishbeth, 1991), and the endophytic *Cryptostroma corticale* can rapidly invade vascular and bark tissues in hot dry summers (Dickenson and Wheeler, 1981).

The biggest threat to the durability of host resistance would come from accelerated pathogen evolution. Changes will occur at all stages in the pathogen life cycle under elevated CO_2. Despite initial delays and reduction in host penetration, established colonies grow faster inside host tissues at elevated CO_2 (Lupton *et al.*, 1995; Hibberd *et al.*, 1996a; Chakraborty *et al.*, 2000a). Fecundity of both biotrophs and necrotrophs (Chakraborty *et al.*, 2000b) studied so far has increased under elevated CO_2. A combination of increased fecundity and a favourable microclimate within enlarged canopies will provide more opportunities for infection. There is evidence of adaptation for increased aggressiveness in some pathogens (Kolmer and Leonard, 1986) within three sexual generations and controlled crossing has shown that aggressiveness is heritable and may be polygenically controlled (Caten *et al.*, 1984). For sexually reproducing pathogen, populations with broad genetic diversity, increased population size and the number of generations in favourable microclimates would increase the probability of more damaging pathotypes evolving more rapidly (Sutherst *et al.*, 1996).

10.15. CONCLUSIONS

If changes in atmospheric composition and global climate continue in the future as predicted, there will be relocation of crops and their diseases and impacts will be felt in economic terms from crop loss. Changes in levels of CO_2, ozone and UV-B will influence diseases by modifying host physiology and resistance. In addition, changes in temperature, precipitation and the frequency of extreme events will influence disease epidemiology. Changes in geographical distribution will potentially alter the relative importance and spectrum of diseases and new disease complexes may arise. Evolution of pathogen populations may accelerate from enhanced UV-B radiation and/or increased fecundity in elevated CO_2. As a result, host resistances may be overcome more rapidly. Disease management will be influenced due to altered efficacy of biological and chemical control options. Information gathered so far has been fragmented and a comprehensive analysis of climate change impacts on diseases is not possible with present knowledge. Experimental research on a diverse range of disease systems is necessary to improve comprehension of climate change impacts. Given the multitude of atmospheric and climatic factors, possible change scenarios and the number of disease systems, modelling approaches to impact assessment need to be strengthened. For instance, changes in both mean temperature and its variability are equally important in predicting the potential impact of climate change (Scherm and van Bruggen, 1994). Given that climate change is a global issue, the focus needs to shift from paddock-based assessment on specific diseases to a more ecologically relevant spatial unit (Scherm *et al.*, 2000) to consider climate with other associated changes in land use and vegetation cover (Luo *et al.*, 1995), among others.

11

Development of Prediction Models for Plant Pathogens

11.1. INTRODUCTION

The use of data series and simulation models could assist projecting future climate change impacts on pathogens and the diseases they cause (Krengel *et al.*, 2014). It is particularly important to collect long-term data sets for weather parameters, crop development, and pathogen/ disease distribution/ and prevalence to be able to develop and validate linked 'disease-crop-climate' models (Madgwick *et al.*, 2011). According to Shaw (2009), disease forecasting models to be used for long-term climatic projections should at least be able to make reliable short-term prediction. However, even the best forecasting models usually cannot address fundamental and yet unknown evolutionary developments of pathogen populations. In addition, technological advancements in breeding, crop management, and crop protection will contribute to altered future disease risks as well (Garrett *et al.*, 2006; Siebold and Tiedemann 2012, 2013; Elad and Pertot, 2014).

11.2. DEVELOPMENT OF PREDICTION MODELS

Predictive models have been developed for a few plant pathogen systems viz., maize (EPICORN – for Southern corn leaf blight), tomatoes and potatoes (EPIDEM, TOMCAST, BLITECAST – for early and late blights), apple (Maryblight, EPIVEN – for fire blight and scab), etc (Tables 11.1 to 11.3).

Powdery mildew, *Erysiphe necator* is one of the most important diseases of grapevine. The European grapevine moth (*Lobesia botrana*) is one of the most noxious vineyard pests in the European and Mediterranean regions. Caffarra *et al.* (2012) combined detailed phenological models of grapevine powdery mildew and the European grapevine moth, applied the models to climate change scenarios for the eastern Italian Alps and considered potential changes in the interactions between these species (Fig. 11.1).

Table 11.1. Crop disease modelling under climate change

Disease	Pathogen	Model components	Comments	Reference
Various fungal diseases on various hosts	*Fusarium oxysporum* f. sp.	Two GCM → CLIMEX	Climate change impacts on global distribution of a pathogenic species complex.	Shabani *et al.*, 2014
Fusarium head blight on wheat	*Fusarium* sp.	One GCM → simulated weather + crop model + disease model.	Climate change impacts in China.	Zhang *et al.*, 2014
	Fusarium culmorum	11 GCM + ensemble + anthesis model + mycotoxin model.	Climate change impacts on mycotoxin levels in Scotland.	Skelsey and Newton, 2015
Brown rust on wheat	*Puccinia recondita*	15 GCM → simulated weather + disease model.	Climate change impacts in Luxembourg.	Junk *et al.*, 2016
Six soil-borne fungi: three affecting cereals, three affecting spring-sown herbaceous crops	*Fusarium nivale*, *F. culmorum*, *Bipolaris sorokiniana*, *Pythium ultimum*, *Sclerotinia minor*, *Macrophomina phaseolina*	One GCM + soil conditions model + disease model.	Climate change impacts in Europe.	Manici *et al.*, 2014
Leaf blast on rice Leaf blight on rice	*Magnaporthe oryzae* *Xanthomonas oryzae* pv. *oryzae*	One GCM → simulated weather + crop model + disease model.	Climate change impacts in Tanzania. Same disease can increase in severity in some areas and decrease in others.	Duku *et al.*, 2015

Disease	Pathogen	Model components	Comments	Reference
Leaf blast on rice Leaf blight on rice	*Magnaporthe oryzae* *Xanthomonas oryzae* pv. *oryzae*	11 GCMs and ensemble → simulated weather + disease model.	Climate change impacts in South Korea.	Kim and Cho, 2015
Phoma stem canker on oilseed rape Brown rust on wheat Net blotch on barley Downy mildew on grape Potato late blight	*Leptosphaeria maculans*, *Puccinia recondite*, *Pyrenophora teres* *Plasmopara viticola* *Phytophthora infestans*	One GCM → simulated weather + infection model.	Climate change impacts in France for five foliar pathogens.	Launay *et al.*, 2014
Downy mildew on grape	*Plasmopara viticola*	One GCM → simulated weather + crop model + disease model.	Climate change impacts in France.	Caubel *et al.*, 2014
Potato late blight	*Phytophthora infestans*	Three GCM → monthly means + crop model + disease model.	Global climate change impacts.	Sparks *et al.*, 2014
	Phytophthora infestans	Weather data + 3 disease models.	Not strictly a climate change impact paper but a comparison of different disease models.	Baker *et al.*, 2015

Table 11.2. Controlled environment (CE), glasshouse and free-air CO_2 enrichment (FACE) experiments for parameterization of crop and disease models through temperature, CO_2 and ozone effects (Challinor *et al.*, 2014).

Crop/ Disease and pathogen	Climate change aspects			Experimental facility	Comments	Reference
	Temp	CO_2	Ozone			
Coffee rust (*Hemileia vastatrix*), Cercospora leaf spot (*Cercospora coffeicola*)		✓		FACE	No effect of elevated CO_2 on disease incidence	Ghini *et al.*, 2015
Barley powdery mildew (*Blumeria graminis*), Spot blotch (*Bipolaris sorokiniana*)	✓	✓	✓	CE	Elevated CO_2, O_3 and temperature, when applied in isolation, gave different effects on the two diseases. Unexpected interactions between elevated CO_2, O_3 and temperature	Mikkelsen *et al.*, 2015
Barley yellow dwarf virus	✓			CE	Elevated temperature increased symptoms	Nancarrow *et al.*, 2014
Maize (*Fusarium verticillioides*)		✓		CE	Elevated CO_2 increased maize susceptibility and fungal biomass. Mycotoxin levels unaltered	Vaughan *et al.*, 2014
Wheat crown rot (*Fusarium pseudograminearum*)	✓	✓		Glasshouse	Elevated CO_2 increased disease severity	Melloy *et al.*, 2014
	✓			Glasshouse	Elevated temperature reduced disease severity	Váry *et al.*, 2015

Table 11.3. Studies on crop disease simulation models in several crops under climate change

Crop/disease	Projected crop disease risk	Geographic Scope/Time horizon of projection	Adaptation options	Reference
Apple fire blight, *Erwinia amylovora*	Number of infection days is projected to remain similar in several sites investigated, except one site where the number of infection days is projected to slightly increase	Switzerland 2060s	Not discussed	Hirschi *et al.*, 2011
Banana black Sigatoka disease, *Mycosphaerella fijiensis*	Area favourable is projected to move from the north to the central region of Brazil. Area favourable is projected to decrease	Global/Brazil 2020s, 2050s, 2080s	Not discussed	Ghini *et al.*, 2007; Jesus Junior *et al.*, 2008; Alves *et al.*, 2011
Coffee rust, *Hemileia vastatrix*	Area favourable is projected to move toward the south of Brazil, the favourable area will decrease. Duration of the incubation period is projected to decrease, suggesting that disease progress will be favoured	Brazil 2020s, 2050s, 2080s	Defining public policies targeted at mitigating and/or adapting to a new climate scenario	Alves *et al.*, 2011 Ghini *et al.*, 2011
Grapevine downy mildew, *Plasmopara viticola*	Severity is projected to increase, earlier start of epidemics also projected	Italy 2030s, 2050s, 2080s	More attention to early downy mildew management, improved fungicide spray timing and frequency	Salinari *et al.*, 2006

Crop/disease	Projected crop disease risk	Geographic Scope/Time horizon of projection	Adaptation options	Reference
	Disease outbreak is projected to be earlier in the future, this could also result to more severe epidemics due to the polycyclic nature of the respective pathogen	Global (e.g. Argentina, Australia, Chile, USA, South Africa, Spain) 2030s, 2050s, 2080s	Timing and frequency of improved fungicide treatments	Salinari *et al.*, 2007
	Number of infection events are projected to increase	Europe 2030s, 2050s	Improved fungicide application (e.g. new active compounds)	Bregaglio *et al.*, 2013
	Time of appearance and probability of attack will change (earlier appearance in vulnerable regions due to temperature increase, disease probability during the growing season will decline due to reduced humidity	Serbia 2030s, 2050s	Fertilization, irrigation, use of hail nets, weed control, permanent shoot cutting, removing of diseased plant debris	Lalic *et al.*, 2013
	Infection risk will decrease	France 2080s	Host plant resistance, timing of sowing, decision support tools	Launay *et al.*, 2014
Grapevine powdery mildew, *Uncinula necator*	Disease severity and number of infection cycles is projected to decrease	Italy 1990–2070	Not discussed	Caffarra *et al.*, 2012
Oilseed rape stem canker, *Leptosphaeria maculans*	Range and severity is projected to increase	United Kingdom 2020s, 2050s	Improved disease resistance, improved fungicide application, under worst case	Evans *et al.*, 2008 Barnes *et al.*, 2010 Butterworth *et al.*, 2010

Crop/disease	Projected crop disease risk	Geographic Scope/Time horizon of projection	Adaptation options	Reference
			scenario: shifting crop production from south to north of the UK	
Potato early blight, *Alternaria* sp.	Infection risk on potato will slightly increase in four out of six locations in North- Rhine-Westfalia	Germany 2001–2050	Crop rotation, soil tillage, host plant resistance, sowing time, fertilization, fungicide application frequency	Volk *et al.*, 2010
Potato diseases	Cumulative relative development rate of late blight is predicted to decrease in already warm growing areas, early blight (*Alternaria solani*) and brown spot (*Alternaria alternata*) are projected to increase in areas with major precipitation, soft rot and blackleg (caused by *Pectobacterium carctovorum* subsp. *brasiliensis*) are predicted to increase particularly in summer rainfall areas, Potato virus Y (PVY) and Potato leaf roll virus (PLRV) are predicted to increase on potato due to higher vector aphid numbers (*Myzus persicae*, vector of PVY and PLRV)	South Africa 1961–2050	Breeding of durable and innate disease resistance, fungicide application frequency, shifting planting dates, pesticide efficacy, monitoring, record keeping, decision support systems, efficient and cost effective plant protection strategies,	Van der Waals *et al.*, 2013

Crop/disease	Projected crop disease risk	Geographic Scope/Time horizon of projection	Adaptation options	Reference
Rice blast, *Magnaporthe oryzae* and *Bipolaris oyzae*	Number of infection events are projected to increase	Europe 2030s, 2050s	Improved fungicide application (e.g. new active compounds)	Bregaglio *et al.*, 2013
Wheat Karnal bunt, *Tilletia indica*	Area favourable for Karnal bunt is predicted to extend northward	Europe 2050s	Not discussed	Baker *et al.* 2000

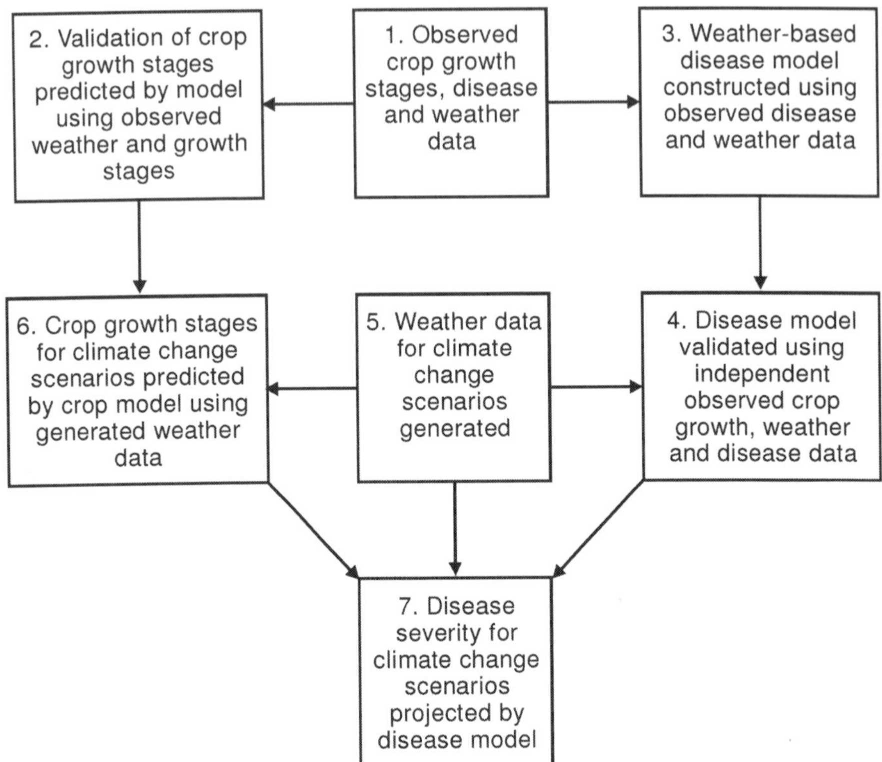

Fig. 11.1. An illustration of how climate, crop growth and disease models can be combined to produce projections of crop growth stages and disease incidence/severity for different climate change scenarios (*Source*: Current Opinions in Plant Biology).

Observed data for weather [e.g. daily minimum and maximum temperature (°C), total rainfall (mm) and solar radiation (MJ day^{-1})], crop growth stage and disease incidence can be collated from a number of sources for different regions for a period of years (Fig. 11.1, 1). The crop growth stages predicted using the crop growth model can be validated by comparing predicted crop growth stages, generated by the model using observed weather data, with observed crop growth stages for the same sites for a given period (Fig. 11.1, 2). A disease model can be developed from data for disease incidence from sites within a certain distance of the site for which there is observed weather for a given period (Fig. 11.1, 3). Predictions of disease incidence can be validated by comparing predictions made using observed weather to observed disease incidence data for a given period for different regions (Fig. 11.1, 4). Weather data can be generated for each of the sites for each climate scenario (Fig. 11.1, 5). The crop growth stages can be projected for each site for each climate scenario using the crop growth model, allowing maps to be generated to show the effect of climate change on crop growth (Fig. 11.1, 6). Using the weather generated and crop growth stage projected using the crop growth model for each of the sites for each of the climate scenarios, the disease model can be used to project disease incidence for each site for each of the climate scenarios (Fig. 11.1, 7).

Simulations of future powdery mildew epidemics predicted a decrease in disease severity, especially in years in which disease symptoms first appear at a later date and in the presence of significantly increased temperatures. They also suggested that in the warmer, more profitable viticulture areas of the region, increased temperatures might have a detrimental effect on crop yield because of increased asynchrony between the larval resistant growth stages of grapevine and European grapevine moth larvae. On the other hand, the increase in pest pressure caused by the increased number of generations might not be as severe as expected on the basis of the pest model only, because of the earlier harvest dates, which would limit the damage caused by late-season generations. The current and future impact of downy mildew, *Plasmophora viticola* on grape vineyards was studied in Piedmont, Italy. Two global circulation models (GCMs) predicted increased temperatures and decreased precipitation in this region. The simulations obtained by combining the disease model with the two GCM outputs predicted increased levels of disease pressure in each of the three tested future decades following more favourable temperature conditions during the months of May and June. The increased temperatures are predicted to promote the epidemics in ways that will be only partially counteracted by the effects of reduced precipitation, which alone would have diminished disease pressure (Salinari *et al.*, 2006).

To guide strategies for adaptation by the agricultural industry, it is essential to project the impacts of climate change on severity of crop disease epidemics.

11.3. COMPLEXITY OF PREDICTING THE EFFECT OF CLIMATE CHANGE ON A PATHOGEN

It is difficult to predict the effects of climate change because, to date, experiments in this area have considered only one or a few of the parameters that are predicted to change. Most of these experiments have been performed under controlled conditions, which may vary greatly from those in the field, and have been carried out on a small scale for relatively short periods of time. Interactions with other changing climatic variables may directly influence, mask, or amplify the effects of a specific factor (Shaw *et al.*, 2002). The common approach for predicting the dynamics of a plant disease is based on modelling the behaviour of the pathogen and/or epidemic under known conditions and then using that model to predict the behaviour of the pathogen/pattern of the epidemic under a projected climate time series or in areas in which the pathogen has not yet been introduced. Empirical epidemiological models, in which climatic variables are predictors and epidemic parameters are variables, are generally used. These empirical models are used as simulation models to make predictions for specific scenarios. General circulation models (GCMs) have been used to create scenarios and predict the effects of climate change (Bergot *et al.*, 2004).

Whereas temperature change is relatively easy to project, there is no overall trend for change in the amount of precipitation, although there is clear historical evidence of changed distribution patterns both regionally and seasonally (Chakraborty and Newton, 2011). The selection of the appropriate spatial and temporal scales for weather and disease remains a challenge. In particular, downscaling weather predictions to the mesoscale for forecasting disease and predicting the microclimate at the level of plant organs are key issues. Furthermore, models should include higher-level ecological interactions (Chakraborty, 2005) and take into consideration the complex changes that may occur in both crops and agricultural practices. Because many pathogens occupy any trophic niche not adequately protected by plant resistance mechanisms or crop protection practices, the prediction of when and where such niches will appear is also very important (Chakraborty and Newton, 2011). The availability of long-term data sets is crucial for the construction of reliable disease models (Jeger and Pautasso, 2008).

The main difficulties in prediction include finding enough suitable data on existing epidemics to construct good empirical models; the uncertainty of climatic models; the difficulty of downscaling projected climatic data series to the regional level; and the fact that these models do not consider other changes that may occur in the future, such as possible adaptations of plants and/or pathogens, changes in agronomic practices, and the effects of climate change on a pathogen's antagonists. Moreover, the pathogen models cannot be disassociated from the phenological models of the host plants, which can be affected by climate change (Caffarra *et al.*, 2012). In general, climate change affects the life cycles and biological synchronicity of plants and pathogens, resulting in changes in the distribution and phenology of events such as bud break, the release of spores by pathogens, and the activities of insects that serve as disease vectors. All of these factors may interact to produce the final outcome (disease incidence and severity), which is difficult to simulate with models (Sturrock *et al.*, 2011).

Predictions should be carried out at the regional level, because large differences may be present within individual countries. For example, the productivity of oilseed rape (*Brassica napus*) in the United Kingdom is projected to increase more in the north than in the south. However, it seems that the diseases that affect this crop will be influenced in a differential way. On the one hand, in the northern United Kingdom, Phoma stem canker (*Leptosphaeria maculans*) will not substantially reduce yield and light leaf spot will become less severe to the point that the application of fungicides may no longer be economically justified in this region. On the other hand, Phoma stem canker will become more dangerous in the southern United Kingdom (Evans *et al.*, 2008), necessitating a strict fungicide schedule (Butterworth *et al.*, 2010).

11.4. IMPACT MODELS

Quantitative (modelling) approaches, which allow one to investigate multiple scenarios and interactions simultaneously, will become more important for impact assessment (Coakley and Scherm, 1996). Guidelines for such model-based assessments are needed, and Sutherst *et al.* (1996) and Teng and Yang (1993) have given a framework.

11.4.1. Climate matching

Climate matching involves the calculation of a "match index" to quantify the similarity in climate between two or more locations. The match index is based on variables such as monthly minimum and maximum temperatures, precipitation, and evaporation. Software packages for climate matching include BIOCLIM (Busby, 1991), HABITAT (Walker and Cocks, 1991) or CLIMEX (Sutherst and Maywald, 1985) and WORLD (Booth, 1990). These packages often come with additional useful features such as internal algorithms for generating "climate surfaces" through interpolation between stations. Climate matching may be used for climate change impact assessment by identifying those locations on the globe with a current climate that is most similar to the predicted future climate at the location of interest. An analysis of the plant disease problems at the matching locations, for example based on disease distribution maps (Weltzien, 1972), would allow predictions to be made about future disease risk at the location of interest.

Booth *et al.* (2000) used climate matching to identify regions suitable for *Cylindrocladium* leaf blight on *Eucalyptus* sp. in Southeast Asia and around the world. They first established a simple rule for presence or absence of the disease based on long-term means of temperature and precipitation. This rule was then implemented in a climate matching programme to identify high-risk regions in Africa, Australia, Latin America, and Southeast Asia under current climate. Further, two climate change scenarios were run for locations in Southeast Asia. The results suggested an increase in disease risk in northern Vietnam, southern Laos, and eastern Thailand. These predictions are consistent with limited field observations indicating that severe disease can occur in these regions during years with extreme weather.

Possible effects of climate change on *Phytophthora cinnamomi*, a soil-borne Oomycete with an extremely wide host range, were considered by Brasier (1996) and Brasier and Scott (1994). This pathogen requires warm, wet soils and is hence limited primarily to tropical and subtropical regions (Lonsdale and Gibbs, 1996). More recently, *P. cinnamomi* has been associated with oak declines in southern and Mediterranean Europe. It was hypothesized (Brasier, 1992) that this may be an early indication of climate warming as the pathogen may have become more active because of higher soil temperatures and/or increased host susceptibility caused by stress (e.g. more frequent winter droughts in the region). For a more formal impact assessment, Brasier and Scott (1994) used the CLIMEX climate matching programme (Sutherst and Maywald, 1985) to map regions in Europe favourable or unfavourable for this pathogen under present and future climate scenarios. The climate change simulations suggested that the pathogen could extend its range further north, although it appeared unlikely that it could become established in those regions where winter temperatures are low such as central and Eastern Europe (Brasier, 1996). It was further hypothesized, that the pathogen's host range could increase if spread occurs into regions where it is currently absent.

11.4.2. Empirical models

Four diseases of two major crops in China, wheat and rice, were examined by regression analysis to determine how they have varied through time and whether this may relate to recent increases in mean and minimum temperatures (Yang *et al.*, 1998). Rice blast and wheat scab have increased sharply since the 1970s. The wheat acreage infected with powdery mildew has become more extensive, whereas stripe rust has decreased steadily (Yang *et al.*, 1998). This may be related to the increased spring and early summer temperatures and would be consistent with the changes in stripe rust observed in the Pacific North-west associated with climate variability (Coakley *et al.*, 1988).

Jahn *et al.* (1996) utilized long-term plant disease monitoring records collected by the State Plant Protection Service in the former German Democratic Republic (GDR) to develop empirical climate-disease models for 15 individual host-pathogen combinations. These models were then used with various climate change scenarios to predict possible changes in "infestation levels" in a future climate. Calculations with the most realistic scenario (a temperature increase of 1°C combined with a decrease in precipitation of 30%) indicated that leaf rusts of wheat and barley and powdery mildew of sugar beet could increase substantially, reaching levels between two and five times as high as under the current climate. Infestation levels on small grains by powdery mildews would remain virtually unchanged, whereas those caused by foot rots and leaf blotch diseases would decrease. Most notable was a decrease in potato late blight to a mere 16% of its current level. The authors cautioned against over-interpreting their results, which were based on calculations with data from only 1 of 14 regions in the former GDR.

11.4.3. Population models

A very different conclusion regarding the importance of potato late blight under climate change was reached by Kaukoranta (1996). This author developed degree day models for the emergence of potatoes and the date of late blight outbreaks in Finland. The two models were coupled and extended by including leaf area expansion of the crop as a function of thermal time; calculating radiation interception as a function of leaf area; transforming the intercepted radiation to tuber dry matter; and simulating the effects of late blight on tuber dry matter through a reduction in green leaf area, assuming that disease reduced leaf area to zero within 14 days after the predicted outbreak. Model parameters were obtained and model validation was done using data from a three-year field and greenhouse study. The combined model was then used with various temperature change scenarios to predict possible changes in potato yield and yield losses caused by late blight in a warmer climate. The results suggested that tuber yield could increase by 2 t/ha per 1° C warming in the absence of late blight. This potential yield gain was almost completely offset when late blight was considered, chiefly because late blight outbreaks occurred 4 to 7 days earlier and the period during which the crop was susceptible was lengthened by 10 to 20 days per 1° C warming. This study did not consider possible yield-enhancing effects of elevated CO_2, nor did it incorporate the effects of changes in precipitation on late blight.

11.4.4. Simulation models

Simulation models have been used extensively to predict yields of various crops in different agro-ecological zones under climate change (Riha *et al.*, 1996). Biotic yield reducing factors such as insects, pathogens, and weeds have, however, been largely ignored in these simulations (Teng *et al.*, 1996). Because of this shortcoming, the development of linked disease-crop models is an important objective within the overall goal of developing a predictive capability for agricultural impact assessment and mitigation (Scherm, 2000). For at least one key crop, rice, preliminary analysis considering the combined effects on yield of increased temperature, elevated UV-B radiation, and rice blast disease (*Magnoporthe grisea*) have been done using a coupled simulation model (Luo *et al.*, 1997). The model consisted of a physiological rice growth model and a leaf blast epidemic simulator, linked via the quantitative effects of leaf blast on photosynthesis and biomass production (Luo *et al.*, 1997). Climate change was imposed by increasing mean temperature in fixed increments and by either including or omitting effects of UV-B on the host and pathogen (Luo *et al.*,

1995). The results suggested that elevated UV-B could result in direct yield losses of 10%. Impacts of increased temperature varied by agro-ecological zone, with an increase in blast and associated yield losses in cool, subtropical rice production regions (e.g. Japan) and a decrease in humid tropics and subtropical regions (e.g. the Philippines). The authors cautioned that the results must be considered preliminary as the simulations did not include neck and panicle blast, two other important symptom types caused by *M. grisea.* Further, increased CO_2 was not considered, nor were changes in precipitation as preliminary analysis had indicated that the combined model was insensitive to changes in rainfall (Luo *et al.*, 1995).

11.5. PREDICTIVE MODELS IN HORTICULTURAL SYSTEMS AND THE OPEN FIELD

Crops grown in fields are likely to be affected by climate change either directly through the climate change effects on various stages of the pathogen life cycle, or indirectly through the susceptibility of host plants to pathogen infection, and changes in the physical environment surrounding the host-pathogen system (Coakley *et al.*, 1999). Any patho-system may be affected by climate change, because the organisms in the system naturally react to climate conditions. Nevertheless, the extent of these effects and their nature, whether positive or negative, are still poorly understood. An extended growing season, as is expected in several regions of the world, is one of the major reasons for the expected increase in severity of disease epidemics in crops (Menzel and Fabian, 1999). This is because of the increased number of disease cycles in polycyclic diseases and the greater impact of monocyclic pathogens. Predictive models have been developed for a few plant-pathogen systems, including Phoma stem canker, *Leptosphaeria maculans* of oilseed rape in the United Kingdom (Evans *et al.*, 2008), downy mildew, *Plasmopara viticola* (Salinari *et al.*, 2006), and powdery mildew, *Erysiphe necator* (Caffarra *et al.*, 2012) of grapevine in northern Italy; coffee rust, *Hemileia vastatrix* in Brazil (Ghini *et al.*, 2011); *Cercospora* leaf spot, *Cercospora beticola* of sugar beet in Germany (Richerzhagen *et al.*, 2011); and banana black sigatoka, *Mycosphaerella fijiensis* in Brazil (Ghini *et al.*, 2007).

Leptosphaeria maculans causes Phoma stem canker (blackleg), which is an important disease of oilseed rape and other *Brassica* species that causes severe yield losses worldwide (Fitt *et al.*, 2006; Howlett *et al.*, 2001). Temperature and rainfall affect the development of this pathogen (Huang *et al.*, 2005) and the resistance of its host (Huang *et al.*, 2006). In the United Kingdom, the most severe Phoma stem canker epidemics occur in southern England. In contrast, in cooler Scotland, Phoma leaf spotting is not accompanied by the damaging Phoma stem cankers (Evans *et al.*, 2008). Weather-based models to predict the development of Phoma stem canker epidemics were combined with climate change models to generate scenarios for the future severity of epidemics in the United Kingdom (Evans *et al.*, 2008). Because this disease is monocyclic, the date in autumn when leaf spots first appear is regarded as a crucial factor affecting the severity of epidemics on stems the following summer (West *et al.*, 2001). A model was developed that incorporated data concerning two aspects. The first input data was the relationship between the first appearance of Phoma leaf spot in autumn and the temperature and rainfall during the period between the harvest of the previous crop and the establishment of the new crop, because maturation on crop debris of *L. maculans* ascospores in pseudothecia and infection of the leaves of oilseed rape through their stomata are dependent on both temperature and moisture. The second input data was the date when Phoma canker begins to develop on stems in spring as predicted on the basis of the date on which the first Phoma leaf spots appeared and the accumulated degree days, which affect the rate at which the pathogen progresses along leaf petioles toward the stem in autumn/winter. This model was used to predict the change in severity of Phoma stem canker (Evans *et al.*, 2008).

To predict the effects of climate change on the range and severity of Phoma stem canker epidemics, Evans *et al.* (2008) used UKCIP02 climate change predictions for the years 2020 and 2050. Their results demonstrate how global warming may increase the geographic spread and severity of plant disease epidemics within the next 20 years. The predicted climate change had a strong effect on the timing of the start of epidemics in spring, often 80 days earlier than during 1960–1990. The area affected by the canker disease was predicted to extend northward from England to areas currently unaffected by the disease, and the model predicted much greater severity of this disease at harvest for 2020 and 2050 compared with earlier years. These major effects on the start date and severity of epidemics were related to predicted increases in temperature, especially during winter (Sun *et al.*, 2000).

The obligate parasite *Plasmophora viticola* causes downy mildew in grapevine crops around the world, including northern Italy. This disease affects both the quantity of harvested grapes and the quality of the wine that can be produced from them (Gessler *et al.*, 2011). The organism overwinters as oospores in fallen grape leaves. In spring, it produces zoosporangia that, in the presence of water, are splashed onto leaves or bunches. Moisture allows for infection by zoosporangia, and dry periods kill the pathogen. The pathogen develops at temperatures of 9°C–34°C (Gessler *et al.*, 2011). To assess the current and future potential effects of projected climate changes on downy mildew, an empirical disease model was developed in a wine grape-growing region in Piedmont, Italy (Salinari *et al.*, 2006). This model used climate change scenarios generated by two GCMs, from the Hadley Centre (Pope *et al.*, 2000), Devon, United Kingdom, and the National Aeronautic and Space Administration Goddard Institute for Space Studies (Schmidt *et al.*, 2005), New York, for the decades of the 2030s, 2050s, and 2080s, which were compared with meteorological data collected between 1955 and 2001 (Salinari *et al.*, 2006). Both GCMs predicted increased temperatures and decreased precipitation in this region. The simulations obtained by combining the disease model with the two GCM outputs predicted increased levels of disease pressure in each of the three tested future decades and more favourable temperature conditions during the months of May and June. The increased temperatures are predicted to promote the epidemics in ways that will be only partially counteracted by the effects of reduced precipitation, which alone would have diminished disease pressure. The authors anticipated that two more fungicide sprays would be needed under the most negative climate scenario compared with the current management programme (Salinari *et al.*, 2006).

Powdery mildew, *Uncinula necator* is another important disease of grapevine (Gadoury *et al.*, 2001). The European grapevine moth, *Lobesia botrana* is one of the most noxious vineyard pests in the European and Mediterranean regions (Delbac *et al.*, 2010). Caffarra *et al.* (2012) combined detailed phonological models of grapevine with phenological models of grape powdery mildew and the European grapevine moth, applied the models to climate change scenarios for the eastern Italian Alps, and considered potential changes in the interactions between these species. Simulations of future powdery mildew epidemics predicted a decrease in disease severity, especially in years in which disease symptoms first appears at a later date and in the presence of significantly increased temperatures. Furthermore, the simulations carried out by Caffarra *et al.* (2012) suggested that in the warmer, more profitable viticulture areas of the studied region, increased temperatures might have a detrimental effect on crop yield because of increased asynchrony between the larvae-resistant growth stages of grapevine and the European grapevine moth larvae. On the other hand, the increase in pest pressure caused by the increased number of generations might not be as severe as expected on the basis of the pest model only, because of the earlier harvest dates, which would limit the damage caused by late-season generations.

Coffee rust, *Hemileia vastatrix* is an important disease in tropical countries. The time interval between inoculation and the appearance of symptoms (incubation period) is a critical point in the pathogen's biological cycle. Optimal environmental conditions (i.e., temperatures) result in shorter incubation periods and more severe epidemics (Ghini *et al.*, 2011). Future coffee rust epidemics were simulated across geographical areas using a model that estimated the pathogen incubation period together with the output from three GCMs (CSIRO-Mk3.0, INM-CM3.0, and MIROC3.2 models). A set of meteorological data from 1961 to 1990 was compared with that predicted for the 2020s 2050s, and 2080s using scenarios A2 and B1 from the Intergovernmental Panel on Climate Change. Maps were prepared for 10 coffee-producing states of Brazil. The climate variables used were maximum and minimum monthly temperatures. Maps obtained for the two tested scenarios predicted a tendency toward a shorter incubation period in future years (Ghini *et al.*, 2011).

Cercospora beticola causes one of the most damaging foliar diseases of sugar beet in south central Europe and North America (Wolf and Verreet, 2005). Higher levels of *Cercospora* leaf spot are found in southern and south-western regions in both continents. Meteorological factors strongly affect the development of this disease. Free water or RH levels above 95% are essential for conidial germination and infection; disease development is favoured at temperatures between 20°C and 30°C (Wolf and Verreet, 2005). Richerzhagen *et al.* (2011) analyzed the effect of climate change on the temporal and regional occurrence of *Cercospora* leaf spot of sugar beet in Lower Saxony, Germany, using the CERCBET1 forecasting model (Rossi and Battilani, 1991). Because epidemics of this disease require warm and humid conditions and in light of the fact that climate models for Germany (until 2080) predict a long-term warming trend with a 1.6°C–3.8°C increase in annual mean temperature, these researchers expected that global warming would shift the area of heavy damage northward. Indeed, based on temperature and RH data collected hourly, the model predicts the times at which 1%, 50%, and 100% of the fields in a region will be infected. The input data were results obtained using the REMO regional climate model (Jacob and Podzun, 1997). The effects of climate change were studied for three time periods: a baseline period (1971–2000), a medium-term period (2021–2050), and a long-term period (2071–2100). A comparison of the baseline time period of 1971–2000 with 2021–2050 shows that the date by which 1% of fields are infected is predicted to be reached 5.2 days earlier in the medium term period than it was previously, the date by which 50% of fields are infected is predicted to be reached 6.7 days earlier, and the date by which 100% fields are infected is predicted to be reached 10.6 days earlier. In the 2071–2100 periods, the date by which 1% of fields are infected is expected to be reached 22.9 days earlier than in the baseline period and the dates by which 50% and 100% of fields are infected are expected to be reached 25.2 and 32.7 days, respectively, earlier than in the baseline period. As chemical control is currently an important part of disease management strategies, Richerzhagen *et al.* (2011) has predicted an increase in the frequency of fungicide applications in the future.

Black sigatoka is a leaf spot disease of banana caused by the Ascomycete, *Mycosphaerella fijiensis*. It is an important disease of banana-growing regions of South and Central America. Plants with leaves infected by this disease may have up to 50% lower yields (Marín *et al.*, 2003). This disease is promoted by high humidity, moisture, and temperatures. Ghini *et al.* (2007) estimated the probable effects of climate change on black sigatoka disease in Brazil through the construction of distribution maps using data from six GCMs and, unlike the situations described earlier, predicted a reduction in the size of the area affected by this disease in Brazil.

11.6. POTENTIAL ADAPTATIONS OF CROPPING SYSTEMS

In majority of the disease risk simulation studies, the improved disease resistance and adjusted fungicide application are considered to be the potential most important adaptation options to future climatic conditions. In addition, sowing time, longer crop rotation, improved disease forecasting, irrigation, shifting of cultivation zones, adjustment of agronomic practices, modification of microclimate (e.g. altered sowing density), cultivar choice, sowing date, crop residue management, and targeted advice can also be used as alternative options. Integrated pest management (IPM) should include the full range of direct and indirect crop health management measures including careful monitoring and optimal timing of needed interventions (Strand, 2000).

11.7. CONCLUSIONS

Because most plant diseases are strongly affected by environmental conditions, climate changes will affect pathogens, hosts, and the interactions between them, resulting in changes in epidemics and their effects. In most of the patho-systems that have been examined, increases in disease severity have been predicted. Actually, climate change is expected to be gradual. The gradual change in climate will probably lead to gradual changes in agriculture and forestry. As part of these changes, we expect that the geographical ranges of existing crops will change and that pathogens may play key roles in these shifts. For particular host species, we expect that the onset time of epidemics and levels of disease severity will change. We also expect that some diseases may become less important, whereas others become more important for particular crops in particular regions. Most pathogens are able to migrate to locations where the climate is more suitable for their survival and reproduce more quickly than plant species. Thus, the variety of diseases that affect a plant species in a particular region will continue to change. Moreover, because abiotic factors, such as temperature and moisture, affects not only pathogen's reproduction, development, and infection, but also the susceptibility of host plant to infection, changes in the interactions between abiotic stress factors and plant diseases may, in the future, become more important for the development of epidemics in plants.

Disease modelling results suggest that projections of future pathogen dynamics and ranges should include several climatic factors and several pathogen life cycle stages to become more reliable, although the overall outcome can be complex due to counteracting effects.

A unifying variable such as yield is lacking in the disease risk simulation studies. Yield is a parameter of common interest both in the agro-climatic and agronomy research community involved in climate change research and thus could enable large-scale multidisciplinary international research initiatives such as AgMIP (The Agricultural Model Intercomparison and Improvement Project). However, it should also be kept in mind that crop disease modelling studies may have different purposes and that yield will not always have first priority.

Interestingly, most simulation studies related to the putative future risks of crop diseases are not taking potential crop protection adaptation options into account (Barnes *et al.*, 2010). Hence, there is a need for simulation studies which account at least for the basic IPM tools such as cultivar choice and timing of sowing in order to obtain a more relevant estimate on required strategies in future crop protection. In some cases, adaptation in crop protection measures such as adjustment of sowing date, use of crop cultivars with superior resistance, and effective chemical and biological plant protection products would be useful even without considering future changes in climate, as they would help addressing current annual weather variability (Olesen *et al.*, 2011).

As uncertainty of projections and cropping system responses is by far the largest challenge in these simulations, it will be useful if modelling studies would estimate and include the uncertainty level of disease risk simulation such as performed by Gouache *et al.* (2013). This study is the first to systematically analyse and quantify different sources of uncertainty in the simulation of climate change impact on plant pathogens and diseases. This could be extremely helpful in determining how meaningful a projection is, and in validating the accuracy of future modelling. In this regard, uncertainty should not only be viewed as the absence of knowledge, because quantified uncertainty may still represent a gain of information (Gouache *et al.*, 2013). It indicates the reliability of results and can also assist in identifying major sources of uncertainty, which could serve as a guideline for future research. Consequently, novel simulation approaches to project future disease risks should be developed even though they are and will be uncertain by nature.

Combining climate change, crop growth and crop disease models to predict impacts of climate change on crop diseases can guide planning of climate change adaptation strategies to ensure future food security.

12

Climate Change Adaptation and Mitigation Strategies for Disease Pathogens

12.1. INTRODUCTION

The climate is changing, resulting in changes in the plant diseases that we need to prevent or manage. The challenges of climate change on disease pathogens have so far been adapted in the past to changing regional production and market factors through introduction of new cultivars, new production systems and new market niches for specialist agricultural products. As the climate change time frame is well within the time frame for breeding new crop cultivars, if plant breeders were able to focus on developing cultivars with lower disease susceptibility, rather than restricting efforts to improved fruit quality, the impact climate change may have on disease risk could be greatly reduced. Unless climate change causes more catastrophic changes than those predicted in the MfE (2008) report, it seems unlikely that agriculture will have difficulty in adapting to changes in plant disease risk.

Climate change can affect disease management by altering efficacy of biological and chemical control options (Chakraborty *et al.*, 1998; Coakley *et al.*, 1999). For example, heavy rains reduces fungicide residue. Crop plants growing under elevated CO_2 could be altered morphologically or physiologically, affecting uptake, translocation, and metabolism of systemic fungicides. For example, increased thickness of the epicuticular wax layer on leaves could result in slower and/or reduced uptake by the host, while increased canopy size could negatively affect spray coverage. Conversely, if higher temperatures increase plants' metabolic rates, they may take up chemicals more quickly, which may result in greater toxicity (Coakley *et al.*, 1999).

New dimensions of climate change may add extra uncertainty in management strategies for diseases caused by different pathogens. Impacts of climate change on plant patho-systems would occur chiefly through influences on host resistance or chemical and biological control agents. Delayed planting to avoid a pathogen may become less reliable. Particular attention is needed to identify cases where the efficacy of disease management may be reduced under climate change.

Applications of conventional crop protection products often disrupt the many processes keeping such organisms in some sort of benign balance in non-epidemic situations. These factors or processes can be categorized in terms of risk mitigation and risk enhancement (Fig. 12.1). These rate determining processes are the result of complex interactions between these 'remediating' and 'enhancing' influences. Each process itself is a complex biological system with multiple components, each influenced by climatic variables in different ways. The challenge is to rank the influences of both the processes and the key environmental/ climatic influences in parallel, in order to construct influence models to predict the likely effects of climate change on production systems. Garrett *et al.* (2011) offers one approach to improve understanding of these complexities.

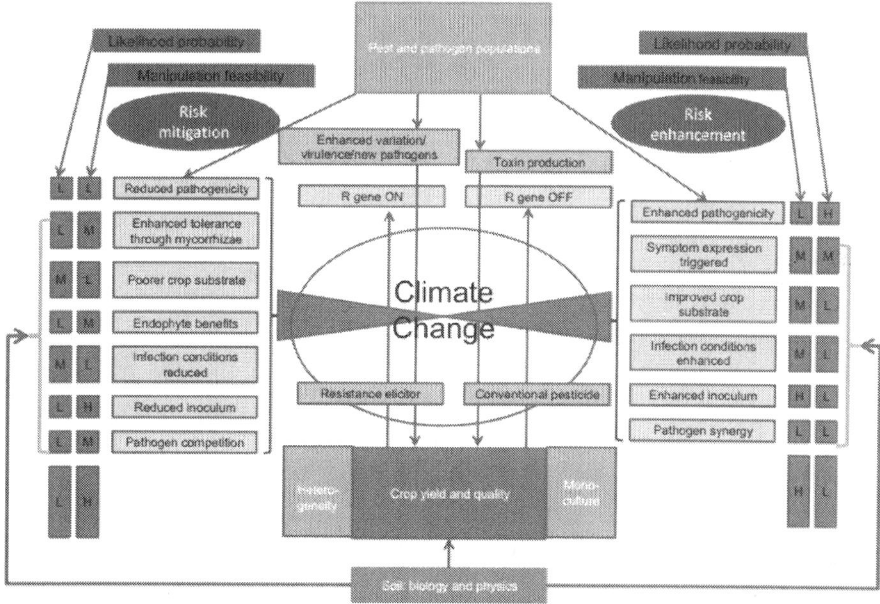

Fig. 12.1. Influence of climate change on rate-determining processes that are the result of the complex interaction between the 'enhancing' (right) and 'mitigating' (left) influences on plant and pest/pathogen interactions. Rankings for likelihood probabilities and manipulation feasibility are initial approximations requiring critical review (L – Low, M – Moderate, H - High).

Evaluating the efficacy of current physical, chemical and biological control methods under changing climatic conditions and research concerning new tools and strategies (including plant breeding) for coping with the predicted changes will be of great strategic importance.

12.2. PLANT DISEASE MANAGEMENT UNDER CLIMATE CHANGE

Plant disease control through host-plant resistance and fungicide treatment makes a major contribution to both climate-change mitigation and sustainable crop production systems to ensure global food security (Berry *et al.*, 2008; Mahmuti *et al.*, 2009; Legreve and Duveiller, 2010). Particularly in high-input production systems, greenhouse gas emissions can be decreased by greater nitrogen use efficiency by a healthy crop (Mahmuti *et al.*, 2009). Disease management strategies may require adjustments under climate change (Garrett *et al.*, 2006). Influences on host-plant resistance and on biological control agents (BCAs) have been the most frequently used management options (Table 12.1). Some selected plant disease management methods in relation to climate change are presented in Table 12.2. Selective preventive plant protection methods, such as delaying planting/sowing dates in order to escape periods of enhanced disease risk, are used by farmers in any climate zone worldwide.

Table 12.1. Most frequently used plant disease management options in agriculture

Plant disease management options	Reference/s
Cultivar breeding for host resistance and tolerance, novel selection techniques, diagnostic tools to detect newly introduced pathogens on new crops, appropriate irrigation technologies, careful selection of crop species and site to minimize disease pressure	Boland *et al.*, 2004
Fungicides, biological control agents, host resistance	Chakraborty *et al.*, 2000
Host resistance, fungicides, disease suppression by soil organic matter, biological control agents, induced systemic resistance	Chakraborty and Pangga, 2004
Host resistance, chemical control, microbial interactions, quarantine and exclusion	Coakley et al., 1999
Biological control, risk models for movement of invasive pathogens, host resistance and its durability, uncertainty in decision making	Garrett *et al.*, 2006
Preventing spread of exotic pathogens, host landscape structure, proximity to higher risk areas, in situ and ex situ conservation of genetic resources for maintaining resistance and tolerance	Garrett, 2008
Fungicides, non-chemical methods, host resistance, biological control	Ghini *et al.*, 2008
Resistance breeding, agricultural practices, rotation, planting time, avoidance, chemical control, forecasting models, quarantine regulations, phytosanitary measures, IPM	Legreve and Duveiller, 2010
Choice of crop and suitable location, appropriate cultivar, crop rotation, tillage practices, manipulation of planting date, intercropping, crop management such as suitable water and nutrient management, disease forecast/prediction, timing of plant protection products application, biological control agents, modification of microclimate, habitat management	Strand, 2000
Host resistance and tolerance, biological control agents, procedures to prevent invasion of pathogens	Sutherst *et al.*, 2007
Crop management, increase of crop diversity, development of new fungicides, increase of range of preventive plant protection methods, reduction of the risk of invasive species, development of IPM tools e.g. models to forecast disease occurrence	Tiedemann and Ulber, 2008

Table 12.2. Some selected plant disease management strategies/tools in relation to climate change

Control strategy	Tool	Expected effects of changed climate	Potential of tool for adaptation
Avoidance	Barrier to entry (quarantine)	Climate-mediated change in pathogen dispersal – frequency, abundance, distance, speed	Altered efficacy of quarantine practices
Preventive	Crop rotation	No direct effect, diversity in cropping systems will remain important to reduce risks of diseases	Crop species better adapted to local climatic conditions may be required
Preventive	Crop residue management	Potential increase in crop biomass through the CO_2 fertilizing effect, unless high temperature and drought counterbalance the fertilizing effect	Innovative approaches needed to reduce inoculum level and saprotrophic colonization
Preventive	Sowing/planting date	Adjustments likely to be necessary, simple and cheap method to escape biotic and abiotic stress; however, disadvantages also possible	Apparently a powerful tool, but maybe limited in overly warm winters (e.g. late sowing in autumn under temperate conditions)
Preventive	Host-plant resistance	Temperature dependent resistance may be overcome by pathogens, changes in plant morphology and physiology with effects on resistance, potentially accelerated pathogen evolution may erode disease resistance prematurely	Altered efficacy of host-plant resistance [higher, same or lower efficacy, depending on resistance (R) gene, pathogen population, etc.]
Preventive	Cleaning machinery and tools	Presumably no major effects	Phytosanitary methods will remain important
Preventive	Use of healthy seeds and plantlets	Presumably no major effects	Preventive methods will remain important
Preventive	Input levels e.g. of irrigation	Presumably higher temperature will promote irrigation in more crops and regions	Water conservation may demand efficient technologies such as drip irrigation, thereby reducing risk of foliar diseases

Control strategy	Tool	Expected effects of changed climate	Potential of tool for adaptation
Preventive and/or curative	Field monitoring and use of decision support systems (DSS)	Presumably no major effects	Field monitoring and DSS will remain or become more important
Preventive and/or curative	Soil solarization	Global warming may facilitate the use (effective in more plant–pathogen systems and regions, heat may reach deeper soil layers, shorter duration of mulching period)	Altered efficacy, generally positive effects, unless drought counterbalances temperature effects
Preventive and/or curative	Antagonists, biological control agents (BCAs)	Presumably, vulnerability of BCAs will be higher as a result of climate variability	Altered efficacy (higher, same or lower, dependent on product, environment, management, etc.)
Preventive and/or curative	Contact fungicides	If rainfall occurs more frequently, more applications may be triggered, faster/slower crop growth may shorten/ lengthen the time between applications	Altered efficacy (higher, same or lower, dependent on product, environment, management, etc.)
Preventive and/or curative	Systemic fungicides	More knowledge on foliar uptake process of systemic fungicides must be gained to make reliable predictions	Altered efficacy (higher, same or lower, dependent on product, environment, management, etc.)

12.3. ADAPTATION AND MITIGATION

According to Olesen *et al.* (2011), there are different types of adaptation options available which can avoid or at least reduce potential negative effects of climate change in agriculture or even exploit the possible positive effects. Autonomous adaptations are mostly short-term adjustments to optimize crop management within the existing cropping system such as variation in the grown varieties, sowing dates, fertilizer application, and pesticide use. These production factors can be optimized by targeted advice in order to improve resource use efficiency on farm-level (Barnes *et al.*, 2010). Planned adaptations refer to major structural farm management changes including alterations in land allocation, crop species, and implementation of new tillage systems to conserve soil water. Long-term adaptations may include breeding efforts for improved disease resistance.

An adaptation option may have both agronomic and plant pathological aspects. For example, sowing time can directly influence both the yield potential of a crop and the incidence and severity of diseases. Delayed sowing to reduce disease risks can be an important tool, particularly if resistant varieties and effective fungicides are not available (Salam *et al.*, 2011). For example, in Southern Australia grain yields of field pea (*Pisum sativum*) can be penalized, if flowering and grain filling are subjected to temperature and drought stress due to delayed sowing (Salam *et al.*, 2011).

Srivastava *et al.* (2010) concluded that more low-cost adaptation strategies such as changing the sowing date and choice of cultivar should be explored to reduce the vulnerability of crop production to climate change, although they can have trade-offs. The practicability of changing planting and/or harvesting dates is dependent on the potential yield penalty and on the region, cultivar preferences of farmers and consumers, magnitude of climate change, and the market situation (Wolfe *et al.*, 2008). However, other potential adaptation options in plant protection to changing climatic conditions which might be effective but more expensive should also be exploited in the future. For example, crop residue management for disease control has gained importance with the expansion of conservation agriculture (Melloy *et al.*, 2010). Therefore, more powerful methods should be developed and combined with already known management techniques such as crop rotation in order to impede saprophytic colonization of crop residues by pathogens and to decrease the carry-over of inoculum between cropping seasons (Melloy *et al.*, 2010).

Finally, considering strategic planning, monitoring and forecasting the future distribution of diseases is very important to decide where to grow perennial agricultural crops such as date palms (Shabani and Kumar, 2013). With the knowledge where economically important crop diseases of particular perennial agricultural crops might occur in the future, low risk locations could be identified in order to avoid or minimize the future impact of these diseases (Shabani and Kumar, 2013). For annual crops such as oilseed rape, shifting of cultivation zones was one of the suggested adaptations under a worse-case scenario.

New cropping systems which are resilient to changed climate conditions are required. It is highly probable, that the changes of climate and CO_2 will occur sufficiently slowly so that changes to sowing date, cultivar, crop and other management practices will allow at least some adaptation of the production system by farmers. Several adaptations are conceivable in the timescale available including:

- Crop selection to determine mechanisms and sources of durable disease resistance.
- Crop selection to identify mechanisms and sources of resistance/resilience to abiotic stresses including drought and cold.
- Genetic enhancement to cope with more variable growing conditions.
- Development of new crops to take advantage of more favourable growing conditions.

There are many possible intervention points in the crop–pest/ pathogen interaction, but decisions on which are to be prioritized will depend on a combination of their likely effects and the feasibility of manipulating them in a beneficial way that is both practical and acceptable. Many treatments require initial investment in capacity and resource building referred to above, but if the potential benefits are great then this should guide investment.

12.3.1. Enhanced surveillance

Diagnostic tools and personnel are needed to detect new diseases. These newly introduced diseases can have devastating effects on plant communities. Increased surveillance by these and other agencies will be necessary if climate change results in more disease introductions.

12.3.2. Quarantine and exclusion

Management of climate change will put additional pressure on agencies responsible for exclusion as a plant disease control strategy (Kahn, 1991). In some regions, certain diseases of economic concern do not currently occur because the climate has precluded the causal agents from becoming established. Use of climate matching tools and geographical information systems may assist quarantine agencies in determining the threat posed by a given pathogen in different regions under current and future climate shifts (Coakley *et al.*, 1999; Legreve and Duveiller, 2010). This approach was used by Sansford of the European Union.

Certainly, quarantine measures to control emerging pathogens, for example, would be very important in order to prevent the spread of the pathogens into new areas, because of the alterations in disease geographical and temporal distribution resulting from climate changes.

Exclusion of pathogens and quarantines through regulatory means may become more difficult for authorities as unexpected pathogens might appear more frequently on imported crops.

12.3.3. Physical methods

Typical target pathogens are soil-borne and include *Verticillium dahliae, Sclerotium rolfsii, Rhizoctonia solani* and *Fusarium* species (Katan, 1981). Soil solarization prior to planting is a non-chemical, physical disease-control method for soil disinfection using plastic mulches (Garibaldi, 1987) that fits the concept of IPM. The longer the treatment period, the higher the pathogen-killing rates and the deeper the effectiveness of soil solarization (Garibaldi, 1987). Climate warming may facilitate the use of soil solarization, both in the greenhouse and open field, because it could be used successfully in more pathogen–plant systems and more regions, with the heat penetrating to deeper soil layers and the mulching periods (usually 4 weeks or longer) becoming shorter. Thus, global warming may improve the efficacy and handling of certain disease control methods, particularly those requiring high temperature, such as soil solarization.

12.3.4. Cultural methods

Agronomic practices, such as crop rotation, soil tillage, fertilization, liming, irrigation, selection of the production site, use of resistant/tolerant varieties, and sanitation to reduce the amount of overwintering inoculum, can be used to prevent or reduce the increased disease risks associated with the predicted climate change (Heitefuss, 1989; Juroszek and von Tiedemann, 2011). In general, farmers first of all have to select the optimal site for their crop species and may introduce new crop species that would benefit from climate change and elevated levels of atmospheric CO_2 (Wolfe *et al.*, 2008). Moreover, preventive measures such as the use of cultivar mixtures in fields and the preservation of tree species diversity in forests are still likely to make sense also in the presence of novel climates. Adding diversity to our fields, plantations, forests, and landscapes appears as a commendable insurance policy which may increase the adaptation potential of a range of managed ecosystems.

Enhanced diversity in cropping systems may reduce the risks of diseases that otherwise, in monoculture, would become more severe as a result of climate change. For example, Hannukkala *et al.* (2007) concluded that increased and earlier occurrence of late blight (*Phytophthora infestans*) epidemics in potato (*Solanum tuberosum*) were probably associated with both climate change and lack of rotation. They detected that the epidemics of late blight started 9 days earlier when the preceding crop was potato than when it was another crop. This emphasizes the importance of crop rotation, including the integration of cover crops and intercropping, in reducing specific disease risks associated with expected climate change.

Changing planting and/or harvest dates of annual crops and short-lived perennials can be an effective, low-cost option to render the crop less vulnerable to diseases or adverse abiotic conditions.

Planting of cultivar mixtures and intercropping (e.g. potatoes and faba beans, *Vicia faba*) might be another way to reduce disease risks, for example, by slowing down epidemic rates, thereby facilitating disease control with fungicides, such as shown for late blight in potatoes (Garrett *et al.*, 2001). Under high disease pressure, these methods have to be combined with other integrated pest management (IPM) measures, such as genetic resistance and fungicide application (Wolfe, 1985; Garrett *et al.*, 2001).

Many overwintering necrotrophic pathogens in crop residues left on the field plant can be managed through crop rotation and other agronomic practices designed to reduce inoculum levels (Melloy *et al.*, 2010). Crop residue management for disease control has gained importance with the expansion of conservation agriculture. An obvious disease management strategy in the context of climate change is conservation agriculture, because it benefits crop resilience in stressed environments (Ortiz *et al.*, 2008). To run conservation agriculture successfully for the long term, cultivars are needed with better water use efficiency, improved root health and durable resistance to the economically important pathogens that emerge from crop residues or result from adoption of conservation practices (Ortiz *et al.*, 2008).

Pathogen aggressiveness may also be favoured as a result of increased plant canopy density (due to increased water use, fertilizer application and enriched CO_2 fertilization effect), which could result in increased fungicide applications (Bloomfield *et al.*, 2006). Hence, water conservation may require efficient technologies such as drip irrigation, as already used under arid climatic conditions (Boland *et al.*, 2004). These systems may also contribute to reduced transfer of plant protection products (PPPs) to soil and groundwater. In addition, efficient technologies such as drip irrigation may contribute to reduced leaf wetness and humidity in the crop canopy, resulting in reduced foliar disease occurrence and severity compared with more traditional practices such as overhead irrigation systems.

Agronomic management practices for the control of diseases under climate change conditions includes:

- Monitor the incidence of diseases.
- Identify the life cycles of diseases, as well as carry out phytosanitary applications, with the aim of breaking their life cycles.
- Carry out an integrated management that allows to conserve or to maintain the populations of the biological control agents.
- As management practices against heavy rains, provide drainage channels and control weeds to avoid increase in relative humidity of the crop.
- Polyculture or agro-biodiversity within a lot allows beneficial fungi to be present that reduce the disease incidence.
- To manage adequate seed density that which favours microclimate conditions in the crop.
- Management of weeds, to reduce host plants for pathogens. However, it should be considered to maintain a vegetative cover on the ground, to avoid high temperatures under drought condition which favours the acceleration of biological cycles of pathogens.

12.3.5. Chemical methods

Fungicides are used to control pathogenic fungi and Oomycetes. They may have additional beneficial side effects on the physiology of the plant, resulting in enhanced general stress resistance and higher yield (Wu and Tiedemann, 2001). Predominantly, fungicides are applied through foliar application or as seed dressings.

Changes in temperature, rainfall, wind speed, soil/air moisture and light conditions can influence the effectiveness of fungicide applications (Bedos *et al.*, 2002; Ziska and Runion, 2007). These environmental factors may alter fungicide dynamics in the soil (Monkiedje *et al.*, 2007) and on/in the foliage, including uptake, degradation and volatilization (Bedos *et al.*, 2002). According to Wolfe *et al.* (2008), more frequent rainfall events projected for the northern latitudes during winter and spring may trigger more frequent fungicide applications, because of the difficulty of keeping contact fungicides on the plant canopy. The introduction of fungicides with a greater rain-fastness might help to reduce this problem (Hannukkala *et al.*, 2007). Optimizing the timing of fungicide application can be a simple way to increase fungicide efficacy (Bedos *et al.*, 2002). For example, Augusto *et al.* (2010) reported that applying fungicide early in the morning to folded, wet leaves in peanut (*Arachis hypogaea*) improved spray deposition in the lower canopy, thereby increasing stem rot control (caused by *Sclerotium rolfsii*) and crop yield.

In general, high temperature and humidity seems to favour degradation and volatilization of the active ingredients. Nonetheless, warm and humid conditions may increase the efficacy of most fungicides, because penetration through plant tissues may be easier and the uptake of active ingredients faster (Garcia, 2004).

More fungicide sprays could be necessary to control diseases as effectively as under present management practices. For example, simulation studies showed that under the most severe climate-change scenario, two additional fungicide treatments might become necessary to manage downy mildew (*Plasmopara viticola*) in grapevine (*Vitis vinifera*) as a result of the projected temperature rise in the northwest of Italy by the end of this century.

Climate change could affect the efficacy of crop protection chemicals in one of two ways. First, changes in temperature and precipitation may alter the dynamics of fungicide residues on the crop foliage. Globally, climate change models project an increase in the frequency of intense rainfall events (Fowler and Hennessy, 1995), which could result in increased fungicide wash-off and reduced control. Data from field experiments and modelling studies suggests precipitation during the post-application period (Schepers, 1996) is critical. Precipitation following fungicide application may improve its distribution (Schepers, 1996) but an increase in rainfall intensity can deplete fungicide residue on the foliage (Neuhaus *et al.*, 1974). The interactions of precipitation frequency, intensity, and fungicide dynamics are complex, and for certain fungicides precipitation following application may result in enhanced disease control because of a redistribution of the active ingredient on the foliage (Schepers, 1996). Neuhaus *et al.* (1974) applied simulated rain to potato foliage at two intensities (6 and 30 mm/ha) and found that the higher rate significantly reduced the fungicide residue that could be measured with a chemical assay, but that there was no difference in disease between the two treatments when the leaves were challenged in a bioassay with *Phytophthora infestans.*

Second, morphological or physiological changes in crop plants resulting from growth under elevated CO_2 could affect uptake, translocation, and metabolism of systemic fungicides. For example, increased thickness of the epicuticular wax layer on leaves (Wolfe, 1995) could result in slower and/or reduced uptake by the host, whereas increased canopy size could negatively affect spray coverage and lead to a dilution of the active ingredient in the host tissue. Both factors would suggest lowered control efficacy at higher concentrations of CO_2. Conversely, increased metabolic rates because of higher temperatures could result in faster uptake by and greater toxicity to the target organism. Despite the potential for important interactions, no similar studies evaluating the impacts of climate change variables on physiological aspects have been published for fungicides.

The more frequent rainfall events predicted by climate change models could result in farmers finding it difficult to keep residues of contact fungicides on plants, triggering more frequent applications. Systemic fungicides could be affected negatively by physiological changes that slow uptake rates, such as smaller stomatal opening or thicker epicuticular waxes in crop plants grown under higher temperatures.

Changes in temperature and precipitation can alter fungicide residue dynamics in the foliage, and the degradation of products can be modified. Alterations in plant morphology or physiology, resulting from growth in a CO_2-enriched atmosphere or from different temperature and precipitation conditions, can affect the penetration, translocation and mode of action of systemic fungicides. Besides, changes in plant growth can alter the period of higher susceptibility to pathogens which can determine a new fungicide application calendar (Coakley *et al.*, 1999; Chakraborty and Pangga, 2004; Pritchard and Amthor, 2005).

Fungicides are an effective method of controlling plant diseases, despite negative public perceptions. Fungicides may continue to serve as common disease suppression agents, although alternative measures, such as cultural methods and biological control, should be developed. The persistence of plant protection in the phyllosphere is highly dependent on weather conditions. Changes in duration, intensity, and frequency of precipitation events will affect the efficacy of chemical pesticides and how quickly the active molecules are washed away. Temperature can directly influence the degradation of chemicals and alter plant physiology and morphology, indirectly affecting the penetration, translocation, persistence, and modes of action of many systemic fungicides (Coakley *et al.,* 1999).

Under worst-case scenarios, several crops may require more fungicide spray treatments or higher application rates, thus increasing costs for farmers, prices for consumers and the likelihood of the development of fungicide resistance (Juroszek and von Tiedemann, 2011). The per acre pesticide usage average cost for corn, cotton, potatoes, soybeans and wheat were found to increase as precipitation increases. Similarly, the pesticide usage average cost for corn, cotton, soybean and potatoes also increases as temperature increases, while the pesticide usage cost for wheat decreases (Chen and McCarl, 2001).

Extreme temperatures may affect the efficacy of pesticides, including fungicides, or increase their phytotoxicity (Strand, 2000). Martin and Edgington (1980) showed that higher temperature (15° C vs. 10° C) reduced the efficacy of fungicides (seed treatment) to control loose smut (*Ustilago nuda*) of barley (*Hordeum vulgare*). Such studies may help to elucidate the potential effects of global warming on fungicide efficacy.

For crops such as potato, economic production is often impossible without the application of fungicides. Fungicide usage may increase if changing crop physiology interferes with the uptake and translocation of fungicides or change in other climatic factors (e.g. more frequent rainfall, washing away residues of contact pesticides) mean that there is a need for more frequent applications. Faster crop development at increased temperature could also increase the need for application of fungicides.

12.3.6. Biological methods

There is almost no information on the effects of climate change on biological control of plant diseases (Ghini *et al.,* 2011a). Some microorganisms have been developed for use as bio-control agents of plant diseases (Elad and Freeman, 2002). Bio-control agents can serve as alternatives to chemical fungicides when applied alone or in combination with other control methods. Because they are living organisms, these bio-control agents are also affected by the abiotic environment. For example, the bio-control of the fungal foliar grey mould disease (*Botrytis cinerea*) in greenhouse cucumber crops is affected by climate. Suppression of grey mould by the bio-control agent, *Trichoderma harzianum* T39 is more pronounced at higher temperatures and lower RH levels (Elad *et al.,* 1993).

The efficacy of bio-control treatments can vary under different environmental conditions. Late blight (*Phytophthora infestans*) development depends on the presence of water on the surface of the plant, temperature and RH, and the management of this disease is facilitated by suitable climate management practices. In growth chambers, the duration of the wet period also increased the severity of late blight. In a bio-control trial involving tomato plants inoculated with *P. infestans* sporangia, a control efficacy of 24% was observed among plants subjected to a 24-hour wet period and 39% efficacy among plants subjected to an 8-hour wet period, which is the minimum period for infection. Among the plants incubated at low temperatures

(10° C–15° C), which delayed the appearance of symptoms and slowed disease development, only *T. harzianum* T39 significantly decreased disease severity relative to the untreated control. Among the plants incubated at temperatures of 20° C or 25° C, most of the examined bio-control treatments (various bacterium and yeast isolates) decreased disease severity by 30%–60%. Thus, when environmental conditions are favourable for late blight development, the examined bio-control agents are less effective.

Grey mould is promoted under conditions of moderate temperatures, high humidity, and water on the surface of plants and can be suppressed by rendering the crop's environment less conducive to disease development. Unlike the late blight bio-control, the bio-control treatments (various bacterium and yeast isolates) provided grey mould control even when environmental conditions favoured disease development, with only a slight decrease in efficacy. However, combining efforts to shorten the wet period and lower the RH in the crop with the application of these treatments will optimize their efficacy. These findings suggest that climate change is likely to affect microbial suppression of some plant diseases.

An analysis of the effects of climate change on commercialized bio-control agents of plant diseases in Brazil indicated that *Bacillus subtilis* and *Trichoderma* sp. will be less affected than some other bio-control agents, whereas the efficacy of *Coniothyrium minitans* and *Clonostachys rosea* may be reduced in some areas (Ghini *et al.*, 2011a). Soil-borne pathogens will remain more difficult to control than foliar pathogens, because some soil-borne pathogens can survive in soil for several years and effective pesticides for the control of these pathogens are not yet available (Ghini *et al.*, 2011a).

There may be problems with applications of bio-control agents in the field because of the vulnerability of bio-control agent populations to environmental variations and environmental extremes. If appropriate temperature and moisture are not consistently available, bio-control agent populations may reach densities that are too small to have important effects, and may not recover as rapidly as pathogen populations when congenial conditions reoccur (Garrett *et al.*, 2006).

In general, the impact of antagonists may be modified by various interaction factors, such as plant species, soil type, soil temperature, moisture and nutrient availability. Ghini *et al.* (2008) argued that in spite of potential problems when applying BCAs research efforts must be strengthened to develop biocontrol measures with more tolerance to variable conditions.

12.3.7. Microbial interactions

Climate change may alter the composition and dynamics of microbial communities in aerial and soil environments sufficiently to influence the health of plant organs (Gunasekera *et al.*, 1997). Changed microbial population in the phyllosphere and rhizosphere may influence plant disease through natural and augmented biological control agents. A direct effect of elevated CO_2 is unlikely in the soil environment as the microflora there is regularly exposed to levels 10 to 15 times higher than atmospheric CO_2.

In the rhizosphere, elevated CO_2 would interact with nitrogen and other soil factors to modify the number and type of Arbuscular Mycorrhizal Fungi (AMF) to influence root health and nutrient uptake. Some short-term studies under controlled conditions have shown that elevated CO_2 can stimulate AMF colonization (Staddon and Fitter, 1998) due to faster plant growth. It is not clear if the increases in soil carbon storage due to greater root and AMF growth under high CO_2 will influence AMF colonization.

Trees grown in soils with poor nutrient status (especially nitrogen) are favoured by colonization of roots with AMF (Klironomos *et al.*, 1997). The relationship between elevated CO_2 and AMF is not well understood (Singer, 1996), and there are conflicting reports on how it may be influenced by the nutrient status of the plant and soil. If a lower nitrogen status of plant tissue under increased CO_2 results in more AMF colonization, this could improve plant health through improved nutrient uptake. Similar confusion exists on the potential role of AMF and ectomycorrhizae in the suppression and biological control of plant pathogens. AMF can have positive, negative, or neutral effects on plant disease, and their role is not well understood despite numerous studies on the subject (Pfleger and Linderman, 1994). Clearly, the influence of AMF on plant health under climate change requires further research.

Enhancing endophytic colonization of plants offers prospects of enhanced abiotic and biotic tolerance, thus addressing multiple consequences of climate change in some plants. However, the magnitude of the responses is likely to be limited and for many plants appropriate endophytes may not be available. There are also many practical issues in establishing and maintaining colonization that have not yet been determined.

Indigenous microbial communities play an important role in maintaining plant health. There is a need to promote these beneficial communities. Recent technological advances, such as metagenomic analysis will increase our understanding of microbial dynamics in soil and other environments and further advance the establishment of plant pathogen suppressive microbial populations. One adaptation measure that can be imagined is the introduction of beneficial microorganisms (bio-control agents) to plant surfaces, so that this niche can be occupied in a way that tilts the plant microorganism interaction in a healthy direction. Among the beneficial microorganisms that have been examined, there are some that have been found to persist in stressful microclimates. The selection of such microorganisms and the development of preparations for agricultural use may help growers cope with both abiotic and biotic plant stresses.

The addition of biochar to soil alters microbial populations in the rhizosphere, and may cause a shift toward beneficial microorganism populations that promote plant growth and resistance to biotic stresses. Interestingly, recent findings demonstrate that the application of biochar to soil results in the induction of systemic resistance to foliar pathogens, such as the necrotroph *Botrytis cinerea* (grey mould) in tomato, pepper, and strawberry; the biotrophs *Podosphaera aphanis* (powdery mildew of strawberry) and *Oidiopsis sicula* (teleomorph: *Leveillula taurica*, powdery mildew of pepper and tomato); and the semi-biotroph *Colletotrichum acutatum* (anthracnose) in strawberry (Elad *et al.*, 2010, 2011; Graber *et al.*, 2010; Kolton *et al.*, 2011; Meller Harel *et al.*, 2012). Biochar induces responses along both systemic acquired resistance and induced systemic resistance pathways (Meller Harel *et al.*, 2011a, b; 2012). Thus, soil applications of biochar can be used not only as a climate change mitigation tool but also as a way to help plants cope with biotic and abiotic stresses.

12.3.8. Host resistance

The ultimate solution for crop adjustment to climate change is breeding for desired characteristics associated with future needs. Breeding programmes for crop plants and forest trees can promote genetic diversity, disease resistance and tolerance of environmental stresses. These breeding goals should, of course, be coupled with traditional breeding goals, such as yield, quality and proper shelf-life.

Host-plant resistance involves the use of cultivars that are able to resist or tolerate pathogen attack (Legreve and Duveiller, 2010). The use of resistant or tolerant cultivars is easy, cheap, environmentally sound and effective (Dodds and Rathjen, 2010), unless pathogens overcome the resistance. Expression of plant resistance is determined by the interaction between genetic factors in the pathogen and the plant, and the environmental impact represented by cultural practices and climate.

Cultivar resistance to pathogens may become more effective because of increased static and dynamic defences from changes in physiology, nutritional status, and water availability. Durability of resistance may be threatened, however, if the number of infection cycles within a growing season increases because of one or more of the following factors: increased fecundity, more pathogen generations per season, or a more suitable microclimate for disease development. This may lead to more rapid evolution of aggressive pathogen races. In a pilot study, evolution of *Colletotrichum gloeosporioides* on *Stylosanthes scabra* under elevated CO_2 was monitored. A susceptible cultivar was grown in a controlled environment under 1x or 2x CO_2 and inoculated with three isolates of the pathogen. For each isolate, conidia collected from infected host tissue were used to inoculate a second group of plants of the same cultivar. Successive groups of plants were inoculated with conidia arising from the previous infection cycle to simulate polycyclic disease development and pathogen evolution over time. After each cycle, measurements were made on components of pathogen aggressiveness, such as fecundity, lesion size, lesion number, and disease severity. Preliminary results suggested a significant trend towards increased disease severity (Fig. 12.2); further, two of the three isolates showed a gradual increase in fecundity under elevated CO_2 after eight infection cycles.

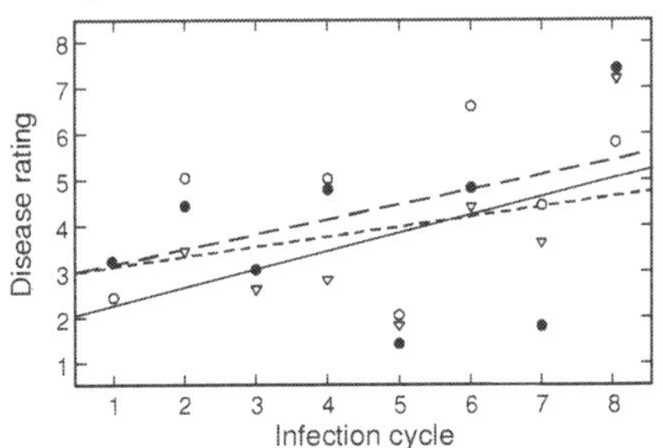

Fig. 12.2. Disease severity (on a scale from 0 to 9) caused by *Colletotrichum gloeosporioides* on susceptible *Stylosanthes scabra* plants under 2x CO_2 over eight cycles of infection. For each of the three isolates (indicated by different symbols and lines), successive groups of plants were inoculated with conidia arising from the previous infection cycle to simulate polycyclic disease development and pathogen evolution over time. The regression lines for the three isolates were significantly different from zero ($P < 0.05$).

Deployment of a major gene for resistance effective under a range of environmental conditions delivers high efficacy for a narrow target disease control with limited duration and high vulnerability. Enhanced efficacy can be delivered through incorporating heterogeneity into both the crop and risk mitigation processes, effectively spreading risk, albeit at the expense of maximum gain from implementation of less durable options, such as deployment of resistant cultivars in extensive monoculture or extensive use of a single fungicide (Newton *et al.,* 2009). However, if the challenge of increasing food production by 70% by 2050 can be met only by deploying cultivars with single or multiple resistance genes or by use of fungicides, it will be difficult to argue for an alternative approach that may not produce the highest attainable yields.

Elevated temperature may cause the breakdown of temperature-sensitive resistance in oat cultivars with Pg3 and Pg4 genes (Martens *et al.,* 1967). Exposure to unsuitable environmental conditions can cause loss of resistance (Strand, 2000). For example, disease resistance can be temperature-dependent (Hartleb and Heitefuss, 1997), as in the case of Phoma stem canker (*Leptosphaeria maculans*) of oilseed rape, where resistance is expressed at 15°C but not at 25°C (Huang *et al.*, 2006). There is a need for improved resistance of oilseed rape to Phoma stem canker, which is favoured at increased temperatures (Stonard *et al.,* 2010). Therefore, strategies for breeding cultivars with improved resistance to pathogens may need to include trials in countries with a warmer climate that represent predicted climates in currently cooler latitudes (Butterworth *et al.,* 2010). There is evidence for increased aggressiveness at higher temperatures of isolates of *Puccinia striiformis* demonstrating that wheat rust fungi can adapt to and benefit from warmer temperatures to cause severe disease in previously unfavourable environmental conditions (Milus *et al.,* 2009). These adapted isolates are also challenging the effectiveness of disease resistance genes which operate at higher temperatures. Disease resistance is also affected by other environmental factors, such as duration of leaf wetness, nutrient status (e.g. nitrogen fertilization), soil type and availability of water. Therefore, breeders need advice on which plant pathogens might become economically important in the future (Chakraborty *et al.,* 1998).

Inversely, a rice bacterial blight resistance gene was reported to benefit from high temperature (Webb *et al.*, 2010). This is also true for some resistance genes in wheat against the rust fungi *Puccinia recondita* and *P. striiformis* (Uauy *et al.,* 2005; Legreve and Duveiller, 2010).

Changes in atmospheric composition and climate may influence host–pathogen interactions and host resistance through numerous types of morphological and physiological alterations, such as stomatal morphology and physiology, increases in rates of net photosynthesis, papillae formation, silicon accumulation at appressorial penetration sites, carbohydrate accumulation in leaves, increased cuticular waxes, additional epidermal cell layers, increased fibre content, reduced nutrient concentration and modified production of resistance-related enzymes such as rubisco (Chakraborty *et al.*, 2000).

Many necrotrophic pathogens that have broad host ranges do not follow gene-for-gene specificity when host resistances are available, they generally rely upon multiple defence mechanisms each offering a partial reduction in disease severity, but not complete protection. A necrotrophic pathogen is able to grow saprophytically after the crop starts to senescence, producing large quantities of inoculum that can infect subsequent crops, thereby often losing the advantage of using a partially resistant variety to reduce inoculum (Melloy *et al.*, 2010). Under climate change, increased biomass of crops and alternative host plants will further increase inoculum production. To be effective, partial resistance has to be combined with agronomic and other practices to develop robust integrated crop protection strategies, which will not suffer such boom–bust cycles. Knowledge of pathogen biology and epidemiology in farming systems must improve significantly to account for changes in geographical distribution of crops to better manage necrotrophs under climate change. As croplands move to match climatic suitability, breeding targets themselves will change with changing pathogen spectra, disease dynamics and relative economic values (Ortiz *et al.*, 2008).

In summary, future plant disease management strategies should include resistance breeding approaches for broad adaptation to multiple environments. Breeding for resistance against several pathogens should be combined with breeding for tolerance to abiotic stresses such as drought and heat (Legreve and Duveiller, 2010). However, when several plant diseases occur at the same time, multiple host-plant resistance may not be sufficient and needs to be supported by integration of other plant protection methods, such as crop rotation and the use of spatial and temporal crop diversity (Tilman *et al.*, 2002).

12.3.9. Transgenic disease resistant varieties

For rust and other biotrophic pathogens that follow a 'gene-for-gene' model of host-pathogen specificity, more sustainable disease management will come from combinations of resistance genes, assembled using marker assisted selection or transgenic approaches. These approaches are being used with wheat rust (Bariana *et al.*, 2007; Ellis *et al.*, 2007). Once developed, resistance sources can be evaluated using facilities mimicking future climate scenarios including increased temperature (Huang *et al.*, 2006) to ascertain their longevity. Pre-emptive breeding can also start in these facilities to identify and replace the most vulnerable genes/ gene combinations.

Transgenic ringspot virus-resistant papaya has been genetically engineered to contain a virus gene that encodes for the production of the coat protein of the virus. As a major component of viruses, the coat protein's primary function is to protect viral genetic information. Expression of this gene in the resulting papaya line renders the plants resistant to the virus (Fig. 12.3).

Fig. 12.3. Papaya ring spot virus. Yellow plants on the left are non-transgenic papaya showing severe symptoms; plants on the right are transgenic 'Rainbow' papaya resistant to the virus.

The transgenic plum called C5 (variety Honey Sweet) express a plum pox virus coat protein, the plant produces the coat protein mRNA and it is processed by a system called post transcriptional gene silencing (PTGS), which functions like the plant's immune system and is mechanistically similar to RNAi (Hily *et al.*, 2004). C5 provides a unique source of germplasm for future breeding programme worldwide (Fig. 12.4).

Fig. 12.4. *Left* - C5 (variety Honey Sweet) genetically modified plum resistant to plum pox virus. *Right* - severe symptoms of *plum pox virus* (PPV) infection on susceptible plum fruits

Transgenic rice lines with inducible production of ethylene (ET) were generated by expressing the rice ACS2 (1-aminocyclopropane-1-carboxylic acid synthase, a key enzyme of ethylene biosynthesis) transgene under control of a strong pathogen-inducible promoter. The transgenic lines exhibited increased resistance to a field isolate of sheath blight (*Rhizoctonia solani*), as well as different races of blast (*Magnaporthe oryzae*). Similarly, transgenic rice plants showing obvious resistance to rice bacterial blight disease have been developed (Fig. 12.5).

Fig. 12.5. *Left* - Transgenic rice IR72 with *Xa21* showing bacterial blight resistance (T) with excellent agronomic performance under field conditions in the Philippines. *Right* – Susceptible check (C)

Approaches that have been used to produce transgenic sweet potato include expression of viral replicase genes, anti-sense RNAs and viral coat protein genes. Transgenic sweet potato is resistant to feathery mottle virus (FMV) and has the potential of increasing yields of sweet potato roots and foliage.

Exploitation of the plant immune system against cassava mosaic disease by expression of hairpin RNA homologous to viral sequences has proven effective to generate virus-resistant cassava (Yadav *et al.*, 2011).

Specifically in summer squash, coat protein-mediated resistance is used against viruses. Transgenic summer squash plants (CZW-3, Liberator III and Destiny III) with resistance to three viruses (Cucumber Mosaic Virus, Zucchini Yellow Mosaic Virus and Watermelon Mosaic Virus 2) produce as many or more marketable fruit than non-transgenic squash. A Cornell University study found that transgenic squash with resistance to three viruses produced a 50-fold increase in marketable yield over non-transgenic varieties (Fuchs *et al.*, 1998).

The GM potato variety Desiree has been transformed with an R or resistance gene (Rpi-vnt1.1) along with its native promoter and terminator intact, using GMO technology. The R gene confers the GM potato line with resistance to the late blight fungus.

The RNA interference (RNAi) gene was introduced in pinto bean (*Phaseolus vulgaris*). This GM bean is resistant to the golden mosaic virus. The transgenic bean could increase production by 10–20%.

The coat protein (CP) gene of cucumber mosaic virus (CMV) was cloned into sweet pepper plants to confer resistance to CMV.

Table 12.3. Commercially available genetically modified crops for disease resistance (Castle *et al.*, 2006)

Crop	Trait phenotype	Target trait gene(s)	Trait designation	Originating company	Year of first commercial sale	Trade name
Squash	Resistance to CMV, WMV2 and ZYMV	Coat protein genes of CMV, WMV2 and ZYMV	CZW3	Asgrow; Semini Vegetable Seeds (now Monsanto)	1998	Destiny III, Conquerer III, Liberator III
	Resistance to WMV2 and ZYMV	Coat protein genes of WMV2 and ZYMV	ZW-20	Asgrow; Semini Vegetable Seeds (now Monsanto)	1995	Preclude II, Patriot II, Declaration II, Independence II
Papaya	Resistance to PRSV	Coat protein gene of PRSV	55-1, 63-1	Cornell university; University of Hawaii, USDA	1998	SunUp, Rainbow

CMV - Cucumber mosaic virus; PRSV - Papaya ringspot virus; WMV2 - Watermelon mosaic virus 2; ZYMV - Zucchini yellow mosaic virus.

Commercially available genetically modified crops for disease resistance are presented in Table 12.3 (Castle *et al.*, 2006).

12.3.10. Integrated disease management (IDM)

IDM is an ecosystem-based strategy that focuses on long-term prevention of diseases or their damage through a combination of methods such as biological control, use of resistant cultivars, habitat management and cultural practices (Strand, 2000). Plant protection products are used following thorough monitoring and based on established economic damage thresholds. Integrated crop management (ICM) adds to IDM components such as soil tillage, nutrient supply and water management in order to develop sustainable production systems (Juroszek *et al.*, 2008). A frequent review of IDM recommendations is necessary to minimize negative environmental and economic impacts associated with the likely increase of plant protection products use, including fungicides (Wolfe *et al.*, 2008).

12.4. FUTURE LINES OF WORK

12.4.1. Enhanced research and development

To date, little attention has been given to the effects of climate change on plant health. For plant diseases, the rapidity of the predicted climate change will necessitate more research into alternative crops and cultivars with increased stress tolerance and disease resistance, enhanced cultural practices, and climate-based site selection. Developing information and research networks will play a vital role by linking researchers and practitioners across the country to allow information to be gathered, reviewed, and redistributed to stakeholders.

12.4.2. Enhanced public and professional awareness

Public awareness is increasing about the plant health hazards. Continuing education on climate-related disease issues would help plant disease scientists to gain the skills to detect and identify newly emerging problems.

12.4.3. Integrated and adaptive policy development

Mitigating and adapting to climate change will only be successful in a supportive and encouraging policy environment in many varied fields, including agriculture and natural resource management.

Under worst-case scenarios, several crops may require more fungicide spray treatments or higher application rates, thus increasing costs for farmers, prices for consumers and the likelihood of the development of fungicide resistance (Juroszek and von Tiedemann, 2011). Some agricultural systems may be more flexible than others in the adoption of new cultivars and cultural practices to cope with the increased risk of certain diseases. Annual crops will have an advantage over perennials, as they provide more flexibility when it comes to adopting new cultivars and cultural practices. Potential adaptation strategies must be accompanied by cost–benefit analysis. Evaluating the efficacy of current physical, chemical and biological control methods under changing climatic conditions and research concerning new tools and strategies (including plant breeding) for coping with the predicted changes will be of great strategic importance.

Fungicides may continue to serve as common disease suppression agents, although alternative measures, such as cultural methods and biological control, should be developed.

Plant diseases are a major problem not only for food production but also for the quality and safety of important food stuffs. In Europe, mycotoxins and pesticide residues are among the top food safety concerns associated with a changing climate. For example, the concentration of mycotoxin produced by Fusarium head blight in grain generally increases with the number of rainy days and days with high RH but decreases with low and high temperatures. Changes in both temperature conditions and atmospheric composition may influence the severity of outbreaks of Fusarium head blight and the production of mycotoxins. The most significant effects of mycotoxin presence generally occur during the production phase, but the entire wheat value chain can be affected (Chakraborty and Newton, 2011). Shifts in any of the components of the disease triangle can dramatically affect the magnitude of disease expression in a given pathosystem. Therefore, it is not at all surprising that disease patterns have already changed and will continue to change in response to the effects of climatic changes on pathogens and hosts. The ultimate solution for crop adjustment to climate change is breeding for desired characteristics associated with future needs. Breeding programmes for crop plants and forest trees can promote genetic diversity, disease resistance and tolerance of environmental stresses. These breeding goals should, of course, be coupled with traditional breeding goals, such as yield, quality and proper shelf-life.

It is not advisable to depend on one 'high input variety', or one breed of crop variety. Varieties should be mixed and changed. A broad genetic variability serves as a foundation for robust crops. In addition, it seems more recent traditional breeding has not selected for CO_2 responsiveness, which simply means newer breeds do not benefit from elevated CO_2 as much as older breeds (Ainsworth *et al.,* 2008). The simple agronomic measures such as mixing varieties reduced rice blast severity by 94% and increased yield by 89%.

12.4.4. Need for adoption of novel approaches

Changing disease scenario due to climate change has highlighted the need for better agricultural practices and use of eco-friendly methods in disease management for sustainable crop production (Boonekamp, 2012). In the changing climate and shift in seasons, choice of crop management practices based on the prevailing situation is important. In such scenarios, weather-based disease monitoring, inoculum monitoring, especially for soil-borne diseases and rapid diagnostics would play a significant role. There is a need to adopt novel approaches to counter the resurgence of diseases under changed climatic scenario. Integrated disease management strategies should be developed to decrease dependence on fungicides (Gautam and Bhardwaj, 2011). Other multipronged approaches include healthy seeds with innate forms of broad and durable disease resistance, and intercropping systems that foster refugees for natural bio-control organisms. In addition, monitoring and early warning systems for forecasting disease epidemics should be developed for important host–pathogens which have a direct bearing on the earnings of the farmers and food security at large. A diversified crop protection strategy has been highlighted in a comprehensive study on an integrated approach to control all foliar diseases in barley (Boonekamp, 2012). Use of botanical pesticides and plant-derived soil amendments such as neem oil, neem cake and karanj seed extract also help in mitigation of climate change because it helps in the reduction of nitrous oxide emission by nitrification inhibitors such as nitrapyrin and dicyandiamide (Pathak, 2010).

12.4.5. Need for biodiversity based approach

It will be important to develop disease management strategies that minimize the risk of severe crop loss under the future climatic conditions. Several Integrated Pest Management (IPM) approaches can be strengthened and used to control diseases through environmental manipulation that put in benign equilibrium between host and pathogen. More breeding can be used to develop resistant varieties. But this will take time and research and, as it happened in the past, the results might not reach smallholder farmers in marginal areas. For these millions of farmers an important answer can come from a wise use of agricultural biodiversity. These farmers are the custodians of this diversity and they are able to manage production systems that are rich in agricultural biodiversity. The idea is that different varieties have different sources of resistance to different diseases and therefore identifying these differences can be very useful to develop mixtures that will include varieties that differ in their resistance to diseases. The fact that diversity can reduce disease pressure and that it works in practice at least in the case of Ugandan beans. But the question is that we also need to see if those systems are also resilient to future changes because we cannot predict how a future pathogen will look like. There is diversity of pathogen populations and also of the overall resistance of different varieties to the diseases. Some of the varieties were resistant to different strains and some were susceptible to all strains, but what is important is that with this information we now know more in detail which varieties are resistant and also to which strain, the strains coming from different sites.

12.5. CONCLUSIONS

The climate is changing, resulting in changes in the plant diseases that we need to prevent or manage. Because climate change predictions are based on uncertain information, especially at regional and local levels, our response must be in the form of adaptive strategies that are reviewed and adjusted as new information and improved climate models become available. Developing effective strategies will require the involvement of government agencies, academia, and the general public.

Systematic quantitative analysis of climate change effects will be necessary for developing future disease management plans, such as plant breeding, altered planting date schedules, chemical and biological control methods and increased monitoring of new disease threats. The existing preventive crop protection measures, such as use of diversity of crop species in the cropping systems, adjustment of sowing or planting dates, use of crop cultivars with superior resistance and/or tolerance to diseases and abiotic stress, use of reliable tools to forecast disease epidemics, application of IPM strategies, and effective quarantine systems, may become important in the future. Effective crop protection technologies are available and will provide appropriate tools to adapt to altered conditions. Therefore, real-time disease monitoring and surveillance have to be in priority to adopt measures against any unforeseen event that might happen due to climate change and/or global change as well as shifts in seasonality.

Significant contributions can derive from better field monitoring of diseases and pests, improved timing of plant protection product (PPP) application and better systems for delivering PPPs to their targets (Strand, 2000). This includes *in situ* and *ex situ* conservation of genetic diversity in crop species, including their wild relatives, in order to increase the availability of genes to breed for resistance to existing and new pathogens and to abiotic stresses (Garrett, 2008). The requirement to breed new cultivars with more durable resistances within shorter time periods will possibly give biotechnological approaches a key role in crop adaptation to changing climate. Accumulation of minor genes and race-non-specific resistance to pathogens by conventional breeding programmes will remain equally important to achieve durable resistance in diverse environments. Preventive plant protection measures may become particularly important under climate change. These may include greater heterogeneity in cropping systems (Christen, 2008) that reduces disease risks, use of superior cultivars resistant and/or tolerant to abiotic and biotic stress (Ordon, 2008), and reliable tools for forecasting pathogen occurrence in order to respond in a timely manner to plant pathogens (Tiedemann and Ulber, 2008).

Disease-forecasting models based on weather data can help to identify the meteorological factors (and the time period) which are significantly correlated with disease (Coakley *et al.*, 1988). These types of disease-forecasting models can be combined with general circulation models in order to simulate future scenarios of disease epidemics, although most general circulation models operate on larger scales of resolution. Down-scaling climate models can contribute to bridge this gap (Soussana *et al.*, 2010). Based on future scenarios of disease epidemics, disease management practices can be suggested and/or improved (Salinari *et al.*, 2006).

In contrast, it should be noted that besides potential risks, there are also potential gains associated with expected climate change. Agricultural systems have shown considerable adaptability to climate through management changes (Chakraborty, 2005). This adaptability also provides a high likelihood that significant climate shifts may be buffered. Moreover, there may be some gains from fewer frosts, altered precipitation patterns, CO_2 fertilization, and longer growing seasons (Chakraborty, 2005). For example, global warming may favour the use of soil solarization practices in horticulture, using plastic mulches over moist planting beds, known to control soil-borne pathogens such as *Verticillium dahliae* and *Fusarium* sp. (Strand, 2000).

The on-going, yet accelerating progress in agricultural technologies, mainly new cultivars, novel PPPs and agro-technological innovations, has constantly required significant adaptations of the production systems by farmers, with increased yields and higher revenues being the main drivers. There is a clear requirement for consistently utilizing, adapting and improving crop protection strategies and tools and enhancing the relevant knowledge base. Improved understanding of the drivers of disease cycles, epidemic development and host responses will continue to be a fundamental prerequisite to enable farmers and advisors to predict and manage diseases under changing local conditions (Chakraborty *et al.*, 2000; Melloy *et al.*, 2010). Consequently, diverse, flexible and resilient crop production systems will be needed (even more than today) that can cope more readily with conditions in a changing environment (Garrett, 2008).

Section IV
NEMATODE PATHOGENS

13

Climate Change and Nematode Pathogen Scenarios

13.1. INTRODUCTION

Plant pathogenic nematodes are one of the important biotic constraints in crop production. They will be affected by climate changes in a complex way. Beneficial or negative effects are expected, depending on species, regional situations, crops, climate and geography. Significant changes induced by variations in climate extremes (mainly, but not uniquely, related to temperature and moisture regimes) are expected within the next decades. Among other environments and natural systems, they will also affect crops as well as plant parasites and biological control agents, with different outcomes, depending on geographic regions and agricultural systems (Olfert and Weiss, 2006). Nematodes by virtue of their trophic diversity act directly as pests and indirectly as vectors of other plant pathogens, and play a significant role in regulating plant growth in agro-ecosystems.

Scientific research on climate change and its impact on herbivorous nematodes are very limited. However, based upon their environmental requirements some assumptions are possible. Severe droughts resulting in a reduction of soil water will most likely negatively affect soil nematodes. Higher average temperatures will probably have little effect, since thermal conductivity of soils is low.

Apparently, a prediction of how climate change will affect herbivorous soil nematodes and thus yields cannot be made. There is some evidence that population dynamics may change, but so far no trend is clear. Basically, and most likely true for all ecological research, the impacts of climate change are specific to crop/plant, region and interacting species.

Majority of plant-pathogenic nematodes spend part of their lives in soil, and therefore, soil is the source of primary inoculum. Life cycle of a nematode can be completed within 2–4 weeks under optimum environmental conditions. Temperature is the most important factor, and development is slower with cooler soil temperatures. Warmer soil temperatures are expected to accelerate nematode development, perhaps resulting in additional generations per season. The drier temperatures are expected to increase symptoms of water stress in plants infected with nematodes such as the soybean cyst nematode. Overwintering of nematodes is not expected to be significantly affected by changes in climate, although for some, such as the soybean cyst nematode, egg viability may be reduced in mild winters.

Climate change due to increased emission of greenhouse gases is posing a serious challenge to sustainability of crop production by interfering with biotic and abiotic components and their interactions with each other. Global warming resulting in elevated carbon dioxide (CO_2) and temperature in the atmosphere may influence plant pathogenic nematodes directly by interfering with their developmental rate and survival strategies and indirectly by altering host plant physiology. Available information on effect of global warming on plant pathogenic nematodes though limited, indicates that nematodes show a neutral or positive response to CO_2 enrichment effects with some species showing the potential to build-up rapidly and interfere with plant's responses to global warming. Studies have also demonstrated that the geographical distribution range of plant pathogenic nematodes may expand with global warming spreading nematode problems to newer areas. Besides plant parasites, other trophic groups (microbial feeders, predators and insect parasites) of soil nematodes also shown to influence the plant productivity indirectly by regulating the key ecosystem processes including decomposition, nutrient mineralization, biological pest suppression and energy transfer in food webs. These findings underline the importance of understanding the impact of climate change on soil nematodes and its implications to crop production while developing mitigation and adaptation strategies to address impact of climate change on agriculture.

13.2. EFFECTS OF CLIMATE CHANGE ON NEMATODES

13.2.1. Increased temperatures

Temperature is the most important factor influencing the biology of nematodes. It affects nematode survival, distribution, embryogenesis and hatching, migration and penetration, development, and symptom expression in the plants. Temperature requirements vary among root-knot nematode populations (thermo types) and with each host-parasite combination. Non-efficient hosts of root-knot nematodes become progressively efficient as temperature rises. The plants become efficient hosts at high temperatures.

Nematode developmental rate is directly influenced by the temperature with slower development at cooler and faster growth rate at warmer soil temperatures. Therefore, increase in atmospheric temperature due to global warming is expected to result in more number of generations per season and expansion of their geographical distribution range. Other potential effects of elevated temperature on parasitic nematodes include altered sex ratio, host defence responses and interference in their survival strategies like dauer juveniles or egg diapauses in extreme environments.

Drier temperatures are expected to increase symptoms of water stress in plants infected with nematodes such as the soybean cyst nematode. Overwintering of nematodes is not expected to be significantly affected by changes in climate, although for some, such as the soybean cyst nematode, egg viability may be reduced in mild winters.

Warming will generally cause a pole-ward shift of the risk of damage from potato nematodes which would increase in all regions and may become a serious problem with additional generations per year.

Data from lab and field experiments demonstrated that higher *Radopholus similis* nematode population densities and greater root damage occurs to banana as the projected temperature increases.

13.2.2. Elevated CO_2 levels

Herbivorous nematodes showed neutral or positive response to CO_2 enrichment effects with some species showing the potential to build up rapidly and interfere with plant's response to global warming. The number of herbivore, bacterivore and fungivore nematodes was significantly higher under winter wheat and sugar beets grown under elevated CO_2 (550 ppm), while the number of carnivores were not changed. The total numbers of herbivore, bacterivore and fungivore nematodes were\ higher under elevated CO_2 wheat than under elevated CO_2 sugar beet, most likely due to the very different root system of both plant species (Sticht *et al.*, 2009).

Available information on effect of global warming on soil nematodes though limited, indicates that abundance of soil nematodes in general is either increased or unaffected by elevated CO_2 levels while individual species and trophic groups differ considerably in their response to climate change. Nematode's reaction to rising CO_2 levels is complex, depending on trophic groups: population increased in forest (Hoeksema *et al.*, 2000), grassland (Hungate *et al.*, 2000) and pasture soils (Yeates and Orchard, 1993; Yeates et al., 1997; 2003).

13.2.3. Precipitation patterns

Increasing rainfalls may induce changes in the selection of varieties or cultivated species that may increase the incidence of nematodes, switching either their species composition (replacement) or even increasing natural antagonists and prevalence levels, due to higher moisture, affecting i.e. the spreading of antagonists or predators (i.e. *Pasteuria* sp. parasites). Higher amounts of water in soil will either increase the incidence of nematodes dispersion and the probability of a local extinction.

Changes in the water regime will influence the water availability for plants and will affect yields, increasing densities and levels of nematode parasitism, due to positive effects on total plant biomass. Changes in water regimes may also influence the duration of a nematode parasitic event, i.e. the infection phase or the progeny produced in a season, with final outcomes on total numbers.

13.2.4. Drought

Soil nematodes are dependent on the continuity of soil water films for movement. Their activities are largely controlled by soil biological and physical conditions (Yeates and Bongers, 1999). Severe droughts resulting in a reduction of soil water will most likely negatively affect soil nematodes. Increased water stress due to climate change diminishes plant vigour and alters C: N ratios, lowering plant resistance to nematodes.

13.2.5. Air pollutants

Air pollutants affect phytoparasitic nematodes through changes in the physiology of the host plant. Synergistic interactions between ozone (O_3) or sulphur dioxide (SO_2) and *Meloidogyne incognita* were observed on tomato, as higher levels of foliar damage were found on nematode infested plants. Galls on roots were higher in nematodes infested plants exposed to 100 ppb O_3. Ozone, however, reduced the reproduction performance of *M. incognita*, with lower numbers of eggs and egg masses (Khan and Khan, 1997).

13.3. IMPACTS OF CLIMATE CHANGE ON NEMATODES

Climate warming contributes to the increase in the range of distribution, colonization of new hosts and modification of their development cycles in biologically diverse nematodes. The effects of climate changes on the epidemiology and management of the main phytoparasitic nematode species occurring includes the alteration of the reproductive cycles due to plants productivity, the geographic dispersion by more northern or higher altitude shifts, and the spread of vectors.

13.3.1. Expansion of geographical range

Studies have demonstrated that the geographical distribution range of plant pathogenic nematodes may expand with global warming spreading nematode problems to newer areas. Before 1970, the soybean cyst nematode was mainly distributed in the Mississippi River Delta area, northern Arkansas, southern Missouri, southern Illinois, and western Kentucky. It is now distributed throughout the main soybean production area and has become the number one soybean pest in the U.S. In Iowa alone it caused an estimated yield loss of 201 million bushels (worth about US$ 1.2 billion) during the 1998 growing season. *Radopholus similis*, currently absent from cooler, higher altitude areas is likely to expand its range.

13.3.2. Increase in additional generations

Warmer soil temperatures are expected to accelerate nematode development, perhaps resulting in additional generations per season.

13.3.3. Impact on overwintering

Overwintering of nematodes is not expected to be significantly affected by changes in climate, although for some, such as soybean cyst nematodes, egg viability may be reduced in mild winters.

13.4. PREDICTION MODELLING FOR NEMATODES

Predictive models may help in evaluating a number of different outcomes and situations, i.e. providing a basis to identify the best among different options, to select in case of invasive species (eradication, suppression, no action) (Fraser *et al.*, 2006). In general, an insight on the possible outcomes of climate changes may be facilitated by the application of simulation models. Their sensitivity should be checked, however, with real scenarios, to verify the level of uncertainty and affordability, as well as their effectiveness as informative tools (Kickert *et al.*, 1999).

One key issue in modelling is the level of resolution achieved and the forecasting precision. The asymmetric increase of temperatures (with different shifts for minima and maxima), as well as the non-linearity in growth or development responses, require sub-daily resolutions when modelling climate changes affect fungal plant pathogens, due to a bias introduced when only mean temperatures are used, without considering the amplitude of the fluctuations (Scherm and Van Bruggen, 1994). Nematodes, however, due to a longer time required for completion of the different life-stages (eggs, juveniles, adults), may be modelled with different scenarios of daily mean temperature changes, aiming at providing a first rough insight about possible field situations and outcomes.

Although modelling necessarily represents a simplification of the more complex, non-linear relationships occurring in soil among different organisms, it provides an insight on dynamics otherwise difficult to trace, due to the amount of experimental data and time required to investigate these relationships. At small scales, generic dynamic crop modelling may result informative about the effects of weather, variety, pests, soil, and management practices on crop growth and yield, as well as on soil N and organic carbon dynamics in aerobic as well as anaerobic conditions. Data generated included pest induced yield losses, allowing a comparative analysis with different greenhouse gas emission situations (Aggarwal *et al.*, 2006). Weather variables introduced in modelling result fundamental to explore the different outcomes of climate changes and hence estimate the spatial and seasonal dynamics of pests, in the mid-term (Yonow *et al.*, 2004).

A further tool is given by probability distribution maps (PDM). Their use allows the identification of potential risks related to the distribution of nematode pests, vectors or plant diseases. PDMs are calculated on the basis of the combined use of local records of pest occurrence, climate data and subsequent statistical analysis. They show the areas susceptible for colonization or of endemism, for a given organism (Morales and Jones, 2004). This tool has several practical advantages, including the possibility of early identification of invasion areas by a pest or the possibility to anticipate the insurgence of epidemics for secondary pests, already present in areas with sub-optimal conditions for life-cycle. PDMs require the monitoring of different climatic variables at the regional scale and a given resolution level, as well as the implementation of an early monitoring and detection support system.

13.5. ADAPTATION AND MITIGATION STRATEGIES

Any research programme aiming at introducing new varieties or improving the long-term application of biological, genetic or agronomic tools (i.e. irrigation programmes, soil labour techniques, forest management or land use) should take into account the effects of climate and environment changes. Particular attention must be paid in the management of nematodes through the introduction of resistance genes in commonly used varieties, since an evaluation is needed about the persistence of the genetic pool introduced, on a scale of decades or years. Hence, protection of biodiversity of plant genetic pools has a practical and immediate justification. The term "adaptive strategy" reflects, in this sense, the need for a quick and flexible response, since some climate changes are expected to occur on a short temporal scale. Some changes are already in action and newly introduced varieties may display in a few years unsuitable agronomic traits, i.e. higher susceptibility due to increased air moisture or rainfall, or may become susceptible to newly colonizing or substitution pests, thus vanishing long term research investments.

Climate change will cause alterations in the spatial and temporal distribution of nematodes and consequently the control methods will have to be altered to suit these new situations. Assessments of the impact of climate change on nematode infestations and in crops provide a basis for revising management practices to minimize crop losses as climate conditions change (Ghini *et al.*, 2008).

Recent observations suggest that nematode pressure on plants may increase with climate change (Ghini *et al.*, 2008). As a result, there may be substantial rise in the use of nematicides in both temperate and tropical regions to control them. Non-chemical nematode management methods (green manuring, crop rotation, intercropping, mulching, application of organic manures, etc.) assume greater significance under changing climate scenario.

Climate change mediated changes in physiology can alter the expression of resistance genes. The most serious threat to genetic resistance to nematodes may be posed by the increased selection pressure resulting from acceleration of nematode developmental rate and increase in number of generations per season due to global warming (Ghini *et al.*, 2008).

Nematode, bacterial and fungal based bio-pesticides are highly vulnerable to environmental stress. Increase in temperature, UV radiation and a decrease in relative humidity may reduce the efficacy of these bio-agents.

Therefore, there is a need to develop appropriate strategies for nematode management that will be effective under situations of global warming in future.

13.5.1. Biotechnological approaches to nematode resistance

Biotechnology offers several benefits for nematode control in an integrated management strategy, such as reducing risks to the environment and to human health, accessibility for food producers in the developing world, and the possibility of achieving durable, broad-spectrum nematode resistance (Thomas and Cottage, 2006). A number of genes that mediate nematode resistance have now been or soon will be cloned from a variety of plant species. Nematode resistance genes are present in several crop species and are an important component of many breeding programmes including those for tomato, potato, soybeans and cereals (Trudgill, 1991). There are essentially three approaches for engineering resistance: transgenic expression of natural resistance genes in heterologous species; targeting and disruption of the nematode; and feeding site attenuation (Thomas and Cottage, 2006).

Using marker-assisted breeding techniques, Monsanto is introducing root-knot nematode resistance into elite genetics to develop cotton varieties that could potentially increase lint yield by an average of 8 to 10 per cent under root-knot nematode infestations.

A new, high throughput screening method is being utilized to accelerate the incorporation of reniform nematode resistance into elite genetics of cotton. The product could potentially increase lint yield by an average of 10 to 15 per cent under reniform nematode infestation conditions.

13.6. CONCLUSIONS

Research on impact of climate change on soil nematodes has been limited, with most work concentrating on the effects of a single atmospheric constituent under controlled conditions. A few recent experiments have also reported the response of herbivorous nematodes to elevated CO_2 beyond the trophic group level. Nevertheless, findings of these studies give an insight into the response of nematode trophic groups to climate change and its consequences to agricultural production. Responses of herbivorous nematodes to CO_2 enrichment were observed to be either neutral or positive but not negative. Further, studies predicting changes in geographical distribution of plant parasitic nematodes using simulation models give a fairly good idea about future scenarios of nematode diseases of plants. More long-term studies in varied agro-ecosystems under different cropping systems of particularly tropical regions are needed to critically assess the impacts of climate change on soil nematodes. This knowledge is vital for developing appropriate adaptation and mitigation strategies to minimize effect of climate change on agriculture.

14

Effects of Climate Change Variables on Nematode Pathogens

14.1. INTRODUCTION

Nematodes are a morphologically and functionally diverse group of soil organisms (Bongers and Bongers, 1998; Ferris, 2010; Ferris *et al.*, 2001; Neher, 2010; Yeates *et al.*, 1993), so shifts in nematode community composition can provide clues about the sensitivity of soil food webs and soil functions to climate change. However, the impacts of climate change on nematode abundance have been idiosyncratic across studies, ranging from positive to negative for unknown reasons. Even when focusing on specific nematode functional groups, such as plant-feeders, the impacts of climate change on nematode abundance vary considerably across studies (A'Bear *et al.*, 2014; Ayres *et al.*, 2008; Blankinship *et al.*, 2011; Cesarz *et al.*, 2015; Ferris *et al.*, 2012). The lack of predictability of nematode responses to climate change likely reflects three sources of variation among studies:

- Biogeographic variation in how climate change influences soil abiotic conditions (e.g., moisture) and the basal resources of soil food webs, including roots, bacteria, and fungi (García-Palacios *et al.*, 2015).
- Variation in the extent to which various factors regulate soil food web structure (e.g., abiotic vs. biotic factors, top-down vs. bottom-up factors, indirect vs. direct effects of climate change) (Kardol *et al.*, 2010; Shao *et al.*, 2015)
- Variation in the taxonomic and functional resolution of nematode identification (Cesarz *et al.*, 2015; Neher and Weicht, 2013).

Thus, additional studies are required to generate hypothesis and to better understand the effects of climate change on soil nematodes. Multi-factor studies are especially needed because of their rarity and potential for interactions among multiple aspects of climate change (e.g., warming and elevated CO_2) (Eisenhauer *et al.*, 2012). There are only two reports of the combined, *in situ* effects of elevated CO_2 (eCO_2) and warming on nematode communities, both for rice paddies in Asia (Chen *et al.*, 2014; Okada *et al.*, 2014).

14.2. ELEVATED TEMPERATURES

Increasing temperatures are expected to enhance plant growth rates and yields, providing a greater food source for nematode pests but also increasing the whole ecosystem complexity. Temperatures will also affect plant's phenology (earlier germination of seeds, plant flowering or ripening). Both factors will be responsible for a higher carrying capacity of plants, an earlier emergence of diseases/vectors and crop attacks, longer life-cycles and reduced pest/disease generation times.

The fact that plants become efficient hosts at high temperatures is probably due to three factors:

- High temperatures are optimum conditions for nematode activity.
- Stress caused by high temperature makes the plants more vulnerable to nematode attack.
- Chemicals responsible for cell necrosis may not be produced or may be neutralized or counter-acted at high temperatures.

As Goudriaan and Zadoks (1995) pointed out for plant diseases that also applies to nematodes, that they will thrive where host plants grow best. However, the range of genetic variability and adaptability is still unknown thus far, for several nematode species. The identification of these adaptive boundaries is important, since they represent the basis for future expansions and colonization of new areas, following isothermal shifts. The reduction or elimination of natural barriers confining a species in a particular geographic or climatic area should always represent a source of concern: for example, changes are expected in the distribution of insects, due to movements towards higher altitudes or latitudes, as a response to higher temperatures (Walther *et al.*, 2002). Reaching a higher altitude may allow a species to overcome a natural barrier, thus colonizing new geographic regions from which it was previously excluded. Although this factor is a major concern for nematode phoresis or vectoring capacity by insects, i.e. *Monochamus* sp., provides a link to the potential spread of nematode pests (i.e. pine wilt nematodes, *Bursaphelenchus* sp.).

Temperature is the most important factor influencing the biology of nematodes. Nematode developmental rate is directly influenced by the temperature with slower development at cooler and faster growth rate at warmer soil temperatures. In plants, under environmental stress (at high temperatures), nematode reproduction is higher. Hatching and embryogenesis is faster and migration and penetration by the nematode is favoured by high temperatures. The nematode life cycle is also completed faster at high temperature; therefore, more generations are produced. Moreover, at high temperatures, fewer males develop. Therefore, increase in atmospheric temperature due to global warming is expected to result in more number of generations per season and expansion of their geographical distribution range. Other potential effects of elevated temperature on parasitic nematodes includes altered sex ratio, host defence responses and interference in their survival strategies like dauer juveniles or egg diapauses in extreme environments.

Drier temperatures are expected to increase symptoms of water stress in plants infected with nematodes such as the soybean cyst nematode. Overwintering of nematodes is not expected to be significantly affected by changes in climate, although for some, such as the soybean cyst nematode, egg viability may be reduced in mild winters.

Plantain (*Musa* sp. AAB) is both an important staple and cash crop throughout the West/Central African humid forest zone. Major yield constraints are root nematodes, particularly *Radopholus similis*. Data from lab and field experiments demonstrates higher nematode population densities and greater plantain root damage at the projected temperature increases. *R. similis*, currently absent from cooler, higher altitude areas is likely to expand its range.

Climate change may also influence the plant nematode interactions by interfering with host defence mechanisms. Rebetez and Dobbertin (2004) reported that strong climate warming that has occurred in recent years favoured pine wood nematode (*Bursaphelenchus mucronatus*) and bark beetles and increased drought stress reduced tree resistance against these pests. This resulted in rapid tree mortality in pine forests in Switzerland.

The alfalfa stem nematode appeared in Yolo County due to climate change. The stem nematode parasitizes alfalfa crops and can cause severe crop losses. The nematode disperses through several vectors, including waterways and irrigation runoff, contaminated farm equipment, and other anthropogenic means similar to other plant diseases. As a new and increasing threat in Yolo County, stem nematode populations are thought to have appeared due to warmer minimum temperatures moving closer to the reproductive threshold for the nematode.

14.3. CHANGES IN WATER REGIMES

Changes in the water regime will influence the water availability for plants and will affect yields, increasing densities and levels of nematode parasitism, due to positive effects on total plant biomass.

Increased flood rates are expected to raise the frequencies of several water-borne (aquatic fungi) or moisture dependent (powdery mildew) diseases, indirectly affecting plants productivity and nematode densities. Availability of water affects soil nematodes, whereas changes in plants communities are known to indirectly affect their abundance and community structure (Kardol *et al.*, 2010). Changes in water regimes may also influence the duration of a nematode parasitic event, i.e. the infection phase or the progeny produced in a season, with final outcomes on total numbers. Similarly, higher/lower incidence of droughts will affect either survival of field inocula, frequency of parasitism by endemic aquatic or nematophagous fungi, or the time spent in soil by juveniles seeking for an available root penetration sites.

Factors affecting a function's response (i.e. crop productivity, pest or disease prevalence) to a given climatic change are key elements and must be correctly identified, in order to yield reliable information for modelling or preventive measures. They also include physical properties of soil, i.e. texture or water retention capacity which may affect, at different extents, soil nematodes responses to changes of the hydrologic cycle (Wessolek and Asseng, 2006).

The effect of increasing rainfall regimes may vary: in arid climates higher water availability may yield positive consequences on agricultural practices, increasing crops productivity or cultivated surfaces. Increasing rainfalls may also induce changes in the selection of varieties or cultivated species that may increase the incidence of nematodes, switching either their species composition (replacement) or even increasing natural antagonists and prevalence levels, due to higher moisture, affecting i.e. the spreading of antagonists or predators (i.e. *Pasteuria* sp. parasites). On the opposite, in an already moist climate, an increasing rainfall regime of the same magnitude may yield floods, with soil losses due to erosion, forcing the adoption of other plants/cultivars or sowing periods. Higher amounts of water in soil will either increase the incidence of nematodes dispersion and the probability of a local extinction. Catastrophic rainfalls may also affect some management practices, like soil labour, solarization, sowing and fertilization, or vanish the effects of chemicals.

14.4. COMBINED CHANGES IN TEMPERATURE AND MOISTURE REGIMES

Combined changes in temperature and moisture regimes will also affect biological control agents (BCA). As for nematodes, increased numbers due to higher yields and plants carrying capacities will affect BCA densities, an effect also expected to produce earlier emergence times and outbreaks due to a corresponding earlier host nematode emergence. Other effects will be the longer life cycle and/or the reduced generation time, an increased spatial spread to newly colonized areas following nematodes spreading, or the increased spread of water related parasites (i.e. *Catenaria*).

Other indirect mechanisms will favour nematodes insurgence or spreading of invasive species (Goudriaan and Zadoks, 1995; Fuhrer, 2003). Among them, the increased (reduced) leaf moisture in wet (dry) conditions (affecting i.e. foliar nematode species), the changes in the survival of BCA or propagules of endophytes (spores, bacterial cells, mycorrhizae) in soil or other host tissues, the reduced (increased) plants resistance due to physiological adaptations to temperature changes (i.e. prolonged vegetation, lignification, longer lasting resistance-breaking temperatures, insurgence of virulent nematode populations), changes in the nutritive value of host plant tissues, higher densities of alternated or secondary hosts (weeds).

14.5. CO_2 ENRICHMENT

CO_2 concentration plays a crucial role in various aspects - biology of plant and insect parasitic nematodes including host recognition, recovery from dauer stage or diapause, etc. Elevated CO_2 levels may also influence these nematodes indirectly by altering host physiology (defence mechanisms such as production of secondary metabolites and nutrient status such as C: N ratio, etc.). It may also influence microbial feeding nematodes due to changes in quality and availability of food under enriched CO_2 conditions in soil. The impacts of climate change can be positive, negative or neutral, since these changes can decrease, increase or have no impact on nematode abundance, depending on each region or period.

Available information on effect of global warming on soil nematodes though limited, indicate that abundance of soil nematodes in general is either increased or unaffected by elevated CO_2 levels while individual species and trophic groups differ considerably in their response to climate change. Herbivorous nematodes showed neutral or positive response to CO_2 enrichment effects with some species showing the potential to build up rapidly and interfere with plant's response to global warming.

Effects of elevated CO_2 on nematode densities as mediated by the host plant are "plant species-specific" and include negative effects (Niklaus *et al.*, 2003), positive effects (Yeates *et al.*, 2003), and no significant effects (Ayres *et al.*, 2008). Most of these studies proposed that changes in root biomass and C/N ratio were the main factors responsible for the effects of elevated CO_2 on nematode abundance (Niklaus *et al.*, 2003; Yeates *et al.*, 2003; Ayres *et al.*, 2008). However, the mechanisms underlying the effect of elevated CO_2 on the interaction between plant-parasitic nematodes and their host plants are poorly understood.

Systemic acquired resistance (SAR) is considered to be the major induced plant defence that confers long-lasting protection against nematodes (Durrant and Dong, 2004). SAR depends on the salicylic acid (SA) pathway and is associated with accumulation of pathogenesis related proteins, which are considered to contribute to resistance. Researchers have recently suggested, however, that the jasmonic acid (JA) pathway is also an indispensable component of plant resistance to nematodes (Howe and Jander, 2007; Bhattarai *et al.*, 2008). The JA pathway is associated with expression of proteins (including proteinase inhibitors, phenylalanine ammonialyase, and lipoxygenase), up-regulation of secondary metabolites, and induction of plant volatile organic compounds (VOC). Cooper *et al.* (2005) reported that the artificial induction of JA-pathway defences reduced reproduction of root-knot nematodes on tomato plants (Cooper *et al.*, 2005). A tomato genotype (35S::Prosystemin) in which induced defence was dominated by the JA pathway (hereafter referred to as a "JA defence-dominated genotype") has stronger resistance to nematodes than a JA defence-recessive genotype (spr2, a jasmonate-deficient mutant) and that the specific responses of these isogenic tomato genotypes to elevated CO_2 requires further investigation (Sun *et al.*, 2010).

Based on several works referring to plant induced defence under elevated CO_2 (Zavala *et al.*, 2008; 2009; Sun *et al.*, 2010), it has been hypothesized that elevated CO_2 would reduce the resistance of a JA defence dominated genotype against *M. incognita* by altering the JA pathway but enhance the SA-pathway defence of a JA-defence-recessive genotype infected by *M. incognita.* In this study, it is determined whether elevated CO_2 affects the regulation of genes and the production of secondary metabolites and the emission of VOC associated with the JA pathway of isogenic tomato genotypes. It has also been determined whether elevated CO_2 affects the regulation of genes associated with the SA pathway. Finally, it is determined whether the changes in these pathways and genes are associated with the performance of *M. incognita* under elevated CO_2.

Similar to other organisms which feed on plants, increased CO_2 levels are believed to have an impact on herbivorous nematodes (Ayres *et al.*, 2008) and several studies have been conducted, where the above ground plant community was exposed to elevated CO_2. Almost all of these studies were done in different grasslands and forests, and thus results have been variable and contradictory. Research results regarding nematodes, from experiments conducted on agricultural crops in arable soils, are very limited. Basically, all kinds of results were determined: increase, decrease and no change of nematodes populations (Sticht *et al.*, 2009). A recent publication presents results of a long term agricultural experiment conducted in winter wheat and sugar beets in Germany. Winter wheat and sugar beet were grown in rotation under 550 ppm atmospheric CO_2 compared to ambient (380 ppm) atmospheric CO_2. The number of herbivore, bacterivore and fungivore nematodes was significantly higher under wheat and sugar beets grown under elevated CO_2, while the number of carnivore was not changed. The total numbers of herbivore, bacterivore and fungivore nematodes were higher under elevated CO_2 wheat than under elevated CO_2 sugar beet, most likely due to the very different root system of both plant species (Sticht *et al.*, 2009). However, impacts on yields were not determined.

The observations that elevated CO_2 levels often induces increased root production, it can be presumed that herbivorous nematode communities will be relatively more affected by increases in atmospheric CO_2 concentration. Positive effects of CO_2 enrichment on the abundance of herbivorous nematodes have been reported in some studies (Yeates *et al.*, 2003). The abundance of *Tylenchus* and *Longidorus* increased after 5 years of CO_2 enrichment, but there was no effect on the abundance of *Paratylenchus, Trichodorus* and members of Hoplolaimidae in pasture plots (Yeates *et al.*, 2003). Yeates *et al.* (1997) reported increase in abundance of *Meloidogyne* in response to CO_2 enrichment in grassland turfs while 7 other herbivorous nematode taxa were not affected. The abundance of *Pratylenchus* was positively associated with CO_2 concentration in gley, but not in organic soil around a natural CO_2 vent in New Zealand (Yeates *et al.*, 1999).

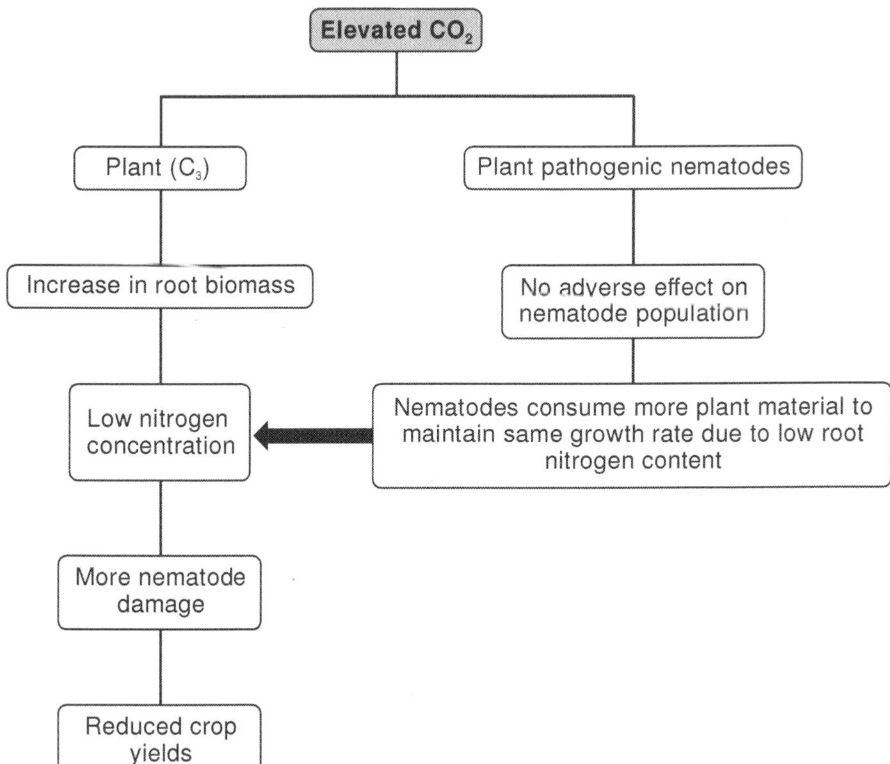

Fig. 14.1. Consequences of elevated CO_2 levels on plant-nematode interaction and crop productivity (Somasekhar and Prasad, 2012)

Although root biomass increased under elevated CO_2, the damage due to root-feeding nematodes was more under elevated CO_2 compared to the ambient levels in a grass species (Wilsey, 2001). Similarly, neutral responses of herbivorous nematodes to CO_2 enrichment were observed despite increase in root production by 3-32% in different locations (Ayres *et al.*, 2008). This may be due to decrease in root quality (low nitrogen content) or increase in nematode antagonists.

The interaction of elevated CO_2 with Nitrogen fertilization or residue addition significantly affected the soil nematode community indices. The residue addition stimulated structure index and inhibited plant-parasites response to the elevated CO_2 in a wheat field (Li *et al.*, 2009).

Experiments with rice have showed no adverse effects of elevated CO_2 levels up to 700 ppm on the abundance of soil nematodes and penetration of rice root-knot nematode, *M. graminicola* (Fig. 14.1) (Somasekhar and Prasad, 2010).

Critical assessment of available information on responses of herbivorous nematodes to CO_2 enrichment reported from different ecosystems indicates that the responses are either neutral or positive but not negative so far (Table 14.1).

Table 14.1. Response of herbivorous nematodes to CO_2 enrichment

Cropping system	Experimental arena*	Nematode response**	Location	Reference
Grassland	CER	+/N	New Zealand	Yeates *et al.* (1997)
Grassland	Vent	N	New Zealand	Yeates *et al.* (1999)
Grassland	FACE	+/N	New Zealand	Yeates *et al.* (2003)
Grassland	OTC	+/N	California, USA	Hungate *et al.* (2000)
Grassland	OTC	N	California, USA	Ayres *et al.* (2008)
Grassland	OTC	N	California, USA	Ayres *et al.* (2008)
Grassland	FACE	+/N	Germany	Sonnemann and Wolters (2005)
Sugar beet & wheat rotation	FACE	+	Germany	Sticht *et al.* (2009)
Grassland	SACC	N	Switzerland	Niklaus *et al.* (2003)
Grassland	CER	N	Montpellier, France	Ayres *et al.* (2008)
Rice- wheat rotation	FACE	+/N	China	Li *et al.* (2007, 2009)
Rice	OTC	N	India	Somasekhar and Prasad (2010)

* OTC = Open top chambers, SACC = Screen aided CO_2 control, FACE = Free air CO_2 enrichment, CER = Controlled environment room, Vent = Natural CO_2 vent.

** + = Positive or increase in abundance, N = Neutral or not affected

Positive effects of increased CO_2 levels alone are expected on plant's productivity (Olszyk and Ingram, 1993; Goudriaan and Zadoks, 1995; Manning and Thedemann, 1995; Chakraborty *et al.*, 2000). They will also increase the numbers of herbivores or efficiency in water, N use and conversion of radiation (Olesen and Bindi, 2002). However, nematodes reaction to rising CO_2 levels is complex, depending on trophic groups. No effect was reported on nematodes from prairie soil (Freckman *et al.*, 1991), but numbers decreased in cotton rhizosphere (Runion *et al.*, 1994), or increased in forest (Hoeksema *et al.*, 2000), grassland (Hungate *et al.*, 2000) and pasture soils (Yeates and Orchard, 1993; Yeates *et al.*, 1997; 2003). Higher CO_2 levels lowered the numbers of bacterial feeders, increasing fungal feeders and predators in forest (Neher *et al.*, 2004) or prairie soils (Yeates *et al.*, 2003).

14.6. SEVERE DROUGHTS

Soil nematodes are dependent on the continuity of soil water films for movement. Their activities are largely controlled by soil biological and physical conditions (Yeates and Bongers, 1999). Increased water stress due to climate change diminishes plant vigour and alters C: N ratios, lowering plant resistance to nematodes. Severe droughts resulting in a reduction of soil water will negatively affect soil nematodes.

Infection with plant-parasitic nematodes can exacerbate or counteract the effects of abiotic stress on plants, as their parasitism in roots severely disrupts plant water relations (Bird, 1974; Haverkort *et al.*, 1991; Smit and Vamerali, 1998). Fig. 14.2 shows an example of the positive interactive effect of drought stress and nematode infection on rice plants. In this case, the addition of nematodes ameliorated the severity of the drought stress. Several studies have been carried out to examine the effect of combined nematode and drought stress on plant growth and development. In the Ivory Coast, the nematode *Heterodera sacchari* increased drought-related losses in upland rice by contributing to reduced leaf water potential, stomatal conductance, and leaf dry weight (Audebert *et al.*, 2000). A similar study investigated the effect of drought and the cyst nematode *Globodera pallida* on water use efficiency in potato (Haverkort *et al.*, 1991). Both factors were found to affect growth negatively, although the simultaneous effect of both stresses was non-additive, perhaps because the infected plants used less water, thus reducing drought stress. *Globodera pallida* has also been shown to cause retardation in potato root development, which in turn had the effect of reducing drought tolerance (Smit and Vamerali, 1998).

Fig. 14.2. Growth of rice under combined biotic and abiotic stress conditions. Plants are shown after exposure to drought stress, infection with the plant-parasitic nematode *Meloidogyne graminicola*, or the two stresses in combination.

Increased drought stress reduced pine tree resistance against pine wood nematode (*Bursaphelenchus mucronatus*) and bark beetles. This resulted in rapid tree mortality in pine forests in Switzerland (Robetez and Dobbertin, 2004).

14.7. AIR POLLUTANTS

Air pollutants affect phytoparasitic nematodes through changes in the physiology of the host plant. Synergistic interactions between ozone (O_3) or sulphur dioxide (SO_2) and *Meloidogyne incognita* were observed on tomato, as higher levels of foliar damage were found on nematode infested plants. Galls on roots were higher in nematodes infested plants exposed to 100 ppb ozone at 5 hours intervals every third day. Ozone, however, reduced the reproductive performance of *M. incognita*, with lower numbers of eggs and egg masses at 50 and 100 ppb (Khan and Khan, 1997).

The root galling on eggplants caused by *M. incognita* was severe at SO_2 and NO_2 polluted sites but egg laying and egg mass production was significantly reduced (Khan and Khan, 2007).

The okra plants grown at SO_2 and NO_2 polluted sites exhibited suppressions in plant growth, yield and photosynthetic pigments. The joint effects of the nematode (*M. incognita* race 1) and coal-smoke were synergistic. Egg laying was reduced by 13-27% (Khan and Khan, 1994).

The effects of the increase in UV-B radiation were highly significant for the occurrence of disease epidemics.

14.8. CONCLUSIONS

Plant pathogenic nematodes are one of the important biotic constraints in crop production. Climate change due to increased emission of greenhouse gases is posing a serious challenge to sustainability of crop production by interfering with biotic and abiotic components and their interactions with each other. Global warming resulting in elevated carbon dioxide (CO_2) and temperature in the atmosphere may influence plant pathogenic nematodes directly by interfering with their developmental rate and survival strategies and indirectly by altering host plant physiology. Available information on effect of global warming on plant pathogenic nematodes though limited, indicate that nematodes show a neutral or positive response to CO_2 enrichment effects with some species showing the potential to build up rapidly and interfere with plant's response to global warming. These findings underline the importance of understanding the impact of climate change on soil nematodes and its implications to crop production while developing mitigation and adaptation strategies to address impact of climate change on agriculture.

Research on impacts of climate change on soil nematodes has been limited, with most work concentrating on the effects of a single atmospheric constituent under controlled conditions. More long term studies in varied ecosystems under different cropping systems of particularly tropical regions are needed to critically assess the impact of climate change on soil nematodes. This knowledge is vital for developing appropriate mitigation and adaptation strategies to minimize the impact of climate change on agriculture.

15

Impacts of Climate Change Induced Consequences on Nematode Pathogens

15.1. INTRODUCTION

The climate change can influence host plant growth and susceptibility; pathogen reproduction, dispersal, survival and activity; as well as host-pathogen interaction. Once climate and nematodes are closely related, climate change will probably alter the geographical and temporal distribution of phytosanitary problems. New nematodes may arise in certain regions, and other nematodes may be economically important, especially if the host plant migrates into new areas (Coakley *et al.*, 1999). The host range changes can be extended due to crop migration. A new geographical distribution of potato cyst nematode (*Globodera rostochiensis*) is predicted, with northward expansion of potato in Finland and a higher number of generations per year. The geographical distribution of the virus-vector nematodes (*Xiphinema* and *Longidorus*) in Great Britain and Continental Europe were associated with increase in temperature. Since the nematodes geographical distribution is directly related to temperature, there could be more problems with these microorganisms in northern Europe, caused by the increase in existing populations and to the dissemination of these species from the southern region.

Nematode management strategies depend on climate conditions. Climate change will cause alterations in the nematode geographical and temporal distributions and consequently the control methods will have to be adapted to this new reality. Changes in temperature and precipitation can alter the nematicide residue dynamics, and the degradation of products can be modified. Alterations in plant morphology or physiology, resulting from growth in a CO_2–enriched atmosphere or from different temperature and precipitation conditions, can affect the penetration, translocation and mode of action of systemic nematicides.

Global warming and climate changes will result in:

- Extension of geographical range of nematode pathogens.
- In cooler latitudes, global warming brings new nematode species.
- Increased risk of invasion by migrant nematodes.
- Reduced effectiveness of crop protection technologies.
- Increase in pesticide use.
- Increased probability of nematodes developing faster resistance to nematicides.

- Warmer winter temperatures would reduce winter kill, favouring the increase of nematode populations.
- Rising temperatures extend the crop growing season.
- Overall temperature increases may influence crop nematode interactions by speeding up nematode growth rates which increases reproductive generations per crop cycle.

15.2. EXPANSION OF GEOGRAPHICAL DISTRIBUTION

Boag *et al.* (1991) used data from soil samples collected during the European plant parasitic nematode survey to assess the possible impacts of climate warming on the geographical range of virus-vector nematodes. Initial analysis of nematode presence-absence data suggested a close association between mean July soil temperature and nematode distribution. Based on this result, the authors predicted that climate change could result in increased nematode and virus problems in northern Europe; a 1° C warming would allow the species in study to migrate northward by 160 to 200 km (Neilson and Boag, 1996). Although nematodes migrate very slowly, humans are credited with efficiently disseminating them. Hence, nematode spread into new regions could put a wide range of crops at risk; additionally, introduction of new crops into a region could also expose them to infestation by nematode species already present. Changes in precipitation could influence nematode distribution on a large scale; although previous findings had suggested that soil moisture would not affect nematode distribution in most agricultural soils in northern Europe (Neilson and Boag, 1996).

Studies have also demonstrated that the geographical distribution range of plant pathogenic nematodes may expand with global warming spreading nematode problems to newer areas. The soybean cyst nematode (*Heterodera glycines*) is the cause of great economic losses to soybean producers in the US. The pest has been expanding since the early 1950s, but the increase has been more dramatic since the early 1970s. Before 1970, the soybean cyst nematode was mainly distributed in the Mississippi River Delta area, northern Arkansas, southern Missouri, southern Illinois, and western Kentucky. It is now distributed throughout the main soybean production area and has become the number one soybean pest in the US (Rosenzweig *et al.*, 2000). In Iowa alone, it caused an estimated yield loss of 201 million bushels (worth about $1.2 billion) during the 1998 growing season (USDA, 1999). In the northern production region, the nematode has up to three generations per year, depending on planting and weather conditions during the growing season. A longer growing season, associated with a warmer climate, would result in an increased risk of losses similar to the ones reported during the year 1998. This pest has been monitored and mapped since the 1950s.

Analysis of the distribution of soybean cyst nematode between 1971 and 1998 in the United States showed rapid dispersal and a very high establishment rate in the central and northern US producing regions (Chakraborty and Datta, 2003). This phenomenon was probably accentuated by the adoption of new production zones linked to climate fluctuations (Somasekhar and Prasad, 2012). In addition, the modelling of soybean phenology within the agricultural potential area of Quebec using climatic parameters of the recent past (1971-2000) and the near future (2041-2070) has shown that more northern latitudes offer new areas for growing soybeans

(Gendron St-Marseille, 2013). Similarly, modelling the life cycle of *H. glycines* based on climatic parameters has demonstrated that it can already be established within the agricultural regions where soybean is currently grown (Dong and Opperman, 1997; Koenning, 2004; Niblack, 2005) and that it has the capacity to follow the move of its host to the north (Gendron St-Marseille, 2013).

Warming will generally cause a pole-ward shift of the risk of damage from potato nematodes which may become a serious problem. Using simulated climate change, Carter *et al.* (1996) predicted that warming will expand distribution of the potato cyst nematode, *Globodera rostochiensis* by 2050 in Finland. The predicted effects of climate change on diseases of selected major agricultural and forestry species in Ontario showed that the cyst nematode, *Heterodera glycines*; root lesion nematodes, *Pratylenchus* sp.; and pine wood nematode, *Bursaphelenchus xylophilus* severity increases due to increase in rate of disease development and potential duration of epidemic due to climate change (Boland *et al.*, 2004).

In India, rice root-knot nematode, *Meloidogyne graminicola*; once considered to be serious pest only in upland rice, has made its importance felt in almost all rice growing areas and in all types of rice culture including hill ecosystems in recent years (Prasad and Somasekhar 2009; Pankaj *et al.*, 2010).

Neilson and Boag (1996) assessed the possible effect of climate change on the distribution of some common virus-transmitting *Xiphinema* and *Longidorus* species within Great Britain. They observed that theoretically an increase in 1° C in mean temperature would result in the northward extension of these nematode species by about 160-200 km. Colonization of new areas by virus-vector nematodes has serious implications for agriculture.

Using mathematical models, Carter *et al.* (1996) simulated climate change in Finland and concluded that warming will expand the cropping area for cereals by 2050 (100 to 150 linear km per 1° C increase in mean annual temperature). Furthermore, higher yields are expected with higher CO_2 concentration. In this scenario, potato cropping will also be benefited with an estimated 20 to 30% increase in yield. However, a new distribution of the potato cyst nematode (*Globodera rostochiensis*) is also predicted, with northward expansion in the country.

In coffee crop, the potential impacts of climate change on the spatial distribution of the coffee nematode (races of *Meloidogyne incognita*) in Brazil were determined using a geographic information system (Ghini *et al.*, 2008). Maps obtained considering the A2 scenario indicated that there could be an increase in infestation of the nematode when compared to the climatological normal from 1961-1990. For the B2 scenario, the infestation level will also be higher as compared to the current scenario, but it will be lower than for the A2 scenario for both.

15.3. BREAKDOWN OF NEMATODE RESISTANCE

Genetic resistance to *Meloidogyne* sp. is sensitive to soil temperatures above 28° C. Tomato, bean, and sweet potato lose resistance at elevated soil temperatures (Dropkin, 1969; Fassuliotis *et al.*, 1970; Jatala and Russell, 1972). High soil temperatures appear to be the main reason that root-knot nematode resistance is not effective in Florida, USA (Walter, 1967), and in many tropical countries. Results by Araujo *et al.* (1983) indicates that race 4 of *M. incognita* reproduces better on resistant tomato genotypes than race 1.

The resistance to root-knot nematode in tomato (cv. 'Sanibel') has often failed as a result of the heat instability or apparent temperature sensitivity of the resistant Mi gene (Fig. 15.1). For example, previous research has demonstrated threshold soil temperatures and incremental reductions in nematode resistance with each degree above 78° F, such that at 91° F tomato plants are fully susceptible. This would suggest that in Florida, use of these varieties may have to be restricted to spring plantings when cooler soil temperatures prevail.

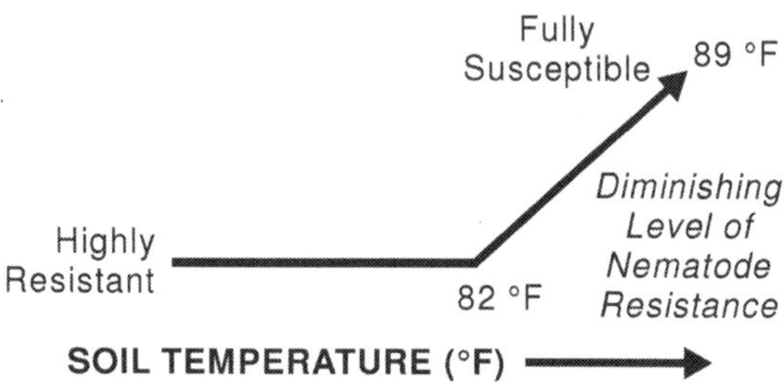

Fig. 15.1. Complete loss of root-knot nematode resistance conferred by the Mi gene in tomato due to increasing soil temperature.

Susceptibility of bell pepper (Charleston Belle, Carolina Wonder, Keystone Resistant Giant, and Yolo Wonder B) to *M. incognita* increased as temperature increased from 24 to 32 °C. Reproduction factor of *M. incognita* and root galling increased ($P < 0.05$) for all cultivars as temperature increased (Table 15.1). Overall, reproduction of *M. incognita* and severity of root galling on the resistant isolines Charleston Belle and Carolina Wonder were less ($P < 0.05$) than on susceptible Keystone Resistant Giant and Yolo Wonder B, and the two cultivars within each group did not differ. However, temperature x cultivar interaction was found ($P < 0.05$) for reproduction index and root galling (Thies and Fery, 1998). Further research is necessary to characterize the usefulness of these varieties under the high soil temperature conditions of Florida. Like tomato, use of these varieties may have to be restricted to spring plantings when cooler soil temperatures prevail.

Table 15.1. Effect of bell pepper cultivars differing in resistance to *Meloidogyne incognita* grown at different temperatures on reproductive index and root galling (eight weeks after inoculation)

Temperature/cultivar	Reproductive index*			Gall index**		
	24° C	28° C	32° C	24° C	28° C	32° C
Charleston Belle	0.3 a	4.9 b	24.7 c	1.53 a	2.46 b	3.61 c
Keystone Resist. Giant	6.1 c	32.2 de	29.9 f	4.79 d	7.62 e	8.74 f
Carolina Wonder	0.1 a	4.0 b	22.9 c	1.57 ab	2.37 ab	3.55 c
Yolo Wonder B	4.7 c	29.4 de	125.8 ef	4.37 cd	7.82 ef	8.61 ef

Mean separation within columns by Duncan's multiple range test at $P \leq 0.05$. Means were compared across all temperature and cultivar combinations.

* Reproduction index (final population/initial population) of *M. incognita*

** Gall index: 1 = no galls, 2 = 1-3%, 3 = 4-16%, 4 = 17-25%, 5 = 26-35%, 6 = 36%-50%; 7 = 51-65%, 8 = 66-80%, and 9 = greater than 80% of root system galled.

In grapevine, the number of galls and egg sacs of *M. javanica* increased with time and temperature on both rootstocks, although not significantly in all cases (Table 15.2). This increasing infestation with time can be ascribed to normal population increase with increasing degree days (Loubser, 1988). Furthermore, pathogenicity as measured by the degree of galling, also appeared to increase with increasing temperature. This was more evident on the moderately resistant rootstock 143 B Mgt (*Vitis vinifera* × *V. riparia*) [compared to the susceptible rootstock Jacquez (*Vitis aestivalis* × *V. cinerea* × *V. vinifera*)] where the number of eggs increased significantly (and apparently galling also) between treatments B and C, irrespective of the number of degree days which remained the same. The reason for this was seen as a breakdown in resistance at the higher temperature. Chitambar and Raski (1984) also found that the grapevine rootstock cultivars Harmony and Couderc 1613 lost their resistance at 36°C.

Table 15.2. Effect of temperature on root growth and infestation of grapevine rootstocks by *Meloidogyne javanica*

Treatment	Jacquez			143 B Mgt		
	Galling[1]	Egg sacs[2]	Eggs[3]	Galling[1]	Egg sacs[2]	Eggs[3]
A. 23 °C/44 days ($572DD_{10}$)	16.2 a	5.2 a	460 a	1.2 a	0.0 a	0 a
B. 23 °C/178 days ($1014DD_{10}$)	62.6 ab	22.2 a	560 a	2.0 a	2.0 a	75 a
C. 33 °C/44 days ($1012DD_{10}$)	41.2 ab	14.6 a	464 a	66.6 ab	16.6 a	396 b
D. 33 °C178 days ($1794DD_{10}$)	613.2 b	613.2 b	855 b	230.6 b	220.6 b	825 b

Treatments which differ significantly ($P \leq 0.05$) are marked vertically with different letters

1. Galling is expressed by the number of galls visible per 5 g of new roots under 20x magnification
2. Egg sacs are expressed by the number visible per 5 g new roots under 20x magnification
3. Eggs are the average number calculated per egg sac

DD_{10}: Physiological time expressed as degree-days above a predetermined threshold of 10° C

Various stages in the Soybean Cyst Nematode (SCN) disease cycle are affected differentially by temperature and moisture. The highest winter survival of SCN eggs occurs in the colder areas of the continent. Thus, spring inoculum levels may be highest in the northern range of soybean culture. Optimal soil temperatures for egg hatch, root penetration, and juvenile and adult development are 24° C, 28° C and 28–32° C, respectively. Temperatures below 15° C and above 35°C, little nematode development occurs. Thus, temperature can affect the number of SCN generations per growing season. In theory, with fewer generations, new races will develop less quickly. The more moderate winter temperatures will reduce egg survival, while the higher temperatures in the growing season will increase egg hatch, the rate of nematode development, and perhaps, the number of generations per season. Soil water is important for SCN movement and development, but water is unlikely to be a limiting factor early in the season. More importantly, the drier growing conditions of summer will increase the yield loss due to SCN because of reduced root surface.

Down-regulation of genes for defence signalling (Iox7, Iox8 and acc-s) reduced the production of cysteine proteinase inhibitors (CystPI), which are specific deterrents of nematodes resulting in increased activity by the nematodes (Fig. 15.2).

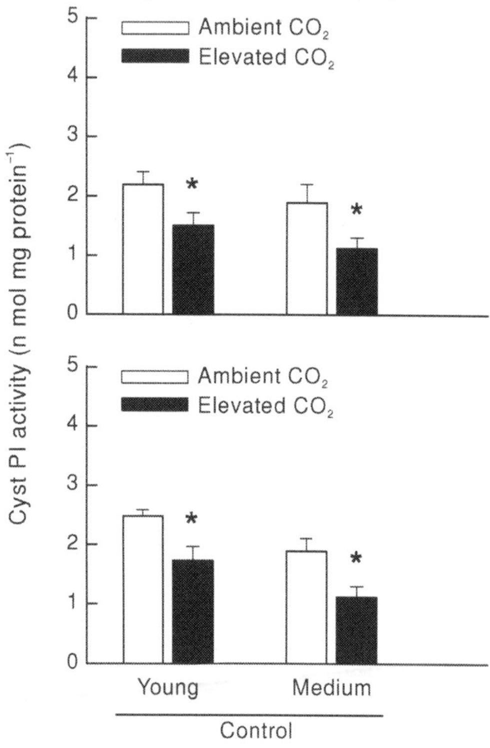

Fig. 15.2. Effect of down-regulation of defence signalling genes on reduced production of cysteine proteinase inhibitors (deterrents of nematodes)

While the exact cause of the recent alfalfa stem nematode outbreak in Yolo County is unclear, an increase in winter temperatures is likely to be an important contributing factor. Stem nematodes do not actively reproduce below 41° F. In Yolo County, average minimum winter temperatures have increased 3°F since 1983 and are currently approaching the lower reproductive threshold (Fig. 15.3). Higher temperatures allow the nematode to complete a larger number of breeding cycles during the winter and thus impact the severity of the infestation. If climate change causes winter temperatures to rise further, outbreaks of alfalfa stem nematode may become more frequent in the region. Additionally, the use of organophosphates and carbamates in alfalfa crops has decreased 50% since 2005. These pesticides are known to suppress stem nematode populations, but are being replaced by pyrethroids, which do not affect the stem nematodes. Consequently, the decreased use of these pesticides may have also played a role in the recent outbreak.

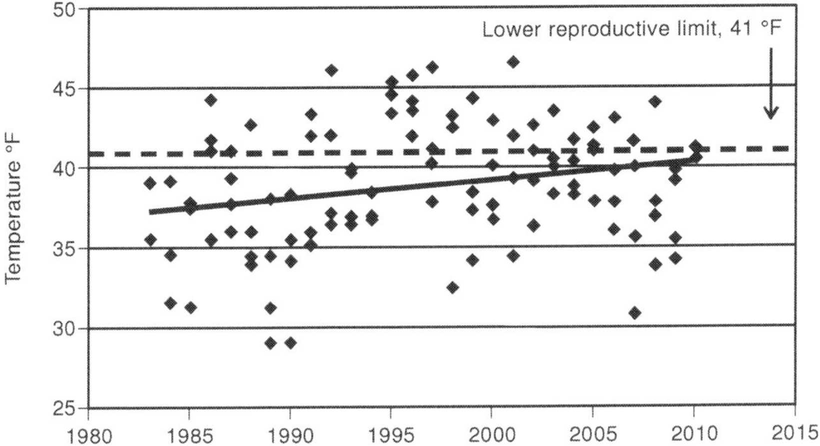

Fig. 15.3. Average minimum monthly ambient temperatures in Davis, California, from November to February 1983-2010. The lower reproductive limit of stem nematode females is 41°F.

15.3.1. Host defence mechanisms

Climate changes could also modify molecular interactions between host and endoparasite by interfering with plant defence mechanisms (Tylka, 2012). The simultaneous action of abiotic stresses (hydric, nutritional, thermal, etc.) and biotic stresses (nematodes, aphids, fungi, etc.) may have an additive, synergistic or antagonistic effect on the various metabolic pathways of phytohormones which confer resistance to stress in plants (Atkinson and Urwin, 2012; Somasekhar and Prasad, 2012). The main metabolic pathway involved in defence against abiotic stress is abscisic acid (ABA) (Atkinson *et al.,* 2013; Atkinson and Urwin, 2012; Studham and MacIntosh, 2012; Suzuki *et al.,* 2014), which alters signalling of the major phytohormones [jasmonic acid (JA), salicylic acid (AS), ethylene (ET)] associated with biotic stresses caused by phytoparasitic nematodes (Rosso *et al.,* 2012). Different stresses can influence plant-pest interactions, such as drought (Atkinson and Urwin, 2012, Atkinson *et al.,* 2013; *Nahar et al.,* 2012; Studham and MacIntosh, 2012; Suzuki *et al., 2014;* Mittler and Blumwald, 2010, Syvertsen and Levy, 2005), deficiencies in essential nutrients (i.e., the increase in salinity), the increase in salinity (Anderson *et al.,* 2004; Atkinson *et al.,* 2013; Prasch and Sonnewald, 2013), Potassium (Dropkin, 1969; Prasch and Sonnewald, 2013; Verdejo-Lucas *et al.,* 2013) and the increase in atmospheric CO_2 (Amtmann *et al., 2008*). Precisely, it has been shown that these environmental stresses, which trigger activation of ABA chemical signals, negatively influence crop resistance, growth and yield due to antagonistic reactions associated with the activation of JA and SA pathways initiated by the presence of the nematode (Sun *et al.,* 2010, 2011). On the other hand, it seems that low temperatures could favour the resistance of certain plant species, including potato (*Solanum* sp.), through cross-host-parasite adaptation (Suzuki *et al.,* 2014; Thaler and Bostock, 2004). Given the expected increase in the prevalence and magnitude of all these stresses due to CC, it is important to question the evolution of resistance to nematodes in various crops. Nevertheless, a limited number of studies have focused on the impact of

CC-associated stresses in genes associated with nematode phytoparasite virulence and host resistance, respectively (Sysoeva *et al.,* 2011, 2012).

15.3.2. Soybean resistance genes and soybean cyst nematode virulence genes

Currently, sources of resistance in soybeans are only effective against some populations of soybean cyst nematode (SCN) (Elad and Pertot, 2014). The relative abilities of different SCN populations to reproduce on seven reference soybean lines led to the identification of HG types (*Heterodera glycines*), quantifying virulence (Niblack *et al.,* 2002). This phenotypic variability of the resistance characteristic can be explained both by the presence of various parasitism genes producing proteins associated with virulence, or effectors in the nematode and resistance genes in the plant (Hogenhout *et al.,* 2009). Since the 1960s, nearly 70 loci of QTL associated with resistance have been identified in soybeans, the most studied being *Rhg1*, *Rhg2* and *Rhg3*, of which the major gene is *recessive,* as well as *Rhg4* and *Rhg5*, of which the major gene has a dominant behaviour (Caldwell *et al.,* 1960; Klink *et al.,* 2009; Matson and Williams, 1965; Rao-Arelli, 1994). Two of these QTLs would be more important in the resistance reaction, *Rhg1* (chromosome 18), which contains about nine genes, and *Rhg4* (chromosome 8), which has at least seven genes (Klink *et al.,* 2013). They are found in cultivars PI 88788 (*Rhg1*) and Peking (*Rhg1* and *Rhg4*), which have divergent cytological responses and form two linkage groups (Klink *et al.,* 2013). The type of resistance called "PI 88788" relies mainly on the genes located on the *rhg1-b* allele of QTL *Rhg1*, which has a dominant behaviour in cultivars of this group (PI 88788, PI 209332 and PI 549316) (Cook *et al.,* 2012; Kim *et al* 2010; Liu *et al.,* 2012; Yuan *et al.,* 2012). The cytological response of group PI 88788 (grouping the cultivar of the same name) is characterized by a rather slow necrosis (between 8 and 10 days) of the syncytium and by the absence of cell wall apposition (Klink *et al.,* 2012). Peking resistance requires the simultaneous presence of co-dominant genes located on the *Rhg1* and *Rhg4* QTL (Kim *et al.,* 2010; Liu *et al.,* 2012; Yuan *et al.,* 2012). Typically, the cytological response associated with the Peking group (Peking, cv. "Forrest", PI 437654, PI 89772, PI 90763) is characterized by rapid necrosis (about 48 hrs.) of the initial cell and increased wall (Cook *et al.,* 2012, 2014). It was suggested that plant resistance was conferred by the number of copies of the *rhg1-b* allele associated with QTL *Rhg1*. Indeed, the lines and cultivars associated with type PI 88788 would have seven to ten copies of the genes located on the *Rhg1* QTL, whereas those located on the same QTL within the Peking-like group would have less than three copies (Brucker *et al.,* 2005; Liu *et al.,* 2012).

A very large genetic variability is found within different populations of soybean cyst nematode and several individuals would carry "*ror*" (resistant host) alleles favouring the bypass of the host resistance and the ability to reproduce by the parasite (Cook *et al.,* 2012; Wang *et al.,* 2014; Zheng *et al.,* 2006). The mechanisms that lead to the bypass of resistance remain unknown. However, due to the extensive research on effectors produced and secreted by the soybean cyst nematode within the host, its parasitic success and the origin of its virulence is better (Liu *et al.,* 2012). To date, different proteins (effectors) secreted by the esophageal glands (two sub-ventral and a dorsal) and injected with the stylet have been identified which allow the soybean cyst nematode to alter the structure and functions of the effectors (Bekal *et al* 2008; Haegeman *et al. 2010*; Klink *et al.*, 2009; Li *et al.*, 1996; Wang *et al.*, 2014). These alterations differ according to the stage of development of the nematode. During the migratory phase, the infectious stage (J2) secretes proteins aimed at dissolving the root cells and dodging its detection by the host, by inhibiting or suppressing the phytohormones associated with the metabolic pathways and the different pathways mechanisms of plant defence (Haegeman *et al.*, 2012). Following the selection by stage J2 of an initial cell used for the development of syncytium, soybean cyst nematode produces and injects other effectors that allow it to complete its life cycle.

CCs are likely to have significant impacts on disease intensity in relation to the interactions between the environment (microclimate), soybean cyst nematode (biology, virulence) and soybean phenology, and resistance that influence the various molecular interactions (Boland *et al.*, 2004; Elad and Pertot, 2014; Gregory *et al.*, 2009; Parmesan, 2006). It has been shown that CC has been shown to modify parasite interactions in other phytoparasitic nematode species (Atkinson and Urwin, 2012; Boland *et al.*, 2004; Elad and Pertot, 2014; Gregory *et al.*, 2009; Newton *et al.*, 2012). High concentrations of atmospheric CO_2 and temperature have an effect on the resistance genes of several crops. For example, rising temperatures limit or even annihilate the effectiveness of R-type resistance genes of different varieties of tomato (*Solanum* sp.), pepper (*Piper* sp.), alfalfa (*Medicago* sp.), sweet potato (*Ipomoea* sp.), cotton (*Gossypium* sp.) and grapes (*Vitis* sp.) known to be resistant to different species of *Meloidogyne* (Ferris *et al.*, 2013; Jablonska *et al.*, 2007; Newton *et al.*, 2012). Moreover, in tomato, rising CO_2 atmosphere would also have an inhibiting effect on the rate of production of plant hormones (JA and SA) that confer different nematode resistance (Cooper *et al.*, 2005; Sun *et al.*, 2010, 2011).

15.3.3. Prospect

At present, the effects of CC on the production of effectors by phytoparasitic nematodes, including soybean cyst nematode, are unknown. It is also impossible to determine whether the loss of resistance in the plant is related to a hypersensitive reaction or the disappearance of the immune response following the construction of a syncytium or giant cell (nematode galls) (Ferris *et al.*, 2013). So, the effects of CC on the soybean and soybean cyst nematode pathosystem should be studied using different methodological approaches in order to deepen current knowledge and to better understand the effects of rising temperatures and CO_2 concentration. First, it would be pertinent to study soybean phenology and the life cycle of the soybean cyst nematode in the light of the various climate scenarios predicted by 2050 (2041-2070) in order to validate the climate is conducive to the development of the nematode and to identify areas of high pest risk. Then, as the use of resistant soybean cultivars is the main means of control, it would be essential to measure the performance of cultivars with maturity groups adapted to the region and resistant to soybean cyst nematode by exposing those to different temperatures and CO_2 concentrations to anticipate their performance against CC. Furthermore, since some populations might include individuals able to bypass resistance, it would be useful to identify molecular markers, for example simple nucleotide polymorphisms (SNPs), which are associated with the genotypes of the different populations of soybean cyst nematode. The discovery of markers capable of differentiating soybean cyst nematode populations and predicting their ability to reproduce on resistant cultivars would allow the development of rapid diagnostic tools in order to optimize control methods by selecting suitable cultivars. Finally, a study of the genes involved in the control of biological mechanisms (e.g., cell cycle) of the host by the parasite could be performed using new transcriptome analysis (RNA-sequence) methods.

15.4. INCREASE IN NUMBER OF GENERATIONS

The expected increase in temperature would positively favour the development of the soybean cyst nematode, *Heterodera glycines*, which would result in the addition of one or two additional generations to reach up to two generations per growing season in the northernmost part of Quebec and up to six generations in the extreme south (Gendron St-Marseille, 2013). This scenario would further weaken the soybean, as higher numbers of nematode infections increase the susceptibility of roots to other pathogens and, on the other hand, restrict the delivery of nutrients and the water to the aerial parts (Gendron St-Marseille, 2013).

Ghini *et al.* (2008) compared climatological norms from 1961-1990 with future scenarios (A2 and B2) of the decades of 2020's, 2050's and 2080's from five General Circulation Models (IPCC, 2001) to predict the changes in spatial distribution of infestation levels of the coffee root-knot nematode (races of *Meloidogyne incognita*) based on number of generations per month in Brazil. They predicted that the nematode infestation will increase in future due to greater number of generations per month. The number of generations of nematodes will increase in both scenarios, but, it will be lower in B2 than A2 scenario.

Warming will generally cause the risk of damage from potato cyst nematode (*Globodera rostochiensis*) that may become a serious problem with additional generations per year. Using simulated climate change, Carter *et al.* (1996) predicted that warming will increase number of potato cyst nematode generations per year by 2050 in Finland.

15.5. CONCLUSIONS

Global warming and climate change will have serious consequences on diversity and abundance of pests, and the extent of losses due to pests, which will impact both crop production and food security. Prediction of changes in geographical distribution and population dynamics of pests will be useful to adapt the pest management strategies to mitigate the adverse effects of climate change on crop production. Pest outbreaks might occur more frequently, particularly during extended periods of drought, followed by heavy rainfall. Some of the components of pest management such as host plant resistance, bio-pesticides, natural enemies, and synthetic chemicals will be rendered less effective as a result of increase in temperatures and UV radiation, and decrease in relative humidity. Climate change will also alter the interactions between the pests and their host plants. As a result, some of the cultivars that are resistant to pests, may exhibit susceptible reaction under global warming. Adverse effects of climate change on the activity and effectiveness of natural enemies will be a major concern in future pest management programmes. Rate of pest multiplication might increase with an increase in CO_2 and temperature. Therefore, there is a need to have a concerted look at the likely effects of climate change on crop protection, and devise appropriate measures to mitigate the effects of climate change on food security.

16

Development of Prediction Models for Nematode Pathogens

16.1. INTRODUCTION

Prediction models may help in evaluating a number of different outcomes and situations, i.e. providing a basis to identify the best among different options, to select in case of invasive species (eradication, suppression, no action) (Fraser *et al.*, 2006). In general, an insight on the possible outcomes of climate changes may be facilitated by the application of simulation models. Their sensitivity should be checked, however, with real scenarios, to verify the level of uncertainty and affordability, as well as their effectiveness as informative tools (Kickert *et al.*, 1999).

One key issue in modelling is the level of resolution achieved and the forecasting precision. The asymmetric increase of temperatures (with different shifts for minima and maxima), as well as the non-linearity in growth or development responses, require sub-daily resolutions when modelling climate change effects on fungal plant pathogens, due to a bias introduced when only mean temperatures are used, without considering the amplitude of the fluctuations (Scherm and Van Bruggen, 1994). Nematodes, however, due to a longer time required for completion of the different life-stages (eggs, juveniles, adults), may be modelled with different scenarios of daily mean temperature changes, aiming at providing a first rough insight about possible field situations and outcomes.

Models predicted that temperature increases coupled with increases in atmospheric CO_2 levels will have a positive effect on soybean root growth, density and length (Lawn and Noel, 1986; Niblack *et al.*, 2006). However, warming of temperatures could exacerbate the negative impacts of early soybean cyst nematode, *Heterodera glycines* on crop yields since increasing root surfaces would allow a greater number of individuals to grow (Rogers *et al.*, 1992; Wang *et al.*, 2003; Ziska, 1998). Also, extreme climatic events, including drought, appear to reduce soybean yields more pronounced in soybean fields infested with soybean cyst nematode (Tylka, 2012). This stress causes the nematode to create its nutrient site more deeply in the vascular tissues of the stele and not in the root cortex area, thus allowing it to continue harvesting the plant's resources (Bonner and Schmitt, 1985; Tylka, 2012). As a result, higher frequencies of drought could have a significant impact on soybean production, which reduce yields more pronouncely in soybean fields infested with soybean cyst nematode (Tylka, 2012).

16.2. MODEL TO SIMULATE THE DENSITY CHANGES OF CARROT CYST NEMATODE, *HETERODERA CAROTAE*

A different model for the carrot cyst nematode, *Heterodera carotae* (Fig. 16.1), was constructed and applied using the daily mean temperatures registered at Manfredonia, a typical carrot growing area in Apulia, during two years (2006-2007) (Fig. 16.2). The life stages considered eggs, second stage juveniles (J2), third stage juveniles (J3), fourth stage females (J4f) and males (J4 male), together with adult females, males and cysts. The model was constructed using published data to derive the daily rates of eggs release and of development, for the other stages of *H. carotae* (Greco and Brandonisio, 1986; Mugniery and Bossis, 1988). The relationship of the daily egg production rate with temperature was provided by a gaussian function, considering the difference between the daily mean value and the optimal temperature for eggs release (20° C) (Fig.16.3, 1). Similarly, the rate of eggs development was calculated using a higher corresponding optimal temperature (24.5° C), whereas the rates of maturation from J2 to J3 and J4 were calculated using the relationship between temperature and length of stage, provided by Mugniery and Bossis (1988). Due to the dependency of *H. carotae* on host plant roots for eggs hatching, the model included also a parameter K, accounting for the presence (K=1) or absence (K=0) of roots, and hence for the production or hatching of eggs during a simulated, two years cropping cycle (August to February). A further parameter was also used to account for a daily minimal egg hatching probability (0.005), in absence of roots. All stages were subjected to natural mortality rates, estimated as the inverse of the corresponding stage average life-length.

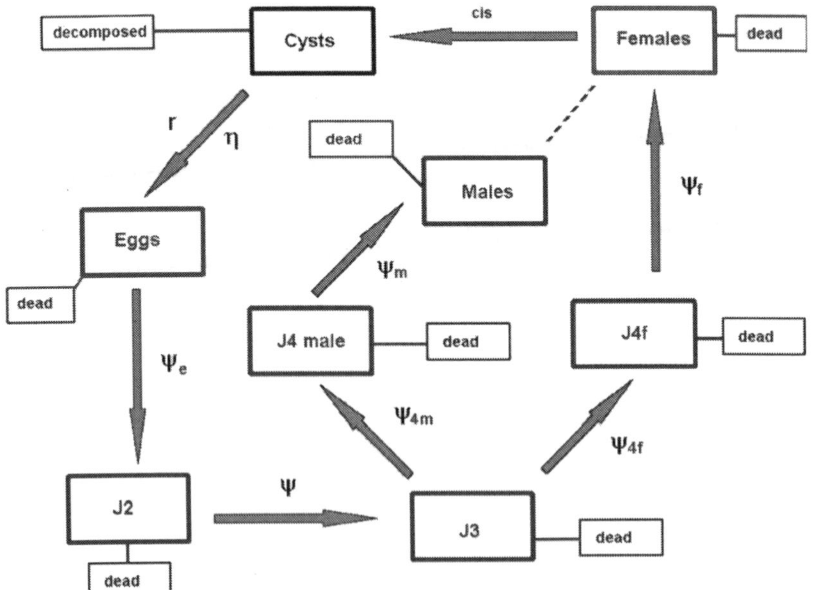

Fig. 16.1. Components of the model applied to simulate the density changes of the carrot cyst nematode, *Heterodera carotae* in function of temperature shifts. Parameters show the maturing rates of life-stages, or the rates of eggs development and release.

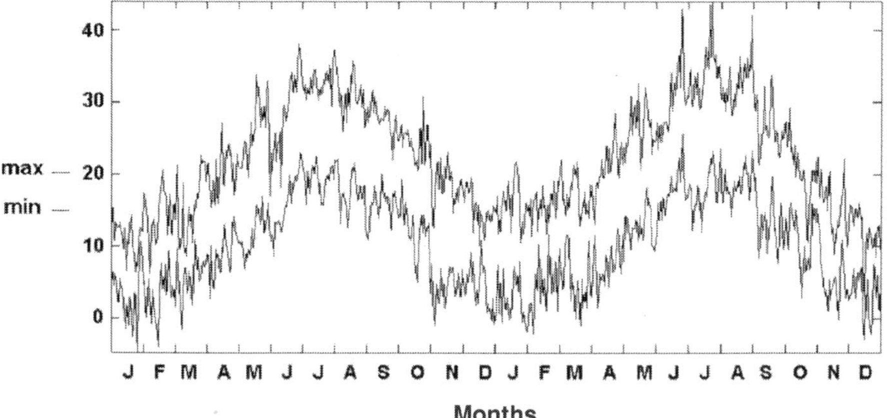

Fig. 16.2. Minimum and maximum temperatures registered at Manfredonia (Foggia, Italy), in the years 2006 - 2007 (source: Protezione Civile).

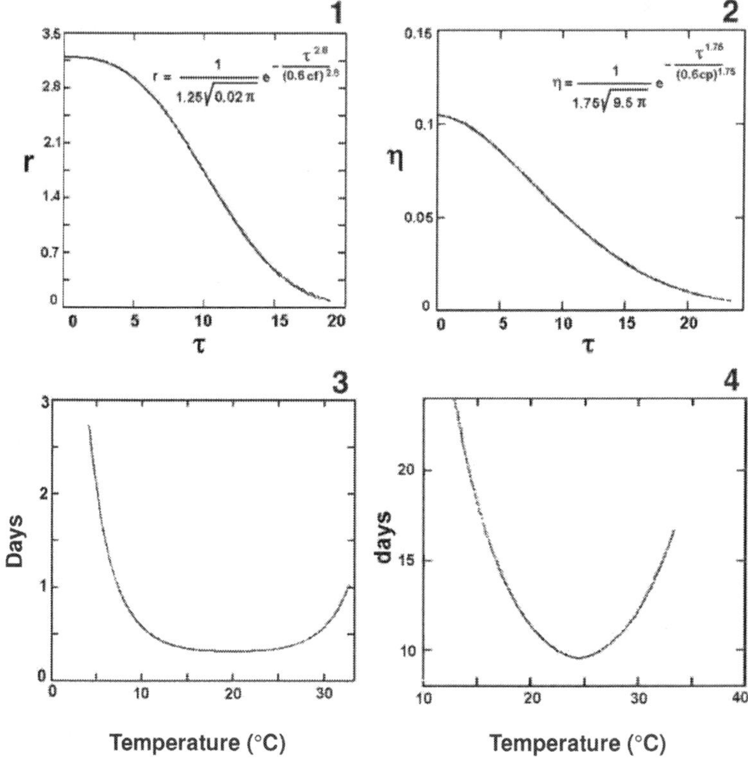

Fig. 16.3. Daily rates of eggs production r (1) and development η (2), estimated in function of the absolute difference τ between mean daily temperature and optimal stage values (cf and cp, respectively), used for modelling *H. carotae*. Inverse values (3 and 4, respectively) show the corresponding stage lengths, in days.

Simulations of an *H. carotae* population dynamics, started from 50 initial eggs, 40 J2-J4f stages, 3 females and cysts and one male per ml of soil, showed the expected numbers of 2-3 nematode generations, observed during the autumn-spring periods (Fig. 16.4, 1). A progressive 0.25° C shift in temperatures, equally applied at each daily mean value to simulate a uniform increase of temperatures, affected the maximum numbers reached by eggs, J2 or cysts in a non-linear way, with highest rates of increase observed at highest (1.5-2° C) increments (Table 16.1, Figs. 16.4 and 16.5). At these temperature shifts, simulations showed further juvenile and eggs generations in winters (Fig. 16.4, 4 and 5).

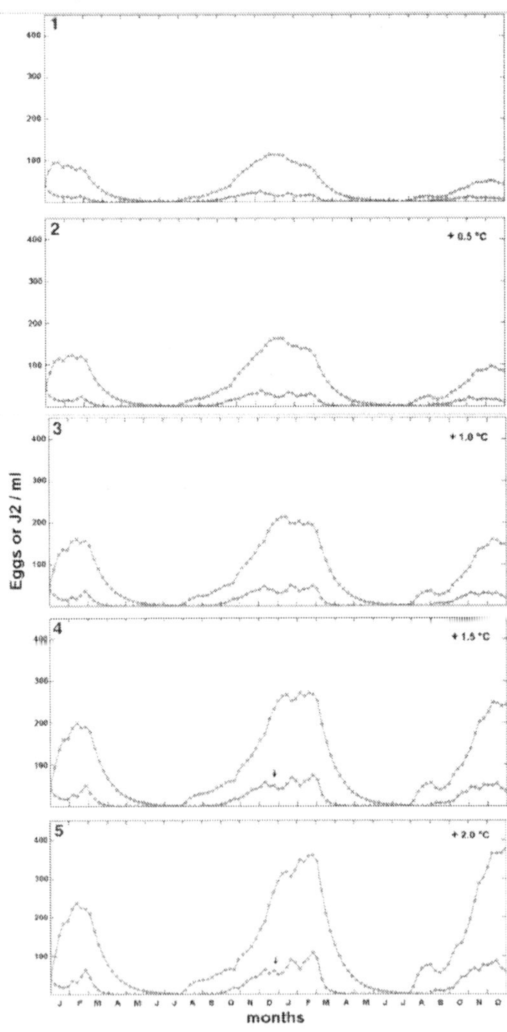

Fig. 16.4. Two years (2006-2007) modelled population dynamics of eggs (squares) and J2 of *H. carotae*, and changes observed with uniform mean daily temperature increases ranging from 0.5 to 2° C. Arrows show an additional J2 generation, at highest temperature shifts.

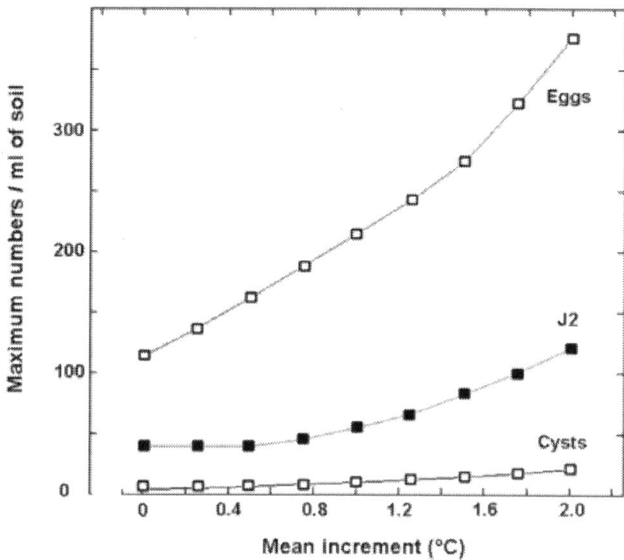

Fig. 16.5. Relationship between mean temperature increments and maximum simulated numbers ml^{-1} of soil reached by eggs, J2 and cysts of *H. carotae*.

Table 16.1. Simulated effects of daily temperature increases (D) on maximum densities reached in soil by *Heterodera carotae*, expressed as ratio with densities at D = 0.

D (° C)	Eggs ml^{-1}	J2 ml^{-1}	Cysts ml^{-1}
0.25	1.19	1.00	1.28
0.50	1.41	1.00	1.60
0.75	1.64	1.14	1.97
1.00	1.88	1.38	2.39
1.25	2.12	1.66	2.90
1.50	2.40	2.05	3.48
1.75	2.81	2.50	4.16.
2.00	3.29	3.02	4.97

Although, modelling necessarily represents a simplification of the more complex, non-linear relationships occurring in soil among different organisms, it provides an insight on dynamics otherwise difficult to trace, due to the amount of experimental data and time required to investigate these relationships. At small scales, generic dynamic crop modelling may result informative about the effects of weather, variety, pests, soil, and management practices on crop growth and yield, as well as on soil N and organic carbon dynamics in aerobic as well as anaerobic conditions. Data generated includes pest induced yield losses, allowing a comparative analysis with i.e. different greenhouse gas emission situations (Aggarwal *et al.*, 2006). Weather variables introduced in modelling result fundamental to explore the different outcomes of climate changes and hence estimate the spatial and seasonal dynamics of pests, in the mid-term (Yonow *et al.*, 2004).

A further tool is given by probability distribution maps (PDM). Their use allows the identification of potential risks related to the distribution of pests, vectors or plant diseases. PDMs are calculated on the basis of the combined use of local records of pest occurrence, climate data and subsequent statistical analysis. They show the areas susceptible of colonization or of endemism, for a given organism (Morales and Jones, 2004). This tool has several practical advantages, including the possibility of early identification of invasion areas by a pest or the possibility to anticipate the insurgence of epidemics for secondary pests, already present in areas with sub-optimal conditions for life-cycle. PDMs require the monitoring of different climatic variables at the regional scale and a given resolution level, as well as the implementation of an early monitoring and detection support system.

Plant-parasitic nematode populations and their damage have been simulated on potato by Ehwaeti *et al.*, (2000), where an initial population, the intrinsic growth rate and the maximum population were used to predict the final population after one crop cycle.

16.3. SIMBA-NEM MODEL TO SIMULATE THE POPULATION DYNAMICS OF *RADOPHOLUS SIMILIS* AND *PRATYLENCHUS COFFEAE* ON BANANA

SIMBA-NEM is a new model developed by Tixier *et al.* (2006a) to simulate the population dynamics of two major plant-parasitic nematode species of banana, *Radopholus similis* and *Pratylenchus coffeae*. For each species, each generation is represented by one cohort. Cohorts of the same species form a chain representing the developmental stages of nematodes. A logistic function describes population growth in relation with: (i) an environmental carrying capacity (K) that depends on available banana root biomass, (ii) an intrinsic growth rate (c) and (iii) competition between nematode species. Soil water content and the quantity of nematicides used are considered to be the main variables influencing the intrinsic population growth rate of each species. The structure of SIMBA-NEM is presented in Fig. 16.6 and Table 16.2. SIMBA-NEM was calibrated and validated using datasets from banana cropping systems in Guadeloupe (French West Indies). By analysing the sensitivity of the model to the main parameters and performing simulations of validation for various cropping systems (banana monoculture with or without nematicide applications use and a banana/sugarcane rotation) they were able to test the ability of the model to predict nematode population dynamics under a range of conditions. SIMBA-NEM is able to predict long-term nematode population size, while taking interspecific competition into account. It also helped to define knowledge gaps in nematology and modelling. SIMBA-NEM was used to optimize the effect of nematicide applications. The best time to apply nematicides is 45 weeks after planting banana, since this best period fits well with the time when nematode populations increase very fast, at the end of the pre-flowering period when root growth is maximum. SIMBA-NEM can be a very helpful tool for designing sustainable and more environment-friendly banana cropping systems. In the SIMBA global crop-modelling environment, SIMBA-NEM is a key sub-model which provides essential information concerning the sustainability of the simulated system and thus permits planning environmentally friendly cropping systems.

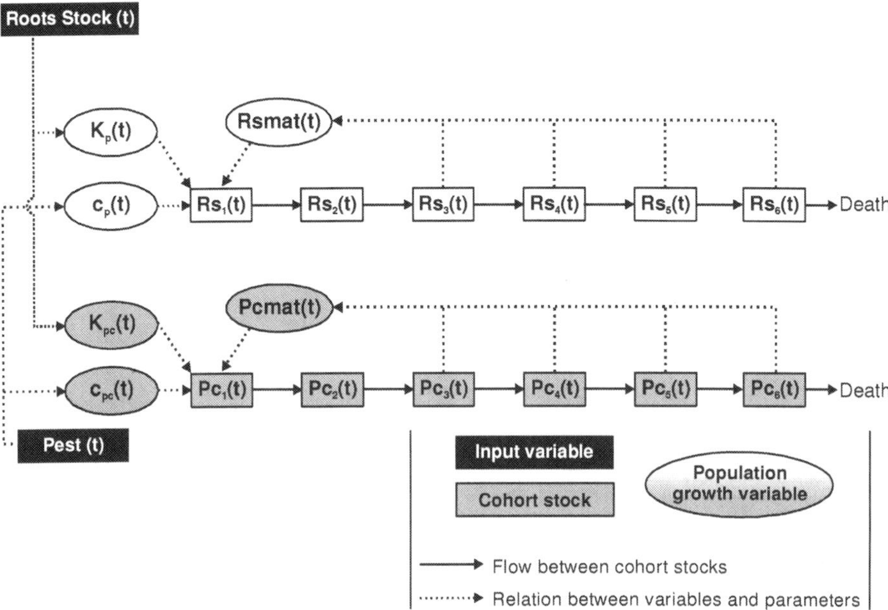

Fig. 16.6. Structure of the SIMBA-NEM model. At each time step t of the model for each species, Rsi(t) and Pci(t) represents the number of nematodes in cohort i, Rsmat(t) and Pcmat(t) represents the number of mature nematodes, KRs(t) and KPc(t) the carrying capacity and cRs(t) and cPc(t) the population growth parameter for *R. similis* (Rs) and *P. coffeae* (Pc), respectively and Rootstock(t), Wat(t) and Pest(t) represents the banana fresh root biomass, per centage of soil water and soil nematicide quantity, respectively.

Table 16.2. Variables and parameters of the SIMBA-NEM model, where 't' is the time step of the model (in weeks) and 'i' is the cohort rank

Variables/ Parameters	Meaning
Input variables	
Roots Stock (t)	Banana root fresh biomass at step t (g ha^{-1})
Wat(t)	Soil water content at step t in per centage of stock (%)
Pest(t)	Grams of nematicide active product in soil per hectare at step t (g ha^{-1})
Rs1(1); Pc1(1)	Initial nematode number for Rs and Pc (in number of nematode per hectare, nb ha^{-1})
Output variables	
Knempot(t)	Total carrying capacity in nematodes per hectare at step t (ha^{-1})
KRs(t); KPc(t)	Carrying capacity per hectare for Rs and Pc at step t (ha^{-1})
cRs(t); cPc(t)	Growth rate at step t for Rs and Pc
FXpest(t)	Nematicide effect at step t

Variables/ Parameters	Meaning
FXwat(t)	Soil water content effect at step t
Rsi(t); Pci(t)	Number of nematodes in cohort i per hectare at step t for Rs and Pc (ha^{-1})
Rsmat(t); Pcmat(t)	Total number of mature nematodes per hectare at step t for Rs and Pc (ha^{-1})
Rstotal(t); Pctotal(t)	Total number of nematodes per hectare at step t for Rs and Pc (ha^{-1})
Rs(t); Pc(t)	Nematode population per gram of root at step t for Rs and Pc (g^{-1})
Parameters	
KnemMax	Maximum carrying capacity per gram of root (g^{-1})
cRspot; cPcpot	Intrinsic growth rate for Rs and Pc
apest, bpest	Parameters of the nematicide effect curve FXpest(t)
awat, bwat, cwat	Parameters of the soil water content effect curve FXwat(t)

Rs for *R. similis* and Pc for *P. coffeae*.

16.4. CONCLUSIONS

Predictive models have been proposed for simulating nematode population dynamics, in relation to the host and to environmental factors. Undoubtedly they need improving, especially by taking into account more environmental factors such as soil temperature or soil physical and chemical characters. These improvements should rise to the capacity to simulate wider range of soil and climate situations. Nevertheless predictive models are very helpful tools for designing sustainable and more environment friendly cropping systems. These can also be used in a global cropping system modelling environment which includes sub-models that simulate plant growth, plant population structure, physical soil properties and water balance (Tixier *et al.*, 2004b). In this global simulation context, models provide essential information on cropping system constraints, functioning and sustainability. In addition to conventional innovations such as cover crops, genetic improvement, etc., the model approach can jointly assess time-course patterns concerning cultural practices and the natural dynamics of the agro-ecosystem in order to optimize yield, preserve soil fertility and minimize environmental risks. Pest dynamic models are rarely included in cropping system models or crop models, due to their complexity. Nevertheless, they are essential for prototyping more sustainable cropping systems. Models will be used in the future to determine the best trade-off between nematicide use and environmental risks when linked to environmental assessment indicators. They could also be used to test biological hypothesis regarding competition between species, the role of root growth in nematode development, or other factors.

Models for invasive species, for possible effects of climatic changes and the integration of a global climate model with local, specific sub-models, represent useful tools, but require a continuous check for their fitness and reliability.

17

Climate Change Adaptation and Mitigation Strategies for Nematode Pathogens

17.1. INTRODUCTION

By 2050, according to new research in the journal *Global Ecology and Biogeography*, those opportunistic viruses, bacteria, fungi, blights, mildews, rusts, beetles, nematodes, flies, mites, spiders and caterpillars that farmers call pests will have saturated the world. Wherever they can make a living, they will. None of this bodes well for food security in a world of nine billion people and increasingly rapid climate change. If crop pests continue to spread at current rates, many of the world's biggest crop-producing nations will be inundated by the middle of the century, posing a grave threat to global food security. Three kinds of tropical root-knot nematodes produce larvae that infect the roots of thousands of different plant species. Predictions such as these are intended to be self-defeating: they present a warning of what might happen if no steps are taken for their management. By unlocking the potential to understand the distribution of nematodes, we are moving one step closer to protecting our ability to feed a growing global population.

To understand how best to progress plant nematode control to improve food security in the context of climate change, we must work with societal change, defining its key processes and influencers to effect change. More specifically, as crop protection professionals, we have a key role to play in improving food security. Plant nematodes could potentially rob humanity of up to 14% of the attainable yield in major crops (Sasser and Freckman, 1987).

The 'fertilization effect' of rising CO_2 increases crop biomass and grain yield (Ainsworth and Long, 2005) raising the possibility of increased food production. But emerging evidence of reduced grain yield from high temperature and water limitations (Anwar *et al.*, 2007; Torriani *et al.*, 2007) make wholesale increase in crop productivity unlikely. Also, impact of plant nematodes, mostly ignored in assessments of global food security under climate change, minimize or reverse any benefit from CO_2 fertilization effect (Butterworth *et al.*, 2010; Fernandes *et al.*, 2004). As world agriculture responds to challenges of securing sufficient, safe, and nutritious food for the ever-expanding human population under changing climate, no doubt pesticide usage will increase even more. Pinpointing key constraints to food security, primarily from a production perspective, this chapter highlights how improving plant nematode management can enhance global food security, and finally highlights key nematode management issues to be addressed in improving food security in a changing climate.

17.2. ADAPTATION AND MITIGATION

Development and application of new diagnostic protocols for accurate identification of nematode species is beneficial and necessary for national and international regulatory and quarantine agencies relative to free trade and economics. Understanding genetic variability and adaptation in nematodes is important for effective nematode management strategies that involve rotation, host plant resistance, cultural manipulations and biological control, because they are influenced directly by genetic variability existing in target nematode field populations. Results from current work indicate that genetic variability and adaptation potential in nematode populations are responsible for the aberrant results of many experiments assessing resistance, crop rotations, host ranges, cover/trap cropping, and biological control.

Implementation of new nematode management tactics that reduce pesticide usage, and thereby benefit human health and the environment; promotion of sustainable farm management practices through new nematode management strategies; increased knowledge base in plant-nematode biology and genetics for use in identifying novel targets for nematode control; economic benefits to producers and consumers through reduction in nematode management costs and food production; and more efficient and effective response and mitigation capabilities for invasive and trade product contamination issues are the needs of the hour.

Bridge (1996) outlines and discusses four different strategies for specific nematodes and some other pests in sustainable and subsistence agricultural systems:

- Preventing the introduction and spread of nematodes by the use of nematode-free planting materials.
- Using non-chemical, cultural, and physical control methods, particularly crop rotation and soil cultivation.
- Encouraging naturally occurring biological control agents by understanding of cultivation methods and appropriate use of soil amendments.
- Maintaining or enhancing the biodiversity inherent in traditional farming systems that use multiple cropping and multiple cultivars to increase the available resistance or tolerance to nematodes.

17.2.1. Alternative management strategies

Management of nematodes in agriculture during the past fifty years has been largely based on the application of broadly efficacious nematicides. It is likely that widely-used and efficacious nematicides will be unavailable or greatly restricted in the future. For example, Phenamiphos (Nemacur), a commonly used nematicide on several crops was cancelled in 2007; Carbofuran (Furadan), commonly used on corn was cancelled in 2009; and Aldicarb (Temik), an important potato nematicide was cancelled in 2011. Increasingly, scientific evidence and public awareness have heightened concerns about environment quality, food quality, and human health and safety relative to pest management in agricultural production. The need for alternative, integrated nematode management has been propelled by the actions triggered by the Montreal Protocol and the Food Quality Protection Act (FQPA) of the 1990's. The agricultural production community (our stakeholders) is scrambling to find viable and agro-ecologically sustainable alternatives to chemical-based soil pathogen and nematode control.

A major shift in nematode-management strategies is occurring, from almost exclusive reliance on soil-applied nematicides, to the use of combinations of alternative strategies such as crop rotation, host plant resistance, cultural manipulations and biological control (Ferris *et al.*, 1992). Greater understanding of nematode genetic response and adaptation to abiotic factors will be important in optimizing the design of cultural management tactics such as manipulations of planting and harvest times, wet or dry fallow, and soil solarization. Wet or dry fallowing starve nematodes while active (wet fallow) or die from extreme moisture stress (dry fallow). Soil solarization involves natural heating of soil under plastic cover to attain the thermal death point of nematodes. Avoidance may include changes in planting and harvest dates, such as delaying planting in the fall to avoid infection activity, and early planting or late harvest of crops to avoid additional nematode generations. Soil amendments such as green manures and various bio-products show promise for nematode suppression in some systems. Some potential biological control agents are known to give effective management of nematodes. Current research is addressing many areas of nematode management, including: development and deployment of nematode-resistant plants; crop rotation to reduce population densities of these pathogens; cover crops, trap crops and soil amendments including green manures to reduce population densities; the role of weed hosts in bringing about phenotypic changes in nematode populations; characterization of resistance genes and resistance responses; the development of molecular diagnostic protocols for nematode identification and the reference databases necessary for their implementation. Efforts to develop improved techniques for nematode diagnostics, understanding of nematode diversity and genetic variability, processes of nematode fitness and adaptation, and the incorporation of this knowledge into the design and analysis of improved nematode management strategies needs to be intensified.

Climate change will cause alterations in the spatial and temporal distribution of nematodes and consequently the control methods will have to be altered to suit these new situations. Assessments of the impact of climate change on nematode infestations and in crops provide a basis for revising management practices to minimize crop losses as climate conditions change (Ghini *et al.*, 2008).

Recent observations suggest that nematode pressure on plants may increase with climate change (Ghini *et al.*, 2008). As a result there may be substantial rise in the use of nematicides in both temperate and tropical regions to control them. Non-chemical nematode management methods (green manuring, crop rotation, mulching, application of organic manures, etc.) assume greater significance under changing climate scenario.

17.3. PHYSICAL METHODS

17.3.1. Solarization

Solarization-heating soil under clear plastic tarps that trap and increase the sun's heat can be an effective means of controlling nematodes in the soil. The soil needs to be moist, well tilled, and heated to at least 50°C for several days, preferably several weeks. This method can be practical for nursery sites and home gardens, but it should be done during the hot months and long days of mid-summer.

Similarly, other heat and steam-based pasteurization methods can be used to prepare potting soil. Healthy plants grown in nematode-free media have a better chance to survive after being transplanted to the field.

17.4. CULTURAL METHODS

17.4.1. Crop rotation

Crop rotation involves growing a crop that is not a host for the nematode present before growing a crop that is susceptible. The non-host or immune crop will cause nematode numbers in the soil to decline, giving the subsequent host crop a chance to establish a good root system. The success of this method depends on growing the non-host crop long enough to reduce the nematode numbers. The rotation crop must be selected carefully because some nematodes (such as root-knot, reniform, and burrowing nematodes) have very wide host ranges. Also, some undesirable species may emerge on the rotational crop and become a pest. Variations on the crop rotation concept include fallowing, multicropping (intercropping), and green manuring.

Crop rotation is a beneficial production practice and has been extensively studied as a more sustainable nematode and pathogen management measure than chemical pesticides. It often reduces nematode populations by reducing reproductive potential and survival, eliminating or reducing population densities. Crop rotations and cover crops also can lead to nematode reduction by increasing the abundance or activity of beneficial soil micro-flora. In such cases, the crop used in the rotation sequence is often selected for the development of a beneficial rhizosphere microbial community (Kloepper *et al.*, 1991; Latour *et al.*, 1996; Smith and Goodman, 1999; Weller *et al.*, 2002). Among the organisms that are most likely favoured by cover crops are fungal egg-parasites, nematode-trapping fungi, endoparasitic fungi, arbuscular mycorrhizal fungi, plant growth promoting rhizobacteria, and obligate bacterial parasites (Sikora, 1992). The use of legume cover or rotation crops provides organic nitrogen while also suppressing plant-parasitic nematodes. Unfortunately, many legumes are susceptible to root-knot nematodes (RKN) (*Meloidogyne* sp.), which are often the key nematode pathogens requiring management in many cropping systems (McSorley, 1998, 1999). A number of legume crops have been identified that are highly resistant to various species and/or races of RKN. Some of these, cowpea (*Vigna unguiculata*), velvet bean (*Mucuna deeringiana*), and sun hemp (*Crotalaria juncea*) have potential in cropping systems (McSorley, 1999; Rodriguez-Kabana *et al.*, 1988, 1989; Sipes and Arakaki, 1997; Weaver *et al.*, 1995). In addition to nematode management and nitrogen provided by nematode-resistant legumes, crop residues may increase levels of soil organic matter, improving water-holding capacity and other soil properties (McSorley and Gallaher, 1995).

Crop rotations can reduce certain nematode densities, but may actually increase populations of other nematode pathogens (LaMondia and Halbrendt, 2003; Noe, 1998). Growers must consider multiple factors including economic returns of rotation crops, additional capital and labour requirements, and grower and market acceptance of the rotation crop. Although crop rotation itself is unlikely to provide nematode management in all crop systems, when integrated with other non-chemical management measures such as biological control, rotation is likely to become increasingly important, as synthetic nematicide options are lost. Several nematode-antagonistic crops including rotation and cover crops such as rapeseed (*Brassica napus*), marigold (*Tagetes* sp.), forage pearl millet (*Pennisetum typhoides*), black-eyed Susan (*Rudbeckia hirta*), Sudan grass (*Sorghum bicolor*) and sorghum-Sudan grass were identified (LaMondia and Halbrendt, 2003). Native prairie plants that are resistant to *M. hapla* (LaMondia, 1996) may be useful as cover crops alone or in mixtures for managing this and other plant-parasitic nematode species. Furthermore, the incorporation of plant shoots as green manures appears to play a significant role in the nematode-antagonistic effects of many plants (Halbrendt, 1996; LaMondia and Halbrendt, 2003).

The RKN resistant pepper Carolina Cayenne was effective as a rotation crop for managing *M. incognita* in susceptible bell peppers (Thies *et al.,* 1998). Yield of squash grown in the spring following a rotation crop of castor, cotton, velvet bean, or crotalaria, produced higher yields than squash grown after RKN susceptible peanut (McSorley *et al.,* 1994). When eggplant was grown following resistant 'Mississippi Silver' southern pea, root galling by *M. incognita* race 1 was less severe than when eggplant was grown following susceptible 'Clemson Spineless' okra (McSorley and Dickson, 1995).

Crop rotation provides for diversity in time and space and is often the preferred means for nematode management. Rotation, however, may be of limited value when several damaging species of nematodes are present or for species with broad host ranges. For growers to accept rotation as a viable tool for nematode management, suitable crops and land must be available. The rotational crop must offer the grower an acceptable return, with similar requirements for labour and equipment. The need to rotate specific crops, however, may vary with location. For example, corn can be grown continuously in some regions with little or no effect on yield (Zawislak and Tyburski, 1992), whereas the parasitic nematodes predominant in the south-eastern United States cause significant yield losses on this crop (Osteen *et al.*, 1982). The highly successful practice of rotating tobacco with fescue (non-host for *Meloidogyne* sp.) has been in place in the south-eastern United States for some five decades (Nusbaum and Ferris, 1973). Periodic incorporation of the dense grass sod improves the soil structure, increases water-holding capacity, and provides control of associated root diseases including root-knot in the primary crop. Other grass fallows that have proven useful in the management of nematodes, especially *Meloidogyne* sp., include Bahia grass (*Paspalum notatum*), Bermuda grass (*Cynodon dactylon*), weeping love grass (*Eragrostis curvula*), Rhodes grass (*Chloris gayana*), Pangola grass (*Digitaria decumbens*), and Guinea grass (*Panicum maximum*) (Bridge, 1996; Rodrıguez-Kabana *et al.*, 1991; Rodrıguez-Kabana and Canullo, 1992). Any significant development of broadleaf weeds in these grass fallows can negate their effects in the control of root-knot nematodes because many are hosts for these pathogens (B´elair and Benoit, 1996; Bridge, 1996). The economic viability of grass or pasture is enhanced when animals are included in the system. An important factor is that many grasses and cereal crops may also support reproduction of many plant-parasitic nematodes, including *Meloidogyne* sp.

Many potential rotations or green manure crops that show promise in nematode management may be antagonistic to some nematode species or even serve as trap crops (Kloepper *et al.*, 1992; Koch and Gray, 1997; Rodrıguez-Kabana and Canullo, 1992; Trivedi and Barker, 1986). These plants may be categorized as being either active or passive, depending on whether they produce some antihelminthic compounds or are simply unsuitable hosts for nematodes. Selected Brassica species, including rapeseed and mustard, may suppress nematode populations, soil-borne pathogens, and weeds in crop rotations (Halbrendt, 1996; Koch and Gray, 1997). These plants produce glucosinolates, and their decomposition products are toxic to nematodes. Nematode-resistant radish is very effective in suppressing *Heterodera schachtii* on sugar beet (Koch and Gray, 1997). In contrast, rapeseed is stunted and supports reproduction of *M. arenaria*. Some antagonistic (or active) plants, including *Crotalaria* sp., mustard, African marigold, asparagus, castor, and sesame, may be grown as commercial crops, used as cover crops, or established in mixed planting with other crops (Bridge, 1996 ; Rodrıguez-Kabana and Canullo, 1992). Caution is needed in selecting from these or other antagonistic plants for cropping systems in case they contain negative features or hazards in addition to providing nematode control. For example, some of the *Crotalaria* species serve as excellent trap crops for root-knot nematodes, but they also synthesize potent toxins that cause primary tumours or suppress growth of swine, cattle, and poultry. Although the African marigold, *Tagetes erecta* and other *Tagetes* sp. may provide effective nematode control under some conditions, the efficacy of the primary nematicidal component (α-terthienyl) is dependent on light activation (Bakker *et al.*, 1993).

The utilization of certain Sudan grass hybrids as a green manure provides excellent control of *Meloidogyne chitwoodi* on potato (Mojtahedi *et al.*, 1993). Sorghum-Sudan grass hybrids also suppress this pathogen, but these plants may contain a higher concentration of dhurrin, a toxin for cattle when these plants are grazed improperly. In addition, the antagonism of these plants may be limited to certain nematode genera or species, e.g. the lesion nematode *Pratylenchus penetrans* is affected little by the use of these plants as green manure crops (MacGuidwin and Layne, 1995).

Earlier work on a selective nematicidal component in decomposing rye residues indicated that nematodes may have differential sensitivity to these products (Sayre *et al.*, 1965). *Meloidogyne incognita* proved to be the most sensitive to the associated decomposition products, *P. penetrans* exhibited intermediate sensitivity, whereas microbivorous nematodes were quite tolerant. Thus, lesion nematodes and the bacterivores probably have developed tolerance to decomposition products, whereas sedentary endoparasites would likely be more sensitive. Butyric acid produced by *Clostridium butyricum* was identified as one of the major toxic components in the decomposing rye (Sayre *et al.*, 1965). However, other compounds may be even more important since only limited amounts of butyric acid were detected in leachates from pots with decaying rye. Rye as a cover crop has both negative and positive aspects. Although highly effective against *Meloidogyne incognita* on cotton, it is much less efficacious against the Columbia lance nematode, *Hoplolaimus columbus*. The impact of rye on the reniform nematode, *Rotylenchulus reniformis*, was likely related to suppression of dicotyledonous winter weeds. In addition, the timing of its incorporation in soil before the establishment of some crops, particularly cotton, is important in that the decomposition products may prevent the normal germination of cotton seeds.

Other legumes including selected clovers, velvet bean, joint vetch, and Cahaba white vetch provide multifaceted contributions to soil health and crop productivity when used in rotation as green manure cover crops. For example, the use of velvet bean, *Mucuna deeringiana* in a soybean rotation enhances the activity of rhizosphere bacteria antagonistic to the soybean cyst nematode, *H. glycines*, and the southern root-knot nematode, *M. incognita* (Kloepper *et al.*, 1992).

17.4.2. Fallowing

Keeping the soil free of plants (fallow) deprives plant-parasitic nematodes of a host, which, over time, reduces their populations. Maintaining good weed control is a critical component of fallowing for nematode control because weeds are hosts of many species of plant-parasitic nematodes.

17.4.3. Multicropping

Multicropping (intercropping) with plants that either are not good nematode hosts or are antagonistic to the nematodes also reduces nematode numbers.

17.4.4. Green manuring

Green manuring—tilling under a crop that grows rapidly and produces a large quantity of biomass-adds organic matter and, depending on the green manure crop used, may add substances that repel or kill nematodes. Sudan grass and corn are excellent green manure crops that provide good nematode control.

17.4.5. Time of planting

Another potential nematode-management strategy involves adjusting the schedules for susceptible crop production to limit nematode reproduction. For example, delayed planting of soybean, which occurs in wheat-soybean double cropping systems, allows for greater nematode attrition in the absence of a host and results in lower at-planting population densities of *Pratylenchus brachyurus* and *H. glycines* (Koenning and Anand, 1991; Koenning *et al.*, 1985). The wheat-soybean double-cropping system was shown to be economically superior, to a rotation with grain sorghum in Arkansas (Dillon *et al.*, 1996). This approach also has been tried for root-knot nematode in carrot. Shifts in the planting and harvest dates of carrot to minimize root-knot development caused by *M. incognita* have produced striking results (Roberts, 1993). Delaying planting to late autumn or early winter clearly restricts root-gall development on carrot. Although the efficacy of this management strategy increases with lateness of planting, some root galling and crop loss occurred even with the best treatments. Thus, this management strategy should be used in concert with available, compatible tactics (Roberts, 1992). In some regions, early harvest of peanut is critical to limiting damage to seeds by *Ditylenchus africanus* (Venter *et al.*, 1992). Although approaches that limit infection by either promoting greater nematode attrition or limiting infection due to physical constraints such as temperature/planting time are useful in some nematode-host interactions, they have not been effective in limiting damage of *Hoplolaimus columbus* to cotton or soybean.

Many crops vary phenotypically for physiological maturity, a factor that can be exploited to suppress final nematode population densities. Soybean cultivars, for example, are classified by maturity groups that range from 000 to IX; each group is separated by 1 to 2 weeks. Late-maturing cultivars support greater reproduction of *H. glycines* in North Carolina (Hill and Schmitt, 1989). The use of early-maturing cultivars suppresses nematode-population increase and benefits succeeding crops (Koenning *et al.*, 1993). Many other cultivated plants including cotton, corn and small grains differ in maturity, and this technique may have wider applicability.

17.4.6. Weed control

Wardle *et al.* (1995) found that cultivation for weed control was an important factor influencing the species diversity of the nematode community. Increased numbers of fungivore nematodes were found in one study in Georgia in reduced tillage plots compared to conventional tillage (Parmelee and Alston, 1986) during the summer, but the reverse was true at other times of year. Similarly, higher numbers of the plant parasites *Helicotylenchus dihystera, Tylenchus*, *Aphelenchoides* sp., dorylamids and mononchids were associated with conventional till systems in North Carolina (Mannion, 1991). Numbers of bacterial feeding and total nematode numbers were greatest in a no-till system in Spain (Lopez-Fando and Bello, 1995). However, Freckman and Ettema (1993) found only small differences in total nematode abundance related to tillage; the trophic diversity was increased, and the ratio of fungivores to bacterivores was decreased in no-till compared to conventional tillage practices. The ratio of fungivores to bacterivores can be regarded as an indicator of the decomposition pathway in detrital food webs (Freckman and Ettema, 1993). The decrease in this ratio associated with no-till may indicate a shift from a bacteria-based food web to a fungus-based food web.

17.5. BIOLOGICAL METHODS

Another non-chemical approach to managing nematodes is biological control-using other organisms against the pest organism. A high level of natural biological control is ordinarily present in the soil. This natural control probably keeps the nematode populations at 10–20 per cent of what they would be in its absence. Nevertheless, the level of natural control is seldom adequate to prevent plant damage from nematodes.

A more realistic strategy for biological control of nematodes is to incorporate soil amendments such as manure (particularly poultry manure) and compost. Such additions of organic matter contributes to biological activity in the soil and enhances the natural activity of organisms antagonistic to nematodes.

In addition to suppressing nematodes, certain plant-growth–promoting rhizobacteria may induce systemic resistance to foliage pathogens such as *Pseudomonas syringae* pv. *lacrymans* and *Colletotrichum orbiculare* on cucumber (Wei *et al.*, 1996). They suggested that these rhizobacteria may control a spectrum of plant pathogens/pests, including fungi, bacteria, nematodes, and insects. In a split-root system, treatments with *Bacillus sphaericus* B43 or *Agrobacterium radiobacter* G12 also induced a significant degree of resistance in potato to *Globodera pallida* (Hoffmann-Hergarten *et al.*, 1997). These rhizobacteria suppressed infection of potato roots by the juveniles, but had no effect on egg production.

Biological control offers an alternative or supplemental management tactic to chemical or cultural management of plant-pathogenic nematodes. *Pasteuria* sp. are obligate, mycelial, endospore forming bacterial parasites of endoparasitic nematodes, such as RKN (*Meloidogyne* sp.), and several ectoparasitic nematodes, such as lesion (*Pratylenchus*) and sting nematodes (*Belonolaimus*) (Chen and Dickson, 1998; Giblin-Davis, 1990). These parasites are promising biological control agents for management of important agricultural species of *Meloidogyne* and *Pratylenchus* (Chen and Dickson, 1998). *Pasteuria penetrans* and *P. thornei* are widespread in soils in the south-eastern USA where RKN and lesion nematode populations occur. However, comparatively little is known about their occurrence or ability to infect nematodes in more northern latitudes. However, *Pasteuria* occurs in northern climates because it has been reported on stunt nematodes from turf soil samples taken from Massachusetts (Wick and Dicklow, 2001), on lesion nematodes from potato fields in Minnesota, on soybean cyst nematodes in Illinois (Atibalentja *et al.*, 1998), and it also has been observed on *M. hapla* in Connecticut, New York and Michigan.

Isolates of *P. penetrans* vary markedly in their specificity towards different RKN species and even to populations within species, however little is known regarding specificity. *P. thornei* and *P. penetrans* can reduce RKN populations below economic threshold levels in several crops and in different environments (Chen and Dickson, 1998). More than 35 publications report that *Pasteuria* may reduce symptoms or disease severity caused by RKN in various crops (Chen and Dickson, 1998). In Florida, RKN and sting nematode suppression has been attributed to *P. penetrans* (Dickson *et al.*, 1994; Chen *et al.*, 1997; Weibelzahl-Fulton *et al.*, 1996) and *Candidatus pasteuria usga*, respectively (Giblin-Davis, 1990).

Nematode, bacterial and fungal based biopesticides are highly vulnerable to environmental stress. Increase in temperature, UV radiation and a decrease in relative humidity may reduce the efficacy of these bio-agents.

17.6. HOST RESISTANCE

The development and deployment of plant resistance to nematodes may be the single most effective means of managing the plant parasitic nematodes. Plant resistance limits reproduction of the nematodes in the target crop and in some cases can be as effective as chemical control (Cook and Evans, 1987; Starr and Roberts, 2004). In one study of the economic benefits, the $1 million cost of developing a soybean cultivar resistant to cyst nematodes was far surpassed by $400 million in benefit.

Any research programme aiming at introducing new varieties, should take into account the effects of climate and environment changes. Particular attention must be paid in the management of nematodes through the introduction of resistance genes in commonly used varieties, since an evaluation is needed about the persistence of the genetic pool introduced, on a scale of decades or years. Hence, protection of biodiversity of plant genetic pools has a practical and immediate justification. The term "adaptive strategy" reflects, in this sense, the need for a quick and flexible response, since some climate changes are expected to occur on a short temporal scale. Some changes are already in action and newly introduced varieties may display in a few years unsuitable agronomic traits, i.e. higher susceptibility due to increased air moisture or rainfall, or may become susceptible to newly colonizing or substitution pests, thus vanishing long term research investments. Programmes on the introduction of resistance genes in plant varieties should consider their durability in a changing environment, and how much they will remain useful.

Climate change mediated changes in physiology can alter the expression of resistance genes. The most serious threat to genetic resistance to nematodes may be posed by the increased selection pressure resulting from acceleration of nematode developmental rate and increase in number of generations per season due to global warming (Ghini *et al.*, 2008).

Therefore, there is a need to develop appropriate strategies for nematode management that will be effective under situations of global warming in future.

17.6.1. Pre-emptive plant breeding

Investment is necessary in pre-emptive crop pest resistance breeding against future high risk pests based on current and future geographic distributions in order to avoid the consequences of shock pest episodes. Predictions of what pests are likely to be a future threat must structure plant breeding programmes more than is currently the case. That a pest is absent from an area only by virtue of not having been introduced, or that a particular pest under a changed cropping system may emerge as a major threat can be anticipated. Pre-emptive breeding, as opposed to the current reactive approach, on crop pests must be seen as a sound investment, if the consequences of shock pest episodes are to be avoided.

17.6.2. Biotechnological approaches

Biotechnology offers several benefits for nematode control in an integrated management strategy, such as reducing risks to the environment and to human health, accessibility for food producers in the developing world, and the possibility of achieving durable, broad-spectrum nematode resistance (Thomas and Cottage, 2006). A number of genes that mediate nematode resistance have now been or soon will be cloned from a variety of plant species. Nematode resistance genes are present in several crop species and are an important component of many breeding programmes including those for tomato, potato, soybeans and cereals (Trudgill, 1991). There are essentially three approaches for engineering resistance: transgenic expression of natural resistance genes in heterologous species; targeting and disruption of the nematode; and feeding site attenuation (Thomas and Cottage, 2006).

Using marker-assisted breeding techniques, Monsanto is introducing root-knot nematode resistance into elite genetics to develop cotton varieties that could potentially increase lint yield by an average of 8 to 10 per cent under root-knot nematode infestations.

A new, high throughput screening method is being utilized to accelerate the incorporation of reniform nematode resistance into elite genetics of cotton. The product could potentially increase lint yield by an average of 10 to 15 per cent under reniform nematode infestation conditions.

Pineapple plants transformed with the modified rice cystatin gene (proteinase inhibitor) have tested positive for transgene expression, particularly in the root tissue. Cystatin has been shown to deter reniform nematode feeding and reduce populations by interfering with their ability to produce digestive enzymes (Atkinson *et al.*, 1996). The planting of the reniform nematode resistant pineapple is assumed to replace the current use of fumigants and nematicides in Hawaiian pineapple. This reduction totals to 1.4 million pounds per year in active ingredient with an associated savings of $2.1 million in expenditures.

Potato plants were developed that transgenically expressed a disulphide-constrained peptide (nAChRbp) capable of binding to nematode acetylcholine receptors and inhibiting chemoreception of cyst nematodes. The results validate the root tip-specific promoter of the *Arabidopsis MDK4–20 gene* (Lilley *et al.,* 2010) as a means of delivering effective root protection from *Globodera pallida* by the peptide under field conditions.

When susceptible sugar beet genotypes are co-transformed with *Agrobacterium rhizogenes* to introduce the (Hs1.sup.pro-1) resistance gene, the hairy roots formed exhibit resistance to sugar beet cyst nematode (Cai *et al.*, 1997). It was reported that the nematode population decreased by 73% with experimental resistant varieties whereas that increased by 35% with a susceptible variety (Werner *et al.*, 1995). A resistant variety called Nematop was developed which showed acceptable performance in nematode-infested soil (Dewar, 2005).

For the first time, it has been shown that a Bt Cry6A protein can confer plant resistance to an endoparasitic nematode (*Meloidogyne incognita*), and that Cry 6A proteins have the potential to control plant-parasitic nematodes in transgenic tomato plants (Xiang-Qian *et al.*, 2007).

The *Mi* gene (true *R* gene) from tomato conferred resistance against a root-knot nematode and an aphid in transgenic potato (Rossi *et al.*, 1998).

The transgenic tobacco plants AtNPR1 high expression line 19-1 recorded the highest shoot and root weight, which is about five folds higher, compared to the wild type plants. The number of root galls and egg-masses developed on the infected roots of the transgenic plants were significantly less (up to 50-60% less) compared to the wild type plants (Bhanu Priya *et al.*, 2011).

Tobacco plants were engineered to produce dsRNA of two essential genes of the parasitic nematode *M. incognita*. The transgenic tobacco plants very effectively resisted *M. incognita* infection and their development was severely impaired. These nematodes were specifically deficient in the mRNA of targeted genes, indicating that the dsRNA produced in plants did indeed trigger RNAi response in the nematode (Yadav *et al.*, 2006).

17.7. INTEGRATED NEMATODE MANAGEMENT

More comprehensive integrated management farming systems that include more restricted tillage, fertilization, pesticide use, the addition of organic manure, and under sowing with clover greatly alter the soil fauna and micro-flora (El Titi and Ipach, 1989). For example, the numbers and biomass of earthworms were six times greater in the integrated plot with limited tillage than in the conventionally managed plot. Predatory mites and microbivorous nematodes (bacterivores and fungivores) also are often greatly increased through this type of integrated management (El Titi and Ipach, 1989). Population densities of *Heterodera avenae* and *Ditylenchus dipsaci* were lower in integrated systems with minimal tillage than in conventional systems with standard tillage practices (El Titi and Ipach, 1989).

17.8. INNOVATIVE TECHNOLOGIES

As sustainable nematode management becomes increasingly based on soil biology–health, new complementary technologies are developing. These new tools undoubtedly will improve the accuracy of nematode diagnosis and assessments of potential problems, and will result in more effective management, reduced pesticides, pesticide usage, and less contamination of groundwater with agricultural chemicals such as nematicides, nitrogen, and fertilizers.

17.8.1. Precision agriculture

Modern computerized harvest-management and data systems offer new opportunities for more precise management of nematodes and general crop production. This technology has the potential to improve water use and limit fertilizer and pesticide application on a spatial and temporal basis as dictated by soil fertility and, more important, differential spatial crop yields (Daberkaw and Christensen, 1996; Evans *et al.*, 1996). Based on early results, this management tool should allow specially prescribed nematode control in high-intensive crop production such as *Radopholus similis* on banana and root-knot nematodes on potato in the north-western United States (Evans *et al.*, 1996). Approaches that focus on a harvest index to locate environmental stress (Copeland *et al.*, 1996) should be able to relate nematode kinds and numbers to poor yield and other stress factors. This approach is now being used in some banana operations in which fruit is harvested in small subunits and yield data are recorded and analysed by computer. Poor-yielding sections can be examined for nematode densities and other potential problems.

17.8.2. Genetically engineered and traditional host resistance

Although almost 100 years elapsed between the appearance of Mendel's rules and the initial discoveries in molecular biology, dramatic progress in the latter area has occurred during the last 30 years (Persley, 1992). The increasing complexity and costs of genetic engineering of plants for pest resistance or altering bio-control agents make it unlikely that significant economic repercussions of molecular biology will be felt on agricultural production in the near future (Persley, 1992). There has been considerable progress made in engineering host resistance to nematodes, genetic mapping, and diagnostics (Cai *et al.*, 1997; Caswell-Chen *et al.*, 1993; Opperman and Conkling, 1998; Williamson *et al.*, 1992). However, genetically engineered resistance to nematodes is still at the developmental stage in contrast to the recently deployed herbicide- and insect-resistant cultivars of cotton, soybean, and other crops. One strategy involves transformation of plants with a transgene(s) encoding a product detrimental to the target nematode or that suppresses the expression of key plant genes involved in the nematode-host interaction (Opperman and Conkling, 1998; Williamson *et al.*, 1992). Candidate genes for this strategy include collagenase, genes expressed in the development of specialized feeding cells induced by species of *Globodera* or *Heterodera* (syncytia) and *Meloidogyne* (giant cells). Constructs of the root-specific TobRB7 gene in tobacco have been used to develop promising root-knot nematode–resistant genotypes (Opperman and Conkling, 1998). Linking this gene with a BARNASE gene resulted in root knot–resistant plants, but difficulties were encountered in recovering resistant lines from progeny of the transformants. Transformed plants with an antisense TobRB7 construct also exhibited root-knot resistance; root-gall development was about 70% less than in susceptible plants (Opperman and Conkling, 1998).

A second approach for engineering nematode-resistant plants involves identifying, cloning, and introducing natural plant-resistance genes into susceptible crop plants. Exciting results with this strategy were recently reported with *Heterodera schachtii* on sugar beet (Cai *et al.*, 1997). In one major development, Cai *et al.* (1997) cloned the cyst-resistant gene in wild *Beta* species. A transformed, normally susceptible sugar beet line exhibited the typical incompatible resistant reaction. Similar progress is being made with the Mi gene, which confers resistance to the common *Meloidogyne* species and populations attacking tomato. With the wide host range of these nematodes, the transfer of the Mi gene to numerous crop species, for which root-knot nematodes affect major crop yields, has great economic promise. Because populations of *M. incognita* may overcome this resistance, much care is needed in developing cropping systems to prolong the durability of this resource in a wider range of genetically engineered resistant crops.

New molecular techniques and markers also have positively affected traditional plant-breeding programmes related to the development of host-resistance to nematodes. Recently, two markers for parasitism in *H. glycines* were identified (Dong and Opperman, 1997) and molecular markers for crop resistance for various cyst nematodes are being investigated. These resistance markers included soybean (*H. glycines*) (Cook, 1991), potato (*G. rostochiensis*) (Pineda *et al.*, 1993), and wheat (*H. avenae*) (Williams *et al.*, 1996). Markers for *M. incognita* races 1 and 3 resistance in tobacco also have been described (Yi, 1997). Undoubtedly, combining markers for parasitism (virulence) within different nematode populations and host-resistance genes should spur advances through traditional plant breeding.

17.8.3. Advisory programmes

Despite the development of nematode advisory programme in some states in the United States in the 1960s and earlier elsewhere (Barker and Imbriani, 1984), low-cost, highly effective nematicides remained in use as a form of insurance until recently. The unreliability of nematode assays, due to difficulties in sampling the contagious infestations, identification of related species, and lack of information on economic thresholds helped to prolong nematicide use. Nevertheless, advisory programmes have successfully contributed to lower pesticide usage and greater farm profits. For peanut alone, growers in Virginia were able to reduce their nematicide use by 35% after a predictive nematode assay programme was established (Phipps, 1993). Savings in production costs for 1989 were estimated at $800,000, primarily through fewer nematicide applications. Currently, about one half of the states in the United States offer their farmers some type of nematode advisory programme, usually through the Extension Service, State Departments of Agriculture, or private consultants. Many growers monitor the relative magnitude of nematode problems in given fields by observing root symptoms and signs of nematodes and through field histories.

The use of hazard indices *in lieu* of damage or economic thresholds has promoted better communication to growers on the relative nematode-damage potential for annual crops in given fields (Barker and Imbriani, 1984). These hazard indices are based on the relative damage potential of the nematode species/races present, their population densities, the cropping history, and the soil type.

Where detailed data on production and nematode populations are maintained, more precise approaches in decision-making are becoming available. Burt and Ferris (1996) developed a sequential decision rule to aid in choosing a rotation crop versus host crop where this practice is the management tactic rather than using a nematicide. The static model used by Ferris (1978) is unsuitable for quantifying the optimal dynamic threshold that would be characterized by population densities lower than where returns from the nematode host and non-host are equal. A dynamic model for this type of crop-nematode management system was recently developed (Burt and Ferris, 1996). Application of this model should allow better economic management of nematodes, but data will still be needed on annual nematode population change under host and non-host crops and the relationships between nematode numbers and crop yields. More comprehensive pest-host simulators and expert systems (Saracino *et al.*, 1995) have bolstered research in recent years (McSorley, 1994).

Management of nematodes, including advisory programmes, poses greater challenges for perennial crops than for annual crops. Control options are limited, and very low population densities often build up to cause severe damage over time. Integrated management, including assays to determine numbers and kinds of nematodes present, and appropriate control tactics such as pre-plant fumigation where necessary, use of nematode-free stock, tolerant cultivars where available, and organic mulches are useful for woody ornamentals (Benson and Barker, 1985).

17.9. FUTURE PROSPECTS

Water-limiting environments, nematode diseases, declining fertility, availability and degradation of the soil resource are among key constraints to increasing production and quality of food. Climate change adds an extra layer of complexity to an already complex agro-ecological system. Crop protection professionals routinely develop and deploy strategies and tools based on well-established principles to manage plant nematodes and many may also be applicable under climate change when projected changes, processes and interactions are factored in. Therefore, research to improve adaptive capacity of crops by increasing their resilience to nematodes may not involve a totally new approach, although managing plant nematodes may have the added advantage of mitigating rising CO_2 levels (Mahmuti *et al.*, 2009). The bulk of any new investment to nematode-proof food crops, therefore, need only to accelerate progress of new and existing promising strategies and approaches and not to 're-invent the wheel' under the guise of climate change research. Such an investment model will ensure that nematode management solutions span the entire range of uncertainties associated with climate change including the 'business as usual' scenario.

There has only been limited empirical research on plant nematodes under realistic field conditions mimicking climate change and this severely restricts development of options to enhance crop adaptation or nematodes management under climate change. In addition, a relatively large body of knowledge has been gathered on potential effects of global climate change using models. First-pass assessments are now available for some countries, regions, crops and particular nematode pathogens. From a food security perspective, emphasis must now shift from impact assessment to developing adaptation and mitigation strategies and options. Two broad areas of empirical investigation will be essential; first, to evaluate the efficacy of current physical, chemical and biological control tactics including nematode resistant varieties under climate change, and second, to factor in future climate scenario in all research aimed at developing new tools and tactics. Transgenic solutions (Huang *et al.*, 2006) must receive serious consideration in integrated nematode management strategies to improve food security.

17.10. CONCLUSIONS

The development of sustainable nematode-management systems is not an option. It is imperative that scientists devise the requisite sustainable tactics as one component of the world's complex food-fibre production system to meet the pressure of the rapid population increase. Management of plant-parasitic nematodes is essential to sustainability, since impaired efficiency of plants' water and nutrient utilization caused by these pathogens limits production and degrades the environment. The proposed strategy of increased use of pesticides and plastics to meet this challenge (Avery, 1995) would likely provide only short-term benefits. For example, the repeated heavy use of chemicals such as methyl bromide essentially sterilizes the soil and eliminates beneficial soil microflora and fauna as well. Many other current crop- and pest-management practices also contribute to the instability of our food production (Ingham *et al.*, 1985). Fortunately, the new technologies forthcoming from molecular and soil biology and truly integrated cropping-nematode-pest management systems are providing new strategies and tactics that can be linked to traditional nematode management for more general integrated and sustainable food and fibre production. In fact, the wide gaps between and within developing and developed countries indicates that global food production still can be increased.

Section V
WEEDS

18

Climate Change and Weed Scenarios

18.1. INTRODUCTION

Weeds are one of the main threats to biodiversity, agriculture and the environment. Under climate change, they pose two challenges; namely, the suite of weed species will change and some weeds will become more invasive.

Agricultural weeds can hurt crop yields or increase costs of production by:

- Competing directly for sunlight, nutrients, moisture, and space.
- Weeds can produce allelopathic substances that are toxic to crop plants.
- Physically hindering crop growth and development, especially climbing vines like morning glories (*Ipomoea* sp.) and hedge bindweed (*Calystegia sepium*).
- Weeds often serve as hosts for many crop diseases and also provide shelter for insect pests and disease pathogens during off-season.
- Promoting disease pathogens by restricting air circulation around the crop.
- Interfering with or contaminating crop harvest.
- Reproducing prolifically, resulting in a greater weed problem in the future.
- Parasitizing crops directly (e.g., dodders, *Cuscuta* sp.; Orabanche; and witch weed, *Striga asiatica*).

18.2. CROP LOSSES

The most recent and comprehensive efforts to provide a measure of global crop losses by weeds are those made by Oerke (2006) (Table 18.1). Weeds produced the highest potential loss (34%). In the US, annual losses in crop production due to weeds have been valued at approximately US$ 12 billion, amounting to some 10% of potential production (Patterson and Flint, 1990). Large efforts are made to limit these damages through a variety of weed control measures.

Table 18.1. Estimated potential of weeds and actual losses due to weeds in six major crops worldwide, in 2001-03

Crop	Attainable production (Million tons)	Crop losses (%) due to weeds	
		Potential	Actual
Wheat	785.0	23.0 (18-29)	7.7 (3-13)
Rice	933.1	37.1 (34-47)	10.2 (6-16)

Crop	Attainable production (Million tons)	Crop losses (%) due to weeds	
		Potential	Actual
Maize	890.8	40.3 (37-44)	10.5 (5-19)
Potatoes	517.7	30.2 (29-33)	8.3 (4-14)
Soybeans	244.8	37.0 (35-40)	7.5 (5-16)
Cotton	78.5	35.9 (35-39)	8.6 (3-13)

Source: Oerke (2006)

Around the world, more human labour is expended in hand weeding than in any other agricultural task, and most cultivation and tillage practices are designed to aid in weed control. The chemical industry manufactures herbicides, which, next to fertilizers account for the largest volume of chemicals applied to crops. Among pesticides used for the management of pests, herbicides account for 65% (IFPRI, 1998; USDA, 1999).

Most analysis concur that in a changing climate, weeds may become even more active than they are currently, thus posing the threat of greater economic losses to farmers (IPCC, 1996; Coakley *et al.*, 1999). While the majority of weeds are invasive species from temperate zones, other weeds in temperate regions originated in tropical or subtropical regions, and in the current climate their distribution is limited by low temperature. Such geographical constraints will be removed under warm conditions. Warmer temperature regimes have been shown to increase the maximum biomass of the grass weeds significantly. In crop monocultures, undesirable competition is controlled through a variety of means, including crop rotations, mechanical manipulations (hoeing), and chemical treatment (herbicides).

18.3. EFFECTS OF CLIMATE CHANGE ON WEEDS

The main drivers for climate change impacts on weeds include increased temperatures, changed rainfall, increased carbon dioxide (CO_2) levels in the atmosphere, altered frequency and intensity of extreme weather events, more frequent frosts, changed phenology and changed land use. Likewise, the risk of negative impacts from weeds due to extreme weather events, such as prolonged drought, heat waves, floods and cyclones that occur under today's climate, are similar to the risks associated with average climate change conditions.

There are concerns that global CO_2 enrichment will affect weeds and crop yields directly or indirectly through global warming and its associated changes in climate such as alteration in precipitation, wind pattern, rise in sea level and more flood and drought. It has been reported that elevated CO_2 concentration cause an increase in photosynthesis (Kendall *et al.*, 1985) and decrease in photorespiration and respiration (Bruce and Caulfield, 1991; Ziska and Bunce, 1993), and different species differ in their response to elevated CO_2 concentrations (Poorter, 1993).

Weeds grown in association with the crops are exposed to increasing concentrations of CO_2 ($\geq$370 ppm) and other active greenhouse gasses (CFCs, CH_4, N_2O, NO_2, etc.) that have led to global warming and the associated greenhouse effect. They are equally and may be differently influenced due to these changes as different species of weeds grow together with a crop. The responses of crops and weeds to increased CO_2 level and temperature are being studied worldwide. The variations in the responses of C_4 and C_3 plants (both crops and weeds) have been noted under such situations. Composite weeds in the crop fields having variable proportions of C_4 and C_3 plants are likely to undergo dynamics in weeds insurgence and shift of weeds in favour of certain species in course of time. Increased CO_2 level may stimulate photosynthesis in some weeds, leading to higher growth of rhizomes and other storage organs in perennial weeds, higher seed production in annual weeds, etc. and make their control difficult concurrently or over the seasons/years. Thus, an appropriate technology towards controlling increasingly more difficult perennial weeds will be the important paradigm in future weeds research. Climate change is likely to trigger differential growth in crop and weeds and may have more implications on weed control across crops and cropping systems.

Changes in temperature and CO_2 are likely to have significant direct (CO_2 stimulation of weed growth) and indirect effects (climatic variability) on weed biology. In spite of the importance of weed biology in both the environment and in farms, very little is known regarding the impact of these environmental changes on either the reproductive success of agronomic or invasive weeds, or the potential consequences for their management. Yet, given what is known, it is clear that the agricultural, environmental and health costs of not understanding the impact of CO_2 on weed biology may be substantial. It is hoped therefore, that the current chapter may serve to both emphasize the critical nature of this topic, and to serve as an initial guide to those who wish to recognize the ramifications of rising CO_2 beyond the polemic of global warming.

An increase in extreme events, such as cyclones, storms and associated floods which are the main characteristics of climate change may increase the dispersal of weed species with efficient seed-dispersal systems (wind, water and birds) will disperse more quickly than weeds that rely on vegetative dispersal move of seeds or pollen which enable them to invade new ecosystems (Singh *et al.*, 2011).

Climate change may affect invasive weeds through:

- Increased disturbance due to fire, floods and other extreme climatic events.
- Higher temperatures and reductions in frost events.
- Changes in rainfall timing, frequency and levels (including humidity and evapotranspiration).
- Reduced stream and river flows (exposing well-watered riparian areas).
- Changes in coastal and estuarine habitat due to rising sea levels.
- Increased CO_2 fertilization (and resultant increases in weed growth).
- Changes to flowering and fruiting times.

18.4. IMPACTS OF CLIMATE CHANGE ON WEEDS

Climate conditions exert a significant influence on the spread, population dynamics, life cycle duration, infestation pressure and the overall occurrence of the majority of agricultural weeds (Šmatas *et al.*, 2008). Climate change is expected to increase the range, or “damage niche” (also called "invasion niche") of many weed species. Research suggests that the composition of invasive weed communities will be fundamentally altered by the end of the century under increasing temperature scenarios, with new weed species entering communities as a result of geographic range shifts (McDonald *et al.*, 2009). For example, the range of the yellow star thistle, a California weed, is expected to increase to more northern parts of California and Nevada due to climate change (Bradley *et al.*, 2009).

It is postulated that competition between plants will change with different responses to elevated CO_2 and temperature (Hunt *et al.,* 1991). Generally, it is reported that plants with C_3 photosynthetic pathways are expected to benefit more than C_4 from CO_2 enrichment but inverse is true with rising temperature. This differential response of C_3 and C_4 plants to elevated CO_2 and temperature can have important implications on crop/weed competition as most of the weeds are C_4. But this fundamental idea that most crops are C_3 and most weeds are C4, and hence weed competition will consequently decrease as CO_2 increases, should not be viewed as universal axiom (Ziska, 2001; 2003). Clearly, there are four major C_4 crops of economic importance (corn, pearl millet, sorghum, and sugarcane), and many important C_3 weeds (e.g. *Chenopodium album, Avena ludoviciana, Phalaris minor, Eclipta prostrata, Ammania baccifera* etc.). Therefore, it is important to examine crop/weed competition case by case to develop effective weed management practice for the emerging species in the changing climate scenario. Climate change may bring changes in weed population and in their phenology. Many weed species may expand their range and spread to new areas. In addition to its impact on agricultural weeds, literature suggests that invasive species may become more of a threat in changing climate because of their strong response to elevated CO_2 and changing climate compared to other native species. Moreover, some weed species which directly impact human health through allergic reactions, skin irritations, or internal poisoning have shown positive response to changing climate particularly to elevated CO_2 concentration by producing higher plant biomass, pollens or poisonous compounds. Climate change may also likely have a direct or indirect impact on the chemical, mechanical and biological control methods by reducing their effectiveness on weeds.

The effects of climate change on weed/plant interactions are likely to vary by region and crop type. Understanding of the underlying physiological mechanisms responses to such factors is needed in order to address these effects. Because the interactions between crops and weeds are "balanced" by various environmental factors, local changes in these factors may tip the scale towards either crop or weed. Furthermore, as the geographic distribution of weed species changes, so will the community composition, posing both challenges and opportunities for invasion control. If the invasion of new weed species can be detected, efforts can be made in advance to prevent and control their establishment.

One of the main impacts of climate change is its influence on species' distributions. The main threat of species migration is from neighbouring regions. The rate of response of invasive plants and weeds is expected to be faster than for other plants, including native species and crops. Secondly, climate change is likely to foster the appearance of a new set of weed species.

Climate change may affect invasive weeds through:

- Potential range shifts (e.g. movement towards cooler latitudes or higher elevations).

- Changes in pathogen pressures, including serious impacts on biological control programmes.
- Changes to species interactions (e.g. between plants and pollinators, weed vectors, etc.).
- Weed shifts and spread of invasive species.

18.5. PREDICTION MODELLING FOR WEEDS

The aim of modelling a species potential distribution is to better understand the geographical area in which that species may be found, survive, thrive, or remain absent now or in the future. Combined with the species present known distribution, this information can be particularly useful for informing strategic management efforts aimed at reducing current and future weed impacts in a rational manner.

Many different factors control where plant species are presently found. These factors can be broadly grouped into four areas: climate, soil, co-occurring species, and their ability to get there from where they were previously (e.g. their native range). Of these factors it is climate that represents the most useful 'outer limit' of where we might need to apply weed management strategies. It is, therefore, climatic suitability that weed models should be first trying to capture in their projections. Once this outer limit has been defined, it can subsequently be partitioned up by then considering non-climatic limitations to weed distributions.

Most modelling techniques for describing species' distributions fall into two groups. Firstly, correlative models relate known distribution records for the species of concern to variables presumed to explain the known range. This produces a projection that represents geographical areas with similar climates to the distribution records, at least according to the variables used in the model. However, we know that species ranges are limited by factors other than climate, so appropriately constructed models based solely on distribution records are likely to underestimate the full potential of distribution of the climatically suitable area for a weed.

The second group, mechanistic models, can also be informed by distribution data, but can also make use of information from other sources, such as physiology experiments and expert knowledge, to understand the true climatic limits of the species in question. The nature of the variables and the form of the functional responses used in these models are usually informed by biological theory.

The general trend that can be drawn out of the many modelled weed distributions is a pole ward shift in modelled suitability (i.e. towards the south in Australia) (Sutherst *et al.*, 2007). Assuming species are able to move from their existing range we may, therefore, see a shift, expansion, or retraction of species distributions with climate change.

18.6. ADAPTATION AND MITIGATION STRATEGIES

Weed management remains a significant financial, logistical and research challenge. Climate change will require revisiting what we deem appropriate for weed control to keep current and future management strategies efficient and effective. Key elements to develop a vision, strategy and weed management plan, and weed management techniques in the context of climate change needs to be outlined.

A major adaptation response to climate change is increased landscape connectivity, but this presents a major opportunity for increased weed invasion. Adaptation responses include quarantine and filtering methods to monitor species displacement. A new management approach will be needed. Adaptation responses for the main weed management techniques of quarantine barriers, eradication and containment include biological control, herbicides and other control methods.

The key weed control methods that require (re)consideration include:

- Quarantine barriers
- Eradication and/or containment
- Biological control
- Herbicides
- Cultivation

18.6.1. Quarantine barriers

The current overall national approach to quarantine of invasive plants is expected to be adequate as a response to climate change. There will be a continued threat of weed invasion from overseas, but the source areas will not change greatly as climate changes. However, the weed risk assessment (WRA) protocol does not have an assessment component specifically for climate change.

The general southwards invasion projected from the various studies of modelling species distributions points to the need to establish new quarantine barriers as an adaptation response. These could be at state boundaries.

18.6.2. Eradication and/or containment

Plants are relatively slow to invade (compared with insects, for example), which gives greater opportunity to instigate eradication or containment. Both control objectives should not be undertaken lightly as considerable time and funds are usually required for eradication (for example, 12 or more years).

A policy of containment implies that control actions against an invasive species will have to go on forever (but this still may be very worthwhile).

An adaptation response is to preferentially target search and control efforts in the area that falls within the region of possible future migration under climate change. Species distribution models should help in delimiting the area where control is needed, as well as the likely direction of species migration.

18.6.3. Biological control

Biological control agents, mainly arthropods and pathogens, have provided permanent and effective control for some major weed species. Part of the process of biological control is to search for and select biological control agents in regions that matches the part of country that has been invaded. Usually climate change is not considered in the assessment of the agent-target association.

The changing dynamics between target weeds and their biological control agents is likely to be complex because the environmental response of the weed may not be the same as the agent's or their interaction. In addition, the changed climate may become unsuitable for the existing biological controls. Interaction between increased CO_2, fast growth and plant nitrogen levels needed to support biological control agents such as herbivores may change.

Some of the main factors that may affect biological control include:

- Increases in temperature leading to increased growth rate of the weed plant that could enable to "escape" the herbivore.
- Changes in rainfall that could make the weed plant and its environment unsuitable for the agent.
- Higher CO_2 concentrations.
- Increased carbon to nitrogen ratio in the plant which could slow herbivore development.
- Changes to plant secondary chemistry which make the plant less palatable.

Two biological control examples show some of the complexity of assessing the interaction between biological control agents, invasive plants and climate change. The growth of the leaf miner *Dialectica scalariella* on Patterson's Curse, *Echium plantagineum*, was mostly related to higher temperatures and not CO_2 concentration, but at the cost of poorer survival of the leaf miner.

A second example concerns the chrysomellids, *Octotoma championi* and *O. scabripennis*, which are leaf miners as larvae and leaf feeders as adults on the widespread invasive shrub, *Lantana camara*. The insects were affected by increased leaf loss at higher temperatures, but insects preferred leaves grown at normal levels of CO_2. In both cases, the plants showed the typical response of increased biomass, and reduced leaf nitrogen and increased carbon - nitrogen ratio when grown under elevated CO_2.

Adaptation responses suitable for the biological control component of weed management includes:

- Monitoring the weed – biological control interaction to determine if additional controls are required.
- Prioritising biological control agent redistribution to areas with a more favourable climate.
- Searching for and assessing new biological control agents from overseas areas with climates that anticipate the future climate of target regions.

18.6.4. Herbicides

Herbicides are an integral part of weed control in crops and are increasingly used to control environmental weeds. Research in overseas has shown that weedy species have an increased tolerance to herbicides when grown under elevated CO_2 conditions. Moisture stress is also likely to decrease the effectiveness of herbicides.

Herbicide resistance is widespread in cropping areas. It can potentially develop within a relatively short period after eight consecutive years of application of the same herbicide. Many invasive plants in agricultural situations have short generation times, mostly with annual generations. It is this combination of high selection pressure (herbicide application), short generation time, and inherent genetic variability that allows for the rapid evolution of herbicide resistance. It is reasonable to assume that these invasive plants could also evolve rapidly in response to climate change. This implies that today's major agricultural weeds (e.g. wild radish, annual ryegrass) are likely to remain as major problems under climate change. This assumes that current agricultural practices continue.

This indicates that the main adaptation response is to manage for herbicide resistance through an integrated approach in controlling agricultural weeds. With a drying environment, this integration is likely to include a move away from continuous cropping to include a greater reliance on pasture phases. Innovative management of weeds, including careful use of genetically modified crops, will be required to avoid a return to weed species favoured by crop-pasture rotations in the past (e.g. doublegee, *Emex australis*), or changed cropping practices (e.g. hoary cress, *Lepidium draba*).

Adaptation responses should consider the following aspects:

- Pasture phases in rotation rather than continuous cropping
- Improved methods of herbicide application
- Cultivation and fire as strategic weed control options
- New innovations in weed control such as weed seed destruction

18.6.5. Other control methods

Cultivation has been the weed control method used over most of agricultural history. Climate change, especially under a drying environment, may make the general use of cultivation problematic given the increased risk of wind erosion and soil moisture loss. The adaptation response might be to use cultivation strategically as part of an integrated weed management approach.

Likewise, increased temperatures will make fire an increasingly risky weed management tool. Frequent fires at certain times of the year might not be the appropriate adaptation response depending on the response of invasive species, which might increase in abundance under frequent fire. Also, consideration needs to be given to the impacts of fire on the vegetation being protected.

It is difficult to see how mechanical weed control (chain pulling, stick raking, blade ploughing, and specialised weeding machines) can be adapted for the increased risk of climate change. Recent developments such as mechanical destruction of seeds as part of the harvesting process will continue to be useful. Managing grazers (goat, sheep, cattle etc.) as part of a pest management adaptation response may also be beneficial to weed control. The management plan for pest animals also provides a model for how to involve the community in a weed management plan.

Considerable attention has been given to robotic means of weed control. These technological innovations for weed control will improve current techniques, such as herbicide use, and may provide specific adaptation responses to climate change, such as enabling highly targeted herbicide spraying during the night, when it is cooler.

18.7. FUTURE THRUSTS

Very limited studies have examined the impact of climate change on weeds, therefore, more emphasis needs to be given to the following aspects:

- To address the anticipatory changes in weed biology and dynamics with climate change in order to develop flexible integrated weed management practices which are based on a foundation of knowledge of weed biology and ecology.

- To stimulate research interest in the area of climate change and weeds to predict possible impacts on weeds.
- Crop-weed competitive interaction under single (CO_2 or temperature) or multiple factors (e.g. CO_2 and temperature or CO_2).
- Invasiveness.
- Weed management.

18.8. CONCLUSIONS

Weeds are a major threat to agriculture and the environment. That threat will continue because of the large number of potentially invasive species that are already present, or that might arrive by new introductions. Climate change, be it through future global change or through the occurrence of extreme events, adds another layer of risk to the existing risk from weeds.

The science of invasion ecology indicates that the suite of weed species will change and that impact may increase under climate change. Increased temperature, changed rainfall, increasing CO_2, changes in extreme weather events and frosts coupled with land use change will be the drivers for the changed weed profile. Species distributions will move, mostly in a southwards direction. This will modify the biosecurity threat, including from native invasive species. Corridors may both aid management or be new conduits for invasion, especially to refugee.

The importance of developing an agreed vision and strategy for addressing weeds issues under climate change needs to be emphasized, from which a more detailed weed management plan can be developed. There is a need to provide a general adaptive management framework to guide weed management planning and emphasize the importance of identifying the main potential weed threats to aid strategic prioritization of resources to key species, locations and control methods.

Research is needed to assess the interactive effects of multiple climate change factors simultaneously to help predicting how weed problems may change in future with changing climate in order to develop flexible integrated weed management practices, which are based on a foundation of knowledge on weed biology and ecology.

19

Effects of Climate Change Variables on Weeds

19.1. INTRODUCTION

Climate change involves rising temperatures (Tubiello *et al.*, 2007; Gillett *et al.*, 2011); altered precipitation patterns, which also increase the probability for summer droughts (Bloomfield *et al.*, 2006; Lobell and Burke, 2008; Robinson and Gross, 2010); increased carbon dioxide (CO2); and higher frequency of extreme weather events such as heavy storms, summer droughts, and extreme cold spells (Diaz *et al.*, 1999; Tubiello *et al.*, 2007; Jentsch *et al.*, 2009; Coumou and Rahmstorf, 2012). Land use and management practices are also influenced by climatic changes (Fleming and Vanclay, 2010). As weeds are closely associated with the cropping system (Pysek et al., 2005; Andreasen and Skovgaard, 2009; Cimalova and Lososova, 2009; Gunton et al., 2011), climate change has an indirect influence on the occurrence of weeds via crop management and land use. Weeds are influenced by these altered abiotic conditions (Dukes *et al.*, 2009; Singer *et al.*, 2013).

Predicted effects of climate changes on weeds include:

- Increased disturbance due to fire, floods and other extreme climatic events.
- Higher temperatures (in excess of 35°C).
- Less incidence of frost.
- More frequent droughts.
- Changes in rainfall timing, frequency and levels (including humidity and evapotranspiration).
- Reduced stream and river flows (exposing well-watered riparian areas).
- Changes in coastal and estuarine habitat due to rising sea levels.
- Increased carbon dioxide (CO_2) fertilization (and resultant increases in weed growth).
- Increased seasonal variability.
- Extreme events such as cyclones can help to spread weeds.

19.2. ELEVATED TEMPERATURES

Higher temperatures can possibly offset some of the benefits of elevated CO_2 for both weeds and crops. High temperatures sometimes limit reproductive development and global warming may decrease reproductive output in such situations despite an increase in CO_2. It is unclear whether this is more likely to occur in C_3 than C_4 species, but if it were, it could alter weed community compositions and affect crop/weed interactions (Bunce and Ziska, 2000).

In temperate regions, global warming will affect the growth and marginally affect phenology, and influence the geographical distribution of weeds. Weed species of tropical and subtropical origins, currently restricted to the southern regions, may expand northward (Patterson, 1995). Warmer seasonal temperatures and milder winters will extend the distribution of invasive weeds (Kudzu and Ragweed).

Increasing temperatures may mean an expansion of weeds into higher latitudes or higher altitudes. Very aggressive weeds that are currently found in the south are limited in the northern states by low temperatures. Many C_4 grass weeds are serious problems in the southern US but do not occur at problem levels in the US Corn Belt. Studies have shown that itch grass (*Rottboellia cochinchinensis*), a profusely tillering, robust grass weed could invade the central Midwest and California with only a 3° C warming trend (Patterson, 1995). Witch weed, a root parasite of corn, is limited at this time to the coastal plain of North and South Carolina. With an increase of temperature of 3° C, it is speculated that this parasite could become established in the Corn Belt with disastrous consequences. The current distribution of both Japanese honeysuckle and kudzu is limited by low winter temperatures. Global warming could extend their northern limits by several hundred miles.

As mean temperatures increase, some weeds will be able to expand their range into new areas. The tropical weed prickly acacia (*Acacia nilotica* s. sp. *indica*) is likely to spread south (Kriticos *et al.*, 2003) and athel pine could spread throughout inland rivers as far south as the Murray River in Victoria.

Some plants have been planted and maintained in locations that are sufficiently warm for them to grow, but too cold for them to reproduce (e.g. certain street trees). Under a warming climate these satellite populations or specimens may become invasive if their seeds are viable and their seedlings can survive to maturity. Monitoring for new invasions can provide the early detection needed for making timely management decisions. However, this monitoring will be of more value in places where early warning signs are most likely to be detected, such as near to environmental assets or agricultural lands.

Atmospheric temperature rise could alter weed proliferation and competitive behaviour in weedy vegetation as well as in crop stands. The indicated likely climate change may favour C_4 over C_3 weeds (Tubiello *et al.*, 2007). As a result, a variation in weed distribution will affect the world's most important cropping systems, for example, rice-based cropping systems through weed shifts to high latitudes and altitudes. In addition, *Striga* sp. might extend their range to moderate climatic zones (Mohamed *et al.*, 2006; 2007). However, *Striga asiatica* is relatively insensitive to temperature, and changes in the geographical range of the host plants seem to play a critical role in its distribution rather than the direct effects of temperature (Patterson et al., 1982). If this concept is generalized for all parasitic weeds in the Orobanchaceae (Phoenix and Press, 2005), which could pose a serious threat to global crop production (especially in fodders) in the near future. Many C_4 weeds, such as *Amaranthus retroflexus*, *Setaria* sp., *Digitaria* sp., *Sorghum halepense*, and *Paspalum dichotomiflorum*, are expected to expand further north (Weber and Gut, 2005; Clements and DiTommaso, 2011), which would have a more pronounced effect in the northern Europe, where the number of weed species is lower than in the south (Fried *et al.*, 2010). Milder and wetter winters would tend to increase the survival of winter annual weeds, while thermophile summer annuals will grow more profusely in areas with warmer summers under prolonged growing seasons, enabling them to grow further north (Walck *et al.*, 2011; Hanzlik and Gerowitt, 2012).

Patterson (1995) predicted that climate change would spread arable weed species. For example, *Datura stramonium*, which needs high temperature for profuse growth (Cavero *et al.*, 1999), would become a more competitive candidate under the climate change scenarios. Warm temperature regimes augmented the abundance of *Hieracium aurantiacum* in Australia through accelerated growth and reproduction (Brinkley and Bomford, 2002).

Under warmer conditions, *Setaria viridis* germinated later in the (August) season (Dekker, 2003). This was a beneficial temporal non-synchrony with emergence of a maize crop, avoiding crop-weed competition. In contrast, a recent study indicated that this species would be a problematic weed in maize-based cropping systems elsewhere, through synchrony with maize emergence, which is probably due to stimulation by increased temperature (Peters and Gerowitt, 2014). Therefore, *S. viridis* would become a competitor of maize at enhanced temperatures at the time of emergence. This has implications for the northern part of the Central Europe, where temperatures are still below the optimum for this species (Walck *et al.*, 2011). Similarly, *Rottboelliia cochinchinensis* could invade the central Midwest of the USA and California from Gulf Coast states, with a 3°C rise in temperature (Patterson, 1995). If new weeds are introduced into a non-native area, new and effective herbicides may be needed.

Lee (2011) opined that increased temperature had a greater effect on plant phenological development than elevated CO_2. The author observed that increasing temperature by 4° C advanced the emergence timing of *C. album* and *S. viridis* by 26 and 35 days, respectively, and flowering time by 50 and 31.5 days, respectively. Increased temperatures strongly affected the biomass accumulation by annual grass species during their reproductive phase as compared with the vegetative phase, and such effects are more pronounced in C_3 than C_4 plant species. However, the increased temperature was believed to offset potential benefits of elevated CO_2 by reducing the reproductive output. The uptake and translocation of herbicides in plants and their persistence in soil will also be affected by rising temperatures (Rodenburg *et al.*, 2011). In addition to these effects, temperature will also affect the rate of water absorption and movement, which affects the rate of leaf development, cuticle thickness, and stomatal number and their aperture, thus indirectly affecting herbicide selectivity and efficacy (Bailey, 2004; Chandrasena, 2009; Rodenburg *et al.*, 2011).

Some weeds will also be able to tolerate the projected increase in temperature. Buffel grass (*Cenchrus ciliaris*) is one of the few weeds in Australia to be extensively assessed for growth response to climate change. There are two types of photosynthesis:

- C_3 plants where the initial steps in photosynthesis builds a three carbon compound. Most plants have C_3 photosynthesis.
- C_4 plants that produce a four carbon compound during photosynthesis. This method of photosynthesis is prominent among the grasses but also present in some other groups of plants.

Buffel grass captures energy from the sun in a way (C_4 photosynthesis) that is more efficient than most plants (that have C_3 photosynthesis) in warmer or drier climates, or at lower levels of atmospheric CO_2. Further, buffel grass plants are able to acclimatize and grow at warmer temperatures (growth at 35° C versus 25° C). This means that the area for management of buffel grass is likely to increase, including both the current range (where they will acclimate and persist) and the spread southwards (as the climate becomes warmer).

The general assumption is that rising temperature will give competitive advantage to C_4 plants than to C_3 plants. In a study, the soybean crop was more adversely affected by rising temperature than prickly sida (*Sida spinosa*) and sickle pod (*Senna obtusifolia*), two weed species (Fig. 19.1).

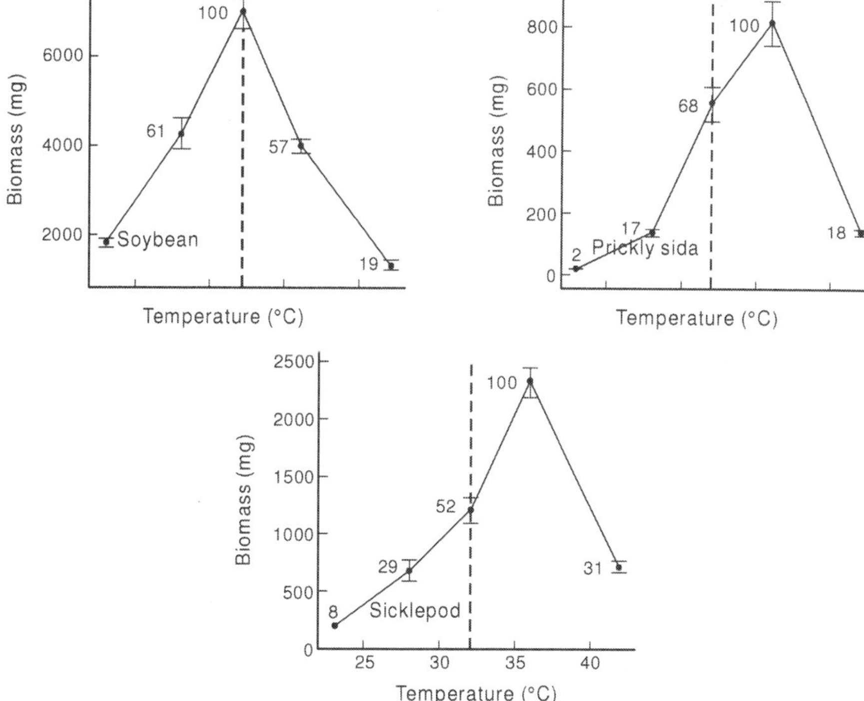

Fig. 19.1. Effects of elevated temperature on soybean crop and prickly sida and sickle pod weed species (*Source:* Tungate *et al.*, 2007).

The impacts of higher temperature include:

- Offset benefits of elevated CO_2 (Bunce and Ziska, 2000).
- Allow sleeper weeds to become invasive.
- Expansion of weeds into higher latitudes or higher altitudes.

- Very aggressive weeds that are currently found in the lower latitudes are limited in the higher latitudes.
- Itch grass, a profusely tillering, robust grass weed could invade the central Midwest and California with a 3° C warming trend (Patterson, 1995).
- A worldwide increase in temperature will make land currently unsuitable for weeds, acceptable places to set up shop in the future. This will expand the potential habitat for weeds species.
- Hotter temperatures earlier in the spring will lengthen the growing season.
- Spraying for weeds will have to be extended over a longer length of time which will lead to more chemicals being leaked into ecosystems.
- Potential for weeds to grow larger, produce more seeds, and disperse farther due to longer growing season.
- Moisture will evaporate from the soil at a faster rate, limiting the water available to both crops and weeds.
- Higher temperatures = longer growing season = more seeds in seed bank, more weeds each season.
- Higher temperature = cool climates become warmer climates = weeds extend their range.
- Negative effects on agriculture (increases weed competition, decreased weed control, weed range shift).

19.3. CO_2 ENRICHMENT

Higher CO_2 will stimulate photosynthesis and growth in C_3 weeds and C_3 crops, and reduce transpiration and increase water use efficiency in both C_3 and C_4 weeds and crops. Due to their different types of photosynthesis C_4 and C_3 plants react very differently to elevated atmospheric CO_2. Basically, elevated CO_2 does not directly stimulate C_4 photosynthesis and growth. Nonetheless, drought stress can be ameliorated at elevated CO_2 as a result of even lower stomatal conductance. Therefore, unlike C_3 crops for which there is a direct enhancement of photosynthesis by elevated CO_2, C_4 crops will only benefit from elevated CO_2 in times and places of drought stress. Fig. 19.2 shows the different response of C_3 in comparison to C_4 plants to CO_2 increase. Crassulacean acid metabolism (CAM) plants react to elevated CO_2 similar to C_3 plants with enhanced growth, if water supply is sufficient. If not they will respond like C_4 plants.

C_3 weeds (using one of two types of photosynthetic pathway, which responds to higher levels of CO_2) such as parthenium (*Parthenium hysterophorus*) may grow more rapidly under higher carbon dioxide levels and become more competitive (McFadyen, 2008; Poorter and Navas, 2003).

CO_2 can affect plant and leaf size, seed size and production, the nutritiousness of leaves to herbivores, plant toxicity and pollen production.

Nitrogen-fixing weeds, such as brooms, gorse and acacias may especially benefit because growth stimulated by CO_2 will not be constrained by low nitrogen levels (Poorter and Navas, 2003).

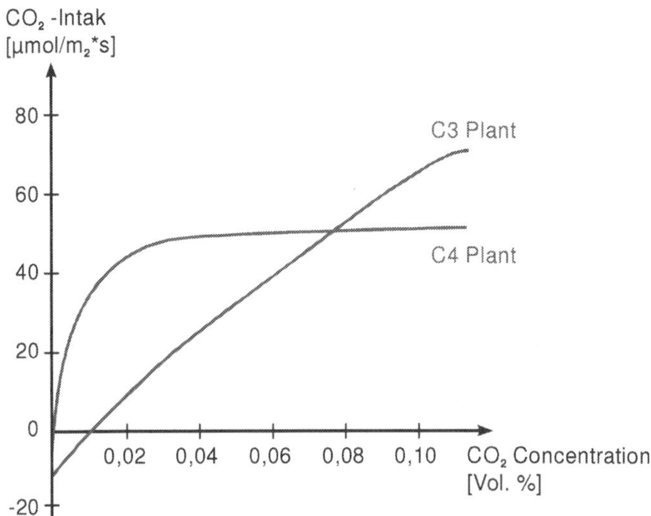

Fig. 19.2. Comparison of CO_2 intake of C_3 and C_4 plants in relation to atmospheric CO_2 concentrations. At current CO_2 levels (380 ppm) CO_2 saturation in C_4 plants is achieved.

Under high CO_2, C_3 plants are likely to become more water-efficient (Ghannoum *et al.*, 2007), potentially allowing weeds such as prickly acacia and rubber vine (*Cryptostegia grandiflora*) to move into drier habitats (Kriticos *et al.*, 2003).

Vines respond strongly to higher CO_2 levels (Gallagher *et al.*, 2006), and there are many highly damaging invasive vines (cat's claw, *Macfadyena unguis-cati* and rubber vine) that could benefit.

Higher CO_2 levels are likely to reduce the effectiveness of glyphosate, the main chemical used to control environmental weeds (Ziska and Goins, 2006; Ziska and Runion, 2007).

However, since climatic change, especially increased CO_2 affects C_3 and C_4 plants differently, and different combinations must be investigated separately:

- C_4 weeds in C_3 crops
- C_3 weeds in C_3 crops
- C_3 and C_4 weeds in C_3 crops
- C_3 weeds in C_4 crops
- C_4 weeds in C_4 crops

When solely looking at the benefit of elevated CO_2, it would be possible to argue that C_4 weeds such as barnyard grass (*Echinocloa crusgalli*) and redroot pigweed (*Amaranthus retroflexus*), which do not react to elevated CO_2 with more biomass production would be less competitive than C_3 crops which grow better under increased CO_2 and *vice versa*. In C_4 crops like millets, sorghum, maize and sugarcane; C_4 weeds may become less competitive than C_3 weeds. But in agricultural setting, weeds with both C_3 and C_4 photosynthetic pathways are present (Table 19.1).

Table 19.1: Important C_3 and C_4 weed species of rice and wheat crops in India.

Crop	C_3 weeds	C_4 weeds
Rice (C_3 crop)	Weedy rice, *Scirpus* sp., *Monochoria, Eclipta prostrata, Ammania baccifera*	*Echinochloa* sp., *Cyperus* sp., *Leptochloa chinensis, Bracharia*
Wheat (C_3 crop)	*Chenopodium album, Phalaris minor, Avena fatua, Convolvulus arvensis, Cirsium arvense*	*Cynadon dactylon*

19.3.1. C_4 weeds in C_3 crops

Among 14 of the world's worst weeds are C_4 plants, while around 76% of the harvested crop area in 2000 was grown with C_3 crops (Monfreda *et al.*, 2008). If the hypothesis is right that C_3 crops would benefit more from elevated CO_2 than C_4 weeds, losses due to C_4 weeds might decrease. In the early 1980s, experiments were conducted to prove this kind of hypothesis (Patterson and Flint, 1980) and basically the hypothesis was supported (Coleman and Bazzaz, 1992; Ziska, 2003). However, more research has been done manipulating CO_2 concentrations alone. Temperature increase or drought in combination with elevated CO_2 was less investigated (Bunce and Ziska, 2000). When including temperature increase, trends are not clear, and will depend on the local conditions. Optimal temperatures for growth in C_4 plants are generally higher than optimal temperatures for C_3 plants, but with higher CO_2 the optimum temperature of many C_3 plants also increases (Bunce and Ziska, 2000).

However, looking at photosynthesis and temperature alone might be insufficient. The barnyard grass (*Echinocloa crusgalli*) in combination with a mycorrhiza also benefits from elevated CO_2 levels. In drought situations, C_4 weeds might also have advantages over C_3 crops under elevated CO_2.

19.3.2. C_3 weeds in C_4 crops

The benefit of elevated CO_2 under sufficient water condition will lead to higher C_3 weed competitiveness in C_4 crops. An experiment with sorghum, and a C_3 and C_4 weed showed what the potential implications of increased CO_2 level may have on the crops. Under ambient CO_2 the presence of the C_3 weed velvetleaf (*Abutilon theophrasti*) had no significant effect on either sorghum seed yield or total above ground biomass; however, at elevated CO_2, yield and biomass losses were significant. The additional loss in sorghum yield and biomass was associated with a three-fold increase in velvetleaf biomass in response to increasing CO_2 (Ziska, 2003).

Elevated CO_2 alone might not only lead to an increase of pure biomass of C_3 weeds. The dandelion (*Taraxacum officinale*) produced more fertile seeds and eventually larger seedlings.

However, C_4 crops might out-compete better growing C_3 weed in drought situations, and at higher temperatures utilizing mycorrhiza.

19.3.3. C_3 weeds in C_3 crops

Logic would imply that the same type of plants (with regards to photosynthesis) in the same ecosystem would react to changes in a similar way. This is only partly true, while C_3 crops and C_3 weeds, both benefit from elevated CO_2 it seems that the magnitude varies. Stimulation of biomass accumulation from CO_2 doubling was estimated by one research team to be + 31% in wheat, + 30% in barley, + 27% in rice, + 39% in soybean, + 57% in alfalfa, and + 84% in cotton. In contrast, a survey of experimental results on 27 non-crop C_3 species revealed that biomass accumulation increased from 79% to 272% compared to ambient CO_2 (Patterson, 1995). An experiment, which investigated seven C_3 crop and three C_3 weeds at 350 ppm and 700 ppm CO_2 showed similar growth rates and mass of C_3 crops and C_3 weeds (Bunce, 1997).

Ziska *et al.* (2010) found that in case of rice (C_3 crop), rice biomass increased with increase in CO_2 from 300 to 400 ppm but did not increase further with increase in CO_2 to 500 ppm, whereas rice yield did not respond to elevated CO_2 (Fig. 19.3). Red rice (C_3 weed) responded linearly in terms of biomass as well as seed production. These results suggest that under elevated CO_2 concentrations, red rice will be more competitive than rice crop, and will produce more seed than at current CO_2 concentration.

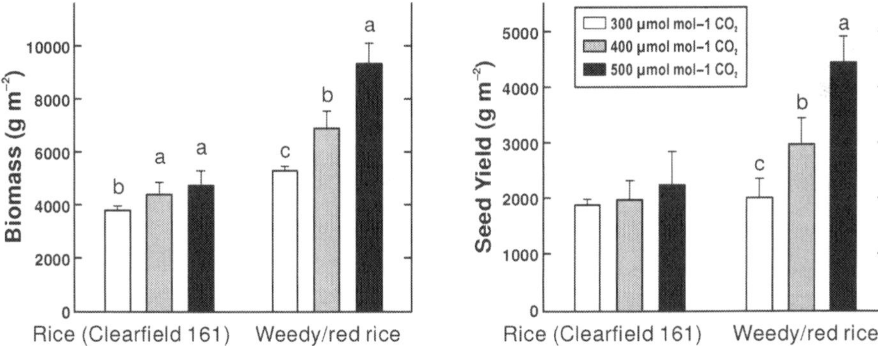

Fig. 19.3. Biomass and seed yield of rice (C_3 crop) and red rice (C_3 weed) under elevated CO_2 conditions (*Source:* Ziska *et al.*, 2010).

19.3.4. C_3 and C_4 weeds in C_3 crops

Ziska (2000) found that soybean (C_3 crop) biomass (32%) and yield (23%) increased at elevated CO_2 (ambient + 250 ppm) when grown in monoculture (Fig. 19.4). But when soybean was grown in competition with *Chenopodium album* (C_3 weed), soybean biomass and yield reduction increased from 23% and 28% at ambient CO_2 to 34 and 39% at elevated CO_2, respectively due to 65% increase in *C. album* dry weight. Conversely, soybean yield diminished from 45% to 30% at elevated CO_2 compared to ambient CO_2 when grown in competition with *Amaranthus retroflexus* (C_s weed). These results suggests that under elevated CO_2, *C. album* would be benefited more than soybean and could become more dominating weed. In contrast, *A. retroflexus* would be less benefitted with rising CO_2 and soybean will likely have competitive advantage when grown in competition with this species.

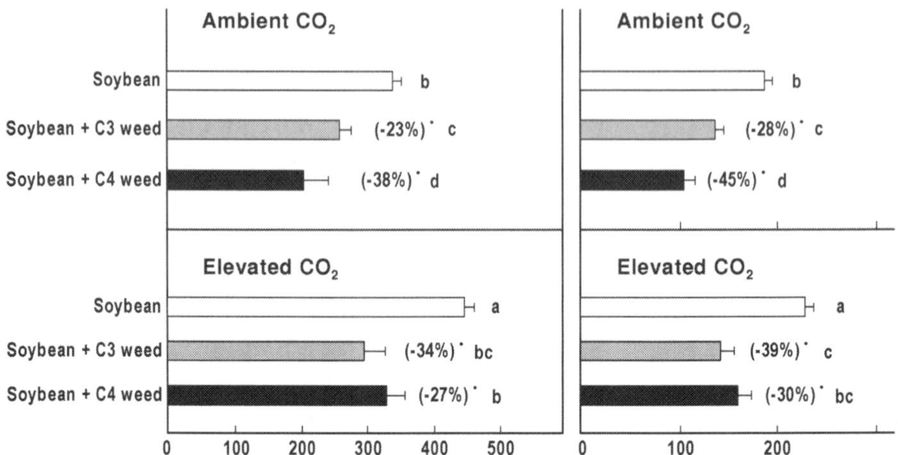

Fig. 19.4. Soybean biomass and yield at ambient and elevated CO_2 when grown in monoculture and in competition with C_3 (*Chenopodium album*) and C_4 weed (*Amaranthus retrofluxus*) (Ziska, 2000).

19.3.5. C_4 weeds in C_4 crops

Since all C_4 plants (weeds and crops) have the same photosynthesis path, they may react to changes in the same ecosystem in a similar way. However, research on impact of climate change in this combination has not been done.

Overall, it can be predicted that, weeds being more genetically diverse than crops have better chances of adaptation to high CO_2 levels than crops. Moreover, increased CO_2 levels may result in not only increased weed growth rate, size and pollen production, but also a change in the plant's chemical composition. In general, under elevated CO_2, it is likely that only when weed is C_4 and crop is C_3, crop is likely benefitted, whereas in all other cases weeds will likely get competitive advantage over crop (Fig. 19.5) (Tables 19.2 and 19.3).

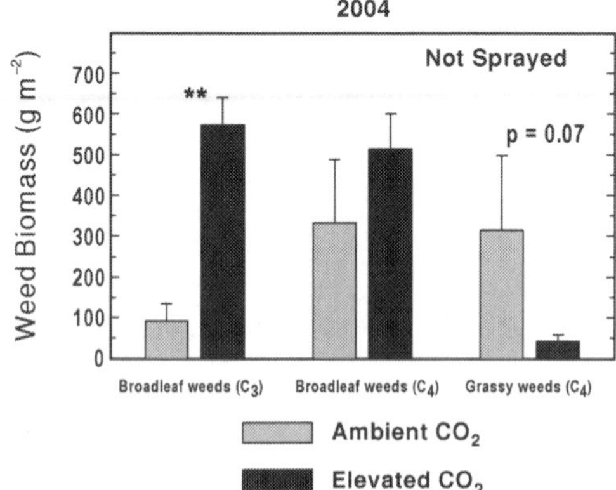

Fig. 19.5. Effect of rising carbon dioxide levels on change in C_3 and C_4 weed populations (some initial evidence suggesting that C_3 weeds could be preferentially selected) (*Source:* Ziska and Goins, 2006)

Table 19.2. Crop/weed competition outcome at elevated CO_2 conditions (Modified from Bunce and Ziska, 2000)

Weed species	Crop	Favoured under elevated CO_2	References
Grasses (C_4)	Lucerne (C_3)	Crop	Bunce, 1995
Amaranthus retroflexus (C_4)	Soyb`ean (C_3)	Crop	Ziska, 2000
Amaranthus retroflexus (C_4)	Sorghum (C_4)	Weed	Ziska, 2003
Chenopodium album (C_3)	Soybean(C_3)	Weed	Ziska, 2000
Taraxacum officinale (C_3)	Lucerne (C_3)	Weed	Bunce, 1995
Albutilon theophrasti (C_3)	Sorghum (C_4)	Weed	Ziska, 2003
Taraxacum and Plantago (C_3)	Grasses (C_3)	Weed	Potvin and Vasseur, 1997
Red rice (C_3)	Rice (C_3)	Weed	Ziska *et al.*, 2010
Echinochloa glabrescens (C_4)	Rice (C_3)	Weed	Alberto et *al.*, 1996

Table 19.3. Effect of elevated CO_2 on invasive species in plant communities

System	Invasive Species	CO_2 Favours?	Reference
Desert	*Bromus madritensis*	Yes	Smith et *al.*, 2000
Prairie	*Prosopis glandulosa*	Yes	Polley et *al.*, 1994
Woods	*Prunus laurocerasus*	Yes	Hattenschewiler and Korner, 2003

Herbicide efficacy may also decrease as CO_2 concentrations increase (Ziska and Teasdale, 2000; Ziska *et al.*, 2004). An increase in CO_2 induces morpho-physiological and anatomical changes in plants affecting the rate of uptake and translocation of herbicides (Ziska and Teasdale, 2000; Manea *et al.*, 2011). C_3 plants showed a decrease in stomatal number and conductance and increased leaf thickness, which might interfere with foliar uptake of herbicides (Nowak *et al.*, 2004; Ainsworth and Long, 2005), as well as an increase in starch accumulation on the leaf surface (Patterson, 1995). Moreover, perennial weeds may become even more noxious, if vegetative growth is stimulated as a result of increased photosynthesis in response to elevated CO_2. These changes are expected to reduce the efficacy of the most commonly used herbicides, such as glyphosate, due to a dilution effect, although the precise mechanism conducive to increased tolerance to glyphosate remains elusive (Fig. 19.6) (Manea *et al.*, 2011). This could be due to less translocation as the root system becomes vigorous. In addition, an increase in the root-shoot ratio may play a critical role in herbicide efficacy (Ziska *et al.*, 2004). It is concluded that CO_2-induced increase in root biomass could make perennial weeds harder to control in a higher CO_2 environment. Thus, research is needed to assess the comparative growth and physiological response of C_3 and C_4 weeds of different age groups (seedlings or mature plants), and their molecular and biochemical bases to herbicide tolerance under ambient and elevated CO_2 concentrations. Allocation of resources to below ground parts, source-sink relationships, and mitochondrial respiration also needs to be reassessed in the wake of climate change scenarios.

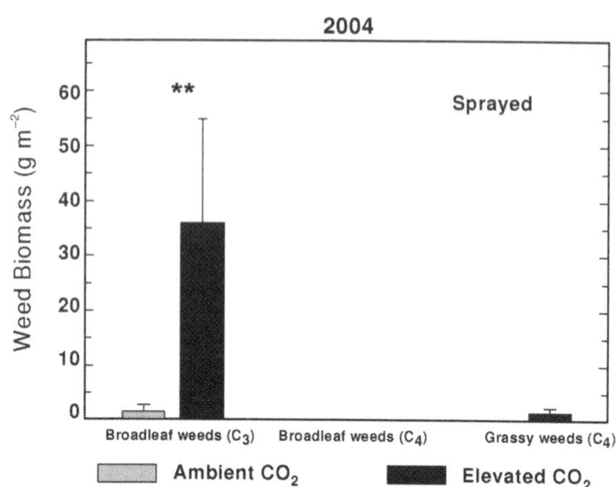

Fig. 19.6. Effect of carbon dioxide concentration on glyphosate efficacy.

19.4. VARYING PRECIPITATION PATTERNS

Plant species have a range of soil moistures that are suitable for survival and growth. A variation in rainfall pattern and increased aridity consistent with a warming climate could alter weed distribution and their impact on crop production. A decrease in annual rainfall will lead to a retraction of plant species' distribution from the dryer inland regions. Under more erratic rainfall and consequently, drought and flood scenario, the distribution and prevalence of weeds will be problematic in crop ecosystems, and in particular summer droughts will affect weed management in spring-sown crops (Peters and Gerowitt, 2014). Rodenburg *et al.* (2010) postulated that under prolonged drought spells, C_4 and parasitic weeds like *Striga hermonthica* will thrive better. Under excess water environments, weeds such as *Rhamphicarpa fistulosa* will be favoured. A change in rainfall patterns would favour hydromorphic weeds while prolonged drought spells will benefit C_4 over C_3 weeds.

Some agricultural areas will receive higher than normal amounts of rain. Weeds will be especially prevalent in these areas due to increased access to water, with more weed management required. Some agricultural areas will receive lower than normal amounts of rain (drought). Weeds will have a tough time in these areas due to lack of water, with less weed management required (but crops will do poorly). It may lead to more drought-tolerable weed species while crop plants remain sensitive to drought. Experiments have shown that water stress affects the degree of competition between native plants and bitou bush, *Chrysanthemoides monilifera* sub sp. *rotundata*, possibly allowing for increased dominance of local ecosystems by this alien invader with further climate change.

Weeds constrained by rainfall may also find new habitats under new climate conditions. Lantana and mist flower (*Eupatorium riparium*), for example, could expand if rainfall increased in some areas (McFadyen, 2008).

The C_4 photosynthetic pathway provides its greatest advantage under hot, arid high light conditions. C_4 plants have higher water use efficiency than C_3 plants. Competition between C_3 and C_4 weeds has been examined in relation to soil moisture regime (Matsunaka, 1983), C_3 plants are dominant in submerged soils; C_4

plants are dominant in dry land soils. Submergence protects rice plants from severe competition with C_4 weeds. On the other hand, upland rice and rain-fed lowland rice with limited precipitation face severe competition with C_4 weeds. Under imposed drought, Patterson (1986) found that the effects of water stress significantly increased leaf area and total dry weight of the three C_4 grasses: *Echinochloa crusgalli, Eleusine indica* and *Digitaria ciliaris*. The CO_2 enrichment can increase the growth of both C_3 and C_4 plants under water stress, but growth stimulation can be expected to be greater in C_3 plants.

Under rain-fed or dry land environments, little or no rainfall will hamper adequate land preparation for wet season rice as a result of limited water availability for flooding, especially early in the season when the rice is most susceptible to weed competition. This will limit traditional weed management in flooded rice and necessitate the use of herbicides. Rice yield losses are expected to be higher under such circumstances. Asch *et al.* (2005) emphasized that drought-tolerant rice cultivars would be required to prevent water stress-induced yield losses and to increase rice competitiveness against weeds under rain-fed conditions. A change from transplanting to direct seeding of rice, in relation to water saving in South Asia, has already resulted in increased weed competition and changed weed dynamics (Matloob *et al.*, 2015a).

Competition of cotton with *Abutilon theophrasti* and *Anoda cristata* increased under drought conditions (Patterson and Highsmith, 1989). A yield reduction due to *Xanthium strumarium* was more pronounced in well-watered soybeans compared with water-stressed soybeans (Mortensen and Coble, 1989). An increase in rainfall provided greater competition to wheat growth and yield against *Cirsium arvense* (Donald and Khan, 1992). According to Patterson (1995), weed competition had little effect on crops under water deficit conditions, as the potential crop yield was already reduced by water stress. This was confirmed by Chauhan and Abugho (2013), who showed that rice could not survive under water stress conditions. By contrast, *Amaranthus spinosus* and *Leptochloa chinensis* survived under water stress conditions and produced a significant number of tillers/branches and leaves even at the lowest soil water content.

Increased rainfall frequency and intensity will have an adverse effect on uptake, retention, and activity of soil-applied herbicides (Bailey, 2004; Rodenburg *et al.*, 2011). Increased cuticle thickness and leaf pubescence in response to drought, will reduce herbicide entry into leaves (Patterson, 1995). These attributes can also affect growth and recovery of crops and weeds following herbicide application. Increasing aridity and drought will increase herbicide volatilization, and, moreover, frequent rain showers will reduce the "rain safe periods" available for herbicide application in a given cropping system posing multidimensional challenges for weed management. An unprecedented increase in rainfall (either as a single rain event or annually) may promote leaching of soil-applied herbicides, and subsequent ground water contamination (Froud-Williams, 1996). A general conclusion that can be drawn from the above discussion is that an increase in rainfall would lead to additional weed pressure, thus increasing the herbicide costs and overall cost of production of major crops.

Weeds of rangelands like cheat grass (*Bromus tectorum*) and yellow star thistle (*Centaurea solstitialis*) depend largely on available soil moisture for seed germination.

Prolonged or heavy winters that greatly enrich the soil moisture favour increased seed production of both the species (Patterson, 1995). Interestingly, both species are also drought adapted. The adoption of cheat grass for drought evidence is shorter life span whereas, deeper root system to endure drought helps star thistle, compared to other native species. Increase in snowfall or a change in snowfall variability is suggested to exacerbate the invasion of forbs in mixed-grass prairie ecosystem which in turn could influence the forage availability (Blumenthal *et al.*, 2008). Soil moisture stress promoted greater herbivore tolerance for the invasive *Alternan theraphiloxeroides* and decreased it for the native congener *A. sessilis* (Sun *et al.*, 2010).

19.5. EXTREME WEATHER EVENTS

Extreme events including cyclones, droughts, floods, heat waves and frosts can sometimes create ideal conditions for weeds to extend their range and invade new areas or out-compete native species in their existing range. Fire is a common extreme event in Australia and its frequency and severity is often influenced by extreme weather. All of these factors can cause great changes in plant abundance, competition between weeds and other plants, and weed distributions at a local scale.

19.5.1. Drought

Dry soil conditions caused by drought prolong the longevity of weed seed banks; and importing fodder and grain into drought areas can bring new weed problems to the region. Drought can reduce the competitiveness of native vegetation, providing new opportunities for weed invasion.

19.5.2. Floods

Floods can spread weeds along water courses into areas that were previously free of weeds. By washing away vegetation and exposing areas of disturbed soil, floods can also provide opportunities for new weed invasions by reducing competition from existing plants. Eleven of the thirty two Weeds of National Significance (WoNS), including tamarisk (*Tamarix aphylla*), and willows (*Salix sp.*) are considered to benefit from flooding events. The major weed lippia (*Phyla canescens*) is also sensitive to altered flooding regimes. When native vegetation is stressed or destroyed by droughts, fires, floods or severe storms, weeds gain new opportunities to replace native species. There is a huge pool of invasive plants available to colonize bare spaces left by drought, fire and storm damage, and wind and flooding waters help spread weeds. Athel pine (*Tamarix aphylla*), for example, spread along 600 km of the Finke River in central Australia after severe flooding in the 1970s and 1980s, replacing river red gums. It could spread much further under climate change.

19.5.3. Cyclones

Cyclones can create new opportunities for weed invasion through associated flooding, soil movement and damage to native vegetation communities. During clean-up activities, it is important to limit further spread by minimizing the movement of soil or plant material from one area to another. Climate change science indicates that tropical regions will experience an increased intensity of cyclones in the future. Cyclones have been shown to accelerate plant invasions and alter the abundance of weeds in tropical forests. Intense cyclones provides a large spatial and temporal window of opportunity for invasion, due to greater disturbance. Empirical research demonstrates that invasive alien species have increased growth rates and recruitment and spread readily following such events. It is the vines, scrambling shrubs and woody invasive alien species that are somewhat shade-tolerant, that are likely to constitute the greatest threat to tropical forests following cyclones.

19.5.4. Severe frosts

An increase in frequency and severity of frosts seems likely for some areas. The increase in frosts is related to more clear skies in winter due to rainfall reduction. Its impact is most evident in the cropping region of south west of Western Australia where crops are affected by less rainfall, late season starts and increased frost, especially during flowering. This may be a temporary phenomenon, as the increased frequency and severity of frost is not likely to persist as the climate warms further. In the interim, frosts will remain a significant driver of plant range limits, including for weeds. Some areas may become more or less suitable and where frosts increase, weeds may be killed or prevented from spreading due to frost damage. This effect may also confer an advantage on plant species adapted to more continental climates. A management option could be to be aware of areas susceptible to frosts, determine what weeds are susceptible and reduce weed management accordingly.

19.5.5. Fire

Fire is a major disturbance in many Australian landscapes and certain frequencies and intensities are often required for ecosystem health. Fire height, intensity and frequency, however, are likely to be altered by weeds through changes in fuel load and fuel structure. The best known example is the ecosystem transforming impact of gamba grass (*Andropogon gayanus*) in northern Australia, where increased fuel loads increase the severity of fires. Climate change, either through warmer and drier environments, or increased fuel loads from CO_2 efficiencies or increased rainfall, could create greater challenges for fire management in Australia.

19.6. PHENOLOGY

The timing of plant growth and reproduction (phenology) will be influenced by changes in seasonal cues, such as temperature and rainfall. Very few observations on changed phenology have been made in Australia, although it is a common study focus for climate change research in the northern hemisphere (e.g. earlier flowering due to rising temperatures). In Australia, the influence of climate change on plant phenology has been investigated for native alpine plants. In contrast, a field based experiment on Tasmanian grassland ecosystems found eCO_2 and temperature did not change the flowering patterns of invasive grasses. Understanding the timing of plant growth and development, however, is critical for guiding the timing of weed control efforts, making this topic an important focus for future research or monitoring and adaptive management.

19.7. LAND USE CHANGE

Human responses to climate change, including deliberate adaptation choices, will be a major contributor to weed problems with climate change. Potential responses include:

- Changed farming practices as farmers implement climate change adaptation measures.
- Pastoralism replacing agriculture under a drier climate in some parts of south-eastern Australia. Shifting the distribution of crop and pasture species will likely shift the distribution of their associated weeds.
- Opportunistic cropping to take advantage of years with good rain.
- The introduction of new crop and pasture species better able to tolerate extreme conditions.
- Changes in choice of sown crops (e.g. increased area of canola in previously wetter areas).
- Development of new cropping regions (in particular in the north of Australia, aided by irrigation).
- The widespread planting of fast-growing alien species for biofuels; biomass crops (including short-rotation trees and high biomass grasses) and oil seed species that are currently being investigated for large-scale production of biofuels in northern Australia.

These changes will likely see a new set of weeds that require on-going management and investment of resources.

19.8. INCREASED ATMOSPHERIC OZONE (O_3)

The effect of increased atmospheric ozone (O_3) is likely to have adverse impacts of growth and productivity of all plants.

19.9. EFFECT OF MULTIPLE CLIMATIC FACTORS ON WEEDS

Despite prediction that both atmospheric CO_2 and air temperature will rise together, very limited studies have been conducted to assess the possible interaction effects of these two global change factors on weeds and crop/weed competition. Most studies have focused mainly on the response of plant species to elevated CO_2. The outcome can be totally different when both factors are taken together compared to when consider only one factor. For example, in a study conducted by Alberto *et al.* (1996), rice had competitive advantage over C_4 weed (*Echinochloa glabrescens*) when considered elevated CO_2 only, whereas when considered both elevated CO_2 and elevated temperature, C_4 weed (*E. glabrescens*) had competitive advantage over rice crop.

Nonhebel (1996) found that although enhanced CO_2 was predicted to increase wheat (*Triticum aestivum*) yields in Europe, these effects would be neutralized by reduced growth due to an elevated temperature. However, if drought occurred due to limited water availability under climate change, wheat yields would be depressed (Nonhebel, 1996). Similar three-way interactions should apply to other plant species, including many weed species, highlighting the value of studying the impact of multiple climatic factors on weeds and their interactions with crops or native plants.

19.10. CONCLUSIONS

Research is needed to unravel whether the so-called CO_2 fertilization could compensate for other negative effects of climate change on crop-weed competition. Moreover, the response of agricultural and invasive weeds to other climatic factors and associated parameters such as temperature, drought, rainfall, and an extended growing period should be explicitly assessed in conjunction with an anticipated rise in CO_2 concentration to predict a wider picture of competitive outcomes in managed and natural ecosystems. The effect of climate change on the geographic distribution of invasive weeds will be a subject of interest in the near future.

Climate change is a looming global crisis and its impacts on agricultural weeds have not been well explored. Conventional thinking around carbon pathways in plants and nutrient management in crops could partially solve the climate change implications, but weed problems could also be aggravated in the wake of increasing CO_2 concentration, high temperature, and most significantly by water stress. These conditions might necessitate the adoption of new agronomic practices to enhance weed competitiveness. As crops and weeds share the same trophic level, the stimulatory or inhibitory behaviour of the climate variables on crops should generally hold true for weeds. An increase in atmospheric temperature was found to favour weed growth as well as herbicide efficacy. Although there is a dominance of C_4 weeds in agriculture, C_3 and C_3-C_4 intermediate pathways of prominent weeds would pose severe crop-weed competition in the years to come. Importantly, due to species interaction, there is a need to study all possible combinations of plant-weed carbon fixation pathways, C_3 crops and C_3 weeds, C_4 crops and C_4 weeds, C_3 crops and C_4 weeds, and C_4 crops and C_3 weeds, while studying the impact of climate change on crop-weed competitive interactions. Several weeds will exert additional pressure for crop-weed competition under the climate change scenario. More adaptive research studies, including complex research conditions, could yield useful solutions for managing yield reduction in the ensuing decades.

20

Impacts of Climate Change Induced Consequences on Weeds

20.1. INTRODUCTION

The impact of climate change on weedy vegetation may be manifested in the form of geographic range expansions (migration or introduction to new areas), alterations in species life cycles, and population dynamics. Migration of weeds will subsequently result in a differential structure and composition of weed communities within natural and managed ecosystems. Through the lens of climate change, Peters *et al.* (2014) outlined three distinct types of shifts in weedy vegetation (range, niche, and trait shifts), occurring at different scales (landscape, community, and population scales), respectively. Changes in weed biology, ecology, and interference potential, in the wake of climate change, will result in complex crop-weed interactions that necessitate alternative adaptive mechanisms.

Uncertainty in agricultural productivity under a climate change scenarios, can be the result of plant-plant interactions through direct effects of a change in temperature and atmospheric CO_2, or indirect effects at the system level through shifts in crop–weed interactions (Fuhrer, 2003) and other biotic stresses.

20.2. EXPANSION OF GEOGRAPHICAL DISTRIBUTION

Under ambient conditions, water availability and temperature are the principal determinants of species distribution (Patterson *et al.*, 1999), but there is the recent addition to this list of CO_2 concentrations through the lens of climate change (Patterson, 1995; Chauhan *et al.*, 2014). The changing climate variables may either increase the distribution range of weed species in response to a change in atmospheric temperature, or allow some non-potent weeds to dominate weed abundance as crop-weed interactions may increasingly favour C_3 weeds (Bazzaz *et al.*, 1985). Other than geographical distribution, the projected climate change might impact their population biology (Patterson *et al.*, 1999; Ziska and Goins, 2006), causing them to move to new areas lying at higher altitudes and latitudes (Patterson, 1995; Ziska and Dukes, 2011). Such effects have been proposed for *Striga* sp., which are expected to extend their geographic range (Mohamed *et al.*, 2006). Climate change will alter the distribution of plant species that undergo shifts in their distribution to more favourable localities.

According to Holm *et al.* (1997), most of the troublesome C_3 and C_4 weeds of the arable land are limited to tropical and subtropical regions, primarily due to low temperatures at higher latitudes. Preliminary data showed an increased tolerance in many weeds to low temperatures under elevated CO_2 (Boese *et al.*, 1997), which suggests the possibility of polar-ward expansion for many weed species (Bradley and Mustard, 2005; McDonald *et al.*, 2009; Ziska and Dukes, 2011).

Lowland species such as lantana (*Lantana camara*) may be able to shift into the uplands (McFadyen, 2008). Weeds moving into alpine areas could have a particularly severe impact because many alpine plant communities are localized with rare endemic species, and there are numerous weed species at lower altitudes (McDougall *et al.*, 2005). On sub Antarctic Heard Island, the weed winter grass (*Poa annua*) has been spreading rapidly on deglaciated sites (Scott and Kirkpatrick, 2005).

Increasing numbers of thermophile weeds such as *Amaranthus retroflexus* and *Abutilon theophrasti* became established in more northern areas, including exotic and potentially invasive species such as *Panicum dichotomiflorum* and *Datura stramonium* (Guillerm *et al.*, 1990; Breitsameter *et al.*, 2014). Some late emergers such as *Chenopodium* sp. and millet weeds such as *Echinochloa* sp., *Setaria* sp., *Digitaria* sp., and *Sorghum halepense* have expanded their distribution range (Mehrtens *et al.*, 2005; Otte *et al.*, 2006).

Weeds may extend their range as the climate gets warmer. This could have quite a big effect as many weeds, particularly invasive species, can reproduce rapidly and therefore unfavourable climate conditions may be important for holding populations in check. This is likely to extend weed ranges northwards and may therefore be problematic in regions like Canada and Northern Europe.

The number of documented examples to promote potential range expansion in invasive plants is increasing and many of these are related to key aspects of climate change, such as northward range expansion in the northern hemisphere (Clements *et al.*, 2004) (Table 20.1).

Table 20.1. Geographical range expansions of weeds or weed genotypes in USA and Canada and associated adaptive traits

Weed species	Description of range expansion	Adaptive traits	Reference/s
Datura stramonium (Jimson weed)	Northward invasion of Canadian and north-eastern US cropland since 1950s	Heavier seeds, earlier growth	Weaver *et al.*, 1985; Warwick, 1990
Hypericum perforatum (St. Johns wort)	Clinal variation across the north–south axis of North American distribution	Leaf size larger in northern populations	Maron *et al.*, 2004
Echinochloa crusgalli (Barnyard grass)	Northward invasion of Quebec from the United States in the 19th century	More rapid maturation at each life cycle stage	Potvin, 1986

Weed species	Description of range expansion	Adaptive traits	Reference/s
Panicum miliaceum (Proso millet)	Northward invasion into Canadian cropland by early 1970s	Seed germination and dispersal characteristics	Bough *et al.*, 1986; McCanny *et al.*, 1988; McCanny and Cavers, 1988
Polygonum cuspidatum (Japanese knot weed)	Range expansion in both Ontario and British Columbia, Canada	Genotypes with different temperature thresholds and potential hybridization	Bourchier and Van Hezewijk, 2010
Setaria faberi (Giant foxtail)	Northward expansion into Canadian cropland by the 1970s	Varied life history traits	Warwick *et al.*, 1987
Setaria viridis (Green foxtail)	Survival at Churchill, Manitoba at nearly 60N latitude (normal range 45–55N)	Leaf production at low temperatures	Douglas *et al.*, 1985; Swanton *et al.*, 1999
Sorghum halapense (Johnson grass)	Northward expansion by 5 latitude between 1926 and 1979	Northern populations annual (versus perennial southern population)	Warwick *et al.*, 1986

Source: Clements *et al.* (2004)

20.3. CHANGE IN FUNCTIONING AND PRODUCTIVITY OF ECOSYSTEMS

Climate change will alter the overall functioning and productivity of ecosystems. For example, increased abundance of woody vines as a consequence of rising CO_2 levels has been associated with increased tree mortality and reduced tree regeneration in forests throughout the world (Phillips *et al.*, 2002). Similarly, an increase in parasitic weeds would become a serious threat to productivity of rice and sorghum crops under rain-fed agriculture (Rodenburg *et al.*, 2011).

In a composite stand of weeds in a cropped field (C_3 and C_4 plants), dynamics in insurgence and shifts of the weed populations in favour of specific species is expected over time (Das *et al.*, 2012). Climate change is likely to trigger differential growth in crops and weeds and will have significant implications for weed management across crops and cropping systems. The abundance, competitive ability, and survival of perennial weeds are expected to be higher, since a rise in CO_2 stimulates tuber and rhizome growth (Chandrasena, 2009).

20.4. INCREASE IN EXOTIC INVASIVE WEED SPECIES

Intentionally or unintentionally, humans have moved many weed species outside of their native ranges to new area where some of these non-native species have become invasive often with detrimental effects on native species and managed ecosystems (Mack *et al.*, 2000; Wilcove, 1998). It is predicted that changing climate (rising CO_2 and temperature, and altered precipitation) will alter the impacts of these invasive plants on native and managed ecosystems (Hellmann *et al.*, 2008; Dukes, 2000). Most of these climate change factors may likely increase the risks of invasion (Bradley *et al.*, 2010). This is mainly because:

- Invasive species are generally well suited to novel environment which is created by climate change.
- Invasive plants are more successful in environment with high resource availability provided by elevated CO_2.

However, recent studies highlighted that it is very difficult to predict the impact of climate change on invasiveness. It is because, on one hand, changes in temperature and precipitation creating novel environment which could benefit invasive species. But on the other hand, they do not consistently increase resource availability, especially elevated temperature and altered precipitation (Bradley *et al.*, 2010). Published results have shown both increased and decreased invasion risks associated with climate change (Table 20.2). It is therefore, critical to assess the response of invasive plants to multiple simultaneous climate change factors (e.g. CO_2, temperature and precipitation) to predict the risks of invasiveness. Rising CO_2 levels by themselves are likely to have a positive effect on the establishment and persistence of invasive species (Cheat grass). Warmer seasonal temperatures and milder winters will extend the distribution of invasive weeds (Kudzu and Ragweed). At regional, state and national levels, biosecurity measures are required to address this threat.

Table 20.2. Likely impact of climate change on risks of invasiveness

Climate change factor	Risks of invasiveness
Elevated CO_2	Increase
Rising temperature	Might increase or decrease
Changing precipitation regime	Might increase or decrease

Source: Bradley *et al.* (2010)

McDonald *et al.* (2009) proposed that if climate change forecasts are realized, damaging endemic weed species of major cropping systems might experience a significant transformation in their host range, besides an overall increase in the chance of invasion by exotic invasive weed species. Besides agronomic weeds, there are also certain non-native weeds whose introduction to new areas can pose ecological and environmental hazards (Mooney and Hobbs, 2000). Several studies have demonstrated that such weeds often benefit from carbonaceous fertilization (Polley *et al.*, 2002; Belote *et al.*, 2003; Ziska and George, 2004). It is believed that under a climate change scenario, these invasive plants would be able to extend their geographic range as well as spread to new areas, including currently agriculturally productive regions (Ziska and Dukes, 2011). An expansion in the geographic range proposed for weeds such as *Lonicera semperviens* and *Pueraria lobata* in the past has now become a reality (Patterson, 1995). Range expansion of arable and invasive weeds in connection with climate change must be studied as an integral part of crop-weed interactions.

20.5. SHIFTS IN COMPETITIVE BALANCE

Climate change will result in a greater frequency of extreme weather events such as frequent droughts and cold spells, so that weeds with less phenotypic plasticity may experience population declines (Peters *et al.*, 2014). Lack of rainfall and prolonged drought will limit growth of arable crops and pastures, resulting in a lack of vegetation cover and bare ground, thus allowing invasion by more resilient drought-tolerant weeds. Increasing CO_2 could alter the competitive balance in a weed-crop mixture through its effect on photosynthesis and stomatal physiology, which is linked with the competitive balance between crops and weeds in a cropping system (Alberto *et al.*, 1996). The range of growth stimulation in response to elevated CO_2 needs to be determined for both crops and weeds with contrasting carbon fixation pathways, growing in variable densities and species compositions. Under conditions of higher temperature and drought, C_4 weeds such as *Amaranthus retroflexus* tend to dominate C_3 crops (e.g., soybean). The infestation of *Phalaris minor* is expected to worsen in wheat fields with CO_2 increase (Mahajan *et al.*, 2012). Likewise, weedy rice will compete more strongly with cultivated rice (Ziska *et al.*, 2010). Exploring differential mechanisms and responses that govern the success of weeds to invade new areas/cropping systems and their ability to utilize growth resources, will be helpful in understanding the implications of rising CO_2 levels on plant-plant interactions. This also requires characterizing their damage niche (McDonald *et al.*, 2009).

20.6. WEED SHIFTS

Shorter distance weed dispersal opportunities have been reduced due to an increase in the size and frequency of barriers, such as habitat fragmentation, changes in land use and the construction of artificial barriers. Climate change is unlikely to directly influence or change dispersal pathways. Instead, changed land use patterns as adaptations to climate change may improve dispersal opportunities and provide new dispersal pathways.

In order to persist in a local habitat, weed species have to respond to the changes of the environment (Woodward and Cramer, 1996). These responses lead to shifts, which act at distinctive scales. Generally, plant species have three options to avoid extinction (Lavorel and Garnier, 2002; Pautasso *et al.*, 2010):

- *Migration* with a favourable climate, which leads to alterations of the distribution of weeds—a process called *range shift*. For migration, weeds need to possess appropriate propagule dispersion mechanisms. In arable ecosystems, this is often also provided by human actions (Kubisch *et al.*, 2013). Range shifts act at the landscape scale (Jump and Peñuelas, 2005).
- *Acclimation* to changes in climate conditions basically refers to the response of species within their phenotypic plasticity without evolutionary adjustments (Pearman *et al.*, 2008). These responses can be divided into tolerance and avoidance of climatic changes that lead to performance beyond the species' ecological optimum (Grime and Hodgson, 1987; Lavorel and Garnier, 2002). As a consequence, the fitness and the competitive ability of the weeds are either reduced or enlarged

(Barrett, 2000). Consequently, the realized niche is being altered, which leads to *niche shifts*. They act at the community scale and can be determined visually as composition shifts.

- *Adaptation* to changes in climate conditions, which is often associated with the evolution of new properties or with the optimization of existing ones (Harlan and de Wet, 1965; Carroll *et al.*, 2007; Tungate *et al.*, 2007). These individual biological adaptations of weeds, which are driven by natural selection, result in *trait shifts*. They become apparent at the population scale, but are brought about by morphological, physiological, and genetic processes at the individual plant scale.

20.6.1. Range shifts

Range shifts represent the transformation of the distribution area of species and occur at the landscape scale, i.e., at a geographical area extending from several arable fields up to few hundred kilometres (Pearson and Dawson, 2003; Petit *et al.*, 2011). With recent climate change, plant species are expected to track the climate favourable to their growth (Jump and Peñuelas, 2005). This is well investigated and documented for Europe (Cimalova and Lososova, 2009; Silc *et al.*, 2009; Walck *et al.*, 2011; Hanzlik and Gerowitt, 2012). Rising temperatures can cause species range boundaries to be moved further toward the poles (Walther *et al.*, 2002). As a result, many C_4 weeds such as *A. retroflexus*, *Setaria* sp., *Digitaria* sp., *Panicum dichotomiflorum*, and *Sorghum halepense* are expected to extend their distribution range to locations further north (Fausey and Renner, 1997; Weber and Gut, 2005; Clements and Ditommaso, 2011). Presumably, increased levels of precipitation during winter will additionally shift the range of many weed species moderately eastward in Europe (Skov and Svenning, 2004; Bergmann *et al.*, 2010). The effect of climate change on the number of weed species is likely to be more pronounced in northern regions of Europe, as the number of weeds is lower there than in southern regions (Holzner and Immonen, 1982; Fried *et al.*, 2008).

Opportunistic weed species possess the ability to track climate change by means of sophisticated dispersal and superior adaptation capabilities (Chapin *et al.*, 1996; Bergmann *et al.*, 2010; Pautasso *et al.*, 2010). Possibly due to an interaction of human actions and climatic changes, the maize weeds of the genus *Setaria* (Douglas *et al.*, 1985; Wang and Dekker, 1995) and *A. theophrasti* (Andersen et al., 1985; Warwick and Black, 1986) have extended their range to Northern America.

In order to sustain a population in a novel habitat, weeds must persist after they have become established (Weiher *et al.*, 1999; Smith *et al.*, 2011). As a consequence, range shifts are often accompanied by natural selection leading to genetic and evolutionary adjustments to the novel environments (Levin, 2009; Lavergne *et al.*, 2010; Richardson *et al.*, 2013). These evolutionary processes result in trait shifts. The level of evolutionary adaptations is highest at the expanding front (the colonizing edge) and at the boundary of the range (Angert *et al.*, 2011; Doxford and Freckleton, 2012).

20.6.2. Niche shifts

Changing climatic conditions lead to a transformation of the size and the shape of the niche pool, which affects the species niches inside the pool and, consequently, causes alterations in the community (Chapin *et al.*, 1996). There are very few examples of climate-mediated niche shifts in arable ecosystems. Broennimann *et al.* (2007) demonstrated that *Centaurea maculosa* actively realized a niche shift. This species was introduced from Europe to Northern America and extended its niche by adapting certain traits. Like in this example, niche shifts are often accompanied by other shifts.

Most agronomic actions are designed to remove weeds from the local species pool (Holzner and Immonen, 1982). Farming practices which vary depending on climate cause disturbance to the system and influence the construction of niches by creating niche gaps (Clements *et al.*, 1994; Eriksson, 2013). Niche gaps also result from disturbance caused by extreme climate events such as extreme wind, frost, rain, and other mechanical disruptions. Niche gaps directly affect the abundance and cover of weeds due to the removal of plants caused by disturbance (Booth and Swanton, 2002; Nogues-Bravo, 2009). This process is often followed by a temporary increase in the availability of nutrients and a reduction in competition (Stohlgren *et al.*, 1999). Ecologically, niche gaps represent empty space in the niche pool. Dekker (2003) describes niche gaps as "opportunity space" for exotics and invaders. Yet, niche gaps as caused by extreme weather events can also be an additional opportunity for weeds to emerge (Hobbs and Huenneke, 1992). Thus, niche gaps most often become apparent by relatively high fluctuations between species or high introduction rates of species (MacArthur, 1955).

20.6.3. Trait shifts

Traditionally, a plant trait represents a measurable attribute of a species (Raunkiaer, 1934; Weiher *et al.*, 1999). In recent decades, efforts have been made to devise functional groups of attributes with relation to ecological factors or resources (Grime, 1977; Craine *et al.*, 2001; Lavorel and Garnier, 2002; Gunton *et al.*, 2011). The ecosystem supplies the plant with nutrients, energy, atmospheric gases, and water. Only those species prevail that possess the appropriate sets of (functional) traits suited for making use of the resources given in a certain area (McIntyre *et al.*, 1999). The trait set of a species is the basis of its functional relationship to the environmental resources and to other species within the ecosystem. A changing climate influences the supply of resources, and as a consequence, it alters the functional connections of traits (Diaz *et al.*, 1998; Jauni and Hyvönen, 2012). Climate change hence selects those species featuring the appropriate set of traits (Keddy, 1992; Grime, 1997) and leads to shifts in the traits of plant populations (Carroll *et al.*, 2007; Tungate *et al.*, 2007). Changing environmental conditions may favour traits of dominant (e.g., invasive) species, which are able to significantly alter ecosystem function (Wilson and Agnew, 1992; Schulze and Mooney, 1994).

The term "trait shift" refers to visible and measurable alterations of morphological or physiological attributes of individual plant species caused by changes in climatic conditions. Trait shifts mainly become apparent at the population scale and result from natural selection of individuals that perform differently under altered environmental conditions. They are usually caused by changes in the genotype and, hence, represent evolutionary adjustments to the altered environmental conditions (Levin, 2009; Lavergne *et al.*, 2010; Richardson *et al.*, 2013). In fragmented populations, climate-mediated trait shifts may result in the formation of climatic ecotypes that fit the local conditions. Trait shifts are a prerequisite to niche shifts and can cause range shifts, if they enable a species to proliferate in areas with different environmental conditions.

Trait shifts are often related to phenology, morphology, physiology, and reproduction. In arable land use systems, they not only occur as a consequence of climatic changes, but also as a result of altered agronomical practice. The adjustment of sowing dates to changing spring and autumn temperature conditions that farmers used to practice in the past decades, is relevant in this context (Otte *et al.*, 2006; Chmielewski *et al.*, 2004). For example, the germination temperature range of *Chenopodium ficifolium* as measured in the 1950s was between 30 and 40° C (Lauer, 1953), whereas in the late 1980s, it was between 0 and 30° C. According to Otte (1991), this trait shift mainly occurred as an adaptation to earlier spring crop sowing dates, which were adopted by farmers during that period of time in Germany. Similar findings were reported for *Chenopodium album* that flowered earlier in northern regions of the UK (Froud-Williams, 1996).

20.6.4. Damage shifts

McDonald *et al.* (2009) introduced the concept of the "damage niche." The latter is defined as "the environmental conditions that make specific weed abundant, competitive, and therefore damaging the production of particular crops." Determining the damage niche of a weed species requires information on the crop-weed interactions and is thus region and cropping system specific.

Climate change is expected to increase the range, or "damage niche" (also called "invasion niche") of many weed species. Research suggests that the composition of invasive weed communities will be fundamentally altered by the end of the century under increasing temperature scenarios, with new weed species entering communities as a result of geographic range shifts (McDonald *et al.*, 2009). For example, the range of the yellow star thistle, a California weed, is expected to increase to more northern parts of California and Nevada due to climate change (Bradley *et al.*, 2009). Precipitation and temperature are likely to define the boundaries of the damage niche for *Chenopodium album*, a summer-annual weed in maize (McDonald *et al.*, 2009). Obviously, it was not a major competitor to maize under the warmer conditions.

20.7. INCREASED NUMBER OF GENERATIONS

Extension of the growing season may lead to an increased number of generations per year for some weed species and the change in species composition of weed communities is important. There may be selective pressures which are exerted by the production of different crops and it is already known that a shift in crop production may cause a shift in weed species prevalence. It is also important that if new crops are introduced then non-indigenous weed species are not accidentally introduced with the crop seed.

Climate change will result in changes in the length of the growing season and this will impact on weed productivity. In northern latitudes, the growing season is likely to increase and this will potentially increase weed productivity and fecundity and, in some cases, may enable there to be more than one generation completed in a single year.

20.8. DEVELOPMENT OF SUPER WEEDS AND SLEEPER WEEDS

Changing climatic conditions can also result in the development of super weeds which are clearly more adaptive and competitive than the crop plants. This has been proved by several lab experiments, where it has been shown that weeds develop better under less favourable conditions when compared to crop plants.

In countries like Australia, there are some 3000 foreign plants now established as weeds. Many of these are ecologically silent and are called sleeper weeds. One possible effect that change in climatic conditions can cause is the awakening of these sleeper weeds into problematic plants.

Super weeds may be more problematic in the future, as may sleeper weeds. This may be influenced by climate change but also through an increase in international trade and intensification of agriculture; opening the possibility for the accidental introduction of exotic species into new areas. The ecological impacts of non-native species which prove to be invasive can be far reaching. This may be exasperated by a general increase in global temperature.

20.9. HERBICIDE-CLIMATE INTERACTIONS

Chemical control is the most common form of weed control. Changes in temperature, wind speed, soil moisture and humidity can influence the efficacy of herbicides. For example, drought can lead to thick cuticle development which in turn can reduce the herbicide entry in the plant. Ziska *et al.* (1999, 2004) found that rising atmospheric CO_2 concentrations can reduce the glyphosate efficacy. Mechanical control of perennial weeds is also likely to be adversely affected by elevated CO_2 conditions. For example, the greater amount of root growth observed at elevated CO_2 in Canada thistle, meaning more asexual propagules (Fig. 20.1). Since Canada thistle propagate asexually, hence, disking/harrowing would result in greater weed infestations. Hence, tillage practices may have to adapt accordingly.

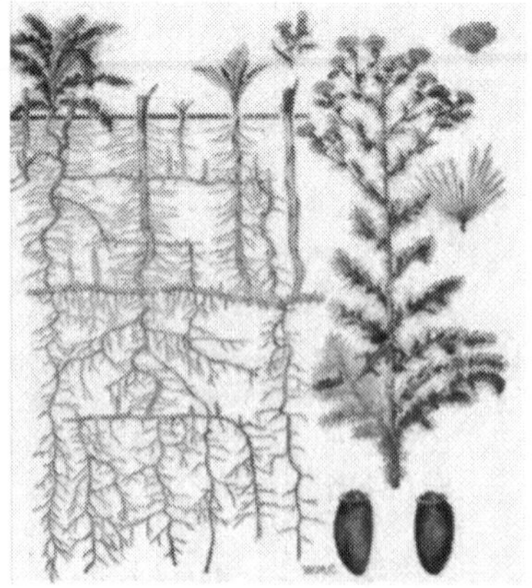

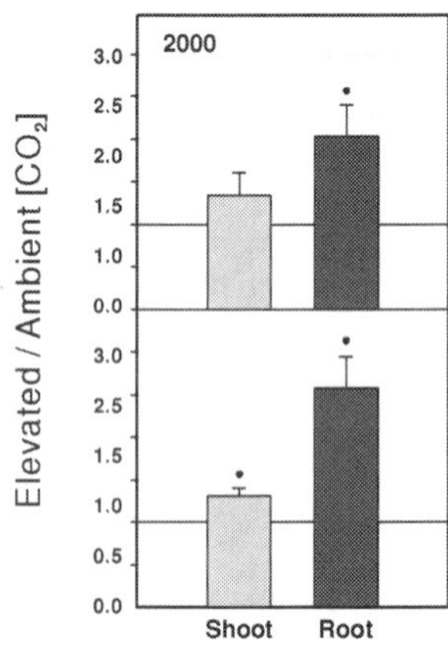

Fig. 20.1. Shoot and root biomass of Canada thistle at elevated CO_2 (brown) relative to ambient CO_2 (yellow) (Source: Ziska *et al.*, 2004)

The effectiveness of foliage applied post-emergence herbicides is dependent on the local climate/microclimate (Kudsk and Kristensen, 1992). A rise in temperature will increase volatility of certain herbicides such as trifluralin, rendering it less efficient. In the 1970s, rapid volatilization of surface-applied alachlor, butachlor, and propachlor occurred from continuously moist soils exposed to a constant 21° C (Beestman and Deming, 1974). Temperatures (day/night 32/22° C and 26/16° C) were not as critical as that of relative humidity in influencing acifluorfen (diphenylether group) phytotoxicity on *Xanthium strumarium* and *Ambrosia artemisiifolia* (Ritter and Coble, 1981). However, temperature had a significant effect on the degradation of imazapyr (imidazolinone group), flumetsulam (sulfonanilide family), and thifensulfuron (sulfonylurea group) in soil (Mcdowell *et al.*, 1997). Glyphosate absorption is dependent on the atmospheric temperature, as evident from *Desmodium tortuosum*, a C_3 weed (Sharma and Singh, 2001). An increase in temperature or relative humidity increased the efficacy of mesotrione on *X. strumarium* and *A. theophrastii* three-fold (Johnson and Young, 2002). The efficacy of the herbicide pyrithiobac (pyrimidinylthiobenzoic acid group) on *Amaranthus palmeri* was reduced at temperatures outside the range of 20-34° C (Mahan *et al.*, 2004). Anderson *et al.* (1993) found that relative humidity had the most significant effect on the phytotoxic action of glufosinate-ammonium, since this is attributed to changes in cuticle hydration and droplet drying (Ramsey *et al.*, 2005). Studies under controlled environmental chambers in Australia using varying night/day temperatures of 5/10, 15/20, and 20/25° C showed that *Raphanus raphanistrum*, grown under cooler temperatures of 5/10° C, was poorly controlled with 1,200 g a.i. ha^{-1} of glufosinate. By comparision, 100% mortality was achieved under 15/20 and 20/25° C for the same dose (Kumaratilake and Preston, 2005), suggesting enhanced efficacy of glufosinate under enhanced atmospheric temperature.

20.10. IMPACT ON WEED MANAGEMENT PRACTICES

It is predicted that climate change can reduce the effectiveness of current weed management practices (Ziska *et al.*, 1999).

20.10.1. Biological control

Climate change could alter the efficacy of weed bio-control agent by affecting the development, morphology, and reproduction of target bio-agent. Overall, it is clear that changing climate or rising CO_2 will likely to alter weed management. Increased environmental fluctuations may have a negative impact on the efficacy of some of the bio-control agents that are currently important weed control agents. If these lose effectiveness, problematic weed species that have previously been largely controlled may begin to cause significant economic damage.

Biological control of pests by natural or manipulated means is likely to be affected by increasing atmospheric CO_2 and climatic change. Climate as well as CO_2 could alter the efficacy of weed bio-control agents by potentially altering the development, morphology and reproduction of the target pest. Direct effects of CO_2 would also be related to changes in the ratio of C: N and alterations in the feeding habits and growth rate of herbivores. As pointed out by Patterson (1995), warming could also result in increased overwintering of insect populations and changes in their potential range. Although this could increase both the biological control of some weeds, it could also increase the incidence of specific crop pests, with subsequent indirect effects on crop-weed competition. Overall, synchrony between development and reproduction of bio-control agents and their selected targets is unlikely to be maintained in periods of rapid climatic change or climatic extremes. Whether this will result in a positive or negative benefit remains unclear.

20.10.2. Mechanical control

A standard means of controlling weed populations, and the one most widely used in developing countries is mechanical removal. Tillage (by animal or mechanical means) is regarded as a global method of weed control in agronomic systems. Elevated CO_2 could lead to further below ground carbon storage with subsequent increases in the growth of roots or rhizomes, particularly in perennial weeds (Rogers *et al.,* 1994). Consequently, mechanical tillage may lead to additional plant propagation in a higher CO_2 environment, with increased asexual reproduction from below ground structures and negative effects on weed control (e.g. Canada thistle) (Ziska *et al.*, 2004).

Overall, there are strong empirical reasons for expecting climate and/or rising CO_2 to alter weed management. Adaptation strategies are available, but the cost of implementing such strategies (e.g. new herbicides, higher chemical concentrations, new bio-control agents) is unclear. Herbicide use is controlled by individual state regulations. If an increase in CO_2 and temperatures allow invasive weed species to expand their geographical locations new herbicides may be needed to combat them. Often it takes a period of time to receive state approval of a new chemical or a chemical that has not been previously used.

20.11. HUMAN HEALTH

Weeds are recognized by the general public as significantly affecting human health either through allergenic reactions, skin irritations, mechanical injury or internal poisoning (Ziska, 2001). For the most part, we are only in the initial stages of quantifying how changes in climate and/or CO_2 may affect those specific weeds associated with public health. One exception has been changes in pollen production and allergenicity in common ragweed (a recognized cause of allergic rhinitis) with changing CO_2 and temperature in both indoor (Ziska and Caulfield, 2000, Wayne *et al.*, 2002) and *in situ* experiments (Ziska *et al.*, 2003). Additional research on how rising CO_2 can affect both poison ivy growth and toxicity is currently on-going. No information is available on how CO_2 could alter the toxicity of secondary compounds associated with mortality in weedy species.

20.12. CONCLUSIONS

Climate change is a looming global crisis and its impacts on agricultural weeds have not been well explored. Weeds are among the agricultural pests that can and will be influenced by climate change. Changes in atmospheric CO_2 levels, rainfall, temperature and other growing conditions will affect weed species distribution and their competitiveness within a weed population and within crop. Climate conditions exert a significant influence on the spread, population dynamics, life cycle duration, infestation pressure and the overall occurrence of weeds. Differential response of C_3 and C_4 plants will alter crop weed interaction because of the fact that majority of weeds are C_4 and most of the food grain crops are C_3. The effect of climate change on the geographic distribution of invasive weeds will be a subject of interest in the near future. Several weeds will exert additional pressure for crop-weed competition under the climate change scenario. More adaptive research studies, including complex research conditions, could yield useful solutions for managing yield reduction in the ensuing decades.

21

Development of Prediction Models for Weeds

21.1. INTRODUCTION

Weeds and climate change are both significant threats to biodiversity and pose considerable challenges to managers of protected areas. Weeds are likely to respond to climate and atmospheric change (e.g. increased temperatures, changed rainfall patterns, and elevated CO_2 levels) in complex ways but knowledge as to the magnitude and direction of these responses is very limited. Invasive plants will continue to spread regardless of climate change, particularly in an era of ever-increasing globalization and movement of goods (Ding *et al.*, 2008). Many plant invasions will progress farther and faster if assisted by climate change, and climate change is likely to increase evolutionary adaptation of weeds to climate extremes.

The combination of high management cost, broad impacts on biodiversity, agriculture and human health, as well as uncertainties about the nature of the impact of climate change on existing and emerging weed species, has led to identify weed research with respect to climate change as a high priority (Natural Resource Management Ministerial Council, 2007). This focus has also led to Key Actions in climate change adaptation strategies.

Thus, identification of emerging weed threats under climate change is an urgent task. It must also be dynamic, adapting to new data on species locations, or new environmental data such as improved future climate projections. A cost-effective monitoring or surveillance approach is needed that can produce useful guidance while research continues into mitigating impacts on biodiversity and prioritization processes (Downey *et al.*, 2010).

21.2. GEOGRAPHICAL DISTRIBUTION MODELS

The aim of modelling a species potential distribution is to better understand the geographical area in which that species may be found, survive, thrive, or remain absent now or in the future. Combined with the species present known distribution, this information can be particularly useful for informing strategic management efforts aimed at reducing current and future weed impacts in a rational manner.

Many different factors control where plant species are presently found. These factors can be broadly grouped into four areas:

- Climate.
- Soil.
- Co-occurring species.
- Their ability to get there from where they were previously (e.g. their native range).

Of these factors it is climate that represents the most useful 'outer limit' of where we might need to apply weed management strategies. It is, therefore, climatic suitability that weed models should be first trying to capture in their projections. Once this outer limit has been defined, it can subsequently be partitioned up by then considering non-climatic limitations to weed distributions.

Most modelling techniques for describing or projecting species' distributions fall into the following two groups.

- Correlative models
- Mechanistic models

21.2.1. Correlative models

Correlative models are known distribution records for the species of concern to variables presumed to explain the known range. This produces a projection that represents geographical areas with similar climates to the distribution records, at least according to the variables used in the model. However, we know that species ranges are limited by factors other than climate, so appropriately constructed models based solely on distribution records are likely to underestimate the full potential distribution of the climatically suitable area for a weed.

21.2.2. Mechanistic models

Mechanistic models can also be informed by distribution data, but can also make use of information from other sources, such as physiology experiments and expert knowledge, to understand the true climatic limits of the species in question. The nature of the variables and the form of the functional responses used in these models are usually informed by biological theory.

Mechanistic models can get much closer to understanding the full climatic limits for weed distributions, but they take longer to build than correlative models. Importantly, modelling weeds and then projecting to future climates are both likely to require an understanding of modelled suitability beyond the range of variables experienced by existing populations (i.e. model extrapolation). Correlative models are poorly equipped to deliver ecologically plausible projections consistently when extrapolating.

Even so, both types of modelling studies have flourished in recent years. All modelling techniques have limitations and the method applied needs to be appropriate both from the spatial scale and question posed. Unfortunately, models are not always applied in this manner, and hence weed managers should understand how to identify and how reliable these model projections are for informing weed management decisions. Carefully considered modelling done appropriately can be highly informative, but poor modelling done carelessly can be extremely misleading.

Depending on factors such as the type of model used, the care taken in cleaning input data and the skill of the modeller, potential distribution models can include substantial errors of omission (areas modelled as unsuitable when they are in fact suitable) and commission (areas modelled as suitable when they are in fact not suitable). Both types of errors can be costly, leading to wasted management effort on surveillance (commission error), or worse, an undetected but preventable invasion (omission error). These errors can be substantial (Kriticos and Randall, 2001; Webber *et al.*, 2011). There is good reason, therefore, to try to gauge the reliability of potential distribution models before they are considered for guiding management. These uncertainties also highlight the need for adopting an adaptive management perspective for managing weed risks under current and future climates.

21.3. SPECIES DISTRIBUTION MODELS

Species distribution models (SDMs) based primarily on climate data provide a first-step tool for assessing potential changes in the geographic range of species under climate change. These models can play an integral part in developing proactive management strategies for current and emerging weed threats. SDMs quantify the relationship between environmental conditions and the observed occurrence of species.

Models are built using baseline climate data and then projected onto a range of climate 'futures' to track the occurrence of climatically suitable habitat. When using SDMs to predict the potential distribution of weed species, data from both the native and invaded range must be used as input to avoid underestimation of the extent of climatically suitable habitat (Beaumont *et al.*, 2009b; Broennimann *et al.*, 2007; Hierro *et al.*, 2005; Mau-Crimmins *et al.*, 2006). Therefore, all available species occurrence data (i.e. native and introduced ranges) should be used to build models that project the distribution models onto the three future times (2020, 2050 and 2080) to provide an indication of medium and long-term trends in favourable climate conditions. The geographical distribution of favourable climate provides an indication of maximum potential species distribution within the country.

There are many methods for modelling the relationship between spatial distribution and the environmental envelope or realized niche of an organism. Elith *et al.* (2006) and Graham and Hijmans (2006) provide an extensive review and comparison of relative performance of many methods. For this preliminary set of models; MaxEnt (Phillips *et al.*, 2006), a very efficient machine-learning method with consistently high performance can be used (Elith *et al.*, 2006; Graham and Hijmans, 2006).

MaxEnt models can be fitted using default settings except that a new random number seed for each model is used and trained on a random selection of 90% of the data for each species, testing model fit on the remaining 10%. Ten cross-validation or replicate models can be made for each species by randomly choosing the location records included in the training and test sets. Averaged maps for the 10 runs can be used in subsequent analysis to represent current climate suitability.

It is important to account for variation in SDM output due to differences in climate predictions produced by GCMs (Beaumont *et al.*, 2008). For each species, each replicate model fitted to current climate can be projected onto climate data for each GCM at the three future times creating an ensemble of models. These future models can be used to produce mean predicted climate suitability maps for 2020, 2050 and 2080 for each species. That is, the 2020 mean map can be created by averaging predicted suitability in each grid cell for the forty maps (four GCMs and 10 projected replicate MaxEnt models for each GCM) for 2020. These mean maps indicate a consensus of probable conditions given the variation in predictions across the four GCMs, particularly for rainfall distribution. This can be referred to as the habitat suitability.

The general trend that can be drawn out of the many modelled weed distributions is a pole-ward shift in modelled suitability (i.e. towards the south in Australia) (Sutherst *et al.*, 2007). Assuming species are able to move from their existing range, a shift, expansion, or retraction of species distributions with climate change can be seen.

An important implication for this southerly range shift is that Australia does not extend far enough south to include the majority of regions of potential future suitability for current southern weeds. The implication may be that the southern displacement of species will result in more species ranges contracting to the southern coastline.

The modelling of weed distributions for the north and centre of Australia is particularly uncertain. While temperature is projected to increase with climate change, in common with other regions, the global climate change models are not clear on the level and direction of rainfall change. Decreasing uncertainty for climate projections is likely to improve our ability to interpret model projections in northern Australia, and this remains an area that requires further research and that requires a cautious approach to weed management.

21.4. ISSUES TO BE CONSIDERED WHEN USING WEED DISTRIBUTION MODELS

There are many bioclimatic modelling platforms available for estimating the potential distribution of weeds. The results of these models are often used to plan and make resource allocation decisions on eradication, containment and control of weeds. In Australasia, these maps often fed into the Post-Border Weed Risk Management System, identifying areas at risk of invasion (Standards Australia, 2006). Thus, they have an important role in strategic weed management. Given importance, their usefulness for intended purpose of gauging the potential distribution of invasive weeds under climate change can be predicted?

There are many species distribution models available in the scientific literature and on the web. However, preparing a reliable species' distribution model requires considerable skill and effort. There is extensive discussion in the scientific literature on these and other relevant issues.

A brief guide to some of the issues to be considered, without necessarily dealing with more fundamental methodological problems that may exist is presented below:

- Does the modelled suitable area cover all the known distribution points of the weed under historical climatic conditions? If not, can the presence of outliers be explained by microclimates?
- Does the potential distribution map make ecological and biological sense? A model of a tropical species that shows high suitability for both the north east of Australia and the Nullarbor, but not other desert regions, is likely to

be a result of a modelling artefact. For Mediterranean climate species does the model show suitability for southern Australia, but not the Perth region despite the weeds' presence there? Any model that is highly fragmented across inland Australia would have to be questioned, given shallow environmental gradients.

- Does the modelling method have a history of being able to project reliably into novel climatic conditions?
- Is it clear whether or not the model projection involves extrapolation beyond the known environmental range of the species?
- Is there a detailed description of how the particular species potential distribution model was developed, including reference to species ecology, physiology and life history stages, and how issues of input data quality and model performance were handled? Modelling species niches accurately and reliably is a skilled operation, and doing so into novel environments doubly so. Highly automated modelling typically fails to detect even gross errors in input data and modelling variables, and makes projections that are sometimes highly erroneous, spanning multiple NRM regions.
- Are there taxonomic issues that may compromise the results?

If any of these quality indicators are compromised for a model, the results should be treated with extra caution. Given the cost implications of incorrect risk assessments, it is often better to proceed with the caution you would apply if you had no knowledge, rather than rely on dubious information.

For example, where there is doubt, you may wish to adopt an adaptive management approach to dealing with a potential invader, sharing surveillance information across NRM regions in the zone of modelling uncertainty.

21.5. EARLY WARNING SYSTEMS MODELS

Although much research has been done to gain a better understanding of the impact of climate change on weeds, there are still many gaps in our knowledge. As with climate change in general, the use of predictive models is critical and as climate change modelling is undergoing continual fine-tuning, predictive models for weed responses must be concomitantly updated. However, improved predictive modelling requires better empirical understanding of weed evolution and ecology (Mortensen *et al.*, 2000). Weed ecology is a multifaceted discipline requiring extensive knowledge specific to the individual weed species. Taken together, predictive modelling and improved understanding of weed biology and ecology should provide more effective early warning systems to track changes in weed distributions and their impact under climate change.

An increasing number of modelling approaches and techniques have been developed in recent years to predict and map the expected ranges of habitat suitability of various invasive weeds. These models have helped to assess the potential geographic distributions of these species in response to different factors, such as climate change and land-use type. Among these modelling approaches, niche-based species distribution models (Guisan, 2000; Guisan and Thuiller, 2005; Peterson, 2003) have been used for assessing and identifying regions with a high invasion potential. These models use known species distributions combined with a set of environmental variables to develop a correlative model of the environmental conditions that meet a species' ecological requirements (Peterson, 2003). This approach makes it possible to project modelled niches into new regions and under future climate change scenarios, and ultimately to estimate the geographical distribution of suitable conditions.

One concern with distribution predictions for invasive species is the type of range information (native, invaded, or full ranges) that is used to develop the niche-based models. A comprehensive list of concerns with modelling plant species' distributions is provided in Thuiller *et al.* (2008). Models based on data from the native range assume that the same environmental factors determine the distribution of the species in the adventive range. However, these predictions may not adequately reflect the distribution of an invasive species in its introduced range (Estrada-Peña *et al.*, 2007). On the other hand, "climatic niche shifts are rare among terrestrial plant invaders" giving support for the use of ecological niche models (Petitpierre *et al.*, 2012).

Models using data from a previously invaded range may be more appropriate and accurate because the fundamental niche is likely to be more fully realized in invaded ranges than in the native range, where the species may be constrained by such factors as competition and dispersal barriers. Combining native and alien distribution records in models (i.e., discriminative correlative models) may be most insightful (Welk, 2004), but has been shown not to consistently improve model projections (Webber *et al.*, 2011). However, given niche differences across native and introduced ranges, distribution models using introduced range data alone may be more valid in some cases (Mau-Crimmins *et al.*, 2006). Despite contrasting views about the robustness of these methods, niche-based models and the presence of species in a region can provide insight into those factors that may favour or restrict the expansion of invasive species over a large scale. For example, Qin *et al.* (2013) used ecological niche maximum entropy (MaxEnt) modelling based on occurrence records of *Ambrosia artemisiifolia* and *A. trifida* in their native and introduced ranges to predict the potential distribution of these two invasive congeners in China under current climate conditions and under future climate projections.

The recent development of a large range of modelling tools for predicting potential weed distributions and to determine what areas are vulnerable to further weed invasion with or without climate change is very promising (Bradley *et al.*, 2010). However, this ability to predict possible future changes in weed distribution must be accompanied by sufficient resources, to develop these tools, utilize them, and then take active measures to prevent new weed infestations.

21.6. GENERAL LIMITATIONS TO MODELLING

Most of the invasive plant species that have been the subject of species distribution models are those on national priority lists, such as weeds of national significance and sleeper species. The major gaps in the types of species modelled include those of tropical Australia, agricultural weeds, and native invasive species.

Looking beyond the many potential pitfalls in modelling methodology, as well as the appropriateness of the methodology for the question being addressed, there are other more general limitations to modelling insight.

For example, species distribution models are mostly based on climate averages and rarely consider climatic extremes or other environmental factors such as soil type, fire and flood regimes, changing land use patterns and altered competitive and trophic (e.g. food chain) interactions. Modelling may also be inappropriate for species with strong associations to microhabitats, where the environment may be very different from that shown by climate averages. An example of the latter would be a plant species only found on rocky outcrops where rainfall collection gives a raised moisture level not captured by the resolution of the variables driving the model.

21.7. RECENT DEVELOPMENTS

Recent modelling has sought to address the issue of combining models using climate and biology with other environmental factors that influence establishment and persistence of invasive plant populations. For example, fundamental changes in land use and altered flood regimes with future climate change in the Murray Darling Basin are likely to impact on the potential distribution of invasive plants such as lippia (*Phyla canescens*).

21.8. CONCLUSIONS

Models that integrate the available quantitative knowledge can be used to design preventive measures, to develop long-term and short-term strategies for weed management, to assist in decision making to determine if, when, where and how weeds should be controlled and to identify new opportunities for weed control. Eco-physiological simulation models for crop-weed competition simulate growth and production of species in mixtures, based on eco-physiological processes in plants and their response to the environment. Such models help improve insight into the crop-weed system and can be used for purposes such as the development of simple predictive yield-loss models, threshold levels or the design of competitive crop plant types. For strategic weed management decisions, preventive measures and the identification of new opportunities for weed control, quantitative insight into the dynamics and spatial patterns of weed populations is also required. The complexity of the process and the long-term character of weed population dynamics make the use of models necessary. Different modelling approaches have been developed. Opportunities to use the available knowledge and models to improve weed management should be worked out. Weeds occur in patches and their sensitivity to herbicides changes strongly with developmental stage, making precision techniques for herbicide application in time and space an option for reducing herbicide use. Limitations related to insight in biological processes as well as the state of technological development needs to be looked into.

22

Climate Change Adaptation and Mitigation Strategies for Weeds

22.1. INTRODUCTION

As the geographical profile of weed communities shift with climate change, local weed management strategies will have to show adaptability. Weed profiles will also be affected by climate change and this may have an influence on what methods are adopted. Effectiveness of weed management is also hypothesized to change along with environmental conditions. Weeds that are under moisture stress can respond by thickening their leaf cuticles, reducing stomatal number and conductance, slowing down vegetative growth and flowering rapidly. Greater increase in biomass will result in dilution of herbicide applied, making weed control more difficult and costly (Patterson, 1995). Drought stressed weeds are more difficult to control with post-emergent herbicides than plants that are actively growing. Moving away from purely environmental factors, issues such as the limited and fast-diminishing oil resources and the reduction in available herbicides will have significant effects on how weeds are managed. A strategy growing in recognition and approval is Integrated Weed Management (IWM). This involves using a variety of weed management tools to provide a more flexible and adaptable management system, often underscored by a more sensitive approach to the natural environment in which the management is occurring.

Weeds are naturally adaptive and have huge genetic diversity. With such a large gene pool, weeds have the potential to adapt to any changes in the environment extremely quickly. Management of weeds will become more complex, with the changes in the environment resulting in changes in selective pressures. This could possibly lead to changes in physiology, which will require redeveloping methods of control, for example, development of new herbicides and rethinking biological control methods. In addition crop-weed interactions could be altered, which would also require the development of new management practices.

To assure harvest in the face of climate change, farmers adapt cropping systems and management measures, for instance, by implementing different crop rotations, crop sowing dates, irrigation, and tillage methods (Kaukoranta and Hakala, 2008; Fleming and Vanclay, 2010; Daccache *et al.*, 2012). Farmers are also likely to choose new, climatically suited crops or cultivars that are better adapted to warm and dry conditions (Bloomfield *et al.*, 2006; Tokatlidis, 2013). Extreme weather events in the future will probably set yield at higher risks, which may lead to an increase in pesticide usage and fertilizer input (Baessler and Klotz, 2006; Lososova *et al.*, 2006). Pre-emergent herbicides or herbicides absorbed by plant roots need soil moisture and actively growing roots to reach their target sites. Occurrence of drought has the potential to reduce the effectiveness of pre-emergent herbicides (Singh *et al.*, 2011). To compensate for any loss of herbicide effectiveness under warmer and drier conditions, producers may need to review the reliance on herbicide use in any integrated weed management approach.

The crop breeding is able to rapidly develop new crop varieties that will have increased tolerance to climate change variables (increased temperature, reduced water availability, increased CO_2 and O_3). Breeding of CO_2-efficient crops such as wheat, maize, or soybean is likely to advance in the future (Ziska *et al.*, 2005; Tokatlidis, 2013).

Overall, there are strong empirical reasons for expecting climate and/or rising CO_2 to alter weed management. Adaptation strategies are available, but the cost of implementing such strategies (e.g. new herbicides, higher chemical concentrations, new biological control agents) is unclear.

22.2. ADAPTING WEED MANAGEMENT

Three potential types of strategies are considered to guide management of invasive weed plants under climate change:

- Continuing current control options.
- Adopting control options suitable for weed management under current extreme weather conditions.
- Developing new weed management techniques adapted for climate change.

22.2.1. Continuing current control options

These may be the most appropriate methods for controlling weeds under the current climate. Progress in weed control tends to be incremental; there has been no radical change in methods in recent times. However, continued investment in improving weed control could take into account climate change as it progresses. The rate of incremental evolution needs to be appropriate for the observed changes. An example is the management of annual weeds for herbicide resistance.

22.2.2. Adopting weed management options suitable under current extreme weather conditions

This is the "no regrets" approach that aims at developing management techniques that are useful under today's climate (and weather extremes) and by consequence these are methods likely to be suitable for use under future climates.

22.2.3. Developing new weed management techniques adapted for climate change

This will require new approaches that take into account projected future climates. An example is the new biological control agents or a different approach to control might be needed in the face of increasing CO_2 concentrations.

Within these various management approaches, we also need to consider the most appropriate responses to climate change for quarantine, eradication and containment. Likewise, how existing methods, herbicides, biological control and other technologies are best adapted for climate change.

22.3. ADAPTATION AND MITIGATION STRATEGIES

The key adaptation and mitigation strategies that require (re)consideration include:

- Plant quarantine
- Physical methods
- Cultural methods
- Chemical methods
- Biological control
- Precision weed management
- Transgenic herbicide tolerant crops
- Integrated weed management

22.3.1. Plant quarantine

The current overall national approach to quarantine of invasive plants is expected to be adequate as a response to climate change. There will be a continued threat of weed invasion from overseas, but the source areas will not change greatly as climate changes. However, the weed risk assessment (WRA) protocol does not have an assessment component specifically for climate change.

The general southwards invasion projected from the various studies modelling species distributions points to the need to establish new quarantine barriers as an adaptation response. These could be at state boundaries.

22.3.2. Physical methods

22.3.2.1. Soil solarization: Soil solarization plays a big role in the management of weeds, nematodes and pathogens under the conditions of increased temperature. It can be incorporated in the schedule of integrated weed management (IWM) and integrated pest management (IPM). However, the thickness and colour of polyethylene, time and duration of solarization, effect on beneficial micro-organisms and nutrient mineralization, and supplementary treatments to be superimposed or followed after solarization across situations need to be more refined under the changing climate.

22.3.3. Cultural methods

22.3.3.1. Mechanical methods: Cultivation has been the weed control method used over most of agricultural history. Climate change, especially under a drying environment, may make the general use of cultivation problematic given the increased risk of wind erosion and soil moisture loss. The adaptation response might be to use cultivation strategically as part of an integrated weed management approach.

A standard means of controlling weed populations, and the one most widely used in developing countries is mechanical removal. Tillage (by animal or mechanical means) is regarded as a global method of weed control in agronomic systems. Under heavy frequent rainfall carrying out of tillage and other mechanical weed control methods will be very difficult due to water-logging of soil and under such conditions undertaking of cultivation will help to reduce soil moisture which may decrease the growth of crop plant and make less competitive with weeds. Elevated CO_2 could lead to further below ground carbon storage with subsequent increases in the growth of roots or rhizomes, particularly in perennial weeds. Consequently, mechanical tillage may lead to additional plant propagation in a higher CO_2 environment, with increased asexual reproduction from below ground structures and negative effects on weed control (e.g. Canada thistle) (Ziska *et al.*, 2004). Examples of weeds that may respond in a similar manner are skeleton weed (*Chondrilla juncea*), and silver leaf nightshade (*Solanumelae agnifolium*) (Kriticos *et al.*, 2010).

It is difficult to see how mechanical weed control (chain pulling, stick raking, blade ploughing, and specialized weeding machines) can be adapted for the increased risk of climate change. Recent developments such as mechanical destruction of seeds as part of the harvesting process will continue to be useful. Managing grazers (goats, sheep, cattle, etc.) as part of a pest management adaptation response may also be beneficial to weed control.

22.3.3.2. Allelopathy: Studies involving allelopathic crop residue mulches (maize, sorghum, wheat, barley or rye) may help to mitigate adverse climatic effects. Effective utilization of such crop residue mulches can act as self-supporting weed management (rice) for the concurrent as well as rotational crops.

22.3.3.3. Other cultural methods: Nonchemical controls include crop rotations, altering planting dates and fertilizer and irrigation applications, and soil management and tillage. Crop rotations should be followed to increase biodiversity. Pasture phases in rotation rather than continuous cropping is beneficial. Noxious weeds establish slower (grassy weeds in cereals), because specific relationships between weeds and host plants are interrupted (Dhawan and Peshin, 2009). These technologies could help minimize projected weed losses and thereby help maintain crop yields.

22.3.4. Chemical methods

Clearly, any direct or indirect impacts from a changing climate will have a significant effect on chemical management. Overall, pesticides are most effective when applied to plants that are rapidly growing and metabolizing, i.e. those free from environmental stress. Environmental factors such as temperature, precipitation, wind and relative humidity influence the efficacy of herbicides (Archambault *et al.*, 2001).

Most of the herbicides required optimum temperature for their activity (function). But prolonged high temperature after application (pre-plant, pre- and post-emergent herbicides) may affect the efficiency and selectivity. That means at high temperature, selective herbicides may become non-selective. There are increasing number of studies (Ziska and Teasdale, 2000) that demonstrate a decline in chemical efficacy with rising CO_2. But scientists suggests that as CO_2 increases the growth and development of some weeds (C_3) will increase and plant can develop immunity that can tolerate or detoxifying mechanism (high tissue volume). Recent work with Canada thistle grown in monoculture under field conditions suggested a greater root to shoot ratio and subsequent dilution effect of glyphosate when grown at elevated CO_2 (Ziska *et al.*, 2004). The scientists pointed out that CO_2

does reduce efficacy of the herbicides. However, then additional work is needed to determine herbicide specificity, concentration and application rates as possible means of adaptation.

Furthermore, there is the possibility that increased CO_2 and temperature will reduce the effectiveness of herbicides, requiring an increase in frequency of application and concentration of herbicide to obtain adequate control (Darren *et al.*, 2010). Similarly, although the effects of elevated atmospheric CO_2 concentrations on weed/crop/herbicide interactions are not well known, a recent study (Archambault *et al.*, 2001) showed that elevated CO_2 levels diminished the efficacy of the widely used herbicide glyphosate (Fig. 22.1). If these effects are common and widespread, they will have a significant impact on agriculture. Moreover, according to Singh *et al.* (2011) weeds that are under moisture stress can respond by thickening their leaf cuticles, slowing down vegetative growth and may flower rapidly. Drought stressed weeds are more difficult to control with post-emergent herbicides than plants that are actively growing for example, systemic herbicides that are translocated within the weed need active plant growth stage to be effective. Pre-emergent herbicides or herbicides absorbed by plant roots need soil moisture and actively growing roots to reach their target sites. Occurrence of drought has the potential to reduce the effectiveness of pre-emergent herbicides.

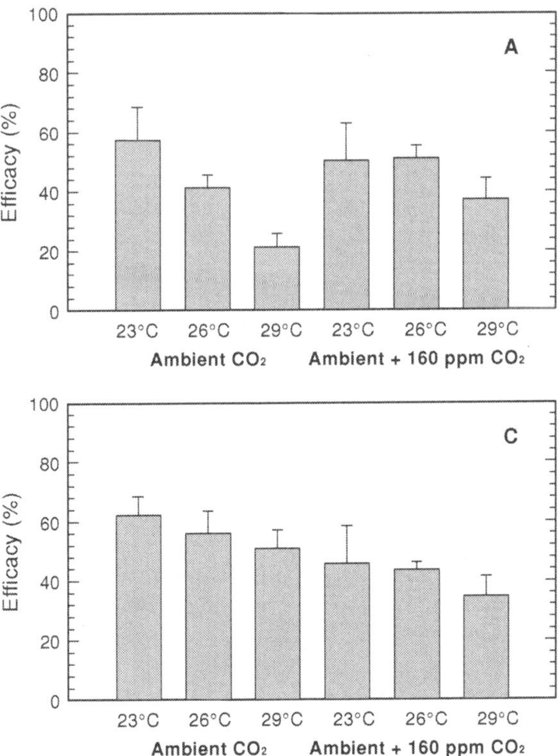

Fig. 22.1. Herbicide efficacy of Fusion (A), and Liberty (C) applied to wild oats grown at different temperatures at ambient or ambient + 160 ppm CO_2 (Archambault et *al.*, 2001).

In general different scientists at different times suggests their own ideas on the possible effects of climate change on chemical weed control methods. These are:

- Greater root to shoot ratio leading to subsequent dilution effect on glyphosate when grown at elevated CO_2 (e.g. Canada thistle) (Ziska *et al.*, 2004).
- Decreased stomatal aperture or number, cuticle thickening, development or increased leaf pubescence with subsequent reductions in herbicide entry into the leaf.
- Altered transpiration leading to reduced uptake of soil herbicides.
- Changes to the wetting and drying of soil profiles affecting microbial activity and herbicide degradation.
- Increased temperatures affecting the efficacy and selectivity of herbicides (Madafiglio *et al.*, 2000; Medd *et al.*, 2001).
- Increased growth rate implying narrower temporal windows in which to apply herbicides before crop yield losses from weed competition and before weed set.
- High temperatures, low relative humidity and unsuitable wind conditions reducing the window of spraying opportunity (Howden *et al.*, 2007).

Herbicides are an integral part of weed control in crops and are increasingly used to control environmental weeds. Research in overseas showed that weedy species have an increased tolerance to herbicides when grown under elevated CO_2 conditions. Moisture stress is also likely to decrease the effectiveness of herbicides.

Herbicide resistance is widespread in cropping areas. It can potentially develop within a relatively short period after eight consecutive years of application of the same herbicide. Many invasive plants in agricultural situations have short generation times, mostly with annual generations. It is this combination of high selection pressure (herbicide application), short generation time, and inherent genetic variability that allows for the rapid evolution of herbicide resistance. It is reasonable to assume that these invasive plants could also evolve rapidly in response to climate change. This implies that today's major agricultural weeds (e.g. wild radish, annual ryegrass) are likely to remain as major problems under climate change. This assumes that current agricultural practices continue.

This indicates that the main adaptation response is to manage for herbicide resistance through an integrated approach to controlling agricultural weeds. With a drying environment, this integration is likely to include a move away from continuous cropping to include a greater reliance on pasture phases. Innovative management of weeds, including careful use of genetically modified crops, will be required to avoid a return to weed species favoured by crop-pasture rotations in the past (e.g. double gee, *Emex australis*), or changed cropping practices (e.g. hoary cress, *Lepidium draba*).

22.3.4.1. Reduction in herbicide effectiveness: Ziska *et al.* (1999) and Ziska and Teasdale (2000) have shown that herbicides (e.g., glyphosate) will be rendered less effective against weeds under CO_2 levels anticipated in the near future (Fig. 22.2). Increased tolerance to glyphosate under elevated CO_2 has been recorded for both agricultural and invasive weed species (Ziska and McConnell, 2015). These alarming findings revealed a sustained increase in photosynthesis and growth of perennial weeds such as *Elymus repens,* with a concurrent decrease in herbicide efficacy and increased potential of invasion and competition (Ziska and Teasdale, 2000). Differential tolerance to glyphosate exhibited by certain weeds under elevated CO_2 is also an issue. Whilst the response of weeds such as *A. retroflexus* was not affected by elevated CO_2, *Chenopodium album* and *Cirsium arvensis* manifested a significant glyphosate tolerance (Ziska *et al.*, 1999; Ziska *et al.*, 2004). A variable response to glyphosate was observed even for invasive grass species possessing the same carbon fixation pathway (Manea *et al.*, 2011). Hence, it can be inferred that some weeds will be more problematic in the near future in glyphosate tolerant crops or under conservation agriculture. Another difficulty will be the knockdown of perennial weeds if glyphosate efficacy is reduced due to climate change. An increase in rhizome and tuber growth, coupled with an increase in biomass, would cause a dilution effect on any herbicide application, causing an increase in weed control costs. Direct effects of climate change on plant physiology, anatomy, and morphology will indirectly affect herbicide efficacy by influencing uptake, translocation, and metabolism. Changes in physical environments, such as drought spells or prolonged rainy seasons, may limit the field conditions necessary for herbicide applications. Climate change will have implications for all dimensions of chemical weed management including application, spray drift, persistence, metabolism, and herbicide efficacy. This justifies diversifying current weed management tactics as well as the urgency of a sound knowledge regarding the ecology and biology of weeds in a changing climate.

Fig. 22.2. Reduction of glyphosate efficacy at increased concentration of CO_2

22.3.4.2. Chemical weed control studies using old and new molecules of herbicides: Climate change, particularly increased temperature may have profound impact on the control of composite, parasitic, aquatic, and invasive weeds using herbicides in crops and non-cropped situations. Studies on the ED_{50} or GR_{50} value and efficiency of herbicides are of paramount importance under this situation. Monitoring of residues in soil, water and plants may add another dimension of research. The following studies may have priority:

- Herbicide bio-efficacy, biochemical selectivity and non-target toxicity of the old and new herbicides and their formulations.
- Herbicide resistance management.
- Herbicide resistant crops (transgenics) and their performance evaluation.

22.3.5. Biological control

Biological control involves the introduction of host-specific agents, usually insects or fungi, for the control of a weed species. The biological control agents have evolved to feed only on the target plant (Medd *et al.*, 2001). Most bio-control systems perform best under a stable environment. However, in addition to the forecast increase in mean temperatures, a CO_2 and shift in rainfall distribution climate variability is expected to increase.

Extreme weather events such as droughts, flood and even unseasonal frosts are predicted to occur more frequently. While many biological control species have mechanisms to cope with extremes, they require time to acclimatize and/or enter the resistant state. The relative vulnerability of the host plants, and bio-control agents to extremes of temperature, desiccation or flooding will determine whether the drought or flood will be followed by a weed or pest outbreak (Gerard *et al.*, 2010). Moreover, according to Gerard *et al.* (2010) increases in CO_2, changes in water availability and increases in temperature will alter plant phenology, growth and distribution, all of which will have flow on effects on the plant herbivores and those that prey on them. Aspects of plant life cycle events controlled by temperature, such as the timing and duration of seed germination, bud burst, and flowering, are likely to change with warmer temperatures, subject to photoperiod and water availability. Similarly, the degree that individual species benefit will depend on how well the life cycle events are synchronized to ensure optimum survival. Changes in host plant quality in response to elevated CO_2, drought or flooding may also cause shifts in herbivore and natural enemy fitness. In general, elevated CO_2 leads to a decrease in the nutritional value of plants. There is a tendency for foliage feeders to consume more plant material (greater damage), but to grow and develop more slowly, and have reduced fecundity and survival.

The direct impacts of climate change will be either on the biology of the biological control agent and/or on the ability of the host plant to tolerate or compensate for the presence of the herbivore or plant pathogen. Increased temperature would be expected to increase the rate of life cycles of both the biological control agents and the weeds. Increased water stress will affect the host plant's development, and through this, the development of biological control agents, so they might be less effective in drier situations. Increase in greenhouse gases will also affect the herbivore plant relationship so impacts at both the spatial and temporal scales are possible (Kriticos *et al.*, 2010). Changes in temperature can influence production of plant defence compounds against plant herbivores. It is known that levels of many insect-resistance allelochemicals increase during drought (Gerard *et al.*, 2010).

At the spatial scale, the effects of moisture and temperature are likely to play out through differential changes in the preferred distribution of the agents and target weed. For example, an increase in temperature may enable biological control agents from subtropical areas to survive on weeds in more temperate regions. Some examples are the agents of aquatic weeds such as salvinia (*Salvinia molesta*) and alligator weed (*Alternanther aphiloxeroides*), and the agents of terrestrial weeds such as prickly pear cactus (*Opuntia* sp.) which may be favoured in southern regions of their distribution. Understanding of the agent's and the weed's physiology, capacity for evolution, and dispersal is needed to determine the potential for a weed to migrate under climate change and "escape" their control agents (Kriticos *et al.*, 2010).

Biological control of pests by natural or manipulated means is likely to be affected by increasing atmospheric CO_2 and climatic change. Climate as well as CO_2 could alter the efficacy of weed bio-control agents by potentially altering the development, morphology and reproduction of the target pest. Direct effects of CO_2 would also be related to changes in the ratio of C: N and alterations in the feeding habits and growth rate of herbivores. As pointed out by Patterson (1995), warming could also result in increased overwintering of insect populations and changes in their potential range. Although this could increase both the biological control of some weeds, it could also increase the incidence of specific crop pests, with subsequent indirect effects on crop-weed competition. Overall, synchrony between development and reproduction of bio-control agents and their selected targets is unlikely to be maintained in periods of rapid climatic change or climatic extremes. Whether this will result in a positive or negative benefit remains unclear.

According to Kriticos *et al.* (2010) who list the range of possible impacts (based on generally polyphagous invertebrate herbivores) identified in a meta-analysis of CO_2 induced changes on plant chemistry and herbivore performance. The impacts included:

- Significantly decreased herbivore abundance (-21.6%).
- Increased relative consumption rates by herbivores (+16.5%) development time (+3.87%); and total consumption (+9.2%).
- Significantly decreased relative growth rate (-8.3%); conversion efficiency (-19.9%); and pupal weight (-5.03%).
- No difference between herbivore associations.
- Plants showed significantly larger biomass (+34.4%), increased C: N ratio (+26.57%), decreased N concentration (-16.4%), increased tannins (+29.9%), and other phenolics.

22.3.5.1. Classical biological control: Biological control agents, mainly arthropods and pathogens, have provided permanent and effective control for some major weed species. Part of the process of biological control is to search for and select biological control agents in regions that match the part of country that has been invaded. Usually climate change is not considered in the assessment of the agent-target association.

The changing dynamics between target weeds and their biological control agents is likely to be complex because the environmental response of the weed may not be the same as the agent's or their interaction. In addition, the changed climate may become unsuitable for the existing biological controls. Interaction between increased CO_2, fast growth and plant nitrogen levels needed to support biological control agents such as herbivores may change.

Some of the main factors that may affect biological control include:

- Increases in temperature leading to increased growth rate of the plant that could enable to "escape" the herbivore.
- Changes in rainfall that could make the plant and its environment unsuitable for the agent.

- Higher CO_2 concentrations.
- Increased carbon to nitrogen ratio in the plant which could slow herbivore development.
- Changes to plant secondary chemistry which make the plant less palatable.

Two biological control examples show some of the complexity of assessing the interaction between biological control agents, invasive plants and climate change. The growth of the leaf miner *Dialectica scalariella* on Patterson's Curse, *Echium plantagineum*, was mostly related to higher temperatures and not CO_2 concentration, but at the cost of poorer survival of the leaf miner.

A second example concerns the chrysomellids, *Octotoma championi* and *O. scabripennis*, which are leaf miners as larvae and leaf feeders as adults on the widespread invasive shrub, *Lantana camara*. The insects were affected by increased leaf loss at higher temperatures, but insects preferred leaves grown at normal levels of CO_2. In both cases the plants showed the typical response of increased biomass and reduced leaf nitrogen and increased carbon: nitrogen ratio when grown under elevated CO_2.

22.3.5.2. Adapting biological control: Adaptation responses suitable for the biological control component of weed management include:

- Monitoring the weed – biological control interaction to determine if additional controls are required.
- Prioritizing biological control agent redistribution to areas with a more favourable climate.
- Searching for and assessing new biological control agents from overseas areas with climates that anticipate the future climate of target regions.

22.3.6. Precision weed management

Combined use of remote sensing, Geographic Information System (GIS) and Global Positioning System (GPS) is a powerful tool in detecting, mapping and monitoring the spread of weeds over inaccessible areas. Identification of weed species using remote sensing in range or wild range lands can be achieved. Site-specific weed management (SSWM) may be undertaken using these tools. Computer-analysed video images of the digital data can provide the area estimates of weeds.

22.3.7. Transgenic herbicide tolerant crops

Another important issue regarding pest management in the future centers on the role of biotechnology in crop protection. The next 20 years will likely see a substantial increase in the use of genetically engineered plants. Some of these plants have been engineered so that the application of herbicides destroys weeds but not the economic crop. Other genetically engineered plants have been designed to resist pests such as stem borers and nematodes without the need for pesticides. Others are expected to combine both herbicide resistance and insect/disease resistance in one seed.

The Roundup Ready (RR) technology incorporates genetic resistance to glyphosate into crop plants by inserting a single bacterial gene that modifies 5-

enolpyruvylshikimate-3-phosphate (EPSP) synthase, an enzyme essential for plant growth.

GTS 40-3-2 (also known as Roundup Ready Soybean) is a genetically engineered variety of glyphosate-resistant soybeans produced by Monsanto. The major problem in soy farming is weeds thus GTS 40-3-2 is revolutionary. The most widely utilized herbicide-tolerant crop in production today is the Roundup Ready soybean resistant to the broad spectrum herbicide Roundup that contains the active ingredient glyphosate.

Liberty Link soybeans were introduced in 2009. These soybeans allow the use of Ignite herbicide as a post-emergence treatment. Ignite is a newer, more concentrated formulation of Liberty herbicide. This gives soybean producers an option to Roundy Ready soybeans.

Roundup Ready Corn (RR Corn) is genetically engineered corn that has had its DNA modified to withstand the herbicide glyphosate (the active ingredient in Monsanto's herbicide Roundup). It is also known as "glyphosate tolerant corn."

LibertyLink corn is genetically engineered for tolerance to over-the-top applications of the non-selective herbicide Liberty (glufosinate-ammonium).

Dicamba, glufosinate and glyphosate-tolerant corn is designed to build on the Roundup Ready Xtend Crop System and provide farmers with additional herbicide-tolerance options. This product would contain multiple herbicide-tolerance traits enabling use of herbicides with different modes of action, expanding growers' options to protect their crop from weed infestations.

Using modern biotechnology, Monsanto Company has developed Roundup Ready cotton plants that confer tolerance to glyphosate, the active ingredient in Roundup agricultural herbicides.

Liberty Link Cotton has been genetically modified to be tolerant to Liberty herbicide, allowing for effective post-emergence herbicide management system in cotton.

Roundup Ready canola has been modified using gene technology to tolerate glyphosate, the active ingredient in Roundup agricultural herbicides.

The LibertyLink trait (glufosinate resistance) is available in top-performing InVigor canola hybrids.

Roundup Ready sugar beets have been enhanced through biotechnology and contain the Roundup Ready gene to tolerate applications of labelled Roundup agricultural herbicides, which contain the active ingredient glyphosate.

Ribas *et al.* (2006) introduced the *bar* gene in *Coffea canephora* and *C. arabica.* The *bar* gene inactivates the herbicide ammonium glufosinate, which is normally used as a non-selective pre-emergence herbicide and also for pre-harvest desiccation.

Commercially available genetically modified herbicide tolerant crops are presented in Table 22.1(Castle *et al.*, 2006).

Table 22.1. Commercially available genetically modified herbicide tolerant crops

Crop	Trait phenotype	Target trait gene(s)	Trait designation	Originating company	Year of first commercial sale	Trade name
Cotton	Resistance to glyphosate herbicides	CP4 epsps	MoB1445/ 1698	Monsanto	1996	Roundup Ready
	Resistance to phosphinothri-cin herbicides	bar	LLCotton25	Bayer Crop Science	2005	LibertyLink
Corn	Resistance to glyphosate	Maize epsps	GA21	DeKalb (now Monsanto)	1998	Roundup Ready
		Two CP4 epsps expression cassettes	NK603	Monsanto	2001	Roundup Ready Corn 2
	Resistance to phosphinothri-cin herbicides	pat	T14, T25	Aventis (now Bayer Crop Science)	1996	LibertyLink
Soybea-n	Resistance to glyphosate herbicides	CP4 epsps	GTS-40-3-2	Monsanto	1996	Roundup Ready
Canola	Resistance to glyphosate herbicides	CP4 epsps, gox v247	GT73	Monsanto	1996	Roundup Ready
	Resistance to phosphinothri-cin herbicides	pat	Topas 19/2	Agrevo (now Bayer Crop Science)	1995	LibertyLink
Alfalfa	Resistance to glyphosate herbicides	CP4 epsps	J101, J163	Monsanto	2005	Roundup Ready

Source: Castle *et al.* (2006)

22.3.8. Integrated weed management

Integrated weed management (IWM) is the recent and more acclaimed aspects of long-lasting weed management programme in crops and non-cropped situations. Integrated weed management lies at the center of weed control. The combination of farming strategies, biological control agents, and necessary herbicide use has helped California farmers address weed problems using a variety of methods.

22.3.9. Non-conventional weed management strategies for modern agriculture

Weeds are a significant problem in crop production and their management in modern agriculture is crucial to avoid yield losses and ensure food security. Intensive agricultural practices, changing climate, and natural disasters affect weed dynamics and that requires a change in weed management protocols. The existing manual control options are no longer viable because of labour shortages; chemical control options are limited by eco-degradation, health hazards, and development of herbicide resistance in weeds. Therefore, reviewing some potential non-conventional weed management strategies for modern agriculture is necessary that are viable, feasible, and efficient. They include:

- Improvement in tillage regimes has long been identified as an impressive weed-control measure.
- Harvest weed seed control and seed predation have been shown as potential tools for reducing weed emergence and seed bank reserves.
- Development in the field of allelopathy for weed management has led to new techniques for weed control.
- The remarkable role of biotechnological advancements in developing herbicide-tolerant crops, bio-herbicides, and harnessing the allelopathic potential of crops is also worth mentioning in a modern weed management programme.
- Thermal weed management has also been observed as a useful technique, especially under conservation agriculture systems.
- Precision weed management has to be elaborated with sufficient details.
- The role of remote sensing, modelling, and robotics as an integral part of precision weed management has to be highlighted in a realistic manner.

All these strategies are viable for today's agriculture; however, site-specific selection and the use of right combinations will be the key to success. No single strategy is perfect, and therefore an integrated approach may provide better results. Future research is needed to explore the potential of these strategies and to optimize them on technological and cultural bases. The adoption of such methods may improve the efficiency of cropping systems under sustainable and conservation practices.

22.4. CONCLUSIONS

After catastrophic climatic events such as drought or flood, weeds will have a greater chance to colonize and invade disturbed habitats. Chemical control measures may become less effective due to a change in the external environment (drier and warmer conditions) or changes in anatomy, growth physiology, and phenology of the target weed flora (Chauhan *et al.*, 2014; Ziska and McConnell, 2015). Asexual reproduction through below-ground parts is always conducive to spread, irrespective of water availability. Extremes of moisture availability, viz., flood as well as drought, hinder physical management methods such as hoeing, inter-cultivation, etc. It seems that growers will have to carefully synchronize the timing of their control measures with the weed life cycle, since these will also respond to climate change. The opinion of Chandrasena (2009) that adaptive responses should be based on a better knowledge on how plant communities will respond to climate change rather than adhoc responses, is therefore valid in the current context.

While the perennial problem of limited resources may make the task ahead appear daunting, applying the principles of adaptive weed management planning should help NRM groups be prepared for most eventualities. One of the most cost effective responses to climate change will be monitoring the consequences of shifts in distributions of species, in order to recognize negative impacts when they arise, so that timely action can be taken. Not every weed species needs to be the focus of intensive management, rather it will be the early identification and targeting of priority weed threats and learning from past efforts that will ensure NRM groups are best situated to respond to future weed challenges.

Section VI
FUTURE THRUSTS AND CONCLUSIONS

23

The Way Forward

23.1. INTRODUCTION

The earth's climate is changing in response to several human activities which markedly contribute to global climate changes (Pachauri and Reisinger, 2007). Agriculture, forestry, and other land uses are responsible for approximately 25 % of anthropogenic greenhouse gas (GHG) emissions mainly from deforestation and agricultural emissions (Bustamante *et al.*, 2014; IPCC, 2014). Over the last century, there have been consistent increases in CO_2 emissions which led to an increase in global mean temperature of 0.74° C, and this is projected to continue at a pace between 0.15 and 0.3° C per decade, based on the emission scenario (IPCC, 2007; Watson, 2001). By the end of the twenty-first century, temperature is projected to rise by 3.4° C while atmospheric CO_2 concentration will be almost four times higher than that of today (Pachauri and Reisinger, 2007; IPCC, 2014). Such a scenario raises questions on the sustainability of current cropping systems, which are likely to be impacted under climate change. Concomitantly, global food production must increase by 70 % to meet the projected demand of the world's population of 9.5 billion by 2050 (Chakraborty and Newton, 2011; Foley *et al.*, 2011). This difficult challenge is further complicated by climate change which influences the distribution of crop pests (i.e., insects, disease pathogens, nematodes, and weeds) and pest severity. Although pest management has played an important role in the marked rise in food production in the last 50 years (Oerke, 2006), most existing management strategies have been based on the indiscriminate use of pesticides in agriculture. Increasing awareness of health and environmental issues as well as increasingly stringent legislation on pesticide use will pose a particular compounded challenge to twenty-first century agriculture. Agriculture needs to meet food demands, satisfy sustainability goals, and adapt to climate change. Indeed, maintaining, or even increasing, crop yield while simultaneously reducing reliance on pesticides represents a challenging task (Hossard *et al.*, 2014) that is further complicated against the background of climate change scenarios.

In the current situation, two factors increase the risk of potential pest pressure in agriculture. While increasing anthropogenic impact, in particular increasing global movement of people and trade of plant commodities, continues to lead to the introduction of exotic pests into new regions, climate change provides suitable conditions for such pests to adapt across the areas which were previously detrimental for their survival (Chakraborty, 2013; Fletcher *et al.*, 2010). In this way, climate change further accelerates pest pressure both in space and time and may have dramatic consequences throughout specific regions of the globe. In some regions, excessively dry, wet, or warm conditions may directly hamper crop production and consequently current crop protection methods may become ineffective (Chakraborty and Newton, 2011). Recent examples of pest adaptation to warmer temperatures have given rise to escalating pest epidemics in new areas where, in retrospect, the current cropping systems are excessively vulnerable to unforeseen threats. Even less direct effects, such as changes in crop or pest phenology can also have significant consequences in crop production.

Crop losses due to pests can account for up to more than 40 % worldwide (Oerke, 2006). These values might significantly rise under changing climatic conditions whereby new and more aggressive pests affect levels and stability of crop yield thereby threatening food security. However, it is not possible to make estimates on overall potential crop losses due to climate change. This is in part because of the unpredictable adaptation of certain pests to a changing environment which creates situations of uncertainty (Chakraborty and Newton, 2011). Indeed, the main consequence of climate change and accelerated globalization is a heightened level of unpredictability regarding spatial and temporal interactions between weather, cropping systems, and pests (Chakraborty, 2013). This is the on-going crop protection challenge that farmers, advisers, researchers, and policy makers face.

In light of this, the reduction of pest impact has become more relevant than ever not only to produce sufficient food and agricultural commodities but also to reduce excessive input and unnecessary CO_2 emissions. Interactions between the changing environment, crops, cropping systems, and pests demand a new multi-disciplinary and integrative knowledge base for the development of viable adaptive strategies. The capacity to adapt is needed to reduce the impact of both invasive alien species as well as rapidly evolving resident pest populations (Chakraborty, 2013; Roy *et al.*, 2014). The anticipatory strategies, enhanced adaptation capacity, and the development of more resilient cropping systems require new insights and innovative organizational change. There is a need to identify relevant research needs and policy implications in order to understand how cropping systems and integrated pest management (IPM) strategies can be designed to better deal with invasive as well as existing but evolving pest populations under climate change. For emerging invasive alien pests that, should they evade biosecurity measures, are likely to become established, their impact needs to be reduced; while for indigenous pests, prevention, escape, or control strategies needs to be employed to manage plant-pest evolutionary processes.

23.2. DIVERSIFICATION OF CURRENT CROP PROTECTION STRATEGIES TO MITIGATE CLIMATE CHANGE IMPACTS

Current agricultural practices may need to be revisited because of climate change. IPM is a widely used strategy that integrates cultural, biological, and chemical controls to reduce harmful pest populations below a threshold which otherwise will cause economic losses. Crop damage could occur more rapidly than expected, because many pests develop more rapidly in response to rising temperatures (Trumble and Butler, 2009). In addition, as temperatures increase, the frequency of spring frosts will decline and the resulting extended frost-free periods will increase the duration and intensity of pest outbreaks. Indeed, farmers may be tempted to take advantage of the changing climate by planting crops earlier in the season.

Taking into account the uncertainty of local and temporal climate change effects in different regions of the world on plant characteristics and crop production, pest and disease epidemiology, and infection probabilities (Chakraborty, 2013), designing appropriate and reliable crop protection strategies will be a real challenge. Hence, the possible effects of climate change on pest evolution, as described earlier, raise several questions concerning the consequences for our current crop protection tools and approaches. In order to address these questions, there is a need to relook into the following aspects:

- Need to study new resistance genes with optimal expression under variable climate change situations.
- Have to collect more data on the behaviour of pests, diseases, and weeds in future circumstances and the efficacy of biological control organisms.
- Have to assess the role for alternative plant hosts and intercropping and landscape management for pests, diseases, and weeds as well as for beneficial organisms.
- Need to develop pest risk analysis models to predict the risk of new invasive pests, their chances of establishment in new cropping areas, and how to counteract disease disasters through local adaptive and resilient cropping and crop protection systems.

All these questions we currently address in order to develop robust IPM less reliant on chemical crop protection can *mutatis mutandis* be raised to develop locally adaptive and diversified IPM systems which are resilient enough to meet extreme weather fluctuations as a result of climate change.

Crop protection *per se* has an important role to play in mitigating climate change. Crop protection practices can contribute to the reduction of GHG emissions. The conventional crop production systems coupled to reduced tillage is generally the best for producing high yields to minimize GHG emissions and to contribute to global food security. Optimum control of pests is insurance for optimum exploitation of all other investments in crop production. Hence, time has come to elaborate new crop protection strategies that are robust and reliable enough for the challenges of changing climatic conditions and more stringent human and environmental protection. Some priorities and recommend action plans which are required to mitigate climate change effects in future crop protection of global relevance are presented.

23.2.1. Resilient cropping systems

Promising avenues to cropping systems resilient to extreme, variable, and unpredictable situations include genetic diversification and combinations of crop protection systems. The pest dynamics and natural enemy populations, as well as levels of pollination services for pollinator-dependent horticultural crops are influenced by spatial and temporal arrangement of crops. An example developed for late blight in potato crops relies on varietal resistance coupled with new methods to monitor loss of resistance. Likewise, the combination of native arthropod natural enemies with entomopathogens against invasive species as those reported for pin worm, *Tuta absoluta* in tomato crops could be effective (Molla *et al.*, 2011). The lack of information is acute regarding what constitutes robust or resilient cropping systems and, more generally, how cropping systems can be managed to reduce the risks associated with invasive alien as well as rapidly evolving indigenous pest species under climate change. This reflects a serious knowledge gap within a priority research area and is one which needs urgent attention. Socio-economic researchers should work on these issues to ensure that the solutions proposed are appropriate for stakeholders.

23.2.2. Breeding for and sustaining resistance

A priority area should be breeding for resistance and devising strategies to sustain that resistance. Crop varieties resistant to pests and tolerance to weeds can be developed using conventional breeding and genetic engineering. This includes *in situ* and *ex situ* conservation of genetic diversity in crop species, including their wild relatives, in order to increase the availability of genes to breed for resistance to existing and new pathogens and to abiotic stresses (Garrett, 2008).

The requirement to breed new cultivars with more durable resistances within shorter time periods will possibly give biotechnological approaches a key role in crop adaptation to changing climate. Accumulation of minor genes and race-non-specific resistance to pathogens by conventional breeding programmes will remain equally important to achieve durable resistance in diverse environments. It requires breeding for varieties that are both adapted to changing climates and those containing temperature-stable resistance genes.

23.2.3. Biological control

Research and development in biological control is another priority area which will help to adapt cropping systems to the appearance of new pests and changing climatic conditions. In particular, biological control studies in arable crops are lacking which need to be addressed to reduce the reliance on pesticides on these species and consequent development of pest resistance to pesticides. Making use of opportunistic native natural enemies against new exotic pests can also be seen as a promising avenue. For example, promoting bio-control services by using generalist natural enemies (Lu *et al.*, 2012), i.e., natural enemies able to attack a wide range of prey including those potential exotic and invasive ones, may increase the robustness of cropping systems when compared to systems relying mainly on pesticides which shows generally very poor bio-control services. There is a need for biological control agents specifically adapted to warmer and drier conditions; pro-active studies maybe developed to identify key natural enemies as possible bio-control agents for management of potential invasive pest species.

23.2.4. Pest risk analysis

Even though there are recognized limits to pest risk analysis in terms of its power to predict establishment, it remains an important early warning tool. Pest risk analysis can help to identify those pest species which present the highest risk for cropping systems. With pest risk analysis, it becomes possible to propose adaptations to cropping systems in advance or to conduct research on the ecological requirements and/or control options related to pest origins. Greater progress could be achieved with pest risk analysis, if it could be made more spatially explicit or based on "hotspots" or case studies so that site-specific conditions affecting the likelihood of establishment are taken into account. Furthermore, enhanced international information exchange with regards to completely new and exotic disease threats and pressures are foreseen as a critical element of an effective, and often urgently required pest risk analysis, e.g., under an emergency incursion situation.

23.2.5. Soil management

Agro-ecological approaches like mulching can provide refugee for natural enemies of pests. The alternate food resources for generalist natural enemies and antagonists of plant disease are provided by building organic matter in soil to enhance pest-regulating populations early in the cropping cycle. Addressing soil salinity problems can render crops less susceptible to pests such as the rice stem borer.

23.2.6. Landscape management

The impact of new forms of landscape management on crop production, and especially on limited water resources, can lead to losses if this is not designed with pests in mind. Pests can bring greater reductions in crop water use efficiency under poorly designed systems. New or modified crop and landscape management approaches such as 'push and pull'; intercropping, relay and sequential planting, border strips, and living mulches can be used for the management of pests while simultaneously conserving water resources. The incorporation of living mulches is an example of an innovative cropping system for integrated soil and pest management in cereal-based farming systems, minimizing pest infestation, sustaining permanent soil cover, and increasing soil fertility (Chabi-Olaye *et al.*, 2005). Similarly, the intercropping of trees and coffee with banana can alter pest pressure and spread (Staver *et al.*, 2001). Intercropping and mulching are effective in the management of soil-borne pests in perennial crops, by stimulating beneficial microorganisms that regulate densities of pests such as plant parasitic nematodes (Pattison *et al.*, 2003).

23.2.7. Ecosystem-based strategy

The world's leading holistic strategy for crop protection for over five decades include the agro-ecological intensification of crop protection (AICP), that is based on ecology, the concept of ecosystems and the goal of sustaining ecosystem functions (FAO, 1966; Smith and Doutt, 1971). The biological processes that underpin production are protected, encouraged and enhanced by IPM which is the first and most fundamental line of defence against pests and diseases in a healthy agro-ecosystem. It reduces input costs, increase yields and sustainability through ecological processes.

The general principles for using integrated pest management in the design of programmes for agro-ecological pest management are as follows:

23.2.7.1. Factors for pest outbreaks: A combination of factors may be involved for the pest outbreaks, and strategies may be developed accordingly to solve the problems. For instance, inappropriate plant density or ploughing that disperses weed seeds can arise due to intensification of agricultural practices. In such cases the practices will need to be modified accordingly. Similarly, the invasive pests like locusts can be managed by methods of biological control or disease suppression used in the place of origin.

23.2.7.2. Productivity risk: The appropriate scale of pest control campaigns or activities can be established based on how much production is at risk. An outbreak that demands a rapid policy response includes infestation (not loss) of more than 10 per cent of the cropped area. However, crops can to some extent compensate the risk physiologically for pest damage. Hence, the estimation of risk should not be disproportionate.

23.2.7.3. Contingency planning: The contingency planning should be implemented when a significant pest threat emerges. Some of the contingency planning includes crop-free periods to prevent the carryover of pest populations to the following season, investment in seed systems to support deployment of resistant varieties, and use of selective pesticides.

23.2.7.4. Surveillance to track pest patterns: The data from fixed plots, along with roving survey data and mapping and analysis tools, may be used for geo-referenced systems for plant pest surveillance, and the response to be adjusted.

23.2.7.5. Ecosystem approach: The intensified crop production system is likely to attract potential pest problems. Hence, a diverse production system needs to be adopted using a diverse range of pest-resistant crop varieties, crop rotations, inter-cropping, optimized planting time and weed management. The eco-friendly approaches such as beneficial species of pest predators, parasites and competitors, along with bio-pesticides and selective, low risk synthetic pesticides should be employed to overcome pest problems.

23.2.8. Human resource development

Lack of relevant expertise in the field of crop protection, both in terms of the numbers of people and their technical back ground; have been reported to severely affect prevention-based pest management. This can be seen in the area of extension services and other areas of crop protection. Action needs to be taken to reverse the decline in skilled crop protection specialists.

Advisory services currently focus on optimizing existing systems, and faced with climate change, this approach will have to change in order to concentrate on making systems more resilient. Diversifying cropping systems in general with strategies such as more varied rotations, as recommended for tackling *Diabrotica* in maize (Gray *et al.*, 2009), can be regarded as major leverage points which can be addressed via extension services. Changes will also be required regarding the role of extension within the innovation process. The involvement of multiple stakeholders, including public-private partnerships, and new, more collective approaches to extension are required to meet the challenges of building more robust cropping systems.

One approach which could be usefully employed to address shortcomings in human and financial resources is to join forces across borders, pool resources, and develop new synergistic collaborations for pest management. Examples are EUROWHEAT and EUROBLIGHT which are internet-based platforms on diseases of wheat and potato, respectively. These platforms bring together all the information available to date on these pathosystems in order to ensure the ready use of the data. In particular, new disease outbreaks, resistance data are updated on a regular basis to support effective disease control, deployment of host resistances, and breeding programmes.

23.3. POTENTIAL BENEFITS

The impacts on our natural resources such as the preservation of energy conservation and non-renewable resources, flora and fauna, and quality and diversity of landscape, are given due importance in agro-ecological pest management. It also gives boost to employment, public health, and wellbeing of persons associated with agriculture, as a long- term sociological benefits (Wijnands and Kroonen-Backbier, 1993). Agro-ecological intensification also gives emphasis on durable systems for pest resistance through natural control by the enriched natural enemy fauna, the use of weed-competitive or disease- and pest-resistant varieties, while reducing pesticide use over 90% on the integrated farms (Fig. 23.1).

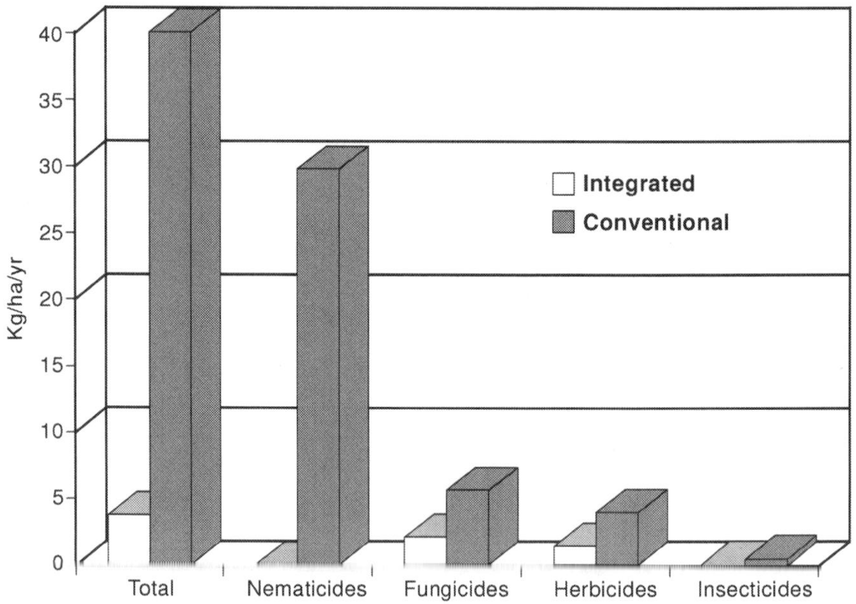

Fig. 23.1. Effect of climate change on pest management and reduction in pesticide use in conventional and integrated farming demonstrations in The Netherlands (1986–1990) (Wijnands and Kroonen-Backbier, 1993)

Significant yield increases are obtained by way of improvements in the management of herbivores, disease pathogens, and weeds. Plant breeders have successfully developed resistance to biotic stresses (pests and diseases) and deployed these defences in space and time, and maintained yield stability despite low crop diversity in continuous cereal systems (wheat, rice and corn) that provide 60% of human food. However, there is no guaranty whether such conventional breeding approaches can work indefinitely. Hence, both agro-ecological pest management and genetic engineering that identifies durable resistance through multiple gene sources should play increasingly important roles (DeVries and Toenniessen, 2001; Ortiz, 1998). The improvement in crop resistance to a pathogen is likely to be transitory, since the pathogens develop new pathogenic races to overcome the resistance (Palumbi, 2001). Likewise, pests are also known to develop resistance to pesticides. For example, within a decade of introduction of new insecticides, insects often evolve resistance to insecticides. Similarly, within 1 to 3 years of the release of many antibiotics, bacterial pathogens develop resistant strains. Likewise, herbicide-resistant weeds were observed within about one or two decades of the introduction of each of seven major herbicides (Palumbi, 2001). Use of agro-ecological practices such as crop rotation and spatial or temporal crop diversity, can overcome the problems of breeding for new disease resistance and discovering newer pesticides. In China, Zhu *et al.* (2000) reported that blast disease-susceptible rice varieties planted in mixtures with resistant varieties had 89% greater yield and blast was 94% less severe than when they were grown in monoculture. The effective elimination of a pathogen can also be achieved through varietal mixtures/multiline cultivars by intermingled planting of crop genotypes that have different disease-resistance profiles.

The insects and other beneficials living in nearby habitats or buffer strips can provide both crop pollination, and parasitoids for effective management of many agricultural pests. The weeds and other agricultural pests can also be managed through buffer strips. Landscape-level management is helpful in providing such ecosystem services.

The basic components, technologies and services, and implementation of pest management are presented in Table 23.1.

Table 23.1. The basic components, technologies and services, and implementation of pest management

Basic components	Technologies and services	Implementation of pest management
Prevention • Location • Crop rotation • Cropping pattern • Seed selection • Crop husbandry and hygiene • Fertilization • Irrigation • Habitat management • Inter- cropping • Harvesting and storage • Tillage practice	***Research and development*** • Low-dose products • Selective action • IPM positioning of broad spectrum products • Safety to people and the environment • Resistance management • Need-directed optimum use recommendations • Application technology • Biopesticides	***Training and education*** • Staff including company, government, distributors and retailers • Universities, colleges and schools • Users: plantations, food processing companies, commercial growers and smallholders • Topics include: pest and beneficial recognition, appropriate IPM strategies, product knowledge, Product safety
Observation • Crop monitoring • Decision support systems • Area-wide management	***Crop variety selection*** • Improved varieties with disease and pest resistance through genetic engineering and traditional breeding	***Multi-stakeholder partnerships*** • Corporate sector • Public sector • Scientific community

Basic components	Technologies and services	Implementation of pest management
Intervention • Cultural and physical control • Biological control • Chemical control	***Disease control*** • Fungicide technology • Diagnostics	***Technology transfer and capacity building*** • Farmers • Government research and extension • Non-government organisations • Industry
	Insect control • Insecticide technology • Pheromone • New modes of action • Band treatment	
	Weed control • Herbicide technology • Band treatment • Weed control in conservation areas	

23.4. POLICY AND REGULATORY ISSUES

Climate change is increasingly recognized as a considerable risk to agriculture in particular with respect to direct impacts on crop production and the stability of yields. It is becoming increasingly central mainly with respect to food security. Potential adaptation of existing pest populations to new environments, as well as exotic incursions of new, better-adapted pest genotypes or species, will further add to the lack of predictability and stability of current cropping systems. However, indirect losses in cropping systems due to climate change, through the probable changing impact of pest and diseases, have received very little attention or recognition compared to human or animal health issues.

In order to reduce crop losses due to either invasive or rapidly evolving resident pests under changing climate, the further integration of the roles of plant health and crop protection specialists for the creation of more resilient cropping systems appears increasingly relevant. Such recognition, although still rather tentative and subject to further policy integration and elaboration, is consistent within the framework of adaptive strategy needs (European Commission, 2009a). Nevertheless, specific actions remain to be outlined.

The impact of climate change, globalization, and increased intra-community movement on the introduction and spread of invasive alien and existing pests, respectively, is considered as one of the central concerns of the current revision of the common plant health regime. In particular, there is an emphasis on the sustainability with regard to the potential increased pressures on cropping systems and yields and overall food security. Recent and potentially future legislation responding to increasingly stringent environmental and human health concerns (European Commission, 2009b,c) and their respective implications, while fuelling innovation toward environmental and health sustainability may add further practical and operational challenges with regard to maintaining the same level of crop yields. It is unclear, however, how both pest evolution and legislative scenarios will be changed in the coming future given the uncertainty on their medium- and longer-term individual and interactive impacts. Of equal concern is also their possible impact on cropping system resilience and the capacity to adapt.

23.5. FUTURE LINES OF WORK

The implementation of crop protection is limited by the "business as usual" approach for pest management, still followed in many countries and by many farmers. The indigenous pest outbreaks due to pest invasions and risks from pesticides to both human health and the environment can be prevented by improvements in agro-ecosystem pest management. Hence, development of a pest management strategy that can be sustained and does not produce adverse side effects is the need of the hour. The policy and institutional changes are needed in the form of removal of pesticide subsidies, tighter enforcement of pesticide regulations, and incentives for local production of agro-ecological inputs, such as insectaries for natural predators, to support agro-ecological intensification in the long term.

In the registration processes, less hazardous pesticides should be given preference by countries, ensuring that the farmers only apply ecologically informed decision-making to determine which pesticides may be sold and used, by whom and in what situations. The subsidies for development and adoption of alternative pest management practices should be financed from pesticide-use fees or pesticide taxes.

There is a scope for development of small scale local industries in view of large-scale adoption of agro-ecological pest management, which is expected to increase demand for bio-control agents such as predators, parasitoids or sterile organisms, biopesticides, commercial monitoring tools, microorganisms, and pollination services. The expansion of local industry should take place with a shift to a more ecosystem-centric approach by producing quality bio-products based on bacteria, viruses, fungi, protozoa and nematodes (Ruttan, 1999).

A more consistent and reliable supply of agricultural produce free of pesticide residues obtained from more stable and sustainable agro-ecosystems should be the perspective for the food processing industry. The farmers' can access new markets by labelling their food products to enhance additional farm income.

The effective advisory services, links to research that respond to farmers' needs, support to the provision of IPM inputs, and effective regulatory control of chemical pesticide distribution and sale, are the pre-requisites for sustaining agro-ecological strategies. The farmer field school (FFS) approach (that supports local learning and encourages farmers), is one of the most effective means of promoting agro-ecological pest management at local level. The transfer of technology can be accelerated by supplying inputs, and using cellular phones to supplement traditional methods of outreach, such as extension, media campaigns and local inputs dealers.

23.5.1. Research and development

The following research and development aspects need to be emphasized:

- Research should be focused on impact of climate change on range shift of the crop pests and extent of crop damages/losses caused by them, the pest may be native or new or invasive. Such studies should be carried out under both simulated (climate chamber) as well as field conditions as long term experiments.
- High yielding crop varieties resilient to climate changes, having notable tolerance/ resistance to biotic stresses (pests) need to be developed/ improved for the targeted crop production/ productivity through multidisciplinary team of researchers.
- There is a need to study the crop-pest-natural enemy-weather relationship in important food crops (agricultural/ horticultural) under changing climatic conditions and develop pest forewarning modules for specific key pests of agricultural importance, using both the historical and experimentally generated data.
- Due to shifts in crop cultivation practices (low inputs production technologies) as well as changing climate, crop pest scenario is being changed, therefore, it is important to work out/develop effective pest surveillance procedures (visual, light trap, pheromone trap or e-pest surveillance) for timely pest management of the crops.
- There is a need to develop effective and climate resistant strains of bio-control agents for accretive and inundative releases. Laboratory reared bio-control agents should be made more heat/temperature tolerant and virulent.
- Research on botanical pesticides especially identification of the active components of the effective botanical pesticides, developing their active formulations and their evaluation against target pests, and also validation of the botanicals being used under indigenous technological knowledge (ITKs) may be strengthened. Exploitation for the biofumigants for the management of insects infesting stored food grains as synthetic fumigants have been recommended only under expert supervision.
- Economic threshold levels (ETLs) of the pests need to be reset taking care of the scenario of climate change as well as kinds of pesticides to be used. ETLs in case of mites, nematodes and diseases should not be considered and prophylactic measures should be given priority.
- Generation of better data and models related to dispersal of the pathogens, current levels of interspecific diversity, strength of selection under different climate change scenarios and heritability of traits are needs of the hour for a better output.
- Pest control measures have to be focused on resistant/tolerance genotypes, locally adoptable, more virulent biocontrol agents and reconstruct of IPM modules are required having integrated and holistic approach using minimum but cost effective interventions.
- Build capacity of plant health organisations, as key partners at the front-line against crop pests, in support of food security, trade and policy implementation.

- Improve data gathering, centred on better understanding of critical metrics of crop pest impacts, extending beyond yield loss and encompassing economic and investment factors to improve prediction capabilities.
- Enhance the understanding of risk behaviours of food chain stakeholders and their willingness to invest and adapt, to support increased adoption of technologies.
- Improve fusion of disparate datasets and risk modelling of crop pest consequences for projections on farmer choices and cropping system at the landscape scale, informed by markets and policy.
- Invest in pre-emptive crop pest resistance breeding against future high risk pests based on current and future geographic distributions in order to avoid the consequences of shock pest episodes.
- Take cognizance of the boundary-less nature of crop pests, and develop regional supra-governmental capabilities for the analysis and articulation of horizon scanning and crop pest risk concerns under climate change.
- Predict and map trends of potential changes in geographical distribution, and study how climatic changes will affect development, incidence, and population dynamics of pests.
- Understand the influence of global warming and climate change on species diversity and cropping patterns, and their influence on the abundance of pests and their natural enemies.
- Understand the changes in expression of resistance to pests, and identify stable sources of resistance, and pyramid the resistance genes in commercial cultivars.
- Study the effect of global warming on the efficacy of transgenic crops in pest management.
- Assess the efficacy of various pest management technologies under diverse environmental conditions, and develop appropriate strategies for pest management to mitigate the effects of climate change.

23.5.2. Training and education

- Development of appropriate programmes and media for in-house agro-ecological intensification training for management, marketing, sales, technical and other relevant staff.
- Support and promotion of practical agro-ecological intensification training programmes for advisory services, agricultural schools, dealers, distributors, farmers, and school children.

23.5.3. Policy support

The support of policymakers is essential to support agro-ecological pest management programmes at a local, regional or national scale. However, the success of agro-ecological intensification for the management of pests and diseases ultimately rests with farmers who make key management decisions. The following aspects need the support of policymakers in implementing agro-ecological intensification:

- The development of ecologically based pest management practices and their adoption by the farmers based on their local knowledge, social learning networks and conditions; require technical assistance from researchers and extension support for transfer of technology.
- The areas like biological control, host plant resistance to pests and diseases, innovative approaches to field pest management, practical monitoring and surveillance methods, and the use of selective pesticides (including biopesticides) need targeted research.
- International Code of Conduct on the Distribution and Use of Pesticides require private sector regulation, including effective systems of governance for the registration and distribution of pesticides.
- Policy and institutional changes are needed for removal of perverse incentives such as pesticide price or transport subsidies, preferential tariffs for pesticides, and unnecessary maintenance of pesticide stocks which encourages their use.

23.6. CONCLUSIONS

Today's challenge is to 'produce more from less'. In the era of climate change, diagnostics of pests and diseases and capacity building of farmers, extension and even research personnel for adaptation to changed pest scenario under future climates assumes significance, wherein integrated crop management in private public partnership mode could be very effective. Farmers' decisions are of vital importance for good yields of crops. Forecasted weather products and area-wide weather networks are becoming more prevalent. Now, the challenge is to bring continuous improvement in productivity, profitability, stability and sustainability of major farming systems, wherein scientific management of plant pests holds a pivotal role (Swaminathan, 1995). Crop loss models, representing a dynamic interaction between pests and host, are essential for forecasting losses thereof. Accurate information concerning possible yield losses due to occurrence of a pest is needed by growers or crop protection specialists to decide on cost-effective control measures. With an increasing concern for cleaner environment and discouragement for pesticide use, there is need to approach pest management through knowledge on their dynamics as an art of living with them (Zadoks, 1985). Thus, future research and education in crop protection does need to address the issue of future climates in pest management, for which fund requirement would certainly be lesser than many ambitious ones.

Hence, more emphasis should be given for the development of farming practices that are compatible with ecological systems and designing cropping systems that naturally limit the elevation of an organism to pest status is the need of the hour. The sustainable and profitable pest management strategies work more in harmony with nature's checks and balances, which are beneficial to all participants in the ecosystem, including humans.

References

A'Bear, A.D., Jones, T.H. and Boddy, L. (2014). Potential impacts of climate change on interactions among saprotrophic cord-forming fungal mycelia and grazing soil invertebrates. *Fungal Ecology* 10: 34-43.

Abolmaaty, S.M., Khalil, A.A. and Amna, M.H. (2011). Using degree- day unit accumulation to predict potato tuber worm incidence under climate change conditions in Egypt. *Nat. Sci.* 9(4): 156-160.

Abril, A., Oliveras, J. and Gomez, C. (2008). Effect of temperature on the development and survival of the Argentine ant, *Linepithema humile*. *Journal of Insect Science* 10: 97.

Adamczyk, J.J. Jr., Adams, L.C. and Hardee, D.D. (2001). Field efficacy and seasonal expression profiles for terminal leaves of single and double *Bacillus thuringiensis* toxin cotton genotypes. *J Econ Ento* 94: 1589–1593.

Adaros, G., Weigel, H.J. and Jäger, H.J. (1991). Growth and yield of spring rape and spring barley as affected by chronic ozone stress. *J. Plant Dis. Destruc* 98: 513–525.

Aerts, M., Cockrell, P., Nuessly, G., Raid, R., Schueneman T. and Seal D. (1999). *Crop Profile for Corn (Sweet) in Florida*. http://www.ipmcenters.org/CropProfiles/docs/FLcorn-sweet.html.

Agbamu, J.U. (2006). *Essentials of Agricultural Communication in Nigeria*. Malthouse Press Ltd., Lagos.

Aggarwal, P.K. (2008). Climate change and Indian agriculture: impacts, adaptation and mitigation. *Indian J. Agric. Sci.* 78: 911-919.

Aggarwal, P.K., Kalra, N., Chander, S. and Pathak, H. (2006). InfoCrop: a dynamic simulation model for the assessment of crop yields, losses due to pests, and environmental impact of agro-ecosystems in tropical environments. I. Model description. *Agricultural Systems* 89: 1-25.

Agrios, G.N. (2005). *Plant Pathology*, 5th edn. Elsevier, London, UK.

Ahmed, S., Naseer, A. and Fiaz, S. (2005). Comparative efficacy of botanicals and insecticides on termites in sugarcane at Faisalabad. *Pak Entomol* 27(1): 23-25.

Ainsworth, E.A., Beier, C., Calfapietra, C. *et al.* (2008). Next generation of elevated CO_2 experiments with crops: a critical investment for feeding the future world. *Plant, Cell and Environment* 31(9): 1317-1324.

Ainsworth, E.A. and Long, S.P. (2005). What have we learned from 15 years of free-air CO_2 enrichment (FACE)? A meta-analytic review of the responses of photosynthesis, canopy properties and plant production to rising CO_2. *New Phytol.* 165: 351–372.

Alam, G. (1994). *Biotechnology and Sustainable Agriculture: Lessons from India*. Technical paper 103, OECD Development Centre, Paris.

Alberto, A.M., Ziska, L.H., Cervancia, C.R. and Manalo, P.A. (1996). The influence of increasing carbon dioxide and temperature on competitive interactions between a C3 crop, rice (*Oryza sativa*) and a C4 weed (*Echinochloa glabrescens*). *Australian Journal of Plant Physiology* 23: 795-802.

Almeida, R.P.P., Wistrom, C., Hill, B.L., Hashim, J. and Purcell, A.H. (2005). Vector transmission of *Xylella fastidiosa* to dormant grape. *Plant Disease* 89: 419-424.

Altieri, M.A., Ponti, L. and Nicholls, C. (2005). Enhanced pest management through soil health: Toward a belowground habitat management strategy. *Biodynamics* (*Summer*), pp. 33–40.

Aluja, M., Guillén, L., Rull, J. *et al.* (2011). Is the alpine divide becoming more permeable to biological invasions? Insights on the invasion and establishment of the Walnut Husk Fly, *Rhagoletis completa* (Diptera: Tephritidae) in Switzerland. *Bull Entomol Res* 101: 451–465.

Alves, M.C., Carvalho, L.G., Pozza, E.A., Sanches, L. and Maia, J.C.S. (2011). Ecological zoning of soybean rust, coffee rust and banana black sigatoka based on Brazilian climate changes. *Procedia Environ Sci* 6: 35-49.

Amerika Singh, Trivedi, T.P., Sardana, H.R. *et al.* (2004). Integrated pest management in horticultural crops – A wide area approach. In: K.L. Chadha, B.S. Ahluwalia, K.V. Prasad and S.K. Singh (eds.) *Crop Improvement and Production Technology of Horticultural Crops*. Hort Soc of India, New Delhi, pp. 621-636.

Amthor, J.S. (1995). Terrestrial higher-plant response to increasing atmospheric CO_2 in relation to the global carbon cycle. *Glob. Change Biol* 1: 243–274.

Amtmann, A., Troufflard, S. and Armengaud, P. (2008). The effect of potassium nutrition on pest and disease resistance in plants. *Physiol. Plant.* 133: 682-691.

Anand, T. and Pereira, G.N. (2008). *Butterflies-The Flying Jewels of the Western Ghats.* http://www.daijiworld.com/chan/exclusive_arch.asp?ex_id=1003, Accessed on 28/12/2010.

Anderson, D.M., Swanton, C.J., Hall, J.C. and Mersey, B.G. (1993). The influence of temperature and relative humidity on the efficacy of glufosinate-ammonium. *Weed Res.* 33: 139–147.

Anderson, J.P., Badruzsaufari E., Schenk P.M. *et al.* (2004). Antagonistic interaction between abscisic acid and jasmonate-ethylene signaling pathways modulates defense gene expression and disease resistance in arabidopsis. *Plant Cell* 16: 3460-3479.

Anderson, P., Cunningham, A.A., Patel, N.G. *et al.* (2004). Emerging infectious diseases of plants: pathogen pollution, climate change and agrotechnology drivers. *Trend Ecol Evol* 19: 533–544.

Andersen, R.N., Menges, R.M. and Conn, J.S. (1985). Variability in velvetleaf (*Abutilon theophrasti*) and reproduction beyond its current range in North America. *Weed Sci* 33: 507–512.

Andreasen, C. and Skovgaard, I.M. (2009). Crop and soil factors of importance for the distribution of plant species on arable fields in Denmark. *Agric Ecosyst Environ* 133: 61–67.

Andrew, N.R. and Hughes, L. (2005). Diversity and assemblage structure of phytophagous Hemiptera along a latitudinal gradient: predicting the potential impacts of climate change. *Global Ecology and Biogeography* 14: 249-262.

Angert, A.L., Crozier, L.G., Rissler, L.J. *et al.* (2011). Do species' traits predict recent shifts at expanding range edges? *Ecol Lett* 14: 677–689.

Anjorin, S.T. (2008). A survey of crop seed protection with botanicals in the FCT – Abuja, Nigeria. *J. Sustainable Trop. Agric. Res.* 24: 35-40.

Anjorin, S.T., Salako, E.A. and Ndana, R.W. (2004). *In vitro* assessment of some plant leaf extracts for the control of *Meloidogyne* sp. and *Rhizoctonia solani*. *Zuma J. Pure Appl. Sci.* 7(1): 2005.

Anon. (1992). *Neem: A Tree for Solving Global Problems.* Report of an Ad-hoc Panel of the Board of Science and Technology for International Development, National Research Council, National Academy Press, Washington D.C.

Anon. (2008). *Food Matters towards a Strategy for the 21st Century.* http://www.cabinetoffice.gov.uk/media/cabinetoffice/strategy/assets/food/food_matters_es.pdf.

Anwar, M.R., O'Leary, G., McNeil, D., Hossain, H. and Nelson, R. (2007). Climate change impact on rainfed wheat in south-eastern Australia. *Field Crops Research* 104: 139–147.

Archambault, D.J., Li, X., Robinson, D. *et al.* (2001). The Effects of Elevated CO_2 and Temperature on Herbicide Efficacy and Weed/Crop Competition. Agriculture and Agri-Food Canada.

Arujo, M.T., Dickson, D.W., Augustine, J.J. and Bassette, M.J. (1983). Reproduction of two races of *Meloidogyne incognita* in tomato plants grown at high temperature. *Journal of Nematology* 15: 640-641.

Asch, F., Dingkuhn, M., Sow, A. and Audebert, A. (2005). Drought-induced changes in rooting patterns and assimilate partitioning between root and shoot in upland rice. *Field Crops Res.* 93: 223–236.

Asin, L. and Pons, X. (2001). Effect of high temperature on the growth and reproduction of corn aphids (Homoptera: Aphididae) and implications for their population dynamics on the Northeastern Iberian Peninsula. *Environ. Entomol.* 30:1127–1134.

Assad, E.D., Pinto, H.S., Zullo Junior, J. and Avila, A.M.H. (2004). Impacto das mudancas climaticas no zoneamento agroclimatico do cafeno Brasil. *Pesqui Agrop Brasil* 39: 1057–1064.

Asselbergh, B., De Vleesschauwer, D. and Höfte, M. (2008). Global switches and fine-tuning ABA modulates plant pathogen defense. *Mol. Plant-Microbe Interact* 21: 709–719.

Atibalentja, N., Noel, G.R., Liao, T.F. and Gertner, G.Z. (1998). Population changes in *Heterodera glycines* and its bacterial parasite *Pasteuria* sp. in naturally infested soil. *Journal of Nematology* 30: 81-92.

Atkinson, H.J., Urwin, P.E., Clarke, M.C. and McPherson M.J. (1996). Image analysis of the growth of *Globodera pallida* and *Meloidogyne incognita* on transgenic tomato roots expressing cystatins. *Journal of Nematology* 28: 209-215.

Atkinson, N.J. and Lilley, C.J. and Urwin, P.E. (2013). Identification of genes involved in the response of Arabidopsis to simultaneous biotic and abiotic stresses. *Plant Physiol.* 162: 2028-2041.

Atkinson, N.J. and Urwin, P.E. (2012). The interaction of plant biotic and abiotic stresses: from genes to the field. *J Exp Bot.* 63(10): 3523-3543.

Audebert, A., Coyne, D.L., Dingkuhn, M. and Plowright, R.A. (2000). The influence of cyst nematodes (*Heterodera sacchari*) and drought on water relations and growth of upland rice in Cote d'Ivoire. *Plant and Soil* 220: 235-242.

Augusto, J., Brenneman, T.B., Culbreath, A.K. and Sumner, P. (2010). Night spraying peanut fungicides II. Application timings and spray deposition in the lower canopy. *Plant Disease* 94: 683–689.

Avery, D.T. (1995). *Saving the Planet with Pesticides and Plastic.* Hudson Inst., Indianapolis, 432 pp.

Awmack, C.S. and Harrington, R. (2000). Elevated CO_2 affects the interactions between aphid pests and host plant flowering. *Agricultural and Forest Entomology* 2(1): 57 - 61.

Awmack, C.S., Woodcock, C.M. and Harrington, R. (1997). Climate change may increase vulnerability of aphids to natural enemies. *Ecological Entomology* 22: 366–368.

Awmack, C.S., Woodcock, C.M., Harrington, R. and Lether, S.R. (1997). Host plant effects on performance of the aphid *Aulacorthum solani* (Kalt) (Homoptera: Aphididae) at ambient and elevated CO_2. *Global change Biol.* 3: 545-559.

Ayres, E., Wall, D.H., Simmons, B.L. *et al.* (2008). Below-ground nematode herbivores are resistant to elevated atmospheric CO_2 concentrations in grassland ecosystems. *Soil Biol Biochem* 40: 978–985.

Ayres, M.P. and Lombardero, M.J. (2000). Assessing the consequences of global change for forest disturbance from herbivores and pathogens. *Science of the Total Environment* 262: 263- 286.

Baessler, C. and Klotz, S. (2006). Effects of changes in agricultural land-use on landscape structure and arable weed vegetation over the last 50 years. *Agric Ecosyst Environ* 115: 43–50.

Bailey, S.W. (2004). Climate change and decreasing herbicide persistence. *Pest Manage. Sci.* 60: 158–162.

Baker, K.M., Lake, T., Benston, S.F., Trenary, R., Wharton, P., Duynslager, L. and Kirk, W. (2015). Improved weather-based late blight risk management: comparing models with a ten year forecast archive. *J Agric Sci* 153: 245-256.

Baker, R.H.A., Cannon, R.J.C. and MacLeod, A. (2003). Predicting the potential distribution of alien pests in the UK under global climate change: *Diabrotica virgifera virgifera. The Proceedings of BCPC International Congress – Crop Science & Technology Volume 2.* SECC, Glasgow, Scotland, UK, pp. 1201–1208.

Baker, R.H.A., Cannon, R.J.C. and Walters, K.F.A. (1996). An assessment of the risks posed by selected nonindigenous pests to UK crops under climate change. In: R.J. Froud-Williams, R. Harrington, T.J. Hocking, H.G. Smith and T.H. Thomas (eds.) *Implications of "Global Environment Change" for Crops in Europe. Aspects of Applied Biology* 45: 323-330.

Baker, R.H.A., MacLeod, A., Cannon, R.J.C. *et al.* (1998). Predicting the impacts of a non-indigenous pest on the UK potato crop under global climate change: reviewing the evidence for the Colorado beetle, *Leptinotarsa decemlineata. Brighton Crop Protection Conference - Pests and Diseases, Vol. III.* BCPC, Surrey, UK, pp. 979-984.

Baker, R.H.A., Sansford, C.E., Jarvis, C.H. *et al.* (2000). The role of climatic mapping in predicting the potential geographical distribution of non-indigenous pests under current and future climate. *Agric Ecosyst Environ* 82: 57-71.

Bakker, J., Gommers, F.J., Smits, L. *et al.* (1983). Photoactivation of isoflavonoid phytoalexins: involvement of free radicals. *Phytochem. Phytobiol.* 38: 323–329.

Bale, J., Masters, G.J., Hodkins, I.D. *et al.* (2002). Herbivory in global climate change research: direct effects of rising temperature on insect herbivores. *Global Change Biol.* 8: 1-16.

Bale, J.S. and Hayward, S.A.L. (2010). Insect overwintering in a changing climate. *The Journal of Experimental Biology* 213: 980-994.

Balsubramanian, A., Mahadevan, N.R., Venugopal, M.S. and Baskaran, R.K.M. (1998). Influence of intercropping on infestation of early season sucking pests of cotton (*Gossypium hirsutum*). *Indian Journal of Agricultural Sciences* 68: 315-316.

Baltrus, D.A. (2013). Exploring the costs of horizontal gene transfer. *Trend Ecol Evol* 28: 489–495.

Barbehenn, R.V., Karowe, D.N. and Chen, Z. (2004). Performance of a generalist grasshopper on a C3 and a C4 grass: compensation for the effects of elevated CO_2 on plant nutritional quality. *Oecologia* 140: 96–103

Barbetti, M.J., Banga, S.S., Salisbury, P.A. (2012). Challenges for crop production and management from pathogen biodiversity and diseases under current and future climate scenarios—case study with oilseed Brassicas. *Field Crop Res* 127: 225–240.

Bardossy, A. (1997). Downscaling from GCMs to local climate through stochastic linkages. *Journal of Environmental Management* 49: 7-17.

Bariana, H.S., Brown, G.N., Bansal, U.K. *et al.* (2007). Breeding triple rust resistant wheat cultivars for Australia using conventional and marker-assisted selection technologies. *Australian Journal of Agricultural Research* 58: 576–587.

Barker, K.R. and Imbriani, J.L. (1984). Nematode advisory programs—status and prospects. *Plant Dis.* 68: 735–741.

Barnes, A.P., Wreford, A., Butterworth, M.H. *et al.* (2010). Adaptation to increasing severity of phoma stem canker on winter oilseed rape in the UK under climate change. *J Agric Sci* 148: 683-694.

Barrett, S.C.H. (2000). Micro-evolutionary influences of global changes on plant invasions. *In:* H.A. Mooney and R.J. Hobbs (eds.) *Invasive Species in a Changing World.* Island Press, Washington D.C., pp. 115–139.

Bartoli, C., Lamichhane, J.R., Berge, O., Guilbaud, C., Varvaro, L., Balestra, G.M., Vinatzer, B.A. and Morris, C.E. (2015). A framework to gage the epidemic potential of plant pathogens in environmental reservoirs: the example of kiwifruit canker. *Mol Plant Pathol.* 16: 137–149.

Battisti, A., Stastny, M., Netherer, S. *et al.* (2005). Expansion of geographic range in the pine processionary moth caused by increased winter temperatures. *Ecol Appl* 15: 2084–2096.

Baxter, R., Ashenden, T.W., Sparks, T.H. and Farrar, J.F. (1994). Effects of elevated CO_2 on three montane grass species. I. Growth and dry matter partitioning. *Journal of Experimental Botany* 45: 305-315.

Bazzaz, F.A., Garbutt, K. and Williams, W.E. (1985). Effect of increased atmospheric carbon dioxide concentration on plant communities. In B.R. Strain and J.D. Cure (eds.) *Direct Effects of Increasing Carbon Dioxide on Vegetation.* United States Department of Energy, ashington, DC, p. 168.

Beaumont, L.J., Gallagher, R.V., Thuiller, W., Downey, P.O., Leishman, M.R. and Hughes, L. (2009). Different climatic envelopes among invasive populations may lead to underestimations of current and future biological invasions. *Diversity and Distributions* 15: 409-420.

Beaumont, L.J., Hughes, L. and Pitman, A.J. (2008). Why is the choice of future climate scenarios for species distribution modelling important? *Ecology Letters* 11: 1135– 1146.

Bedos, C., Cellier, P., Calvet, R., Barriuso, E. and Gabrielle, B. (2002). Mass transfer of pesticides into the atmosphere by volatilization from soils and plants: overview. *Agronomie* 22: 21–33.

Beed, F., Benedetti, A., Cardinali, G. *et al.* (2011). *Climate Change and Micro-Organism Genetic Resources for Food and Agriculture: State of Knowledge, Risks and Opportunities.* Commission on Genetic Resources for Food and Agriculture, FAO. doi: http://www.fao.org/ docrep/ meeting/022/mb392e.pdf

Beestman, G.B. and Deming, J.M. (1974). Dissipation of acetanilide herbicides from soils. *Agron. J.* 66: 308–311.

Bekal, S., Craig, J.P., Hudson, M.E. *et al.* (2008). Genomic DNA sequence comparison between two inbred soybean cyst nematode biotypes facilitated by massively parallel 454 micro-bead sequencing. *Mol. Genet. Genom.* 279: 535-543.

B´elair, G. and Benoit, D.L. (1996). Host suitability of 32 common weeds to *Meloidogyne hapla* in organic soils of southwestern Quebec. *J. Nematol.* 28(4S): 643–647.

Belote, R.T., Weltzin, J.F. and Norby, R.J. (2003). Response of an understory plant community to elevated CO_2 depends on differential responses of dominant invasive species and is mediated by soil water availability. *New Phytol.* 161: 827–835.

Benson, D.M. and Barker, K.R. (1985). Nematodes—a threat to ornamental plants in the nursery and landscape. *Plant Dis.* 69: 97–100.

Bergmann, J., Pompe, S., Ohlemüller, R. *et al.* (2010). The Iberian Peninsula as a potential source for the plant species pool in Germany under projected climate change. *Plant Ecol* 207: 191–201.

Bergot, M., Cloppet, E., Pérarnaud, V. *et al.* (2004). Simulation of potential range expansion of oak disease caused by *Phytophthora cinnamomi* under climate change. *Glob. Change Biol.* 10:1539–1552.

Berry, P.M., Kindred, D.R. and Paveley, N.D. (2008). Quantifying the effects of fungicides and disease resistance on greenhouse gas emissions associated with wheat production. *Plant Pathology* 57: 1000–1008.

Bezemer, T.M. and Jones, T.H. (1998). Plant-insect herbivore interactions in elevated atmospheric CO_2; quantitative analyses and guild effects. *Oikos* 82: 212-222.

Bezemer, T.M., Jones T.H. and Knight K.J. (1998). Long-term effects of elevated CO_2 and temperature on populations of the peach potato aphid *Myzus persicae* and its parsitoid *Aphidius matricariae*. *Oecologia* 116: 128–135.

Bhanu Priya, D., Somasekhar, N., Prasad, J.S. and Kirti, P.B. (2011). Transgenic tobacco plants constitutively expressing Arabidopsis NPR1 show enhanced resistance to root-knot nematode, *Meloidogyne incognita*. *BMC Research Notes* 4: 231-235.

Bhattarai, K.K., Xie, Q.G., Mantelin, S. *et al.* (2008). Tomato susceptibility to root-knot nematodes requires an intact jasmonic acid signalling pathway. *Mol Plant Microbe Interac* 21: 1205–1214.

Bird, A.F. (1974). Plant response to root-knot nematode. *Annual Review of Phytopathology* 12: 69-85.

Blanford, S. and Thomas, M.B. (1999). Host thermal biology: the key to understanding host–pathogen interactions and microbial pest control? *Agricultural and Forest Entomology* 1(3): 195-202.

Blankinship, J.C., Niklaus, P.A. and Hungate, B.A. (2011). A meta-analysis of responses of soil biota to global change. *Oecologia* 165: 553-565.

Blenkinsop, S. and Fowler, H.J. (2007). Changes in European drought characteristics projected by the PRUDENCE regional climate models. *Int. J. Climatol.* 27: 1595–1610.

Bloomfield, J.P., Williams, R.J., Gooddy, D.C. *et al.* (2006). Impacts of climate change on the fate and behaviour of pesticides in surface and groundwater – a UK perspective. *Science of the Total Environment* 369: 163–177.

Blumenthal, D., Chimner, R.A., Welker, J.M. and Morgan, J.A. (2008). Increased snow facilitates plant invasion in mixed grass prairie. *New Phytologist* 179: 440-448.

Boag, B., Crawford, J.W. and Neilson, R. (1991). The effect of potential climatic changes on the geographical distribution of the plant parasitic nematodes *Xiphinema* and *Longidorus* in Europe. *Nematologica* 37: 312–323.

Boese, S.R., Wolfe, D.W. and Melkonian, J. (1997). Elevated CO_2 mitigates chilling-induced water stress and photosynthetic reduction during chilling. *Plant Cell Environ.* 20: 625–632.

Bokhorst, S., Huiskes, A., Convey, P. *et al.* (2008). Climate change effects on soil arthropod communities from the Falkland Islands and the Maritime Antarctic. *Soil Biology and Biochemistry* 40: 1547-1556.

Boland, G.J., Melzer, M.S., Hopkin, A. *et al.* (2004). Climate change and plant diseases in Ontario. *Canadian Journal of Plant Pathology* 26: 335–350.

Bonan, G.B., Oleson, K.W., Vertenstein, M. *et al.* (2002). The land surface climatology of the community land model coupled to the NCAR community climate model. *J. Clim.* 15: 3123–3149.

Bongers, T. and Bongers, M. (1998). Functional diversity of nematodes. *Appl. Soil Ecol.* 10: 239-251.

Bonner, M.J. and Schmitt, D.P. (1985). Population dynamics of *Heterodera glycines* life courses on soybean. *J. Nematol.* 17: 153-158.

Boonham, N., Glover, R., Tomlinson, J. *et al.* (2008). Exploiting generic platform technologies for the detection and identification of plant pathogens. *Eur J Plant Pathol* 121: 355-363.

Boonekamp, P.M. (2012). Are plant diseases too much ignored in the climate change debate? *Eur. J. Plant Pathol.* 133: 291–294.

Booth, B.D. and Swanton, C.J. (2002). Assembly theory applied to weed communities. *Weed Sci* 50: 2–13.

Booth, T.H. (1990). Mapping regions climatically suitable for particular tree species at the global scale. *Forest Ecology and Management* 36: 47–60.

Booth, T.H., Jovanovic, T., Old, K.M. and Dudzinski, M.J. (2000). Using climatic mapping programs to identify high risk areas for *Cylindrocladium quinqueseptatum* leaf blight on eucalypts in mainland South East Asia and around the world. *Environmental Pollution* 108: 365 – 372.

Bough, M.A., Colosi, J.C. and Cavers, P.B. (1986). The major weedy biotypes of proso millet (*Panicum miliaceum* L.) in Canada. *Canadian Journal of Botany* 64: 1188–1198.

Bourchier, R.S. and Van Hezewijk, B.E. (2010). Distribution and potential spread of Japanese knotweed (*Polygonum cuspidatum*) in Canada relative to climatic thresholds. *Invasive Plant Science and Management* 3: 32–39.

Bowes, G. (1993). Facing the inevitable: plants and increasing atmospheric CO_2. *Annual Review of Plant Physiology and Plant Molecular Biology* 44: 309-332.

Boyer, J.S. (1995). Biochemical and biophysical aspects of water deficits and the predisposition to disease. *Annu Rev Phytopathol* 33: 251–274.

Bradley, B. and Mustard, J.F. (2005). Identifying land cover variability distinct from land cover change: cheat grass in the Great Basin. *Remote Sens Environ.* 94: 204–213.

Bradley, B.A. (2013). Distribution models of invasive plants over-estimate potential impact. *Biol Invasions* 15: 1417–1429.

Bradley, B.A., Blumenthal, D.M., Wilcove, D.S. and Ziska, L.W. (2010). Predicting plant invasions in an era of global change. *Trends in Ecology and Evolution* 25: 310-318.

Bradley, B.A., Oppenheimer, M. *et al.* (2009). Climate change and plant invasions: restoration opportunities ahead? *Global Change Biology* 15(6): 1511-1521.

Bradshaw, W.E. and Holzapfel, C.M. (2011). Genetic shift in photoperiodic response correlated with global warming. *Proc Natl Acad Sci USA* 98: 14509–14511.

Braga, M.R., Aidar, M.P.M., Marabesi, M.A. and de Godoy, J.R.L. (2006). Effects of elevated CO_2 on the phytoalexin production of two soyabean cultivars differing in the resistance to stem canker disease. *Env Exp Bot* 58: 85-92.

Brasier, C.M. (1992). Oak tree mortality in Iberia. *Nature* 360: 539.

Brasier, C.M. (1996). *Phytophthora cinnamomi* and oak decline in southern Europe: environmental constraints including climate change. *Annals of Science Forestry* 53: 347–358.

Brasier, C.M. (2001). Rapid evolution of introduced plant pathogens via interspecific hybridization. *BioScience* 51: 123–133.

Brasier, C.M., Cooke, D.E.L. and Duncan, J.M. (1999). Origin of a new *Phytophthora* pathogen through interspecific hybridization. *Proc. Natl. Acad. Sci. USA* 96: 5878–5883.

Brasier, C.M., Jones, B. and Webber, J.F. (2010). New tree hosts and new disease etiologies beneath larch (Larix) infected by *Phytophthora ramorum* in the UK. *Abstracts: 5th IUFRO Phytophthoras in Forests and Natural Ecosystems*, Auckland and Rotorua, New Zealand, 25 pp.

Brasier, C.M. and Scott, J.K. (1994). European oak declines and global warming: a theoretical assessment with special reference to the activity of *Phytophthora cinnamomi. EPPO Bulletin* 24: 221–232.

Bregaglio, S., Donatelli, M. and Confalonieri, R. (2013). Fungal infections of rice, wheat, and grape in Europe in 2030–2050. *Agron Sustain Dev* 33: 767-776.

Breitsameter, L., Bürger, J., Peters, K. *et al.* (2014). Klimafolgenforschung zu Ackerunkräutern – Daten, Methoden und Anwendungen auf verschiedenen Skalen. *Julius-Kühn-Archiv* 443: 123–132.

Bridge, J. (1996). Nematode management in sustainable and subsistence agriculture. *Annu. Rev. Phytopathol.* 34: 201–225.

Bridge, P., Jeffries, P., Morse, D.R. and Scott, P.R. (eds.). (1998). *Information Technology, Plant Pathology and Biodiversity.* CAB International, Wallingford, UK.

Brinkley, T.R. and Bomford, M. (2002). *Agricultural Sleeper Weeds: What is the Potential Threat?* Bureau of Rural Sciences, Kingston, ACT.

Broennimann, O., Treier, U.A., Müller-Shärer, H. *et al.* (2007). Evidence of climatic niche shift during biological invasion. *Ecol Lett* 10: 701–709.

Brosi, G.B., McCulley, R.L., Bush, L.P. *et al.* (2010). Effects of multiple climate change factors on the tall fescue–fungal endophyte symbiosis: infection frequency and tissue chemistry. *New Phytologist* 189: 797–805.

Browder, L.E. and Eversmeyer, M.G. (1986). Interactions of temperature and time with some *Puccinia recondita: Triticum* corresponding gene pairs. *Phytopathology* 76: 1286–1288.

Bruce, D. and Wilson, A. (1998). The extensive spread of heart water in Zimbabwe since 1975. In L. Coons and M. Rothschild (eds.) *Proceedings of Second International Conference on Tick Borne Pathogens at the Host Vector Interface: A Global Perspective*, Kruger National Park, South Africa, pp. 105–112.

Bruce, J.A. and Caulfield, F. (1991). Reduced respiratory carbon dioxide efflux during growth at elevated carbon dioxide in three herbaceous perennial species. *Annuals of Botany* 67: 325-330.

Bunce, J.A. (1995). Long-term growth of alfalfa and orchard grass plots at elevated carbon dioxide. *Journal of Biogeochemistry* 22: 341-348.

Bunce, J.A. (1997). Variation in growth stimulation by elevated carbon dioxide in seedlings of some C3 crop and weed species. *Global Change Biology* 3: 61–66.

Bunce, J.A. and Ziska, L.H. (2000). Crop ecosystem responses to climatic change: crop–weed interactions. In K.R. Reddy and H.F. Hodges (eds.) *Climate Change and Global Crop Productivity.* CABI Publishing, New York, pp. 333–352.

Burdon, J.J., Thrall, P.H. and Ericson, L. (2006). The current and future dynamics of disease in plant communities. *Annu Rev Phytopathol* 44: 19–39.

Burt, O.R. and Ferris, H. (1996). Sequential decision rules for managing nematodes with crop rotations. *J. Nematol.* 28: 457–474.

Busby, J.R. (1991). BIOCLIM-a bioclimate analysis and prediction system. *Plant Protection Quarterly* 6: 8-9.

Bustamante, M., Robledo-Abad, C., Harper, R. *et al.* (2014). Co-benefits, trade-offs, barriers and policies for greenhouse gas mitigation in the agriculture, forestry and other land use sector. *Glob Chang Biol.* 20(10): 3270-3290.

Butterworth, M.H., Semenov, M., Barnes, A.P. *et al.* (2010). North-south divide; contrasting impacts of climate change on crop yields in Scotland and England. *Journal of the Royal Society Interface* 7: 123–130.

Caffarra, A., Rinaldi, F., Eccel, E. *et al.* (2012). Modelling the impact of climate change on the interaction between grapevine and its pests and pathogens: European grapevine moth and powdery mildew. *Agric. Ecosyst. Environ.* 148: 89-101.

Cai, D.G., Kleine, M., Kifle, S. *et al.* (1997). Positional cloning of a gene for nematode resistance in sugar beet. *Science* 275: 832–834.

Caldwell, B.E., Brim, C.A. and Ross, J.P. (1960). Inheritance of resistance of soybeans to the cyst nematode, *Heterodera glycines*. *Agron. J.* 52: 635-636.

Calvero, S.B., Coakley, S.M., McDaniel, L.R. and Teng, P.S. (1994). *IRRI Discussion Papers Series No. 5*. Int. Rice Res. Inst., Manila, The Philippines.

Cammell, M. and Knight, J. (1992). Effects of climatic change on the population dynamics of crop pests. *Adv. Ecol. Res.* 22: 117–162.

Cannon, R.J.C. (1998). The implications of predicted climate change for insect-pests in the UK, with emphasis on non-indigenous species. *Global Change Biol.* 4: 785-796.

Cardinale, B.J., Harvey, C.T., Gross, K. and Ives, A.R. (2003). Biodiversity and biocontrol: emergent impacts of a multi-enemy assemblage on pest suppression and crop yield in an agroecosystem. *Ecology Letters* 6: 857–865.

Carefoot, G.L. and Sprott, E.R. (1967). *Famine on the Wind: Plant Diseases and Human History*. Angus & Robertson, Sydney.

Carlton, R.R., West, J.S., Smith, P. and Fitt, B.D.L. (2012). A comparison of GHG emissions from UK field crop production under selected arable systems with reference to disease control. *Eur J Plant Pathol* 133: 333–351.

Carroll, S.P, Hendry, A.P., Reznick, D.N. and Fox, C.W (2007). Evolution on ecological time-scales. *Funct Ecol* 21: 387–393.

Carter, T.R., Nurro, M. and Torkko, S. (1996). Global climate change and agriculture in the north. *Agriculture and Food Science in Finland* 5: 223-385.

Carter, T.R., Saarikko, R.A. and Niemi, K.J. (1996). Assessing the risks and uncertainties of regional crop potential under a changing climate in Finland. *Agric Food Sci Finl* 5: 329–350.

Castle, L.A., Wu, G. and McElroy, D. (2006). Agricultural input traits: Past, present and future. *Current Opinion in Biotechnology* 17: 105-112.

Caswell-Chen, E.P., Williamson, V.M. and Westerdahl, B.B. (1993). View point-applied biotechnology in nematology. *J. Nematol.* 25: 719–730.

Caten, C.E., Person, C., Groth, J.V. and Dhali, S.J. (1984). The genetics of pathogenic aggressiveness in three dikaryons of *Ustilago hordei*. *Canadian Journal of Botany* 62: 1209-1219.

Caubel, J., Launay, M., Garcia de Cortazar-Atauri, I. *et al.* (2014). A new integrated approach to assess the impacts of climate change on grapevine fungal diseases: the coupled MILA-STICS model. *J Int Sci Vigne Vin* 48(1): 45-54.

Caulifield, F. and Bunce, J.A. (1994). Elevated atmospheric carbon dioxide concentration affects interactions between *Spodoptera exigua* (Lepidoptera: Noctuidae) larvae and two host plant species outdoors. *Env. Entomol.* 23: 999-1005.

Cavero, J., Zaragoza, C., Suso, M.L. and Pard, A. (1999). Competition between maize and *Datura stramonium* in an irrigated field under semi-arid conditions. *Weed Res.* 39: 225–240.

Cesarz, S., Reich, P.B., Scheu, S. *et al.* (2015). Nematode functional guilds, not trophic groups, reflect shifts in soil food webs and processes in response to interacting global change factors. *Pedobiologia* 58: 23-32.

Chahal, S.K., Bains, G.S. and Dhaliwal, L.K. (2008). Climate change: Mitigation and Adaptation. *Proceedings of International Conference on Climate Change, Biodiversity and Food Security in the South Asian Region*, Punjab State Council for Science and Technology, Chandigarh and United Nations Educational, Scientific and Cultural Organization, New Delhi, 12 pp.

Chakraborty, S. (2005). Potential impact of climate change on plant-pathogen interactions. *Australasian Plant Pathol.* 34: 443–448.

Chakraborty, S. (2013). Migrate or evolve: options for plant pathogens under climate change. *Glob Chang Biol* 19: 1985–2000.

Chakraborty, S. and Datta, S. (2003). How will plant pathogens adapt to host plant resistance at elevated CO_2 under a changing climate? *New Phytol.* 159: 733-742.

Chakraborty, S., Liu, C.J., Mitter, V. *et al.* (2006). Pathogen population structure and epidemiology are keys to wheat crown rot and Fusarium head blight management. *Australasian Plant Pathology* 35: 643–55.

Chakraborty, S., Murray, G.M., Magarey, P.A. *et al.* (1998). Potential impact of climate change on plant diseases of economic significance to Australia. *Australian Plant Pathology* 27: 15-35.

Chakraborty, S. and Newton, A.C. (2011). Climate change, plant diseases and food security: an overview. *Plant Pathology* 60: 2-14.

Chakraborty, S. and Pangga, I.B. (2004). Plant disease and climate change. In: M. Gillings and A. Holmes (eds.) *Plant Microbiology*. BIOS Scientific, London, pp. 163-180.

Chakraborty, S., Pangga, I.B., Lupton, J. *et al.* (2000a). Production and dispersal of *Colletotrichum gloeosporioides* spores on *Stylosanthes scabra* under elevated CO_2. *Environmental Pollution* 108: 381-387.

Chakraborty, S., Tiedemann, A.V. and Teng, P.S. (2000b). Climate change: potential impact on plant diseases. *Environmental Pollution* 108: 317-326.

Challinor A.J., Watson, J., Lobell, D.B. *et al.* (2014). A meta-analysis of crop yield under climate change and adaptation. *Nature Climate Change* 4: 287-291.

Challinor, A.J., Wheeler, T.R. and Craufurd, P.Q. (2007). Adaptation of crops to climate change through genotypic responses to mean and extreme temperatures. *Agric Ecosyst Environ* 119: 190–204.

Champeil, A., Fourbet, J.F., Doré, T. and Rossignol, L. (2004). Influence of cropping system on Fusarium head blight and mycotoxin levels in winter wheat. *Crop Protection* 23: 531–537.

Chand, R. and Raju, S.S. (2009). Instability in Indian agriculture during different phases of technology and policy. *Indian J. Agric. Econ.* 64: 187-207.

Chander, S. (1998). Infestation of root and foliage/earhead aphids on wheat in relation to predators. *Indian Journal of Agricultural Science* 68(11): 754-755.

Chander, S., Reji, G. and Aggarwal, P.K. (2009). Assessing impact of climate change on rice gundhi bug using a population dynamics simulation model. In P.K. Aggarwal (ed.) *Global Climate Change and Indian Agriculture: Case Studies from the ICAR Network Project.* ICAR, New Delhi, pp. 60-65.

Chander, S., Singh, V.S. and Kalra, N. (2003). Aphid infestation on barley in relation to climatic variability. *Proceedings of the National Symposium on Frontier Areas of Entomological Research*, Entomology Division, Indian Agricultural Research Institute, New Delhi, pp. 37-38.

Chandrasena, N. (2009). How will weed management change under climate change, some perspective. *Journal of Crop and Weed* 5(2): 95-105.

Chapin, F.S.I.I.I., Bret-Harte, M.S., Hobbie, S.E. and Zhong, H. (1996). Plant functional types as predictors of transient responses of arctic vegetation to global change. *J Veg Sci* 7: 347–358.

Chapman, R.F. (1998). *The Insects - Structure and Function.* Cambridge University Press, 4th Edition, 788 pp.

Chauhan, B.S. and Abugho, S.B. (2013). Effect of water stress on the growth and development of *Amaranthus spinosus*, *Leptochloa chinensis*, and rice. *Am. J. Plant Sci.* 4: 989–998.

Chauhan, B.S., Kaur, P., Mahajan, G. *et al.* (2014). Global warming and its possible impact on agriculture in India. *Adv. Agron.* 123: 65–121.

Chellappan, P., Vanitharani, R., Ogbe, F. and Fauquet, C.M. (2005). Effect of temperature on geminivirus-induced RNA silencing in plants. *Plant Physiology* 138: 1828-1841.

Chen, C.C. and McCarl, B.A. (2001). An investigation of the relationship between pesticide usage and climate change. *Climatic Change* 50: 475–487.

Chen, F., Wu, G., Parajulee, M.N. and Ge, F. (2007). Long-term impacts of elevated carbon dioxide and transgenic Bt cotton on performance and feeding of three generations of cotton bollworm. *Entomologia Experimentalis et Applicata* 124: 27–35.

Chen, F.J., Wu, G., Lü, J. and Ge, F. (2005). Effects of elevated CO_2 on the foraging behavior of cotton bollworm, *Helicoverpa armigera*. *Insect Science* 12: 359-365.

Chen, J., Chen, F., Liu, M. *et al.* (2014). Short-term effects of CO_2 concentration elevation, warming and transgenic Bt rice cropping on soil labile organic carbon and nitrogen, and nematode communities. *Acta Ecol. Sin.* 34 (6):1481-1489.

Chen, Z.X. and Dickson, D.W. (1998). Review of *Pasteuria penetrans*: Biology, ecology, and biological control potential. *Journal of Nematology* 30: 313-340.

Chen, Z.X., Dickson, D.W., Mitchell, D.J. *et al.* (1997). Suppression mechanisms of *Meloidogyne arenaria* race 1 by *Pasteuria penetrans*. *Journal of Nematology* 29:1-8.

Chippendale, G.M. (1979). *The South-western Corn Borer,* Diatraea grandiosella*: Case History of an Invading Insect.* Research Bulletin 1031, University of Missouri Agricultural Experiment Station, Columbia, MO.

Chitambar, J.J. and Raski, D.J. (1984). Reactions of grape rootstocks to *Pratylenchus vulnus* and *Meloidogyne* sp. *J. Nematol.* 16: 166-170.

Chitwood, D.J. (2003). Research on plant-parasitic nematodes biology conducted by United States Department of Agriculture – Agricultural Research Service. *Pest Management Science* 3359: 748-753.

Chmielewski, F.M., Müller, A. and Buns, E. (2004). Climate changes and trends in phenology of fruit trees and field crops in Germany, 1961–2000. *Agric For Meteorol* 121: 69–78.

Christen, O. (2008). Langfristige Trends und Anpassung der Anbausysteme an den Klimawandel. In: A.V. Tiedemann, R. Heitefuss and F. Feldmann (eds.) *Pflanzenproduktion im Wandel – Wandel im Pflanzenschutz*. Braunschweig, DPG Selbstverlag, Germany, pp. 57–64.

Ciais, P., Reichstein, M., Viovy, N. *et al.* (2005). Europe-wide reduction in primary productivity caused by the heat and drought in 2003. *Nature* 437: 529–533.

CIKS. (2000). Bollworm control in cotton. Center for Indian Knowledge Systems. Chennai, India. *Pesticide Post* 8(6): 1-4.

Cimalova, S. and Lososova, Z. (2009). Arable weed vegetation of the north-eastern part of the Czech Republic: effects of environmental factors on species composition. *Plant Ecol* 203: 45–57.

Cleland, E.E., Chuine, I., Menzel, A. *et al.* (2007). Shifting plant phenology in response to global change. *TREE* 22: 357-365.

Clements, D.R. and Ditommaso, A. (2011). Climate change and weed adaptation: can evolution of invasive plants lead to greater range expansion than forecasted? *Weed Res* 51: 227–240.

Clements, D.R., Ditommaso, A., Jordan, N. *et al.* (2004). Adaptability of plants invading North American cropland. *Agriculture, Ecosystems and Environment* 104: 379–398.

Clements, D.R., Weise, S.F. and Swanton, C.J. (1994). Integrated weed management and weed species diversity. *Phytoprotection* 75:1–18.

Coakley, S.M., Line, R.F. and McDaniel, L.R. (1988). Predicting stripe rust severity on winter wheat using an improved method for analyzing meteorological and rust data. *Phytopathology* 78: 543–550.

Coakley, S.M., McDaniel, L.R. and Line, R.F. (1988). Quantifying how climatic factors affect variation in plant disease severity: a general method using a new way to analyze meteorological data. *Climatic Change* 12: 157–175.

Coakley, S.M. and Scherm, H. (1996). Plant disease in a changing global environment. *Aspects of Applied Biology* 45: 227–238

Coakley, S.M., Scherm, H. and Chakraborty, S. (1999). Climate change and plant disease management. *Annual Review of Phytopathology* 37: 399–426.

Coleman, J.S., Jones, C.G. and Smith, W.H. (1988). Interaction between an acute ozone dose, eastern cottonwood and *Marssonia* leaf spot: Implications for pathogen community dynamics. *Canadian Journal of Forest Research* 66: 863-868.

Coleman, J.S., Kingston, A.C.T. and Bazzaz, F.A. (1992). Effects of CO_2 and temperature on growth and resource use of co-occurring C3 and C4 annuals. *Ecology* 73(4): 1244-1259.

Colhoun, J. (1973). Effects of environmental factors on plant disease. *Annual Review of Phytopathology* 11: 343–364.

Coll, M. and Hughes, L. (2008). Effects of elevated CO_2 on an insect omnivore: A test for nutritional effects mediated by host plants and prey. *Agriculture, Ecosystems and Environment* 123: 271–279.

Collier, R.H., Finch, S., Phelps, K. and Thompson, A.R. (1991). Possible impact of global warming on cabbage root fly (*Delia radicum*) activity in the UK. *Annals of Applied Biology* 118: 261-271.

Conlong, D.E. and Rutherford, R.S. (2009). Conventional and new biological and habitat interventions for integrated pest management systems: Review and case studies using *Eldana saccharina* Walker (Lepidoptera: Pyralidae). In R. Peshin and A.K. Dhawan (eds.) *Integrated Pest Management: Innovation - Development Process*, Vol. 1. Springer, Netherlands, pp. 241-261.

Cook, D.E., Lee, T.G., Guo, X. *et al.* (2012). Copy number variation of multiple genes at Rhg1 mediates nematode resistance in soybean. *Science* 338: 1206-1209.

Cook, R. and Evans, K. (1987). Resistance and tolerance. In R.H. Brown and B.R. Kerry (eds.) *Principles and Practice of Nematode Control in Crops*. Academic, Sydney, pp. 179–231.

Cook, R.J. (1991). Challenges and rewards of sustainable agriculture, research and education. In *Sustainable Agriculture Research and Education in the Field*. Natl. Acad. Press, Washington, DC, pp. 32–76.

Cook, S.M., Khan, Z.R. and Pickett, J.A. (2007). The use of push-pull strategies in integrated pest management. *Annual Review of Entomology* 52: 375-400.

Cooper, W.R., Jia, L. and Goggin, L. (2005). Effects of jasmonate-induced defenses on root-knot-nematode infection of resistant and susceptible tomato cultivars. *J Chem Ecol* 31: 1953–1967.

Copeland, P.J., Malzer, G.L., Davis, J.G. *et al.* (1996). Using harvest index to locate environmental stress. In P.C. Robert, R.H. Rust and W.E. Larson (eds.) *Proc. 3rd. Int. Conf. Precision Agric.*, Madison, WI: ASA/CSSA/SSSA, pp. 531–540.

Corlett, T.T. and LaFrankie, Jr. J.V. (1998). Potential impacts of climate change on tropical Asian forests through and influences on phenology. *Climate Change* 39: 439- 454.

Cornelissen, T. (2011). Climate change and its effects on terrestrial insects and herbivory patterns. *Neotrop. Entomol.* 40: 155–163.

Costanza, R., d'Arge, R., de Groot, R. *et al.* (1987). The Value of the world's ecosystem services and natural capital. *Nature* 387: 253-260.

Coulson, S.J., Leinaas, H.P., Ims, R.A. and Søvik, G. (2000). Experimental manipulation of the winter surface ice layer: the effect on a High Arctic soil microarthropod community. *Ecography* 23: 299-306.

Coumou, D. and Rahmstorf, S. (2012). A decade of weather extremes. *Nat Clim Chang* 2: 491–496.

Coviella, C., Margon, D.J.W. and Trumble, J.T. (2000). Interaction of elevated CO_2 and nitrogen fertilization: Effect on the production of *Bacillus thuringiensis* toxins in transgenic plants. *Env. Entomol.* 29: 281-287.

Coviella, C.E. and Trumble, J.T. (1999). Effects of elevated atmospheric carbon dioxide on insect plant interactions. *Conservation Biology* 13: 700-712.

Coviella, C.E. and Trumble, J.T. (2000). Effect of elevated atmospheric CO_2 on use of foliar application of *Bacillus thuringiensis*. *Biocontrol* 45: 325-336.

Craine, J.M., Froehle, J., Tilman, D.G. *et al.* (2001). The relationships among root and leaf traits of 76 grassland species and relative abundance along fertility and disturbance gradients. *Oikos* 93: 274–285.

Crowl, T.A., Crist, T.O., Parmenter, R.R. *et al.* (2008). The spread of invasive species and infectious disease as drivers of ecosystem change. *Frontiers Ecol. Environ.* 6: 238–246.

Cruz, R.V. *et al.* (2007). Asia. Climate change 2007: Impacts, adaptation and vulnerability. In M.L. Parry (ed.) *Contribution of Working Group 2 to the 4 Assessment Report of the Intergovernmental Panel on Climate Change*. Cambridge University Press, Cambridge.

Cuculeanu, V., Tuinea, P. and Balteanu, D. (2002). Climate change impacts in Romania: vulnerability and adaptation options. *Geol. J.* 57: 203–209.

Cure, J.D. and Acock, B. (1986). Crops response to carbon dioxide doubling; a literature survey. *Agricultural and Forest Meteorology* 38: 127-145.

Currano, D., Wilf, P., Wing, S.L. *et al.* (2008). Sharply increased insect herbivory during the Paleocene–Eocene Thermal Maximum. *PNAS* 1056: 1960–1964.

Czajkowski, R., Pérombelon, M.C.M., van Veen, J.A. and van der Wolf, J.M. (2011). Control of blackleg and tuber soft rot of potato caused by *Pectobacterium* and *Dickya* species: a review. *Plant Pathol* 60: 999–1013.

Czepak, C., Albernaz, K.C., Vivan, L.M. *et al.* (2013). First reported occurrence of *Helicoverpa armigera* (Hu¨bner) (Lepidoptera: Noctuidae) in Brazil. *Pesqui. Agropec. Trop.* 43: 110-113.

Daane, KM. and Caltagirone, L.E. (1989). Biological control of black scale in olives. *Calif Agric* 43: 9–11.

Daberkaw, S. and Christensen, L. (1996). The potential contributions of precision agriculture to IPM. In P.C. Robert, R.H. Rust and W.E. Larson (eds.) *Proc. 3rd. Int. Conf. Precision Agric.*, Madison, WI: ASA/CSSA/SSSA, p. 567.

Daccache, A., Keay, C., Jones, R.J.A. *et al.* (2012). Climate change and land suitability for potato production in England and Wales: impacts and adaptation. *J Agric Sci* 150: 161–177.

Das, D.K., Singh, J. and Vennila, S. (2011). Emerging crop pest scenario under the impact of climate change – a brief review. *J. Agric. Phy.* 11: 13-20.

Das, T.K., Sharma, A.R. and Pathak, H. (2012). Crop-weed balance studies under climate change. In H. Pathak, P.K. Aggarwal, and S.D. Singh (eds.) *Climate Change Impact, Adaptation and Mitigation in Agriculture: Methodology for Assessment and Application.* Indian Agricultural Research Institute, New Delhi, p. 131.

De, U.S. and Mukhopadhyay, R.K. (1999). Severe heat wave over the Indian subcontinent in 1998 in perspective of global climate. *Curr. Sci.* 75: 1308-1315.

DeBach, P. (1964). *Biological Control of Insect Pests and Weeds.* Chapman and Hall Ltd., London, 884 pp.

Deka, M.K. and Singh, S.N. (2001). Neem formulation in the management of sugarcane insects and pests. *Procceding 63rd Ann. Conv. Sugar Technol. Associ.*, Jailpur, India, pp. 33-38.

Deka, S., Byjesh, K., Kumar, U. and Choudhary, R. (2012). Climate change and impacts on crop pests—a critique. *ISPRS Archives* 38: 8.

Dekker, J. (2003). Evolutionary biology of the foxtail (*Setaria*) species group. In: Inderjit (ed.) *Weed Biology and Management.* Kluwer Academic Publishers, The Netherlands, pp. 65–114.

Delbac L., Lecharpentier P. and Thiery D. (2010). Larval instars determination for the European grapevine moth (Lepidoptera: Tortricidae) based on the frequency distribution of head-capsule widths. *Crop Prot.* 29: 623–630.

DeLucia, E.H., Casteel, C.L., Nabity, P.D. and O'Neill, B.F. (2008). Insects take a bigger bite out of plants in a warmer, higher carbon dioxide world. *PNAS* 105 (6): 1781–1782.

Desjardins, A.E. (2006). *Fusarium Mycotoxins: Chemistry, Genetics and Biology. APS Press, St Paul, MN, USA.*

Deslippe, J.R., Hartmann, M., Mohn, W.W.A. and Simard, S.W. (2010). Long-term experimental manipulation of climate alters the ectomycorrhizal community of *Betula nana* in Arctic tundra. *Global Change Biology* 17: 1625–1636.

Desneux, N., Luna, M.G., Guillemaud, T. and Urbaneja, A. (2011). The invasive South American tomato pinworm, *Tuta absoluta,* continues to spread in Afro-Eurasia and beyond—the new threat to tomato world production. *J. Pest. Sci.* 84: 403–408.

Desneux, N., Wajnberg, E., Wyckhuys, K.A.G. *et al.* (2010). Biological invasion of European tomato crops by *Tuta absoluta*: ecology, geographic expansion and prospects for biological control. *J Pest Sci* 83: 197–215.

Deutsch, C.A., Tewksbury, J.J., Huey, R.B. *et al.* (2008). Impacts of climate warming on terrestrial ectotherms across latitude. *PNAS* 105 (18): 6668–6672.

DeVries, J. and Toenniessen, G. (2001). *Securing the Harvest: Biotechnology, Breeding, and Seed Systems for African Crops.* CAB International, Wallingford.

Dewar, A.M. (2005). Beet cyst nematodes – a peril in the soil. *British Sugar Beet Review* 73: 40–46.

Dewer, R.C. and Watt, A.D. (1992). Predicted changes in the synchrony of larval emergence and bud burst under climatic warming. *Oecologia* 89: 557-559.

Dhaliwal, G.S. and Dilawari, V.K. (1993). *Advances in Host Resistance to Insects.* Kalyani Publishers, New Delhi.

Dhaliwal, G.S., Jindal, V. and Dhawan, A.K. (2010). Insect-pest problems and crop losses: changing trends. *Indian J. Ecol.* 37: 1-7.

Dhawan, A.K. and Peshin, R. (2009). Integrated pest management: Concept, opportunities, challenges. In R. Peshin and A.K. Dhawan (eds.) *Integrated Pest Management: Innovation - Development Process, Vol. 1.* Springer, Netherlands, pp. 113-129.

Dhawan, A.K., Singh, K., Saini, S. *et al.* (2007). Incidence and damage potential of mealybug, *Phenacoccus solenopsis* Tinsley, on cotton in Punjab. *Indian J. Ecol.* 34: 110-116.

Dhillon, M.K. and Sharma, H.C. (2008). Temperature and *Helicoverpa armigera* food influence survival and development of the ichneumonid parasitoid, *Campoletis chlorideae. Ind. J. Pl. Protect.* 36: 240-44.

Dhillon, M.K. and Sharma, H.C. (2009). Temperature influence the performance and effectiveness of field and laboratory starins of the ichneumonid parasitoid *Campoletis chlorideae. Biocontrol* 54: 743-50.

Diaz, S., Cabido, M. and Casanoves, F. (1998). Plant functional traits and environmental filters at a regional scale. *J Veg Sci* 9: 113–122.

Diaz, S., Cabido, M., Zak, M. *et al.* (1999). Plant functional traits, ecosystem structure and land-use history along a climatic gradient in central western Argentina. *J Veg Sci* 10: 651–660.

Diaz, S., Fraser, L.H., Grime, J.P. and Falczuk, V. (1998). The Impact of elevated CO_2 on plant herbivore interactions: experimental evidence of moderating effects at the community level. *Oecologia* 117: 177-186.

Dickenson, S. and Wheeler, B.F.J. (1981). Effects of temperature and water stress in sycamore on growth of *Cryotostroma corticale. Transactions of the British Mycological Society* 76: 181-185.

Dickson, D.W., Oostendorp, M., Giblin-Davis, R.M. and Mitchell, D.J. (1994). Control of plant-parasitic nematodes by biological antagonists. In D. Rosen, F.D. Bennett and J.L. Capinera (eds.). *Pest Management in the Tropics, Biological Control-A Florida Perspective.* Intercept Ltd, U.K., pp. 575-601.

Diffenbaugh, N.S., Krupke, C.H., White, M.A. and Alexander, C.E. (2008). Global warming presents new challenges for maize pest management. *Environment Research Letter* 3: 1-9.

Dillon, C.R., Keisling, T.C., Riggs, R.D. *et al.* (1996). Profitability of non-irrigated soybean cropping rotations. *Proc. 5th Annu. South. Soybean Conf.*, pp. 88–97.

Ding, J., Mack, R.N., Lu, P. *et al.* (2008). China's booming economy is sparking and accelerating biological invasions. *BioScience* 58: 317–324.

Dinoor, A. (1974). Role of wild and cultivated plants in the edipemiology of plant diseases in Israel. *Annual Review of Phytopathology* 12: 413-436.

Dixon, G.R. (2012). Climate change – impact on crop growth and food production, and plant pathogens. *Can. J. Plant Pathol.* 34: 362–379.

Dodds, P. and Rathjen, J.P. (2010). Plant immunity: towards an integrated view of plant–pathogen interactions. *Nature Reviews Genetics* 11: 539–548.

Donald, W.W. and Khan, M. (1992). Yield loss assessment for spring wheat (*Triticum aestivum*) infested with Canada thistle (*Cirsium arvense*). *Weed Sci.* 40: 590–598.

Dong, K. and Opperman, C.H. (1997). Genetic analysis of parasitism in the soybean cyst nematode *Heterodera glycines*. *Genetics* 146: 1311-1318.

Dosdall, LM. (1994). Evidence for successful overwintering of diamondback moth, *Plutella xylostella* (L.) (Lepidoptera: Plutellidae), in Alberta. *The Canadian Entomologist* 126: 183–185.

Douglas, B.J., Thomas, A.G., Morrison, I.N. and Maw, M.G. (1985). The biology of Canadian weeds 70. *Setaria viridis* (L.) Beauv. *Canadian Journal of Plant Science* 65: 669–690.

Downey, P.O., Williams, M.C., Whiffen, L.K. *et al.* (2010). Managing alien plants for biodiversity outcomes - the need for triage. *Invasive Plant Science and Management* 3: 1-11.

Doxford, S.W. and Freckleton, R.P. (2012). Changes in the large-scale distribution of plants: extinction, colonisation and the effects of climate. *J Ecol* 100: 519–529.

Dropkin, V.H. (1969). Necrotic reaction of tomatoes and other hosts resistant to *Meloidogyne*: reversal by temperature. *Phytopathology* 59: 1632-1637.

Dropkin, V.H., Helgeson, J.B. and Upper, C.D. (1969). The hypersensitivity reaction of tomato resistance to *Meloidogyne incognita*: Reversal by cytokinins. *Journal of Nematology* 1: 55-61.

Duffield, S.J. and Reddy, Y.V.R. (1997). Distribution and increment of predators of *Helicoverpa armigera* in intercropped sorghum and short duration pigeonpea. *Crop Research* 14: 315-335.

Dukes, J.S. (2000). Will the increasing atmospheric CO_2 concentration affect the success of invasive species? In H.A. Mooney and R.J. Hobbs (eds.) *Invasive Species in a Changing World.* Island Press, Washington, DC, pp. 95–113.

Dukes, J.S. and, Mooney, H.A. (2000). Does global change increase the success of biological invaders? *Trend Ecol Evol* 14: 135–139.

Dukes, J.S., Pontius, J., Orwig, D. *et al.* (2009). Responses of insect pests, pathogens, and invasive plant species to climate change in the forests of north-eastern North America: what can we predict? *Can J For Res* 39: 231–248.

Duku C., Sparks A.H. and Zwart S.J. (2015). Spatial modelling of rice yield losses in Tanzania due to bacterial leaf blight and leaf blast in a changing climate. *Clim Change* 135(3-4): 569-583.

Dunbar, H.E., Wilson, A.C.C., Ferguson, N.R. and Moran, N.A. (2007). Aphid thermal tolerance is governed by a point mutation in bacterial symbionts. *PLOS Biol.* 5: 1006–1015.

Duncan, L.W. and Noling, J.W. (1998). Agricultural sustainability and nematode IPM. In K.R. Barker, G.A. Pederson and G.L. Windham (eds.) *Plant and Nematode Interactions*. Agron. Monogr. 36. Am. Soc. Agron., Crop Sci. Soc. Am. and Soil Sci., pp. 251–287.

Durgapal, J.C. and Singh, B. (1980). Taxomony of pseudomonads pathogenic to horse-chestnut, wild fig and wild cherry in India. *Indian Phytopathol* 33: 533–535.

Durrant, W.E. and Dong, X. (2004). Systemic acquired resistance. *Annu Rev Phytopathol* 42: 185–209.

Dury, S.J., Good, J.E.G., Perrins, C.M. *et al.* (1998). The effects of increasing CO_2 and temperature on oak leaf palatability and the implications for herbivorous insects. *Global Change Biology* 4: 55-61.

Duveiller, E., Singh, R.P. and Nicol, J.M. (2007). The challenges of maintaining wheat productivity: Pests, disease and potential epidemics. *Euphytica* 157: 417-430.

Eastburn, D.M., Cytryn, Y. Meller H. et al. (2011). Influence of atmospheric and climate change on plant pathogen interactions. *Plant Pathol* 60: 54-69.

Eastburn, D.M., Degennaro, M.M., DeLucia, E.H. *et al.* (2010). Elevated atmospheric carbon dioxide and ozone alter soybean diseases at SoyFACE. *Glob Change Biol* 16: 320-330.

Eastburn, D.M., McElrone, A.J. and Bilgin, D.D. (2011). Influence of atmospheric and climatic change on plant–pathogen interactions. *Plant Pathol.* 60: 54–69.

Easterling, W.E., Aggarwal, P.K., Batima, P. *et al.* (2007). Food, fibre and forest products. In M.L. Parry, O.F. Canziani, J.P. Palutikof *et al.* (eds.) *Climate Change 2007: Impacts, Adaptation and Vulnerability*. Contribution of Working Group II to the Fourth Assessment Report of the Intergovernmental Panel on Climate Change, Cambridge University Press, Cambridge, pp. 273-313.

Egli, S. (2011). Mycorrhizal mushroom diversity and productivity - an indicator of forest health? *Annals of Forest Science* 68: 81–88.

Ehwaeti, M.E., Elliott, M.J., McNicol, J.M. *et al.* (2000). Modelling nematode population growth and damage. *Crop. Prot.* 19: 739–745.

Eisenhauer, N., Cesarz, S., Koller, R., Worm, K. and Reich, P.B. (2012). Global change belowground: impacts of elevated CO_2, nitrogen, and summer drought on soil food webs and biodiversity. *Global Change Biology* 18: 435-447.

El Titi, A. and Ipach, U. (1989). Soil fauna in sustainable agriculture: results of an integrated farming system at Lautenbach, F.R.G. *Agric. Ecosyst. Environ.* 27:561–72.

Elad, Y., Cytryn E., Meller Harel Y. et al. (2011). The biochar effect: Plant resistance to biotic stresses. *Phytopathol. Mediterr.* 50: 335–349.

Elad, Y. and Freeman, S. (2002). Biological control of fungal plant pathogens. In: F. Kempken (ed.) *The Mycota XI, Agricultural Applications*. Springer, Heidelberg, Germany, pp. 93-109.

Elad, Y. and Pertot, I. (2014). Climate change impacts on plant pathogens and plant diseases. *J Crop Improv* 28: 99-139.

Elad, Y., Rav David, D., Meller Harel, Y. *et al.* (2010). Induction of systemic resistance in plants by biochar, a soil-applied carbon sequestering agent. *Phytopathology* 100: 913–921.

Elad, Y., Williamson, B., Tudzynski, P. and Delen, N. (eds.) (2004). *Botrytis: Biology, Pathology and Control.* Kluwer Academic Publishers, Dordrecht, Netherlands.

Elad, Y., Zimand, G., Zaqs, Y. *et al.* (1993). Use of *Trichoderma harzianum* in combination or alternation with fungicides to control cucumber grey mould (*Botrytis cinerea*) under commercial greenhouse conditions. *Plant Pathol.* 42: 324-332.

Elith, J., Graham, C.H., Anderson, R.P. *et al.* (2006). Novel methods improve prediction of species' distributions from occurrence data. *Ecography* 29: 129-151.

Ellis, B. and Bradley, F. (1996). *The Organic Gardener's Handbook of Natural Insect and Disease Control.* Rodale Press. Emmaus, Pennsylvania.

Ellis, J.G., Mago, R., Kota, R. *et al.* (2007). Wheat rust resistance research at CSIRO. *Australian Journal of Agricultural Research* 56: 507–511.

Elphinstone, J. and Toth, I.K. (2008). Erwinia chrysanthemi (Dikeya *sp.)* - *The Facts.* Potato Council, Oxford, U.K.

El-Sayed, A.M., Suckling, D.M., Byers, J.A. *et al.* (2009). Potential of "Lure and Kill" in long-term pest management and eradication of invasive species. J. Econ. Entomol. 102(3): 815-835.

Emmerson, M., Bezemer, T.M., Hunter, M.D. *et al.* (2004). Trophic interactions in a changing world: altering of interaction strengths. *Basic and Applied Ecology* 5: 505-514.

EPA. (1989). *The Potential Effects of Global Climate Change on the United States. Vol 2: National Studies.* Review of the Report to Congress, US Environmental Protection Agency, Washington DC, 261 pp.

EPPO. (2014). *EPPO Reporting Service, No. 2014/015.* European and Mediterranean Plant Protection Organization, Paris, France.

Epstein, P.R., Diaz, H.F. and Flias, S. (1998). Biological and physical sign of climate change: Focus mosquito-borne diseases. *Bull. Am. Meteorol. Soc.* 79: 409-417.

Erasmus, B.F.N., Van Jarsveld, A.S. and Chown, S. (2002). Vulnerability of South Africa animal taxa to climate change. *Global Change Boil.* 6: 679- 693.

Eriksson, O. (2013). Species pools in cultural landscapes—niche construction, ecological opportunity and niche shifts. *Ecography* 36: 403–413.

Estrada-Peña, A., Pegram, R.G., Barré, N. and Venzal, J.M. (2007). Using invaded range data to model the climate suitability for *Amblyomma variegatum* (Acari: Ixodidae) in the New World. *Exp Appl Acarol* 41: 203–214.

European Commission. (2009a). White Paper from the European commission to the Council, the European Parliament, the European Economic and Social Committee and the Committee of the Regions - Adapting to climate change: Towards a European framework for action (COM(2009) 147/4) (01.04.20).

European Commission. (2009b). Regulation (EC) No 1107/2009 of the European Parliament and of the Council of 21 October 2009 concerning the placing of plant protection products on the market and repealing Council Directives 79/117/EEC and 91/414/EEC.

European Commission. (2009c). Directive 2009/128/EC of the European Parliament and of the Council of 21 October 2009 establishing a framework for Community action to achieve the sustainable use of pesticides.

Evans, N., Baierl, A., Semenov, A.M. *et al.* (2007). Range and severity of a plant disease increased by global warming. *J. R. Soc. Interface* 5: 525–531.

Evans, R.G., Han, S., Kraiger, M.W. and Schneider, S.M. (1996). Precision center pivot irrigation for efficient use of water and nitrogen. In P.C. Robert, R.H. Rust and W.E. Larson (eds.) *Proc. 3rd. Int. Conf. Precision Agric.*, ASA/CSSA/SSSA, Madison, WI, pp. 75–84.

Eze, S.C., Asiegbu, J.E., Mbah, B.N. *et al.* (2006). Effects of aqueous and ethanolic extracts and concentration of four agrobotanicals and gibberrelic acid (GA3) on the shelf life of the white Guinea yam *Dioscorea rotundata. Bio-Research J* 4: 67-73.

Fabre, F., Plantegenest, M., Mieuzet, L. *et al.* (2005). Effects of climate and land use on the occurrence of viruliferous aphids and the epidemiology of barley yellow dwarf disease. *Agriculture, Ecosystems and Environment* 106: 49–55.

Facknath, S. (2000). Application of neem extract and intercropping for the control of some cabbage pests in Mauritius. *Proc. International Neem Conference*, Queensland, Australia.

Fand, B.B., Kamble, A.L. and Kumar, M. (2012). Will climate change pose serious threat to crop pest management: A critical review? *International Journal of Scientific Research Publications* 2: 2250-3153.

FAO. (1966). Proceedings of the FAO Symposium on Integrated Pest Control, Rome, 1965. FAO, Rome.

FAO. (2002). *Training Modules on Conservation Agriculture.* Land and Water Digital Media Series #

FAO. (2003). Focus on Food Insecurity and Vulnerability – A Review of the UN System Common Country Assessments and World Bank Poverty Reduction Strategy Papers. FIVIMS Secretariat and Wageningen University and Research Centre: www.fao.org/DOCREP/006/ Y5095E/Y5095E00.htm.

FAO. (2003). *Trade Reforms and Food Security. Conceptualizing the Linkages.* Food and Agriculture Organisation of the United Nations, Rome, Italy. http://www.fao.org/ docrep/005/y4671e/ y4671e00.htm. Accessed 12 Nov 2010.

FAO. (2008). Rapid Assessment of Pollinators' Status. FAO, Rome, Italy.

Fassuliotis, G., Deakin, J.R. and Hoffman, J.C. (1970). Root-knot nematode resistance in snap beans: breeding and nature of resistance. *Journal of American Society of Horticultural Science* 95: 640-645.

Fausey, J.C. and Renner, K.A. (1997). Germination, emergence, and growth of giant foxtail (*Setaria faberi*) and fall panicum (*Panicum dichotomiflorum*). *Weed Sci* 45: 423–425.

Feil H., Westerdahl, B.B., Smith, R.J. and Verdegaal, P. (1997). Effects of seasonal and site factors on *Xiphinema index* populations in two California vineyards. *Journal of Nematology* 29(4): 491-500.

Fenton, B., Woodford, J.A.T. and Malloch, G. (1998). Analysis of clonal diversity of the peach-potato aphid, *Myzus persicae* (Sulzer), in Scotland, UK and evidence for the existence of a predominant clone. *Mol Ecol* 7: 1475–1487.

Fernandes, F.R., de Albuquerque, L.C., Giordano, L.D.B. *et al.* (2008). Diversity and prevalence of Brazilian bipartite begomovirus species associated to tomoatoes. *Virus Gene* 36: 251–258.

Fernandes, J.M., Cunha, G.R., Del Ponte, E. *et al.* (2004). Modelling Fusarium head blight in wheat under climate change using linked process-basedmodels. In: S.M. Canty, T. Boring, J. Wardwell and R.W. Ward (eds.) *2nd International Symposium on Fusarium Head Blight; Incorporating the 8th European Fusarium Seminar*, Orlando, FL, USA. Michigan State University, East Lansing, MI., pp. 441-444.

Ferris, H. (1978). Nematode economic thresholds: derivation, requirements, and the theoretical considerations. *J. Nematol.* 10: 341–350.

Ferris, H. (2010). Form and function: metabolic footprints of nematodes in the soil food web. *European Journal of Soil Biology* 46: 97-104.

Ferris, H., Bongers, T. and de Goede, R.G.M. (2001). A framework for soil food web diagnostics: extension of the nematode faunal analysis concept. *Applied Soil Ecology* 18: 13-29.

Ferris, H., Castro, C.E., Caswell, E.P. *et al.* (1992). Biological approaches to the management of plant-parasitic nematodes. In J.P. Madden (ed.). *Beyond Pesticides: Biological Approaches to Pest Management in California*. Divn. Ag. Nat. Res., Univ. California, Oakland, pp. 68-101.

Ferris, H., Griffiths, B.S., Porazinska, D.L. *et al.* (2012). Reflections on plant and soil nematode ecology: past, present and future. *Journal of Nematology* 44: 115-126.

Ferris, H., Zheng L. and Walker M.A. (2013). Soil temperature effects on the interaction of grape rootstocks and plant–parasitic nematodes. *J. Nematol.* 45: 49-57.

Fischer, G., Shah, M., van Velthuizen, H. and Nachtergaele, F.O. (2001). *Global Agro-Ecological Assessment in the 21st Century*. International Institute for Applied Systems Analysis, Laxenburg, Austria.

Fitt, B.D.L., Brun, H., Barbetti, M.J. and Rimmer, S.R. (2006). World-wide importance of phoma stem canker (*Leptosphaeria maculans* and *L. biglobosa*) on oilseed rape (*Brassica napus*). *Eur. J. Plant Pathol.* 114: 3–15.

Fitt, B.D.L., Fraaije, B.A., Chandramohan, P. and Shaw, M.W. (2011). Impacts of changing air composition on severity of arable crop disease epidemics. *Plant Pathology* 60: 44–53.

Fitt, B.D.L., Hu, B.C., Li, Z.Q. *et al.* (2008). Strategies to prevent spread of *Leptosphaeria maculans* (phoma stem canker) onto oilseed rape crops in China: costs and benefits. *Plant Pathol* 57: 652–664.

Fitter, A.H. and Fitter, R.S. (2002). Rapid changes in flowering time in British plants. *Science* 296: 1689- 1691.

Fleming, A. and Vanclay, F. (2010). Farmer responses to climate change and sustainable agriculture. A review. *Agron Sustain Dev* 30: 11–19.

Fleming, R.A. and Volney, W.J.A. (1995). Effect of climate change on insect defoliator population process in Canada boreal forest - some plausible scenarios. *Water Air Soil Pollution* 82: 445-454.

Fletcher, J., Luster, D., Bostock, R. *et al.* (2010). Emerging infectious diseases. In: W.M. Scheld (ed.) *Emerging Infections*. ASM Press, Washington DC, pp. 337–366.

Foley, J.A., Ramankutty, N., Brauman, K.A. *et al.* (2011). Solutions for a cultivated planet. *Nature* 478: 337–342.

Ford, K.A., Casidaa, J.E., Chandran, D. *et al.* (2010). Neonicotinoid insecticides induce salicylate associated plant defense responses. *Proceedings of the National Academy of Sciences* 107(41): 17527–17532.

Fowler, A.M. and Hennessy, K.J. (1994). Potential impacts of global warming on the frequency and magnitude of heavy precipitation. *Natural Hazards* 11: 283-303.

Fowler, H.J., Kilsby, C.G., O'Connell, P.E. *et al.* (2005). A weather-type conditioned multi-site stochastic rainfall model for the generation of scenarios of climatic variability and change. *J. Hydrol.* 308: 50–66.

Fraser, R.S.S. and Loughlin, S.A.R. (1982). Effects of temperature on the Tm-1 gene for resistance to tobacco mosaic virus in tomato. *Physiologial Plant Pathology* 20(1): 109-117.

Fraser, R.W., Cook, D.C., Mumford, J.D. *et al.* (2006). Managing outbreaks of invasive species: eradication versus suppression. *International Journal of Pest Management* 52: 261-268.

Freckman, D.W. and Ettema, C.H. (1993). Assessing nematode communities in agroecosystems of varying human intervention. *Agric. Ecosyst. Environ.* 45: 239–261.

Freckman, D.W., Moore, J.C., Hunt, H.W. and Elliot, E.T. (1991). The effects of elevated CO_2 and climate change on soil nematode community structure of prairie soil. *Bulletin of the Ecological Society of America* 72 (Suppl.): 119.

Freier, B. and Triltsch, H. (1996). Climate chamber experiments and computer simulation on the influence of increasing temperature on wheat-aphid-predator interaction. *Asp. of App. Bio.* 45: 293-98.

French, S. and Appleton, B. (2001). *A Guide to Successful Pruning: Pruning Basics and Tools. Ornamental and Turf.* Virginia Cooperative Extension, Virginia Polytechnic Institute and State University Publication, pp. 430-455.

Fried, G., Norton, L.R. and Reboud, X. (2008). Environmental and management factors determining weed species composition and diversity in France. *Agric Ecosyst Environ* 128: 68–76.

Fried, G., Petit, S. and Reboud, X. (2010). A specialist-generalist classification of the arable flora and its response to changes in agricultural practices. *Bio Med Central Ecol* 10: 1–11.

Froud-Williams, R.J. (1996). Weeds and climate change: implications for their ecology and control. *Asp. Appl. Biol.* 45: 187–196.

Fry, W. (2008). *Phytophthora infestans:* the plant (and R gene) destroyer. *Molecular Plant Pathology* 9: 385–402.

Fuchs, M., Tricoli, D.M., Carney, K.J. *et al.* (1998). Comparative virus resistance and fruit yield of transgenic squash with single and multiple coat protein genes. *Plant Disease* 82: 1350-1356.

Fuhrer, J. (2003). Agro-ecosystem responses to combinations of elevated CO_2, ozone, and global climate change. *Agriculture, Ecosystems and Environment* 97: 1-20.

Fuhrer J. (2006). *Agricultural Systems: Sensitivity to Climate Change.* CAB Reviews: Perspectives in Agriculture, Veterinary Science, Nutrition and Natural Resources 1, No 052.

Gadoury, D.M., Seem R.C., Pearson R.C. *et al.* (2001). Effects of powdery mildew on vine growth, yield, and quality of concord grapes. *Plant Dis.* 85: 137–140.

Gallagher, R., Beaumont, L., Downey, P.R. *et al.* (2006). Assessing the potential impacts of climate change on weeds in New South Wales: establishing priorities. *XV Australian weed conf.*, pp. 35-39.

Gang, W., Chen, F.; Sun, C. and Feng, G. (2007). Response of successive three generations of cotton bollworm, *Helicoverpa armigera* (Hubner), fed on cotton bolls under elevated CO_2. *J. Environ. Sci.* 19: 1318-1325.

Garbelotto, M. (2008). Molecular analysis to study invasions by forest pathogens: examples from Mediterranean ecosystems. *Phytopathologia Mediterranea* 47: 183–203.

Garbelotto, M. and Pautasso, M. (2012). Impacts of exotic forest pathogens on Mediterranean ecosystems: four case studies. *European Journal of Plant Pathology* 133: 101-116.

Garcia, M. (2004). *Effects of Pesticides on Soil Fauna: Development of Ecotoxicological Test Methods for Tropical Regions.* Centre for Development Research, University of Bonn, Bonn, Germany.

García-Palacios, P., Vandegehuchte, M.L., Shaw, E.A. *et al.* (2015). Are there links between responses of soil microbes and ecosystem functioning to elevated CO_2, N deposition and warming? A global perspective. *Global. Change Biology* 21: 1590-1600.

Garibaldi, A. (1987). Soil solarization: possibilities and limitations in Mediterranean countries. In: R. Cavalloro (ed.) *Integrated and Biological Control in Protected Crops.* Proceedings of a meeting of the EC Experts' Group, Heraklion, 24–26 April 1985. Balkema, Rotterdam, the Netherlands, pp. 24–26.

Garrett, K., Forbes, G., Savary, S. *et al.* (2011). Complexity in climate change impacts: an analytical framework for effects mediated by plant disease. *Plant Pathology* 60: 15–30.

Garrett KA. (2008). Climate change and plant disease risk. In: D.A. Relman, M.A. Hamburg, E.R. Choffnes and A. Mack (eds.) *Global Climate Change and Extreme Weather Events: Understanding the Contributions to Infectious Disease Emergence.* National Academies Press, Washington, USA, pp. 143–155.

Garrett, K.A., Dendy, S.P., Frank, E.E. *et al.* (2006). Climate change effects on plant disease: genomes to ecosystems. *Annual Review of Phytopathology* 44: 489–509.

Garrett, K.A., Forbes, G.A., Savary, S. *et al.* (2011). Complexity in climate-change impacts: an analytical framework for effects mediated by plant disease. *Plant Pathol* 60: 15–30.

Garrett, K.A., Nelson, R.J., Mundt, C.C. *et al.* (2001). The effects of host diversity and other management components on epidemics of potato late blight in the humid highland tropics. *Phytopathology* 91: 993–1000.

Gaston, K.J. and Williams, P.H. (1996). Spatial patterns in taxonomic diversity. In: K.J. Gaston (ed.) *Biodiversity*. Blackwell Science, Oxford, pp. 202–229.

Gautam, H.R. and Bhardwaj, M.L. (2011). Better practices for sustainable agricultural production and better environment. *Kurukshetra* 59(9): 3–7.

Ge F., Chen, F.J., Parajulee, N.M. *et al.* (2005). Quantification of diapausing fourth generation and suicidal fifth generation cotton bollworm, *Helicoverpa armigera,* in cotton and corn in northern China. *Entomologia Experimentalis et Applicata* 116: 1–7.

Gendron St-Marseille, A.F. (2013). Le nématode à kyste du soja (*Heterodera glycines*): enjeux des changements climatiques sur sa distribution, sa reproduction et sur les probabilités de synchronisme avec le soja (*Glycines max*) au Québec. Université de Sherbrooke, Sherbrooke, Qc, Canada.

Gerard, P.J., Kean J.M., Phillips C.B. *et al.* (2010). *Possible Impacts of Climate Change on Biocontrol Systems in New Zealand Agri Research.* Report for MAF Pol Project 0910-11689, Client report number: SC110/2010/170.

Gertsson, C.A. (2005). New species and new province records of scale insects from Sweden (Hemiptera: Coccoidea) up to the year 2004. *Entomologisk Tidskrift. Stockholm* 126: 35-42.

Gessler, C. and Pertot, I. (2012). Vf scab resistance of *Malus. Tree Struct. Funct* 26: 95–108.

Gessler, C., Pertot, I. and Perazzolli, M. (2011). *Plasmopara viticola:* A review of knowledge on downy mildew of grapevine and effective disease management. *Phytopathol. Mediterr* 50: 3–44.

Ghannoum, O., Searson, M.J. and Conroy, J.P. (2007). Nutrient and water demands of plants under climate change. In P.C.D. Newton, G. Edwards, A. Carran and P. Niklaus (eds.) *Agroecosystems in a Changing Climate. Advances in Agroecology Series, Vol 12.* CRC Press, Boca Raton.

Ghini, R. (2005). *Mudanças Climáticas Globais e Doenças de Plantas.* Jaguariúna: Embrapa Meio Ambiente, 104 pp.

Ghini, R., Bettiol, W. and Hamada, E. (2011). Diseases in tropical and plantation crops as affected by climate changes: current knowledge and perspectives. *Plant Pathol* 60: 122–132.

Ghini, R., Hamada, E. and Bettiol, W. (2008). Climate change and plant diseases. *Scientia Agricola* 65: 98–107.

Ghini, R., Hamada, E., Goncalves, R.R.V. *et al.* (2007). Análise de risco das mudanças climáticas globais sobre a sigatoka-negra da bananeira no Brasil. *Fitopatologia Brasileira* 32: 197-204.

Ghini, R., Hamada, E., Pedro Junior, M.J. and Goncalves, R.R.V. (2011). Incubation period of *Hemileia vastatrix* in coffee plants in Brazil simulated under climate change. *Summa Phytopathol* 37: 85-93.

Ghini, R., Hamada, F., Pedro Junior, M.J. *et al.* (2008). Risk analysis of climate change on coffee nematodes and leaf miner in Brazil. *Pesquisa Agropecudria Brasileira* 43: 187-194.

Ghini, R., Torre-Neto, A., Dentzien, A.F.M. *et al.* (2015). Coffee growth, pest and yield responses to free-air CO_2 enrichment. *Clim Change* 132: 307-320.

Ghule, B.D., Jagtap, A.B., Dhumal, V.S. and Deokar, A.B. (1989). Influence of weather factors on the incidence of leaf miner (*Aproaerema modicella* Deventer) on groundnut. *Journal of oil seeds Research* 6: 17-21.

Giblin-Davis, R.M. (1990). Potential for biological control of phytoparasitic nematodes in Bermuda grass turf with isolates of the *Pasteuria penetrans* group. *Proc. Fla. State Hort. Soc.* 103: 349-351.

Gillett, N.P., Arora, V.K., Zickfeld, K. *et al.* (2011). On-going climate change following a complete cessation of carbon dioxide emissions. *Nat Geosci* 4: 83–87.

Gleason, M.L., Parker, S.K., Pitblado, R.E. *et al.* (1997). Validation of a commercial system for remote estimation of wetness duration. *Plant Disease* 81: 825-829.

Gonias, E.D., Oosterhuis, D.M. and Bibi, A.C. (2008). Physiologic response of cotton to the insecticide imidacloprid under high temperature stress. *J Plant Growth Regul* 27: 77–82.

Goswami, R.S. and Kistler, H.C. (2004). Heading for disaster: *Fusarium graminearum* on cereal crops. *Molecular Plant Pathology* 5: 515–525.

Gouache, D., Bensadoun, A., Brun, F. *et al.* (2013). Modelling climate change impact on *Septoria tritici* blotch (STB) in France: Accounting for climate model and disease uncertainty. *Agric for Meteorol* 170: 242-252.

Goudriaan, J. and Zadoks, J.C. (1995). Global climate change: modelling the potential responses of agro-ecosystems with special reference to crop protection. *Environmental Pollution* 87: 215-224.

Graber, E.R., Meller-Harel, Y., Kolton, M. *et al.* (2010). Biochar impact on development and productivity of pepper and tomato grown in fertigated soilless media. *Plant Soil* 337: 481–496.

Graham, C.H. and Hijmans, R.J. (2006). A comparison of methods for mapping species ranges and species richness. *Global Ecology and Biogeography* 15: 578-587.

Gray, M.E., Sappington, T.W., Miller, N. *et al.* (2009). Adaptation and invasiveness of western corn rootworm: intensifying research on a worsening pest. *Annu Rev Entomol* 54:303–321.

Greco, N. and Brandonisio, A. (1986). The biology of *Heterodera carotae*. *Nematologica* 32: 447-460.

Greenplate, J.T., Penn, S.R., Shappley, Z. *et al.* (2000). Bollgard II efficacy: quantification of total lepidopteran activity in a 2-gene product. In: P. Dugger and D. Richter (eds.) *Proceedings of Beltwide Cotton Conference.* National Cotton Council of America, Memphis, Tennesse, USA, pp 1041–1043.

Gregory, P.J., Ingram, J.S.I., Campbell, B. *et al.* (1999). Managed production systems. In: B. Walker, W. Steffen, J. Canadell and J.S.I. Ingram (eds.) *The Terrestrial Biosphere and Global Change: Implications for Natural and Managed Systems.* Cambridge University Press, Cambridge, pp. 229–270.

Gregory, P.J., Johnson, S.N., Newton, A.C. and Ingram, J.S.I. (2009). Integrating pests and pathogens into the climate change/food security debate. *Journal of Experimental Botany* 60: 2827-2838.

Grime, J.P. (1997). Climate change and vegetation. In: M. Crawley (ed.) *Plant Ecology.* Blackwell Science, Oxford.

Grime, J.P. and Hodgson, J.G. (1987). Botanical contributions to contemporary ecological theory. *New Phytol* 106: 283–295.

Groninger, J.W.J.R., Zedaker, S.S.M. and Ber-rang, P.C. (1996). Photosynthetic response of loblolly pine and sweet gum seedling stands to elevated carbon dioxide, water stress, and nitrogen level. *Canadian Journal of Forest Research* 26: 95-102.

Groves, D. and Lempert, R. (2007). A new analytic method for finding policy relevant scenarios. *Global Environ. Change* 17: 73–85.

Grundy, P.S.S. (2003). *Potential Alternative to Chickpeas for Trap Cropping.* ANP Technology, Green mount Press, DPI, Queensland. 22(1): 14.

Gu, G., Cevallos-Cevallos, J.M., Vallad, G.E. *et al.* (2013). Organically managed soils reduce internal colonization of tomato plants by *Salmonella enterica* serovar *typhimurium. Phytopathology* 103: 381–388.

Gu, G., Hu J., Cevallos-Cevallos, J.M. *et al.* (2011). *Internal colonization of* Salmonella enterica *serovar* typhimurium *in tomato plants.* PLoS One 6:e27340 PubMedCentralPub MedGoogle Scholar

Guillerm, J.L., Floch, E. Le Maillet, J. and Boulet, C. (1990). The invading weeds within the Western Mediterranean Basin. In: F. di Castri, A.J. Hansen and M. Debussche (eds.) *Monogr Biol 65.* Springer, pp. 61–84.

Guisan, A. and Thuiller, W. (2005). Predicting species distribution: offering more than simple habitat models. *Ecol Lett* 8: 993–1009.

Guisan, A.Z.N.E. (2000). Predictive habitat distribution models in ecology. *Ecol Model* 135: 147–186.

Gunasekera, T.S., Paul, N.D. and Ayres, P.G. (1997). Responses of phyllosphere yeasts to UV-B (290–320) radiation: inter specific differences in sensitivity. *Mycol Res* 101: 779–785.

Gunton, R.M., Petit, S. and Gaba, S. (2011). Functional traits relating arable weed communities to crop characteristics. *J Veg Sci* 22: 541–550.

Guo, H., Sun, Y., Li, Y. *et al.* (2013). Pea aphid promotes amino acid metabolism both in *Medicago truncatula* and bacteriocytes to favor aphid population growth under elevated CO_2. *Glob Chang Biol* 19: 3210–3223.

Gupta, G.P., Mahapatro, G.K. and Puri, S.N. (2009). Contemporary scenario of pests and pesticides in the global climate change: Realities, apprehensions and possibilities. In: V.V. Ramamurthy, G.P. Gupta and S.N. Puri (eds.) *Proc. Natn. Symp. IPM Strategies to Combat Emerging Pests in the Current Scenario of Climate Change*, Pasighat, Arunachal Pradesh, pp. 399-412.

Gurr, G.M., Wratten, S.D. and Luna, J.M. (2003). Multifunction agricultural biodiversity: pest management and other benefits. *Basic and Applied Ecology* 4: 107–116.

Gutierrez, A.P., Ponti, L. and Cossu, Q.A. (2009). Effects of climate warming on olive and olive by *Bactrocera oleae* (Gmelin)) in California and Italy. *Clim Chang* 95: 195–217.

Gutierrez, A.P., Pointi, L., d'Oultremont, T. and Ellis, C.K. (2008). Climate change effects on poikilotherm tritrophic interactions. *Climate change* 87: 67-92.

Hadley Centre. (2007). *Climate Change.* www.metoffice.gov.uk/climatechange. Exeter, Devon, UK

Haegeman, A., Kyndt, T. and Gheysen, G. (2010). The role of pseudo-endoglucanases in the evolution of nematode cell wall-modifying proteins. *J. Mol. Evol.* 70: 441-452.

Haegeman, A., Mantelin, S., Jones, J.T. and Gheysen, G. (2012). Functional roles of effectors of plant-parasitic nematodes. *Gene* 492: 19-31.

Hahn, D.A. and Denlinger, D.L. (2007). Meeting the energetic demands of insect diapause: Nutrient storage and utilization. *Journal of Insect Physiology* 53: 760-773.

Halbrendt, J.M. (1996). Allelopathy in the management of plant-parasitic nematodes. *Journal of Nematology* 28: 8-14.

Hall, M. and Wall, R. (1995). Myiasis of humans and domestic animals. *Advances in Parasitology* 35: 257–334.

Hamada, E. and Ghini. R. (2011). Impacts of climate change on plant pests and diseases in Brazil. *Revista Mexicana de Ciencias Agrícolas* 2: 195-205.

Hamilton, J.G., Dermody, O., Aldea, M. *et al.* (2005). Anthropogenic changes in tropospheric composition increase susceptibility of soybean to insect herbivory. *Environment Entomology* 34: 479-485.

Hampson, G.F. (1908). The moths of India- Supplementary papers to the volume in the fauna of British India. *Ser.III, Pt.X, J.Bombay Nat. Hist.Soc.* 18: 257-271.

Hance, T., van Baaren, J., Vernon, P. and Boivin, G. (2007). Impact of extreme temperatures on parasitoids in a climate change perspective. *Annual Review of Entomology* 52: 107-126.

Hannukkala, A.O., Kaukoranta, T., Lehtinen, A. and Rahkonen, A. (2007). Late-blight epidemics on potato in Finland, 1933–2002; increased and earlier occurrence of epidemics associated with climate change and lack of rotation. *Plant Pathology* 56: 167–176.

Hanzlik, K. and Gerowitt, B. (2012). Occurrence and distribution of important weed species in German winter oilseed rape fields. *J Plant Dis Prot* 119: 107–120.

Hare, J.D. (1992). Effects of plant variation on herbivore-natural enemy interactions. In R.S. Fritz and E.L. Simms (eds.) *Plant Resistance to Herbivores and Pathogens: Ecology, Evolution and Genetics.* University of Chicago Press, Chicago, Illinois, USA, pp. 278-298.

Harlan, J.R. and deWet, J.M.J. (1965). Some thoughts about weeds. *Econ Bot* 19: 16–24.

Harrington, R., Fleming, R. and Woiwood, I.P. (2001). Climate change impacts on insect management and conservation in temperate regions: can they be predicted? *Agricultural and Forest Entomology* 3: 233-240.

Hartleb, H. and Heitefuss, R. (1997). Abiotic and biotic influences on resistance of crop plants against fungal pathogens. In: H. Hartleb, R. Heitefuss and H.H. Hoppe (eds.) *Resistance of Crop Plants Against Fungi.* Gustav Fischer Verlag, Jena, Germany, pp. 298–326.

Harvell, C.D., Mitchell, C.E., Ward, J. *et al.* (2002). Climate warming and disease risks for terrestrial and marine biota. *Science* 296: 2158-2162.

Hasse, V. (1987). *Cotton. No 1-2.* Philippine-German Cotton Project, Department of Agriculture. Manila, Philippines.

Hatfield, J.L., Boote, K.J., Kimball, B.A. *et al.* (2011). Climate impacts on agriculture: implications for crop production. *Agron J* 103: 351–370.

Hattenschwiler, S. and Korner, C. (2003). Does elevated CO_2 facilitate naturalization of the non-indigenous *Prunus laurocerasus* in Swiss temperate forests? *Functional Ecology* 17: 778–785.

Haverkort, A.J., Fasan, T. and Vandewaart, M. (1991). The influence of cyst nematodes and drought on potato growth. 2 Effects on plants water relations under semi-controlled conditions. *Netherlands Journal of Plant Pathology* 97: 162-170.

Hawkes, C.V., Kivlin, S.N., Rocca, J.D. *et al.* (2011). Fungal community responses to precipitation. *Global Change Biology* 17: 1637–1645.

Heagle, A.S. (2003). Influence of elevated carbon dioxide on interactions between *Frankliniella occidentalis* and *Trifolium repens*. *Environmental Entomology* 32(3): 421-424.

Heagle, A.S., Burns, J.C., Fisher, D.S. and Miller, J.E. (2002). Effects of carbon dioxide enrichment on leaf chemistry and reproduction by two spotted spider mites (Acari: Tetranychidae) on white clover. *Environmental Entomology* 31(4): 594-601.

Heagle, A.S. and Key, L.W. (1973). Effect of ozone on the wheat stem rust fungus. *Phytopathology* 63: 397–400.

Heitefuss, R. (1989). *Crop and Plant Protection: the Practical Foundations*. Ellis Horwood Limited, Chichester, UK.

Hellmann, J.J. (2008). Five potential consequences of climate change for invasive species. *Conservation Biology* 22: 534–543.

Hibberd, J.M., Whitbread, R. and Farrar, J.F. (1996a). Effect of elevated concentrations of CO_2 on infection of barley by *Erysiphe graminis*. *Physiological and Molecular Plant Pathology* 48: 37-53.

Hibberd, J.M., Whitbread, R. and Farrar, J.F. (1996b). Effect of 700 μmol^{-1} CO_2 and infection with powdery mildew on the growth and carbon partitioning of barley. *New Phytol.* 134: 309–315.

Hierro, J.L., Maron, J.L. and Callaway, R.M. (2005) A biogeographical approach to plant invasions: the importance of studying exotics in their introduced and native range. *Journal of Ecology* 93: 1-15.

Hilder, V.A. and Boulter, D. (1999). Genetic engineering of crop plants for insect resistance – A critical review. *Crop Protecion* 18: 177-191.

Hill, D.S. (1987). *Agricultural Insects Pests of Temperate Regions and Their Control.* Cambridge University Press, Cambridge, UK, 659 pp.

Hill, M.G. and Dymock, J.J. (1989). *Impact of Climate Change: Agricultural/Horticultural Systems.* DSIR Entomology Division Submission to the New Zealand Climate Change Program, Department of Scientific and Industrial Research, Auckland, New Zealand, 16 pp.

Hill, N.S. and Schmitt, D.P. (1989). Influence of temperature and soybean phenology on dormancy induction of *Heterodera glycines*. *J. Nematol.* 21: 361–369.

Hily, J.M., Scorza, R., Malinowski, T. *et al.* (2004). Stability of gene silencing-based resistance to plum pox virus in transgenic plum (*Prunus domestica* L.) under field conditions. *Transgenic Research* 13: 427–436.

Hirneisen, K.A. and Kniel, K.E. (2013). Norovirus surrogate survival on spinach during pre-harvest growth. *Phytopathology* 103: 389–394.

Hirschi, M., Stoeckli, S., Dubrovsky, M. *et al.* (2011). Downscaling climate change scenarios for apple pest and disease modelling in Switzerland. *Earth Syst Dyn Dissuss* 2: 493-529.

Hobbs, R.J. and Huenneke, L.F. (1992). Disturbance, diversity, and invasion: implications for conservation. *Conserv Biol* 6: 324–337.

Hoeksema, J.D., Lussenhop, J. and Teeri, J.A. (2000). Soil nematodes indicate food web responses to elevated atmospheric CO_2. *Pedobiologia* 44: 725-735.

Hoffmann-Hergarten, S., Hasky, K., Reitz, M. and Sikora, R.A. (1997). Induced systemic resistance by Rhizobacteria toward the cyst nematode, *Globodera pallida*, on potato. In A. Ogoshi, K. Kobayashi, Y. Homma *et al.* (eds.) *Plant Growth-Promoting Rhizobacteria—*

Present Status and Future Prospects. Proc. 4th Int. Workshop Plant-Growth-Promot. Rhizobacteria, OECD Joint Workshop, Sapporo, Japan, pp. 292–295.

Hogenhout, S.A., Van der Hoorn R.A., Terauchi R. and Kamoun S. (2009). Emerging concepts in effector biology of plant-associated organisms. *Mol. Plant-Microbe Interact.* 22: 115-122.

Holden, N., Pritchard, L. and Toth, I. (2009). Plants as an alternative environmental reservoir for human pathogenic enterobacteria. *FEMS Microbiol Rev* 33: 689–703.

Holley, J.D., Hofstra, G. and Hall, R. (1985a). Appearance and fine structure of lesions caused by the interaction of ozone and *Alternaria solani* in potato leaves. *Canadian Journal of Plant Pathology* 7: 277-282.

Holley, J.D., Hofstra, G. and Hall, R. (1985b). Effect of reducing oxidant injury and early blight on fresh weight and tuber density of potato. *Phytopathology* 75: 529-532.

Holm, L.G., Doll, J., Holm, E. *et al.* (1997). *Worlds Weeds: Natural Histories and Distribution.* John Wiley & Sons, New York, 1129 pp.

Holzner, W. and Immonen, R. (1982). The agrestal weed flora and vegetation of the world. In: W. Holzner and M. Numata (eds.) *Biology and Ecology of Weeds.* Dr. W. Junk, The Hague, pp 203–226.

Hossard, L., Philibert, A., Bertrand, M. *et al.* (2014). *Effects of Halving Pesticide Use on Wheat Production.* Sci Rep. 4, Article number: 4405. doi:10.1038/ srep04405PubMed CentralPub MedGoogle Scholar.

Hovmøller, M.S., Sørensen, C.K., Walter, S. and Justesen, A.F. (2011). Diversity of *Puccinia striiformis* on cereals and grasses. *Annu Rev Phytopathol* 49: 197–217.

Hovmøller, M.S., Walter, S. and Justesen, A.F. (2010). Escalating threat of wheat rusts. *Science* 329: 369.

Hovmøller, M.S., Yahyaoui, A.H., Milus, E.A. and Justesen, A.F. (2008). Rapid global spread of two aggressive strains of a wheat rust fungus. *Mol Ecol* 17: 3818–3826.

Howden, S.M., Gifford R.G. and Meinke Grains H. (2010). In C. Stokes, M.H. Haden, M. Howden (eds.) *Adapting Agriculture to Climate Change: Preparing Australian Agriculture for the Future.* CSIRO, Melbourne, Australia, pp. 21-40.

Howden, S.M., Soussana, J.F., Tubiello, F.N. *et al.* (2007). Adapting agriculture to climate change. *Proceedings of the National Academy of Science of the United States of America* 104: 19691–19696.

Howe, G.A. and Jander, G (2007). Plant immunity to insect herbivores. *Ann Rev Plant Biol* 17: 41–66.

Howlett, B.J., Idnurm, A. and Pedras, M.S.C. (2001). *Leptosphaeria maculans*, the causal agent of blackleg disease of Brassicas. *Fungal Genet. Biol.* 33: 1–14.

Hoyt, B. (2001). *Integrated Pest Management (IPM) and Food production.* Communication Services, Branch Office of Pesticide Programmes, US EPA, Washington, DC 20460 (703): 305-5017

Huang, G.Z., Allen, R., Davis, E.L. *et al.* (2006). Engineering broad root-knot resistance in transgenic plants by RNAi silencing of a conserved and essential root-knot nematode parasitism gene. *Proceedings of the National Academy of Sciences,* USA 103: 14302–14306.

Huang, J., Pray, C. and Rozelle, S. (2002). Enhancing the crops to feed the poor. *Nature* 418: 678–84.

Huang, Y.J., Evans N., Li Z.Q. *et al.* (2006). Temperature and leaf wetness duration affect phenotypic expression of Rlm6-mediated resistance to *Leptosphaeria maculans* in *Brassica napus. New Phytologist* 170: 129–141.

Huang, Y.J., Fitt B.D.L., Jedryczka M. *et al.* (2005). Patterns of ascospore release in relation to phoma stem canker epidemiology in England (*Leptosphaeria maculans*) and Poland (*Leptosphaeria biglobosa*). *Eur. J. Plant Pathol.* 111: 263–277.

Huang, Y.J., Pirie, E.J., Evans, N. *et al.* (2009). Quantitative resistance to symptomless growth of *Leptosphaeria maculans* (phoma stem canker) in *Brassica napus* (oilseed rape). *Plant Pathol* 58: 314–323.

Huber, L. and Gillespie, T.J. (1992) Modeling leaf wetness in relation to plant-disease epidemiology. *Annu Rev Phytopathol* 30: 553–577.

Hughes, L. and Bazzaz, F.A. (2001). Effects of elevated CO_2 on five plant-aphid interactions. *Entomologia Experimentalis et Applicata* 99: 87–96.

Huisman, O.C. (1982). Interrelations of root growth dynamics to epidemiology of root-invading fungi. *Annu. Rev. Phytopathol* 20: 303–327.

Hungate, B.A., Holland, E.A., Jackson, R.B. *et al.* (1997). The fate of carbon in grasslands under carbon dioxide enrichment. *Nature* 388: 576-579.

Hungate, B.A., Jaeger, C.H., Gamara, G. *et al.* (2000). Soil microbiota in two annual grasslands: Responses to elevated atmospheric CO_2. *Oceologica* 124: 589-598.

Hunt, R., Hand, D.W., Hannah, M.A. and Neal, A.M. (1991). Response to CO_2 enrichment in 27 herbaceous species. *Functional Ecology* 5: 410-421.

Hunter, M.D. (2001). Effects of elevated atmospheric carbon dioxide on insect-plant interactions. *Agric. For. Entomol.* 3: 153-159.

Hutchinson, G.E. (1957) Concluding remarks. *Cold Spring Harb Symp Quantum Biol* 22: 415–427.

Hyvönen, T., Luoto, M. and Uotila, P. (2012). Assessment of weed establishment risk in a changing European climate. *Agric Food Sci* 21: 348–360

IARI News. (2008). Brown plant hopper outbreak in rice. *IARI News* 24: 1-2.

ICIPE. (2003). *Scientific Report.* International Center for Insect Physiology and Ecology, Nairobi, Kenya.

IFPRI (International Food Policy Research Institute). (1998). *Pest Management and Food Production: Looking into the Future.* 2020 Vision Brief 52.

IMD. (2010). *Annual Climate Summary 2010.* India Meteorological Department, Pune; Government of India, Ministry of Earth Sciences, 27 pp.

Ingham, R.E., Trofymow, J.A., Ingham, E.R. and Coleman, D.C. (1985). Interactions of bacteria, fungi, and their nematode grazers: effects on nutrient cycling and plant growth. *Ecol. Monogr.* 55: 119–140.

Ingram, J.S.I., Gregory, P.J. and Izac, A.M. (2008). The role of agronomic research in climate change and food security policy. *Agric Ecosyst Environ* 126: 4–12.

Ingram, M., Nabhan, G. and Buchmann, S. (1996). Our forgotten pollinators: Protecting the birds and bees. *Global Pesticide Campaigner* 6(4): 1-12.

IPCC. (1990a). Climate change: The IPCC scientific assessment. In J.T. Houghton, G.J. Jenkins and J.J. Ephraums (eds.), *World Meteorological Organization and United Nations Environmental Program.* Cambridge University Press, Cambridge, 365 pp.

IPCC. (1990b). *The Potential Impacts of Climate Change on Agriculture and Forestry.* Intergovernmental Panel on Climate Change, Geneva and Nairobi, Kenya: World Meteorological Organization and UN Environment Program, 55 pp.

IPCC. (1996). Climate Change 1995: The Science of Climate Change. In J.T. Houghton, L.G. Meira Filho, B.A. Callander *et al.* (eds.) *Contribution of Working Group I to the Second*

Assessment Report of the Intergovernmental Panel on Climate Change. Cambridge University Press, Cambridge.

IPCC. (2001). Climate Change 2001. The Scientific Basis. In J.T. Houghton, Y. Ding, D.J. Griggs *et al.* (eds.) *Contribution of Working Group I to the Third Assessment Report of the Intergovernmental Panel on Climate Change.* Cambridge University Press, Cambridge, United Kingdom and New York, USA, 881pp.

IPCC. (2007) Climate change 2007: The physical science basis. In: S. Solomon, D. Qin, M. Manning *et al.* (eds.) *Contribution of Working Group I to the Fourth Assessment Report of the Intergovernmental Panel on Climate Change.* Cambridge University Press, Cambridge, 996 pp.

IPCC. (2007). Climate Change 2007: Impacts, Adaptation and Vulnerability. In M.L. Parry, O.F. Canziani, J.P Palutikof *et al.* (eds.) *Contribution of Working Group II to the Fourth Assessment Report of the Intergovernmental Panel on Climate Change.* Cambridge University Press, Cambridge, United Kingdom and New York, USA.

IPCC. (2007). Climate Change 2007: Synthesis Report. In: R.K. Pachauri and A. Reisinger (eds.) Contribution of Working Groups I, II and III to the Fourth Assessment Report of the Intergovernmental Panel on Climate Change Core Writing Team. IPCC, Geneva, Switzerland.

IPCC. (2007). Climate Change 2007: Working Group II: Impacts, Adaptation and Vulnerability. Cambridge University Press, Cambridge, UK.

IPCC. (2007). Climate Change, *The Fourth IPCC Assessment Report.* Cambridge University Press, Cambridge, UK.

IPCC. (2007a). Appendix I: Glossary. In: M.L. Parry, O.F. Canziani, J.P. Palutikof *et al.* (eds.) *Climate Change 2007: Impacts, Adaptation and Vulnerability.* Contribution of working group II to the fourth assessment report of the intergovernmental panel on climate change. Cambridge University Press, Cambridge, pp. 79–131.

IPCC. (2007b). Summary for policymakers. In: S. Solomon, D. Qin, M. Manning *et al.* (eds.) *Climate Change 2007: The Physical Science Basis.* Contribution of working group I to the fourth assessment report of the intergovernmental panel on climate change, Cambridge University Press, Cambridge, pp. 1–18.

IPCC. (2014). *Climate Change.* http://www.ipcc.ch/pdf/assessment-report/ar5/syr/SYR_AR5_LONGERREPORT.pdf

IRRI News. (2009). Pest outbreaks in India. *Rice Today* 6. www.ricenews.irri.org.

Isebaert, S., De Saeger, S., Devreese, R. *et al.* (2009). Mycotoxin-producing Fusarium species occurring in winter wheat in Belgium (Flanders) during 2002–2005. *J Phytopathol* 157: 108–116.

Ishigami, Y. Shimizu, Y. and Omasa, K. (2005). Evaluation of the risk to natural vegetation from climate change in Japan. *J. Agric. Meteorol.* 61: 69-75.

Isman, M.B. (1997). Neem and other botanical pesticides: Barriers to commercialization. *Phytoparasitica* 25: 339- 344.

IUCN. (1994). *International Union for Conservation Red List Categories.* Prepared by the IUCN Species Survival Commission. IUCN, Gland, Switzerland.

Jablonska, B., Ammiraju, J.S.S., Bhattarai, K.K. *et al.* (2007). The Mi-9 gene from *Solanum arcanum* conferring heat-stable resistance to root-knot nematodes is a homolog of Mi-1. *Plant Physiol.* 143: 1044-1054.

Jacob, D. and Podzun, R. (1997). Sensitivity studies with the regional climate model REMO. *Meteorol. Atmos. Phys.* 63: 119–129.

Jahn, M., Kluge, E. and Enzian, S. (1996). Influence of climate diversity on fungal diseases of field crops—evaluation of long-term monitoring data. *Aspects of Applied Biology* 45: 247–252.

Jatala, P. and Rusell, C.C. (1972). Nature of sweet potato resistance to *Meloidogyne incognita* and the effect of temperature on parasitism. *Journal of Nematology* 4: 1-7.

Jauni, M. and Hyvönen, T. (2012). Interactions between alien plant species traits and habitat characteristics in agricultural landscapes in Finland. *Biol Invasions* 14: 47–63.

Jeger, M.J. and Pautasso, M. (2008). Plant disease and global change—the importance of long-term data sets. *New Phytol.* 177: 8–11.

Jentsch, A., Kreyling, J., Boettcher-Treschkow, J. and Beierkuhnlein, C. (2009). Beyond gradual warming: extreme weather events alter flower phenology of European grassland and heath species. *Glob Chang Biol* 15: 837–849.

Jesus Junior, W.C., Valadares Junior, R., Cecilio, R.A. *et al.* (2008). Worldwide geographical distribution of black sigatoka for banana: predictions based on climate change models. *Sci Agric* 65: 40-53.

Johnson, B.C. and Young, B.G. (2002). Influence of temperature and relative humidity on the foliar activity of mesotrione. *Weed Sci.* 50: 157–161.

Jones, R. and Barbetti, M. (2012). Influence of climate change on plant disease infections and epidemics caused by viruses and bacteria. CAB reviews: perspectives in Agriculture, Veterinary Science. *Nutr Nat Resour* 7: 1–31.

Jones, R.A.C. (2009). Plant virus emergence and evolution: origins, new encounter scenarios, factors driving emergence, effects of changing world conditions, and prospects for control. *Virus Research* 141: 113–130.

Jones, R.L. Lindroth, M.C., Press, I. *et al.* (2002). Herbivory in global climate change research: direct effects of rising temperatures on insect herbivores. *Global Change Biology* 8: 1-16.

Joshi, S. and Viraktamath, C.A. (2004) The sugarcane woolly aphid, Ceratovacuna lanigera Zehntner (Hemiptera: Aphididae): its biology, pest status and control. *Curr Sci* 87: 307–316.

Joutei A.B. Roy L. Van Impe G. and Lebrun P. (2000). Effect of elevated CO_2 on the demography of a leaf-sucking mite feeding on bean. *Oecologia* 123: 75-81.

Jump, A.S. and Penuelas, J. (2005). Running to stand still: adaptation and the response of plants to rapid climate change. *Ecology Letters* 8: 1010–1020.

Jung, T. (2009). Beech decline in central Europe driven by the interaction between *Phytophthora* infections and climatic extremes. *For Pathol* 39: 73–94.

Junk J., Kouadio, L., Delfosse, P. and El Jarroudi, M. (2016). Effects of regional climate change on brown rust disease in winter wheat. *Clim Change* 135: 439-451.

Juroszek, P., Lumpkin, T.A. and Palada, M.C. (2008). Sustainable vegetable production systems. *Acta Horticulturae* 767: 133–149.

Juroszek, P. and Tiedemann, A.V. (2011). Potential strategies and future requirements for plant disease management under a changing climate. *Plant Pathol* 60: 100–112.

Kahn, R.P. (1991). Exclusion as a plant disease control strategy. *Annu Rev Phytopathol* 29: 219-246.

Kaiser, J. (1996). Pests overwhelm Bt cotton crop. *Nature* 273: 423.

Kannan, R. and James, D.A. (2009). Effects of climate change on global diversity: a review of key literature. *Tropical Ecology* 50: 31-39.

Kardol, P., Cregger, M.A., Campany, C.E. and Classen, A.T. (2010). Soil ecosystem functioning under climate change: plant species and community effects. *Ecology* 91: 767-781.

Karnosky, D.F., Percy, K.E., Xiang, B. *et al.* (2002). Interacting elevated CO_2 and tropospheric O_3 predisposes aspen *(Populus tremuloides* Michx.*)* to infection by rust *(Melampsora medusae* f. sp. *tremuloidae). Glob. Change Biol* 8: 329–338.

Karuppaiah, V. and Sujayanad, G.K. (2012). Impact of climate change on population dynamics of insect pests. *World Journal of Agricultural Sciences* 8(3): 240-246.

Katan, J. (1981). Solar heating (solarization) control of soil-borne pests. *Annual Review of Phytopathology* 19: 211–36.

Kaukoranta, T. (1996). Impact of global warming on potato late blight: risks, yield loss and control. *Agriculture and Food Science in Finland* 5: 311-327.

Kaukoranta, T. and Hakala, K. (2008). Impact of spring warming on sowing times of cereal, potato and sugar beet in Finland. *Agric Food Sci* 17: 165–176.

Keddy, P.A. (1992). Assembly and response rules: two goals for predictive community ecology. *J Veg Sci* 3: 157–165.

Kendal, A.C., Turner, J.C. and Thomas, S.M. (1985). Effects of CO_2 enrichment at different irradiances on growth and yield of wheat, I. Effects of cultivar and duration of CO_2 enrichment. *Journal of Experimental Botany* 36: 252-260.

Kennedy, A.D. (1997). Bridging the gap between general circulation model (GCM) output and biological microenvironments. *International Journal of Biometeorology* 40: 119-122.

Kerlin, K. (2002). *As Global Warming Rises, so do Tree-Killing Infestations of Bark Beetles.* E/The Environmental Magazine, 4 April. Available: http://www.enn.com/news/enn-stories/2002/04/ 04042002/s_46495.asp [accessed 8 July 2003].

Khadar, A.B., Prabhuraj, A., Srinivasa Rao, M. *et al.* (2014). Influence of elevated CO_2 associated with chickpea on growth performance of gram caterpillar, *Helicoverpa armigera* (Hüb.). *Applied Ecology and Environmental Research* 12(2): 345-353

Khaderkhan, H., Nataraju, M.S. and Nagaraja, G.N. (1998). Economics of IPM in tomato. In: P. Parvatha Reddy, N.K. Krishna Kumar and, A. Verghese (eds.) *Advances in IPM for Horticultural Crops.* Association for Advancement of Pest Management in Horticultural Ecosystems, Divn. of Entomol and Namatol, Indian Inst of Hort Res, Bangalore, pp. 151-152.

Khalid, S. and Shad, R.A. (2002). Potential advantage of recent allelochemical discovery and agro-ecosystems. *Prog. Farm.* 11: 30-35.

Khan, M.R. and Khan, M.W. (1994). Single and interactive effects of root-knot nematode and coal-smoke on okra. *New Phytologist* 126: 337–342.

Khan, M.R. and Khan, M.W. (1997). Effect of the root-knot nematode, *Meloidogyne incognita*, on the sensitivity of tomato to sulfur dioxide and ozone. *Environmental and Experimental Botany* 38: 117-130.

Khan, M.R. and Khan, M.W. (2007). Smoke pollution and the effects of root - knot nematodes on the growth of eggplants. *Plant Pathology* 43(4): 683 – 693.

Khan, S. (2008). Managing climate risks in Australia: options for water policy and irrigation management. *Aust. J. Exp. Agric.* 48: 265–273.

Khan, Z., Midega, C.A.O., Pittchar, J. *et al.* (2011). Push-pull technology: A conservation agriculture approach for integrated management of insect pests, weeds and soil health in Africa. *Int. J. Agric. Sustain.* 9: 162–170.

Khan, Z.R., Midega, C.A.O., Hutter, N.J. *et al.* (2006). Assessment of the potential of Napier grass (*Pennisetum purpureum*) varieties as trap plants for management of *Chilo partellus*. *Entomol. Exp. Appl.* 119: 15–22.

Khan, Z.R., Midega, C.A.O., Njuguna, E.M. *et al.* (2008). Economic performance of the "push-pull" technology for stem borer and striga control in smallholder farming systems in western Kenya. *Crop Prot* 27: 1084-1097.

Kickert, R.N., Tonella, G., Simonov, A. and Krupa, S.V. (1999). Predictive modelling of effects under global change. *Environmental Pollution* 100: 87-132.

Kim, K.H. and Cho, J. (2015). Predicting potential epidemics of rice diseases in Korea using multi-model ensembles for assessment of climate change impacts with uncertainty information. *Clim Change* 134: 327-339.

Kim, M., Hyten, D.L., Bent, A.F. and Diers, B.W. (2010). Fine mapping of the SCN resistance locus rhg1-bfrom PI 88788. *Plant Genome* 3: 81-89.

Kimball, B.A. (1985). Adaptation of vegetation and management practices to a higher carbon dioxide world. In: B.R. Strain and J.D. Cure (eds.) *Direct Effects of Increasing Carbon Dioxide on Vegetation*. US Department of Energy, Washington, pp. 185-204.

Kiritani, K. (1971). Distribution and abundance of the southern green stink bug, *Nezara viridula*. In: *Proceedings of the Symposium on Rice Insects*, TARC, MAF, Tokyo, pp. 235-248.

Kiritani, K. (2006). Predicting impacts of global warming on population dynamics and distribution of arthropods in Japan. *Popul Ecol* 48:5–12.

Klein, A.M., Vaissiere, B., Cane, J.H. *et al.* (2007). Importance of pollinators in changing landscapes for world crops. *Proceedings of the Royal Society B* 274: 303-313.

Klerks, M.M., Franz, E., van Gent-Pelzer, M. *et al.* (2007). Differential interaction of *Salmonella enterica* serovars with lettuce cultivars and plant-microbe factors influencing the colonization efficiency. *ISME J* 1: 620–631.

Klink, V.P., Hosseini P., MacDonald M.H. *et al.* (2009). Population-specific gene expression in the plant pathogenic nematode *Heterodera glycines* exists prior to infection and during the onset of a resistant or susceptible reaction in the roots of the *Glycine max* genotype Peking. *BMC Genomics* 10: 111.

Klink, V.P., Matsye P.D., Lawrence K.S. and Lawrence G.W. (2013). Engineered soybean cyst nematode resistance. *In:* H. El-Shemy (éd.), *Soybean - Pest Resistance*. InTech, Rijeka, Croatia, pp. 139-172.

Klironomos, J.N., Rillig, M.C., Allen, M.F. *et al.* (1997). Soil fungal-arthropod responses to *Populus tremuloides* grown under enriched atmospheric CO_2 under field conditions. *Global Change Biology* 3: 473-478.

Kloepper, J.W., Rodríguez-Kábana, R., McInroy, J.A. and Collins, D.J. (1991). Analysis of populations and physiological characterization of microorganisms in rhizospheres of plants with antagonistic properties to phytopathogenic nematodes. *Plant and Soil* 136: 95-102.

Kloepper, J.W., Rodnguez-Kabana, R., McInroy, J.A. and Young, R.W. (1992). Rhizosphere bacteria antagonistic to soybean cyst (*Heterodera glycines*) and root-knot (*Meloidogyne incognita*) nematodes: identification by fatty acid analysis and frequency of biological control activity. *Plant Soil* 139: 75–84.

Kobayashi, T., Ishiguro, K., Nakajima, T. *et al.* (2006). Effects of elevated atmospheric CO_2 concentration on the infection of rice blast and sheath blight. *Phytopathology* 96: 425-431.

Koch, D.W. and Gray, F.A. (1997). Nematode resistant oil radish for control of *Heterodera schachtii* I. Sugar beet-barley rotations. *J. Sugar Beet Res.* 34: 31–43.

Koenning, S.R. (2004). Population biology. In D.P. Schmitt, J.A. Wrather and R.D. Riggs (eds.) *Biology and Management of the Soybean Cyst Nematode, 2nd Ed.* Schmitt and Associates of Marceline, Marceline, MI, USA, pp. 73-88.

Koenning, S.R. and Anand, S.C. (1991). Effects of wheat and soybean planting date on *Heterodera glycines* population dynamics and soybean yield with conventional tillage. *Plant Dis.* 75: 301–304.

Koenning, S.R., Schmitt, D.P. and Barker, K.R. (1985). Influence of planting date on population dynamics and damage potential of *Pratylenchus brachyurus* on soybean. *J. Nematol.* 17: 428–434.

Koenning, S.R., Schmitt, D.P. and Barker, K.R. (1993). Cropping systems effects on *Heterodera glycines* population densities and associated soybean yield. *Plant Dis.* 77: 780–786.

Kolmer, J.A. and Leonard, K.J. (1986). Genetic selection and adaptation of *Cochliobolus heterostrophus* to corn hosts with partial resistance. *Phytopathology* 76: 774-777.

Kolton, M., Meller Harel Y., Pasternak Z. *et al.* (2011). Impact of biochar application to soil on the root-associated bacterial community structure of fully developed greenhouse pepper plants. *Appl. Environ. Microbiol.* 77: 4924–4930.

Konvicka, M., Maradova, M., Benes, J. (2003). Uphill shifts in distribution of butterflies in the Czech Republic: Effects of changing climate detected on a regional scale. *Global Ecology and Biogeography* 12: 403-410.

Kranti, K.R., Naidu, S., Dhawad, C.S. *et al.* (2005). Temporal and intraplant variation in Cry1Ac expression of Bt cotton and its influence on the development of cotton bollworm, *Helicovera armigera* (Hubner), (Noctuidae, Lepidoptera), *Current Science* 89: 291-298.

Kremen, C., Colwell, R.K., Erwin, T.L. *et al.* (1993). Terrestrial arthropod assemblages: Their use in conservation planning. *Conservation Biology* 7: 796- 808.

Krengel, S., Klocke, B., Seidel, P. and Freier, B. (2014). Changes in the occurrence of plant diseases, pests and their natural enemies. In: J.L. Lozan, H. Grassel, L. Karbe and Jendritzky (Hrsg.). *Warnsignal Klima: Gefahren für Pflanzen, Tiere und Menschen. 2. Auflage, Elektronische Veröffentlichung (Kap. 4.3)* – (www.warnsignale.uni-hamburg.de).

Krishna Reddy, M. (2010). Climate change and viral diseases of horticultural crops. In *Challenges of Climate Change: Indian Horticulture.* Westville Publishing House, New Delhi, 210 pp.

Kriticos, D.J., Crossman, N.D., Ota, N. and Scott, J.K. (2010). *Climate Change and Invasive Plants in South Australia.* National Research Flagship Climatic Adaptation-Australia.

Kriticos, D.J. and Randall, R.P. (2001). A comparison of systems to analyse potential weed distributions. In: R.H. Groves, F.D. Panetta and J.G. Virtue (eds.) *Weed Risk Assessment.* CSIRO Publishing, Melbourne, Australia, pp. 61-79.

Kriticos, D.J., Sutherst, R.W., Brown, J.R. *et al.* (2003). Climate change and the potential distribution of an invasive alien plant: *Acacia nilotica* s. sp. *indica* in Australia. *Journal of Applied Ecology* 40: 111-124.

Kroschel, J., Alcazar, J. and Pomar, P. (2009). Potential of plastic barriers to control Andean potato weevil *Premnotrypes suturicallus* Kuschel. *Crop Prot* 28: 466-476.

Kubisch, A., Degen, T., Hovestadt, T. and Poethke, H.J. (2013). Predicting range shifts under global change: the balance between local adaptation and dispersal. *Ecography* 36: 873–882.

Kuchlein, J.H. and Ellis, W.N. (1997). Climate-induced changes in the microlepidoptera fauna of the Netherlands and the implications for nature conservation. *J Insect Conserv* 1: 73–80.

Kudela, V. (2009). Potential impact of climate change on geographic distribution of plant pathogenic bacteria in Central Europe. *Plant Prot. Sci.* 45: S27–S32.

Kudo, G., Nishikawa, Y., Kasagi, T. and Kosuge, S. (2004). Does seed production of spring ephemerals decrease when spring comes early? *Ecol. Res.* 19: 255–259.

Kudsk, P. and Kristensen, J.L. (1992). Effect of environmental factors on herbicide performance. In *Proceedings of the First International Weed Control Congress*, Melbourne, VIV: 173–186.

Kumaratilake, A.R. and Preston, C. (2005). Low temperature reduces glufosinate activity and translocation in wild radish (*Raphanus raphanistrum*). *Weed Sci.* 53: 10–16.

Lal, M. (2003). Global climate change: India's monsoon and its variability. *Journal of Environmental Studies and Policy* 6 (1): 1-34.

Lal, R., Follett, R.F. and Kimble, J.M. (2003). Achieving soil carbon sequestration in the US: A challenge to the policymakers. *Soil Science* 168: 827-845.

Lalic, B., Jankovic, D. and Ninkov, M. (2013). Assessment of climate change impact on downy mildew appearance in Serbia using ECHAM5 climate model outputs. *Proceedings of the Conference Environmental Changes and Adaptation Strategies*, Skalica, Slovakia, 4 pp.

Lamichhane, J.R. (2014). *Xanthomonas arboricola* diseases of stone fruit, almond and walnut trees: progress toward understanding and management. *Plant Dis* 98: 1600–1610.

Lamichhane, J.R., Varvaro, L., Parisi, L. *et al.* (2014). Disease and frost damage of woody plants caused by *Pseudomonas syringe*: seeing the forest for the trees. *Adv Agron* 126: 235–295.

LaMondia, J.A. (1996). Response of additional herbaceous perennial ornamentals to *Meloidogyne hapla*. *Supplement to the Journal of Nematology* 28(4S): 636-638.

LaMondia, J.A. and Halbrendt, J.M. (2003). Differential host status of rotation crops to dagger, lesion and rot-knot nematodes. *Journal of Nematology* 35: 349.

Latour, X., Corberand, T., Laguerre, G. *et al.* (1996). The composition of fluorescent pseudomonad populations associated with roots is influenced by plant and soil type. *Appl Environ Microbiol* 62: 2449–2456

Lauer, E. (1953). Über die Keimtemperaturen von Ackerunkräutern und deren Einfluß auf die Zusammensetzung von Unkrautgesellschaften. *Flora* 140: 551–595.

Launay, M., Caubel, J., Bourgeois, G. *et al.* (2014). Climatic indicators for crop infection risk: application to climate change impacts on five major foliar fungal diseases in northern France. *Agric Ecosyst Environ* 197: 147-158.

Lavergne, S., Mouquet, N., Thuiller, W. and Ronce, O. (2010). Biodiversity and climate change: integrating evolutionary and ecological responses of species and communities. *Annu Rev Ecol Evol Syst* 41: 321–350.

Lavorel, S. and Garnier, E. (2002). Predicting changes in community composition and ecosystem functioning from plant traits: revisiting the Holy Grail. *Funct Ecol* 16: 545–556.

Lawn, D.A. and Noel, G.R. (1986). Field interrelationships among *Heterodera glycines*, *Pratylenchus scribneri*, and other nematode species associated with soybean. *J. Nematol.* 18: 98-105.

Lee, J. (2011). Combined effect of elevated CO_2 and temperature on the growth and phenology of two annual C3 and C4 weedy species. *Agric. Ecosys. Environ.* 140: 484–491.

Legrand, M.A., Colinet, H., Vernon, P. and Hance, T. (2004). Autumn, winter and spring dynamics of aphid *Sitobion avenae* and parasitoid *Aphidius rhopalosiphi* interactions. *Annals of Applied Biology* 145: 139–144.

Legreve, A. and Duveiller, E. (2010). Preventing potential diseases and pest epidemics under a changing climate. In: M.P. Reynolds (ed.) *Climate Change and Crop Production*. CABI Publishing, Wallingford, UK, pp. 50–70.

Lever, R.J.W. (1969). Do armyworm follow the rain? *Wild Crops* 21: 351-352.

Levin, D.A. (2009). Flowering-time plasticity facilitates niche shifts in adjacent populations. *New Phytol* 183: 661–666.

Lewin, H.D., Saroja, R., Leela David, A. and Padmanabhan, M.D. (1979). Influence of sowing time and weather on the incidence of groundnut leaf miner. *Indian J. Agric. Sci.* 49: 886-891.

Lewis, T. (1997). *Thrips as Crop Pests*. CAB International, Cambridge University Press, Cambridge, 740 pp.

Li, J., Faghihi J., Ferris, J.M. and Ferris, V.R. (1996). The use of RAPD amplified DNA as markers for virulence characteristics in soybean cyst nematode. *Fundam. Appl. Nematol.* 19: 143-150.

Li, Q., Liang, W.J., Jiang, Y. and Neher, D.A. (2007). Effect of elevated CO_2 and N fertilization on soil nematode abundance and diversity in wheat field. *Applied Soil Ecology* 36: 63-69.

Li, Q., Xu, C., Liang, W.J. *et al.* (2009). Residue incorporation and N fertilization affect the response of soil nematodes to the elevated CO_2 in a Chinese wheat field. *Soil Biology and Biochemistry* 41: 1497-1503.

Licciardi, S., Assogba-Komlan, F., Sidick, I. *et al.* (2008). A temporary tunnel screen as an eco-friendly method for small-scale growers to protect cabbage crop in Benin. *Int J Trop Insect Sci* 27: 152–158.

Lilley, C.J., Wang, D., Atkinson, H.J. and Urwin, P.E. (2010). Effective delivery of a nematode-repellent peptide using a root cap-specific promoter. *Plant Biotechnology Journal* 9: 151–161.

Lincoln, D.E., Couvet, D. and Sionit, N. (1986). Responses of an insect herbivore to host plants grown in carbon dioxide enriched atmospheres. *Oecologia* 69: 556–560.

Lipsitch, M., Singer, R.S. and Levin, B.R. (2002). Antibiotics in agriculture: when is it time to close the barn door. *Proc Natl Acad Sci USA* 99: S572–S574.

Liu, D. and Trumble, J.T. (2007). Comparative fitness of invasive and native populations of the potato psyllid (*Bactericera cockerelli*). *Entomologia Experimentalis et Applicata* 123(1): 35-42.

Liu, S., Kandoth, P.K., Warren, S.D. *et al.* (2012). A soybean cyst nematode resistance gene points to a new mechanism of plant resistance to pathogens. *Nature* 492: 256-260.

Loehle, C. (1996). Forest response to climate change: Do simulations predict unrealistic dieback? *Journal of Forestry* 94(9): 13–15.

Logan, J.A. and Powell, J.A. (2001). Ghost forests, global warming and the mountain pine beetle. *Am Entomol* 47: 160–173.

Logiswaran, G. and Mohanasundaram, M. (1986). Influence of weather factors on the catches of moths of groundnut leaf miner (*Aproaerema modicella* Deventer) in the light trap. *Entomon* 12:147-150.

Lonsdale, D. and Gibbs, J.N. (1996). Effects of climate change on fungal diseases of trees. In J.C. Frankland, N. Magan and G.M. Gadd (eds.) *Fungi and Environmental Change*. Cambridge University Press, Cambridge, United Kingdom, pp. 1–19.

L´opez-Fando, C. and Bello, A. (1995). Variability in soil nematode populations due to tillage and crop rotation in semi-arid Mediterranean agroecosystems. *Soil Tillage Res.* 36: 59–72.

Lososova, Z., Chytry, M., Kühn, I. *et al.* (2006). Patterns of plant traits in annual vegetation of man-made habitats in Central Europe. *Perspect Plant Ecol Evol Syst* 8: 69–81.

Loubser, J.T. (1988). Occurrence and pathogenicity of root-knot nematodes (*Meloidogyne* species) in South African vineyards. *S. Afr. J. Enol. Vitic.* 9(2): 21-27.

Lu, Y.H., Wu, K.M., Jiang, Y.Y. *et al.* (2012). Widespread adoption of Bt cotton and insecticide decrease promotes biocontrol services. *Nature* 487: 362–365.

Luck, J., Spackman, M., Freeman, A. et al. (2011). Climate change and diseases of food crops. *Plant Pathology* 60: 113–121.

Luo, Q., Bellotti, W., Williams, M. and Wang, E. (2009). Adaptation to climate change of wheat growing in South Australia: analysis of management and breeding strategies. *Agric Ecosyst Environ* 129: 261–267.

Luo, Y., TeBeest, D.O., Teng, P.S. and Fabellar, N.G. (1995). Simulation studies on risk analysis of rice blast epidemics associated with global climate in several Asian countries. *Journal of Biogeography* 22: 673-667.

Luo, Y., Teng, P.S., Fabellar, N.G. and TeBeest, D.O. (1997). A rice-leaf blast combined model for simulation of epidemics and yield loss. *Agricultural Systems* 53: 27–39.

Lupton, J., Chakraborty, S., Dale, M. and Sutherst, R. (1995). Assessment of the enhanced greenhouse effect on plant diseases. Case study of Stylosanthes anthracnose. In: *Proceedings of the 10th Biennial Australasian Plant Pathology Conference*, Lincoln University, New Zealand.

MacArthur, R.H. (1955). Fluctuations of animal populations and a measure of community stability. *Ecology* 36: 533–536.

MacGuidwin, A.E. and Layne, T.L. (1995). Response of nematode communities to sudan grass and sorghum-sudan grass hybrids grown as green manure crops. *J. Nematol. Suppl.* 27(4S): 609–616.

Mack, R.N., Simberloff, D., Lansdale, W.F. and Evans, H. (2000). Biotic invasions: causes, epidemiology, global consequences, and control. *Ecological Applications* 10: 689–710.

Macvean, R. and Dixon, A.F.G. (2001). The effect of plant drought-stress on populations of the pea aphid *Acyrthosiphon pisum*. *Ecological Entomology* 26: 440-443.

Madafiglio, G.P., Medd, R.W., Cornish, P.S. and Van de Ven, R. (2000). Temperature mediated responses of flumetsulam and metosulam on *Raphanus raphanistrum*. *Weed Research* 40: 387-395.

Madgwick, J.W., West, J.S., White, R.P. *et al.* (2011). Impacts of climate change on wheat anthesis and Fusarium ear blight in the UK. *Eur J Plant Pathol* 130: 117-131.

Mahajan, G., Singh, S. and Chauhan, B.S. (2012). Impact of climate change on weeds in the rice-wheat cropping system. *Curr. Sci.* 102: 1254–1255.

Mahan, J.R., Dotray, P.A. and Light, G.G. (2004). Thermal dependence of enzyme function and inhibition; implications for herbicide efficacy and tolerance. *Physiol. Plant.* 20: 187–195.

Mahmuti, M., West, J.S., Watts, J. *et al.* (2009). Controlling crop disease contributes to both food security and climate change mitigation. *International Journal of Agricultural Sustainability* 7: 189–202.

Malcolm, J.R., Markham, A., Neilson, R.P. and Garaci, M. (2002). Estimated migration rates under scenarios of global climate change. *J Biogeogr* 29: 835–849.

Malmstrom, C.M. and Field, C.B. (1997). Virus induced differences in the response of oat plants to elevated carbon dioxide. *Plant Cell Environment* 20: 178–88.

Malmstrom, C.M., Melcher, U. and Bosque-Pérez, N.A. (2011). The expanding field of plant virus ecology: Historical foundations, knowledge gaps, and research directions. *Virus Research* 159(2): 84-94.

Malumphy, C. (1997). Laurel scale, *Aonidia lauri* (Bouché) (Homoptera, Coccoidea, Diaspididae), a pest of bay laurel, new to Britain. *Entomologists Gazette* 48: 195-198.

Manea, A., Leishman, M.R. and Downey, P.O. (2011). Exotic C4 grasses have increased tolerance to glyphosate under elevated carbon dioxide. *Weed Sci.* 59: 28–36.

Manici L.M., Bregaglio S., Fumagalli D. and Donatelli M. (2014). Modelling soil borne fungal pathogens of arable crops under climate change. *Int J Biometeorol* 58: 2071-2083.

Manning, W.J. and Thedemann, A. (1995). Climate change: potential effects of increased atmospheric carbon dioxide (CO_2), ozone (O_3), and ultraviolet B (UV-B) radiation on plant diseases. *Environmental Pollution* 88: 219-245.

Mannion, C.D. (1991). *Influence of Tillage Practices and Crop Rotation on Soil Arthropods and Nematodes.* MS thesis. NC State Univ., Raleigh, NC.

Marín, D.H., Romero, R.A., Guzmán, M. and Sutton, T.B. (2003). Black sigatoka: An increasing threat to banana cultivation. *Plant Dis.* 87: 208–222.

Maron, J.L., Vila, M., Bommarco, R. *et al.* (2004). Rapid evolution of an invasive plant. *Ecological Monographs* 74: 261–280.

Martens, J.W., McKenzie, R.H. and Green, G.J. (1967). Thermal stability of stem rust resistance in oat seedlings. *Canadian Journal of Botany* 45: 451-458.

Martin, R.A. and Edgington, L.V. (1980). Effect of temperature on the efficacy of triadimenol and fenapanil to control loose smut of barley. *Canadian Journal of Plant Pathology* 2: 201–204.

Masters, G.J., Brown, V.K., Clarke, I.P. *et al.* (1998) Direct and indirect effects of climate change on insect herbivores: Auchenorrhyncha (Homoptera). *Ecological Entomology* 23(1): 45-52.

Matloob, A., Khaliq, A. and Chauhan, B.S. (2015). Weeds of rice in Asia: problems and opportunities. *Adv. Agron.* 130: 291–336.

Matros, A., Amme, S., Kettig, B. *et al.* (2006). Growth at elevated CO_2 concentrations leads to modified profiles of secondary metabolites in tobacco cv. SamsunNN and to increased resistance against infection with potato virus Y. *Plant, Cell and Environment* 29: 126–137.

Matson, A.L. and Williams, L.F. (1965). Evidence of a fourth gene for resistance to soybean cyst nematode. *Crop Sci.* 5: 477.

Matsunaka, S. (1983). Evolutions of rice weed control practices and research world perspective. In *Weed Control in Rice.* IRRI, P.O. Box 933, Manila, Philippines. pp. 5-18.

Mattson, W.J. and Haack, R.J. (1987). The role of drought in outbreaks of plant-eating insects. *BioScience* 37: 374-380.

Mau-Crimmins, T.M., Schussman, H.R. and Geiger, E.L. (2006). Can the invaded range of a species be predicted sufficiently using only native-range data?: Lehmann lovegrass (*Eragrostis lehmanniana*) in the southwestern United States. *Ecol Model* 193: 736–746.

Maywald, G.F. and Sutherst, R.W. (1991). *User's Guide to CLIMEX, a Computerised System for Comparing Climates in Ecology.* Australian CSIRO Division of Entomology Report No. 48, Canberra, 51 pp.

Mboup, M., Bahri, B., Leconte, M. *et al.* (2012). Genetic structure and local adaptation of European wheat yellow rust populations: the role of temperature-specific adaptation. *Evol. Appl.* 5: 341–352.

McCanny, S.J., Bough, M. and Cavers, P.B. (1988). Spread of proso millet (*Panicum miliaceum* L.) in Ontario, Canada. I. Rate of spread and crop susceptibility. *Weed Research* 28: 59–65.

McCanny, S.J. and Cavers, P.B. (1988). Spread of proso millet (*Panicum miliaceum* L.) in Ontario, Canada. II. Dispersal by combines. *Weed Research* 28: 67–72.

McDonald, A., Riha, S., DiTommaso, A. and DeGaetano, A. (2009). Climate change and the geography of weed damage: analysis of U.S. maize systems suggests the potential for significant range transformations. *Agric. Ecosys. Environ.* 130: 131–140.

McDonald-Madden, E., Runge, M.C., Possingham, H.P. and Martin, T.G. (2011). Optimal timing for managed relocation of species faced with climate change. *Nature Climate Change* 1: 261–265.

McDougall, K., Morgan, J., Walsh, N.G. and Williams, R. (2005). Plant invasions in treeless vegetation of the Australian Alps. *Perspectives in Plant Ecology, Evolution and Systematics* 7: 159-171.

Mcdowell, R.W., Condron, L.M., Main, B.E. and Dastgheib, F. (1997). Dissipation of imazapyr, flumetsulam and thifensulfuron in soil. *Weed Res.* 37: 381–389.

McElrone, A.J., Reid, C.D., Hoye, K.A. *et al.* (2005). Elevated CO_2 reduces disease incidence and severity of a red maple fungal pathogen via changes in host physiology and leaf chemistry. *Global Change Biol.* 11: 1828–1836.

McElrone, A.J., Sherald, J.L. and Forseth, I.N. (2001). Effects of water stress on symptomatology and growth of *Parthenocissus quinquefolia* infected by *Xylella fastidiosa. Plant Dis* 85: 1160–1164.

McFadyen, R. (2008). *Invasive Plants and Climate Change.* Weeds CRC Briefing Notes. CRC for Australian Weed Management.

McIntyre, S., Lavorel, S., Landsberg, J. and Forbes, T.D.A. (1999). Disturbance response in vegetation—towards a global perspective on functional traits. *J Veg Sci* 10: 621–630.

McNeil, D.L., Redden, R. and Patil, S.A. (2010). *Business Media.* Springer Science, Dordrecht, The Netherlands, pp. 115-140.

McSorley, R. (1994). Nematode management in sustainable agriculture. In K.L. Campbell, W.D. Graham and A.B. Del Bottcher (eds.) *Environmentally Sound Agriculture.* Am. Soc. Agric. Eng., St. Joseph, pp. 517–522.

McSorley, R. (1998). Alternative practices for managing plant-parasitic nematodes. *Amer. J. Alternative Agriculture* 13: 98-104.

McSorley, R. (1999). Host suitability of potential cover crops for root-knot nematodes. *Journal of Nematology Supplement* 31: 619-623.

McSorley, R. and Dickson, D.W. (1995). Effect of tropical rotation crops on *Meloidogyne incognita* and other plant-parasitic nematodes. *Journal of Nematology Supplement* 27: 535-544.

McSorley, R., Dickson, D.W., de Brito, J.A. *et al.* (1994). Effects of tropical rotation crops on *Meloidogyne arenaria* population densities and vegetable yields in microplots. *Journal of Nematology* 26:175-181.

McSorley, R. and Gallaher, R.N. (1993). Effects of crop rotation and tillage on nematode densities in tropical corn. *J. Nematol. Suppl.* 25: 814–819.

Mearns, L.O. (2000). Climate change and variability. In K.R. Reddy and H.F. Hodges (eds.) *Climate Change and Global Crop Productivity.* CABI Publishing, New York, pp. 7–35.

Medd, R.W., Van de Ven, R., Pickering, D.I. and Nordblom, T. (2001). Determination of environment specific dose response relationships for clodinafoppropargyl on *Avena* sp. *Weed Res.* 41: 351-68.

Mehrtens, J., Schulte, M. and Hurle, K. (2005). Unkrautflora in Mais - Ergebnisseeines Monitorings in Deutschland. *Gesunde Pflanz* 57: 206–218.

Meller Harel, Y., Elad, Y., Rav-David, D. *et al.* (2012). Biochar-induced systemic response of strawberry to foliar fungal pathogens. *Plant Soil* 357: 245–257.

Meller Harel, Y., Kolton, M., Elad, Y. *et al.* (2011). Induced systemic resistance in strawberry (*Fragaria* × *ananassa*) to powdery mildew using various control agents. *IOBC/WPRS Bulletin* 71: 47–51.

Melloy, P., Aitken, E., Luck, J. *et al.* (2014). The influence of increasing temperature and CO_2 on Fusarium crown rot susceptibility of wheat genotypes at key growth stages. *Eur J Plant Pathol* 140: 19-37.

Melloy, P., Hollaway, G., Luck, J. *et al.* (2010). Production and fitness of *Fusarium pseudograminearum* inoculum at elevated carbon dioxide in FACE. *Global Change Biol.* 16: 3363-3373.

Menendez, R. (2007). How are insects responding to global warming. *Tijdschrift voor Entomologie* 150: 355-365.

Menzel, A. and Fabian, P. (1999). Growing season extended in Europe. *Nature (London)* 397: 659.

Mesaros, N., Nordmann, P., Plésiat, P. *et al.* (2007). *Pseudomonas aeruginosa*: resistance and therapeutic options at the turn of the new millennium. *Clin Microbiol Infect* 13: 560–578.

Metzger, M.J., Leemans, R. and Schroter, D. (2005). A multidisciplinary multi-scale framework for assessing vulnerabilities to global change. *Int J Appl Earth Observ Geoinform* 7: 253–267.

MfE. (2008). *Climate Change Effects and Impacts Assessment: A Guidance Manual for Local Government in New Zealand, 2nd Edition.* B. Mullan, D. Wratt, S. Dean, M. Hollis, S. Allan, T. Williams and G. Kenny (eds.). Ministry for the Environment, Wellington, 149 pp.

Mikkelsen, B.L., Jørgensen, R.B. and Lyngkjær, M.F. (2015). Complex interplay of future climate levels of CO_2, ozone and temperature on susceptibility to fungal diseases in barley. *Plant Pathol* 64: 319-327.

Millennium Ecosystem Assessment. (2005). *Ecosystems and Human Wellbeing: Synthesis.* Island Press, Washington, DC.

Milus, E.A., Kristensen, K. and Hovmøller, M.S. (2009). Evidence for increased aggressiveness in a recent widespread strain of *Puccinia striiformis* f. sp. *tritici* causing stripe rust of wheat. *Phytopathology* 99: 89–94.

Mittler, R. (2006). Abiotic stress, the field environment and stress combination. *Trends in Plant Sci.* 11: 15-19.

Mittler, R. and Blumwald, E. (2010). Genetic engineering for modern agriculture: challenges and perspectives. *Annu. Rev. Plant Biol.* 61: 443-462.

MMSU. (2003). *Seven New Botanical Extracts vs. Garlic Pests.* Mariano Marcos State University, Batac, Ilocos Norte, Philippines.

Mohamed, K.I., Bolin, J.F., Musselman, L.J. and Peterson, T.A. (2007). Genetic diversity of *Striga* and implications for control and modelling future distributions. In G. Ejeta and J. Gressel (eds.) *Integrating New Technologies for Striga Control–Towards Ending the Witch-Hunt.* World Scientific, Singapore, pp. 71–84.

Mohamed, K.I., Papes, M., Williams, R. *et al.* (2006). Global invasive potential of 10 parasitic witch weeds and related Orobanchaceae. *Ambio* 35: 281–288.

Mojtahedi, H., Santo, G.S. and Ingham, R.E. (1993). Suppression of *Meloidogyne chitwoodi* with Sudan grass cultivars as green manure. *J. Nematol.* 25: 303–311.

Molla, O., Gonzalez-Cabrera, J. and Urbaneja, A. (2011) The combined use of *Bacillus thuringiensis* and *Nesidiocoris tenuis* against the tomato borer *Tuta absoluta. Biocontrol* 56: 883–891.

Mondal, K.K., Mani, C., Singh, J. *et al.* (2012). Fruit rot of Tinda caused by *Pseudomonas aeruginosa*—a new report from India. *Plant Dis* 96: 141.

Mondor, E.B., Tremblay, M.N., Awmack. C.S. and Lindroth, R.L. (2005). Altered genotypic and phenotypic frequencies of aphid populations under enriched CO_2 and O_3 atmospheres. *Glob. Change Biol.* 11: 1990-1996.

Monfreda, C., Ramankutty, N. and Foley, J.A. (2008): Farming the planet: 2. Geographic distribution of crop areas, yields, physiological types, and net primary production in the year 2000. *Global Biogeochemical Cycles* 22: GB1022.

Monkiedje, A., Zuehlke, S., Maniepi, S.J.N. and Spiteller, M. (2007). Elimination of racemic and enantioenriched metalaxyl based fungicides under tropical conditions in the field. *Chemosphere* 69: 655–663.

Mooney, H.A. and Hobbs, R.J. (eds.) (2000). *Invasive Species in a Changing World.* Island Press, Washington, DC, 457 pp.

Morales, F.J. and Jones, P.G. (2004). The ecology and epidemiology of whitefly-transmitted viruses in Latin America. *Virus Research* 100: 57-65.

Morecroft, M.D., Masters, G.J., Brown, V.K. *et al.* (2004). Changing precipitation patterns alter plant community dynamics and succession in an ex-arable grassland. *Functional Ecology* 18: 648-655.

Morimoto, N., Imura, O. and Kiura, T. (1998). Potential effects of global warming on the occurrence of Japanese pest insects. *Appl Entomol Zool* 33:147–155.

Mortensen, D.A., Bastiaans, L. and Sattin, M. (2000). The role of ecology in the development of weed management systems: an outlook. *Weed Res* 40: 49–62.

Mortensen, D.A. and Coble, H.D. (1989). The influence of soil water content on common cocklebur (*Xanthium strumarium*) interference in soybeans (*Glycine max*). *Weed Sci.* 37: 76–83.

Mosier, A.R., Duxbury, J.M., Freney, J.R. *et al.* (1998). Mitigating agricultural emissions of methane. *Climatic Change* 40: 39-80.

Mugniery, D. and Bossis, M. (1988). *Heterodera carotae* Jones, 1950. 1. Gamme d'hôtes, vitesse de développement cycle. *Revue de Nématologie* 11: 307-313.

Muñiz, M. and Nombela, G. (2001). Differential variation in development of the B- and Q-biotypes of *Bemisia tabaci* (Homoptera: Aleyrodidae) on sweet pepper at constant temperatures. *Environ. Entomol.* 30: 720–726.

Murugan, K. (2006). Bio-diversity of insects. *Curr. Sci.* 91: 1602-1603.

Musolin, D. (2007). Insects in a warmer world: ecological, physiological and life-history responses of true bugs (Heteroptera) to climate change. *Global Change Biology* 13: 1565-1585.

Musser, F.P. and Shelton, A.M. (2005). The influence of post-exposure temperature on the toxicity of insecticides to *Ostrinia nubilalis* (Lepidoptera: Crambidae). *Pest Management Science* 61: 508-510.

Myers, N., Mittermeier, R.A., Mittermeier, C.G. *et al.* (2000). Biodiversity hot spots for conservation priorities. *Nature* 403: 853-858.

NACCAP. (2008). *Climate Change Impacts on Pest Animals and Weeds. Communicating Climate Change.* National Agriculture and Climate Change Action Plan (NACCAP), Bureau of Meteorology, Department of Agriculture, Fisheries and Forestry, Australian Government, pp. 1-6.

Nahar, K., Kyndt T., Nzogela Y.B. and Gheysen G. (2012). Abscisic acid interacts antagonistically with classical defense pathways in rice-migratory nematode interaction. *New Phytol.* 196: 901-913.

Naimov, S., Dukiandjiev, S. and de Maagd, R.A. (2003). A hybrid *Bacillus thuringiensis* delta-endotoxin gives resistance against a coleopteran and a lepidopteron pest in transgenic potato. *Plant Biotechnology Journal* 1: 51–57.

Nancarrow, N., Constable, F.E., Finlay, K.J. *et al.* (2014). The effect of elevated temperature on barley yellow dwarf virus-PAV in wheat. *Virus Res* 186: 97-103.

Nash, M.A., Thomson, L.J. and Hoffmann, A.A. (2008). Effect of remnant vegetation, pesticides and farm management on abundance of the beneficial predator *Notonomus gravis* (Chaudoir) (Coleoptera: Carabidae). *Biological Control* 46: 83–93.

NATCOM. (2004). *India's Initial National Communication to the United Nations Framework-Convention on Climate Change.* Ministry of Environment and Forests, 268 pp.

Natural Resource Management Ministerial Council. (2007). *The Australian Weeds Strategy: A National Strategy for Weed Management in Australia.* Australian Government Department of the Environment and Water Resources, Canberra ACT.

Neher, D.A. (2010). Ecology of plant and free-living nematodes in natural and agricultural soil. *Annual Review of Phytopathology* 48: 371-394.

Neher, D.A. and Weicht, T.R. (2013). Nematode genera in forest soil respond differentially to elevated CO_2. *J. Nematol* 45: 214-222.

Neher, D.A., Weicht, T.R., Moorhead, D.L. and Sinsabaugh, R.L. (2004). Elevated CO_2 alters functional attributes of nematode communities in forest soils. *Functional Ecology* 18: 584-591.

Neilson, R. and Boag, B. (1996). The predicted impact of possible climatic change on virus-vector nematodes in Great Britain. *Eur J Plant Pathol* 102:193–199.

Neuhaus, W., Stachewicz, H. and Dunsing, M. (1974). UÈ ber den Ein¯uss von NiederschlaÈ gen auf die biologische Wirkung von Fungiziden zur Phytophthora-BekaÈmpfung. *Nachrichtenbl. P¯anzensch. DDR* 28: 149-153.

Newman, J.A. (2004). Climate change and cereal aphids: the relative effects of increasing CO_2 and temperature on aphid pollution dynamics. *Glob Change Biol* 10: 5-15.

Newman, J.A. (2005). Climate change and the fate of cereal aphids in Southern Britain. *Global Change Biology* 11: 940–944.

Newman, J.A. (2006). Using the output from global circulation models to predict changes in the distribution and abundance of cereal aphids in Canada: A mechanistic modeling approach. *Glob Change Biol* 11:1634–1642.

Newman, J.A., Gibson D.J., Parsons, A.J. and Thornley, J.H.M. (2003). How predictable are aphid population responses to elevated CO_2? *J. Animal Ecol.* 72: 556–566.

Newton, A.C., Begg, G.S. and Swanston, S. (2009). Deployment of diversity for enhanced crop function. *Annals of Applied Biology* 154: 309-322.

Newton, A.C., Gravouil, C. and Fountaine, J.M. (2010). Managing the ecology of foliar pathogens: ecological tolerance in crops. *Ann Appl Biol* 157: 343–359.

Newton, A.C., Johnson, S.N. and Gregory, P.J. (2011). Implications of climate change for diseases, crop yields and food security. *Euphytica* 179: 3-18.

Newton, A.C., Torrance, L., Holden, N. *et al.* (2012). Climate change and defense against pathogens in plants. *Adv. Appl. Microbiol.* 81: 89-132.

Newton, A.C. and Toth, I.K. (1999). Helper bacteria and pathogenicity assessments. *New Phytol* 144: 385-386.

Newton, A.C., Toth, I.K., Neave, P. and Hyman, L. (2004). Bacterial inoculum from a previous crop affects fungal disease development on subsequent non-host crops. *New Phytol* 163: 133-138.

Niblack, T.L. (2005). Soybean cyst nematode management reconsidered. *Plant Dis.* 89: 1020-1026.

Niblack, T.L., Arelli, P.R., Noel, G.R. *et al.* (2002). A revised classification scheme for genetically diverse populations of *Heterodera glycines*. *J. Nematol.* 34: 279-288.

Niblack, T.L., Lambert, K.N. and Tylka, G.L. (2006). A model plant pathogen from the kingdom animalia: *Heterodera glycines*, the soybean cyst nematode. *Annu. Rev. Phytopathol.* 44: 283-303.

Niklaus, P.A., Alphei, J., Ebersberger, D. *et al.* (2003) Six years of *in situ* CO_2 enrichment evoke changes in soil structure and soil biota of nutrient-poor grassland. *Global Change Biol* 9: 585–600.

Noe, J.P. (1998). Crop- and nematode-management systems. In K.A. Barker, G.A. Pederson, and G.L. Windham (eds.) *Plant and Nematode Interactions*. Agronomy Monograph No. 36, American Society of Agronomy, Madison, WI, pp. 159-171.

Noel, J.T., Arrach, N., Alagely, A. *et al.* (2010). Specific responses of *Salmonella enterica* to tomato varieties and fruit ripeness identified by *in vivo* expression technology. PLoS One 5:e12406 PubMedCentralPubMedGoogle Scholar.

Nogues-Bravo, D. (2009). Predicting the past distribution of species climatic niches. *Glob Ecol Biogeogr* 18: 521–531.

Nonhebel, S. (1996). Effects of temperature rise and increase in CO_2 concentration on simulated wheat yields. *Eur Clim Change* 34: 73–90.

Norby, R.J., Edwards, N.T., Riggs, J.S. *et al.* (1997). Temperature-controlled open top chambers for global change research. *Global Change Biology* 3: 259–267.

Nowak, R.S., Ellsworth, D.S. and Smith, S.D. (2004). Functional responses of plants to elevated atmospheric CO_2: do photosynthetic and productivity data from FACE experiments support early predictions? *New Phytol.* 162: 253–280.

Nusbaum, C.J. and Ferris, H. (1973). The role of cropping systems in nematode population management. *Annu. Rev. Phytopathol.* 11: 423–440.

Obeng-Ofori, D. and Ankrah, D.A. (2002). Effectiveness of aqueous neem extracts for the control of insect pests of cabbage (*Brassica oleracea* var *capitata* L) in Accra Plains of Ghana. *Agric. and Food Sc. J. of Ghana* 1(1): 83-94.

Oenema, O., Wrage, N., Velthof, G.L. *et al.* (2005). Trends in global nitrous oxide emissions from animal production systems. *Nutrient Cycling in Agroecosystems* 72: 51-65.

Oerke, E.C. (2006). Crop losses to pests. *J. Agric. Sci.* 144: 31–43.

Oerke, E.C., Dehne, H.W., Schonbeck, F. and Weber, A. (1994). *Crop Production and Crop Protection: Estimated Losses in Major Food and Cash Crops*. Elsevier, Amsterdam.

Oerke, E.C., Gerhards, R., Menz, G. and Sikora, R.A. (eds.) (2010). *Precision Crop Protection - The Challenge and Use of Heterogeneity*. Springer, Dordrecht, The Netherlands, 441 pp.

Ofuoku, A.U., Egho, E.O. and Enujeke, E.C. (2008). Integrated pest management (IPM) adoption among farmers in central agro-ecological zone of Delta State, Nigeria. *African J. of Agric Res.* 3 (12): 852 – 856.

Okada, H., Sakai, H., Tokida, T. *et al.* (2014). Elevated temperature has stronger effects on the soil food web of a flooded paddy than does CO_2. *Soil Biology and Biochemistry* 70: 166-175.

Olesen, J.E. and Bindi, M. (2002). Consequences of climate change for European agricultural productivity land use and policy. *European Journal of Agronomy* 16: 239- 262.

Olesen, J.E., Trnka, M., Kersebaum, K.C. *et al.* (2011). Impacts and adaptation of European crop production systems to climate change. *Eur J Agronomy* 34: 96-112.

Olfert, O. and Weiss, R.M. (2006). Impact of climate change on potential distributions and relative abundances of *Oulema melanopus, Meligethes viridescens* and *Ceutorrhynchus obstrictus* in Canada. *Agriculture, Ecosystems and Environment* 113: 295-301.

Olkowski, W., Daar, S. and Olkowski, H. (1995). *The Gardener's Guide to Common Sense Pest Control.* The Taunton Press, USA.

Olmstead, A.L. and Rhode, P.W. (2011). Adapting North American wheat production to climatic challenges, 1839–2009. *Proc. Natl. Acad. Sci. U.S.A.* 108: 480-485.

Olson, D.M. and Wackers, F.L. (2007). Management of field margins to maximize multiple ecological services. *Journal of Applied Ecology* 44: 13–21.

Olszyk, D.M. and Ingram, K.T. (1993). Effects of UV-Band global climate change on rice production. In *The EPA/IRRI Cooperative Research Plan.* International Rice Research Institute, The Philippines, pp. 234-253.

Opperman, C.H. and Conkling, M.A. (1998). Bioengineering resistance to plant parasitic nematodes. In K.R. Barker, G.A. Pederson and G.L. Windham (eds.) *Plant and Nematode Interactions.* Agron. Monogr. 36, Am. Soc. Agron., Crop Sci. Soc. Am., and Soil Sci. Soc. Am., pp. 239–250.

Ordon, F. (2008). Pflanzenzüchterische Möglichkeiten der Anpassung von Nutzpflanzen an zukünftige Produktionsbedingungen. In: A.V. Tiedemann, R. Heitefuss and F. Feldmann (eds.) *Pflanzenproduktion im Wandel – Wandel im Pflanzenschutz.* DPG Selbstverlag, Braunschweig, Germany, pp. 90–121.

Ortiz, R. (1998). Critical role of plant biotechnology for the genetic improvement of food crops: perspectives for the next millennium. *J. Biotechnol.* 1: 1–8.

Ortiz, R., Sayre, K.D., Govaerts, B. *et al.* (2008). Climate change: can wheat beat the heat? *Agriculture, Ecosystems & Environment* 126: 46–58.

Osbrink, W.L.A., Trumple, J.T. and Wagner, R.E. (1987). Host suitability of *Phaseolu lunata* for *Trichoplusia ni* (Lepidoptora: Noctuidae) in controlled carbon dioxide atmosphere. *Envi. Entomol.* 16: 639-644.

Osteen, C., Johnson, A.W. and Dowler, C.C. (1982). *Applying the Economic Threshold Concept to Control Lesion Nematodes on Corn.* Tech. Bull. No. 1670, USDA ERS.

Otte, A. (1991). Agro-ökosysteme und Habitatinseln in der Agrarlandschaft. 1. Agroökosysteme. 1.1 Struktur und Dynamik. Veränderungen im Keimungs- und Auflaufverhalten bei *Chenopodium ficifolium* im Vergleich von 1950 zu 1985-88. *Wissenschaftliche Beiträge der Martin-Luther-Universität Halle-Wittenberg R P Biowissenschaftliche Beiträge* 46: 38–48.

Otte, A., Bissels, S. and Waldhardt, R. (2006). Samen-, Keimungs- und Habitateigenschaften: Welche Parameter erklären Veränderungstendenzen in der Häufigkeit von Ackerwildkräutern in Deutschland? *J Plant Dis Prot Sonderheft* 20: 507–516.

Owensby, C.E. (1994). Climate change and grasslands: ecosystem-level responses to elevated carbon dioxide. In *Proceedings of XVII Iternational Grassland Congress*, pp. 1119-1124.

Paajanen, R., Julkunen-Tiitto, R., Nybakken, L. *et al.* (2011). Dark-leaved willow (*Salix myrsinifolia*) is resistant to three-factor (elevated CO_2, temperature and UV-B radiation) climate change. *New Phytologist* 190: 161–168.

Pachauri, R.K. and Reisinger, A. (2007). *Climate Change 2007: Synthesis Report.* Contribution of Working Groups I, II and III to the Fourth Assessment Report of the Intergovernmental Panel on Climate Change. IPCC, Geneva, Switzerland.

Pachauri, R.K. and Reisinger, A. (eds.). (2007). *IPCC Fourth Assessment Report: Climate Change 2007.* Cambridge University Press, Cambridge.

Padmanabhan, S.Y. (1973). The great Bengal famine. *Annual Review of Phytopathology* 11: 11-26.

Palumbi, S.R. (2001). Humans as the world's greatest evolutionary force. *Science* 293: 1786–1790.

Pande, S. and Shanna, M. (2010). Climate change: potential impact on chickpea and pigeon pea diseases in the rainfed semi-arid tropics (SAT). In *Proceedings of the 5th International Food Legumes Research Conference (IFLRC V) and 7th European Conference on Grain Legumes (AEP VII)*, Antalya, Turkey.

Pangga, I., Hanan, J. and Chakraborty, S. (2013). Climate change impacts on plant canopy architecture: implications for pest and pathogen management. *Eur J Plant Pathol* 135: 595–610.

Pangga, I.B., Chakraborty, S. and Yates, D. (2004). Canopy size and induced resistance in *Stylosanthes scabra* determine anthracnose severity at high CO_2. *Phytopathology* 94: 221-227.

Pankaj, Sharma, H.K. and Prasad, J.S. (2010). The rice root-knot nematode, *Meloidogyne graminicola*: an emerging problem in rice-wheat cropping system. *Ind. J. Nematol.* 40: 1-11.

Parmelee, R.W. and Alston, D.G. (1986). Nematode trophic structure in conventional and no-till agroecosystems. *J. Nematol.* 18: 403–407.

Parmesan, C. (2006) Ecological and evolutionary responses to recent climate change. *Ann Rev Ecol Evol Syst* 37: 637–669.

Parmesan, C., Ryrholm, N., Stefanescu, C. *et al.* (1999). Poleward shifts in geographical ranges of butterfly species associated with regional warming. *Nature* 399: 579-583.

Parmesan, C. and Yohe, G. (2003). A globally coherent fingerprint of climate change impacts across natural systems. *Nature* 421: 37- 42.

Parry, M.L., Canziani, O.F., Palutikof, J.P. *et al.* (eds.). (2007). Climate Change 2007: *Impacts, Adaptation and Vulnerability. Contribution of Working Group II to the Fourth Assessment Report of the Intergovernmental Panel on Climate Change.* Cambridge University Press, Cambridge.

Parry, M.L. and Carter, T.R. (1989). An assessment of the effects of climatic change on agriculture. *Climatic Change* 15: 95-116.

Parsons, M.S. and Davey, P. (2007). The rise of the scarce bordered straw *Helicoverpa armigera* (Hübner) (Lep.: Noctuidae). *Entomologist's Record and Journal of Variation* 119: 185-192.

Paterson, R.R.M. and Lima, N. (2010). How will climate change affect mycotoxins in food? *Food Res Int* 43: 1902–1914.

Pathak, H. (2010). Mitigating greenhouse gas and nitrogen loss with improved fertilizer management in rice: quantification and economic assessment. *Nutr. Cycling Agroecosyst.* 87: 443–454.

Pathak, M.D. and Khan, Z.R. (1994). *Insect Pests of Rice.* International Rice Research Institute, Manila, The Philippines.

Patterson, D.T. (1986). Responses of soybean CO_2 enrichment during drought. *Weed Sci.* 34: 203-210.

Patterson, D.T. (1995). Weeds in a changing climate. *Weed Science* 43: 685-701.

Patterson, D.T. and Flint, E.P. (1990). Implications of increasing carbon dioxide and climate change for plant communities and competition in natural and managed ecosystems. In: B.A. Kimball, N.J. Rosenberg and L.H. Allen Jr. (eds.) *Impact of Carbon Dioxide, Trace Gases, and Climate Change on Global Agriculture*. American Society of Agronomy, ASA Special Publication No. 53, Madison, WI, pp. 83-110.

Patterson, D.T. and Highsmith, M.T. (1989). Competition of spurred anoda (*Anoda cristata*) and velvet leaf (*Albutilon theophrasti*) with cotton (*Gossypium hirsutum*) during simulated drought and recovery. *Weed Sci.* 37: 658–664.

Patterson, D.T., Musser, R.L., Flint, E.P. and Eplee, R.E. (1982). Temperature responses and potential for spread of witch weed (*Striga lutea*) in the United States. *Weed Sci.* 30: 87–93.

Patterson, D.T., Westbrook, J.K., Joyce, R.J.V. *et al.* (1999). Weeds, insects, and diseases. *Climate Change* 43: 711–727.

Paul, N.D., Ayres, P.G., Rasanayagam, S. and Royle, D.J. (1998) Stratospheric ozone depletion, UV-B radiation and *Septoria tritici* infection of wheat. *7th International Congress of Plant Pathology*, Edinburgh, Scotland. British Society for Plant Pathology, Birmingham, UK, Invited papers abstracts 1: 4.2.6.S.

Pautasso, M., Dehnen-Schmutz, K. and Holdenrieder, O. *et al.* (2010). Plant health and global change—some implications for landscape management. *Biol Rev* 85: 729–755.

Pautasso, M., Döring, T.F., Garbelotto, M. *et al.* (2012). Impacts of climate change on plant diseases - opinions and trends. *Eur. J. Plant Patho.* 33(1): 295–313.

Pearman, P.B., Guisan, A., Broennimann, O. and Randin, C.F. (2008). Niche dynamics in space and time. *Trends Ecol Evol* 23: 149–158.

Pearson, R.G. and Dawson, T.P. (2003). Predicting the impacts of climate change on the distribution of species: are bioclimate envelope models useful? *Glob Ecol Biogeogr* 12: 361–371.

Pennypacker, B.W., Leath, K.T. and Hill, R.R. Jr. (1991). Impact of drought stress on the expression of resistance to *Verticillium albo-atrum* in alfalfa. *Phytopathology* 81: 1014–1024.

Persley, G.L. (1992). Beyond Mendel's garden. In A.A.S.A. Kadir and H.S. Barlow (eds.) *Pest Management and the Environment in 2000*. CAB Int., Wallingford, UK, pp. 360–370.

Peters, K., Breitsameter, L. and Gerowitt, B. (2014). Impact of climate change on weeds in agriculture: a review. *Agron. Sustain. Dev.* 34: 707–721.

Peters, K. and Gerowitt, B. (2014). Important maize weeds profit in growth and reproduction from climate change conditions represented by higher temperatures and reduced humidity. *J. Appl. Bot. Food Qual.* 87: 234–242.

Peterson, A.T. (2003). Predicting the geography of species' invasions via ecological niche modeling. *Q Rev Biol* 78: 419–433.

Petit, S., Boursault, A., Le Guilloux, M. *et al.* (2011). Weeds in agricultural landscapes. A review. *Agron Sustain Dev* 31: 309–317.

Petitpierre, B., Kueffer, C., Broennimann, O. *et al.* (2012). Climatic niche shifts are rare among terrestrial plant invaders. *Science* 335: 1344.

Petzoldt, C. and Seaman, A. (2010). *Climate Change Effects on Insects and Pathogens*. Climate Change and Agriculture: Promoting Practical and Profitable Responses, pp. 6-16. Available at http://www.climateandfarming.org/pdfs/FactSheets/III.2Insects. Pathogens.pdf.

Pfender, W.F. and Vollmer, S.S. (1999). Freezing temperature effect on survival of *Puccinia graminis* sub sp. *graminicola* in *Festuca arundinacea* and *Lolium perenne*. *Plant Dis.* 83: 1058–1062.

Pfleger, F.L. and Linderman, R.G. (eds.) (1994). *Mycorrhizae and Plant Health*. APS Press, St. Paul.

Phillips, O.L., Martinez, R.V., Arroyo, L. *et al.* (2002). Increasing dominance of large lianas in Amazonian forests. *Nature* 418: 770–774.

Phillips, S.J., Anderson, R.P. and Schapire, R.E. (2006). Maximum entropy modeling of species geographic distributions. *Ecological Modelling* 190: 231-259.

Phipps, P.M. (1993). IPM in peanuts: developing and delivering working IPM systems. *Plant Dis.* 77: 307–309.

Phoenix, G.K. and Press, M.C. (2005). Effects of climate change on parasitic plants: the root hemiparasitic Orobanchaceae. *Folia Geobotanica* 40: 205–216.

Pimentel, D., Brown, N., Vecchio, F. *et al.* (1993). Ethical issues concerning potential global climate change on food production. *Journal of Agricultural and Environmental Ethics* 5: 113-146.

Pimentel, D., McLaughlin, L., Zepp, A. *et al.* (1991). Environmental and economic impacts of reducing U.S. agricultural pesticide use. In D. Pimentel (ed.) *Handbook on Pest Management in Agriculture*, Vol. I. CRC Press, Boca Raton, FL, pp. 679-718.

Pineda, O., Bonierbale, M.W., Plaisted, R.L. *et al.* (1993). Identification of RFLP markers linked to the *Il*1 gene conferring resistance to the potato cyst nematode *Globodera rostochiensis*. *Genome* 36: 152–156.

Plazek, A., Hura, K., Rapacz, H. and Zur, I. (2001). The influence of ozone fumigation on metabolic efficiency and plant resistance to fungal pathogens. *J Appl Bot* 75: 8-13.

Plessl, M., Heller, W., Payer, H.D. *et al.* (2005). Growth parameters and resistance against *Drechslera teres* of spring barley (*Hordeum vulgare* L. cv. Scarlett) grown at elevated ozone and carbon dioxide concentrations. *Pant Biol* 7: 694-705.

Ploetz, R.C., Thomaz, J.E. and Slabaugh, W.R. (2003). Diseases of banana and plantain. In: R.C. Ploetz (ed.) *Diseases of Tropical Fruit Crops*. CABI Publishing, Wallingford, UK, pp. 73–134.

Podger, F.D., Mummery, D.C., Palzer, C.R. and Brown, M.L. (1990). Bioclimatic analysis of the distribution of damage to native plants in Tasmania by *Phytophthora*. *Australian Journal of Ecology* 15: 281-289.

Pollan, S. (2009). *Effect of Temperature on Development of the Microsporidium* Nosema lymantriae *and Disease Progress in the Host* Lymantria dispar. Master thesis, Institute of Forest Entomology, BOKU University of Natural Resources and Applied Life Sciences, Vienna.

Pollard, E. and Yates, T.J. (1993). *Monitoring Butterflies for Ecology and Conservation*. Chapman & Hall, London.

Polley, H.W., Johnson, H.B. and Mayeux, H.S. (1994). Increasing CO_2: Comparative responses of the C4 grass *Schizachyrium* and grassland invader *Prosopis*. *Ecology* 75: 976–988.

Polley, H.W., Johnson, H.B. and Tischler, C.R. (2002). Woody invasion of grasslands: evidence that CO_2 enrichment indirectly promotes establishment of *Prosopis glandulosa*. *Plant Ecol.* 164: 85–94.

Poorter, H. (1993). Interspecific variation in the growth response of plants to an elevated ambient CO_2 concentration. *Vegetatio* 104/105: 77–97.

Poorter, H. and Navas, M.L. (2003). Plant growth and competition at elevated CO_2: on winners, losers and functional groups. *New Phytol* 157: 175–198.

Pope, V.D., Gallani, M.L., Rowntree, P.R. and Stratton, R.A. (2000). The impact of new physical parametrizations in the Hadley Centre climate model – hadam3. *Clim. Dyn.* 16: 123–146.

Porter, J. (1995). The effects of climate change on the agricultural environments for crop insect pests with particular reference to the European corn borer and grain maize. In: R. Harrington and N.E. Stork (eds.) *Insect in a Changing Environment.* Academic Press, pp. 93-123.

Porter, J.H., Parry, M.L. and Carter, T.R. (1991). The potential effects of climate change on agricultural pests. *Agricultural and Forest Meteorology* 57: 221-240.

Porter, J.R. and Semenov, M.A. (2005). Crop responses to climatic variation. *Phil Trans R Soc B* 360: 2021–2031.

Potvin, C. (1986). Biomass allocation and phenological differences among southern and northern populations of the C4 grass *Echinochloa crusgalli*. *Journal of Ecology* 74: 915– 923.

Potvin, C. and Vasseur, L. (1997). Long-term CO_2 enrichment of a pasture community. *Ecology* 78: 666-677.

Prasad, J.S. and Somasekhar, N. (2009). *Nematode Pests of Rice: Diagnosis and Management.* Technical bulletin No. 38, Directorate of Rice Research (ICAR), Hyderabad, 29 pp.

Prasad, Y.G. and Bambawale, O.M. (2010). Effects of climate change on natural control of insect pests. *Indian Journal of Dryland Agriculture Research and Development* 25: 1-12.

Prasannakumar, N.R., Chander, S. and Pal, M. (2012). Assessment of impact of climate change with reference to elevated CO_2 on rice brown planthopper, *Nilaparvata lugens* (Stal.) and crop yield. *Curr. Sci.* 103: 10.

Prasch, C.M. and Sonnewald, U. (2013). Simultaneous application of heat, drought, and virus to Arabidopsis plants reveals significant shifts in signalling networks. *Plant Physiol.* 162: 1849-1866.

Price, P.W. 1987. The role of natural enemies in insect populations. In: P. Barosa and J.C. Schultz (eds.) *Insect Outbrteaks.* Academic Press, San Diego, CA, USA, pp. 287-312.

Pritchard, S.G. (2011). Soil organisms and global climate change. *Plant Pathol.* 60: 82 99.

Pritchard, S.G. and Amthor, J.S. (2005). *Crops and Environmental Change.* Food Products Press, Binghamton, 421 pp.

Pritchard, S.G., Rogers, H.H., Prior, S.A. and Peterson, C.M. (1999). Elevated CO_2 and plant structure: A review. *Glob. Change Biol* 5: 807–837.

Pysek, P., Jarosik, V., Kropac, Z. *et al.* (2005). Effects of abiotic factors on species richness and cover in Central European weed communities. *Agric Ecosyst Environ* 109: 1–8.

Qin, Z., DiTommaso, A. and Zhang, J.E. (2013). Potential distribution of two *Ambrosia* species in China under projected climate change. *Weed Res* 54: 520–531.

Radcliffe, E.B. and Ragsdale, D.W. (2002). Aphid-transmitted potato viruses: the importance of understanding vector biology. *Am J Pot Res* 79: 353–386.

Rajendran, T.P. (2009). Integrated pest management-Policy directions in the context of climate change. In: V.V. Ramamurthy, G.P. Gupta and S.N. Puri (eds.) *Proc. Natn. Symp. IPM Strategies to Combat Emerging Pests in the Current Scenario of Climate Change.* Pasighat, Arunachal Pradesh, pp. 8-13.

Ramsey, R.J.L., Stephenson, G.R. and Hall, J.C. (2005). A review of the effects of humidity, humectants, and surfactant composition on the absorption and efficacy of highly water-soluble herbicides. *Pesticide Biochem. Physiol.* 82: 162–175.

Rao, G.D., Katyal, J.C., Sinha, S.K. and Srinivas, K. (1995). Impacts of climate change on sorghum production in India: simulation study. In: C. Rosenzweig, L.H. Allen, L.A. Harper *et al.* (eds.) *Climate Change and Agriculture: Analysis of Potential International Impacts.* American Society of Agronomy, Madison, USA, pp. 325-337.

Rao, G.G.S.N., Rao, A.V.M.S. and Rao, V.U.M. (2009). Trends in rainfall and temperature in rainfed India in previous century. In: P.K. Aggarwal (ed.) *Global Climate Change and Indian Agriculture Case Studies from ICAR Network Project.* ICAR Publication, New Delhi, pp. 71-73.

Rao-Arelli, A.P. (1994). Inheritance of resistance to *Heterodera glycines* race 3 in soybean accessions. *Plant Dis.* 78: 898-900.

Raunkiaer, C. (1934). The *Life Forms of Plants and Statistical Plant Geography.* Oxford Univ Press, Oxford.

Rebetez, M. and Dobbertin, M. (2004). Climate change may already threaten Scots pine stands in the Swiss Alps. *Theorotical and Applied Climatology* 79: 1-9.

Reddy, M.K. (2010). Climate change and viral diseases of horticultural crops. In *Challenges of Climate Change: Indian Horticulture.* Westville Publishing House, New Delhi, 210 pp.

Reed, W. and Pawar, C.S. (1982). *Heliothis*: a global problem. In *Proceedings of the International Workshop on* Heliothis *Management,* November 15-20, 1981, International Crops Research Institute for the Semi-Arid Tropics, Patancheru, A.P., India, pp. 9-14.

Regnier, J. (1983). An oviposition model for the spruce budworm, *Choristoneura fumiferana* (Lepidoptera: Tortricidae). *Can Entomol.* 115: 1371-1382.

Reiners, S. and Petzoldt, C. (eds.) (2005). *Integrated Crop and Pest Management Guidelines for Commercial Vegetable Production.* Cornell Cooperative Extension publication #124VG http://www.nysaes.cornell.edu/recommends/.

Reitz, S.R., Karowa, D.N., Diawara, M.M. and Trumble, J.T. (1997). Effects of elevated atmospheric carbon dioxide on the growth and linear furanocoumarin content of celery. *Journal of Agricultural and Food Chemistry* 45: 3642-3646.

Rhoades, D.F. (1985). Offensive-defensive interactions between herbivores and plants: their relevance in herbivore population dynamics and ecological theory. *American National* 125: 205-238.

Ribas, A.F., Kobayashi, A.K., Pereira, L.F.P. and Vieira, L.G.E. (2006). Production of herbicide-resistant coffee plants (*Coffea canephora* P.) via *Agrobacterium tumefaciens*-mediated transformation. *Brazilian Archives in Biology Technology* 49: 11-19.

Richardson, A.D., Keenan, T.F., Migliavacca, M. *et al.* (2013). Climate change, phenology, and phenological control of vegetation feedbacks to the climate system. *Agric For Meteorol* 169: 156–173.

Richerzhagen, D., Racca, P., Zeuner, T. *et al.* (2011). Impact of climate change on the temporal and regional occurrence of Cercospora leaf spot in Lower Saxony. *Journal of Plant Diseases and Protection* 118: 168-177.

Ricketts, T.H., Regetz, J., Steffan-Dewenter, I. *et al.* (2008). Landscape effects on crop pollination services: Are there general patterns? *Ecology Letters* 11: 499-515.

Riha, S.J., Wilks, D.S. and Simoens, P. (1996). Impact of temperature and precipitation variability on crop model predictions. *Climatic Change* 32: 293–311.

Rishbeth, J. (1991). Armillaria in an ancient broad leaved woodland. *European Journal of Forest Pathology* 21: 239-249.

Ritter, R.L. and Coble, H.D. (1981). Influence of temperature and relative humidity on the activity of acifluorfen. *Weed Sci.* 29: 480–485.

Robert, Y. and Woodford, J.A.T. and Ducray-Bourdin, D.G. (2000) Some epidemiological approaches to the control of aphid-borne virus diseases in seed potato crops in northern Europe. *Virus Res* 71: 33–47.

Roberts, P.A. (1992). Current status of the availability, development, and use of host plant resistance to nematodes. *J. Nematol.* 24: 213–227.

Roberts, P.A. (1993). The future of nematology: integration of new and improved management strategies. *J. Nematol.* 25: 383–394.

Robinson, T.M.P. and Gross, K.L. (2010). The impact of altered precipitation variability on annual weed species. *Am J Bot* 97: 1625–1629.

Rodenburg, J., Meinke, H. and Johnson, D.E. (2011). Challenges for weed management in African rice systems in a changing climate. *J. Agric. Sci.* 149: 427–435.

Rodenburg, J., Riches, C.R. and Kayeke, J.M. (2010). Addressing current and future problems of parasitic weeds in rice. *Crop Prot.* 29: 210–221.

Rodrıguez-Kabana, R. and Canullo, G.H. (1992). Cropping systems for the management of phytonematodes. *Phytoparasitica* 20: 211–224.

Rodriguez-Kabana, R., King, P.S., Robertson, D.G. and Weaver, C.F. (1988). Potential of crops uncommon to Alabama for management of root-knot and soybean cyst nematodes. *Journal of Nematology Supplement* 20:116-120.

Rodrıguez-Kabana, R., Robertson, D.G., Weaver, C.F. and Wells, L.W. (1991). Rotations of bahia grass and castor bean with peanut for the management of *Meloidogyne arenaria*. *J. Nematol. Suppl.* 23 (4S): 658–661.

Rodriguez-Kabana, R., Robertson D.G., Wells L. *et al.* (1989). Crops uncommon to Alabama for the managing of *Meloidogyne arenaria* in peanut. *J. Nematology Supplement* 21: 712-716.

Rogers, H.H., Peterson C.M., McCrimmon J.N. and Cure J.D. (1992). Response of plant roots to elevated atmospheric carbon dioxide. *Plant Cell Environ.* 15: 749-752.

Rogers, H.H., Runion, G.B. and Krupa, S.V. (1994). Plant responses to atmospheric CO_2 enrichment, with emphasis on roots and the rhizosphere. *Environ Pollut* 83: 155–189.

Rohrs-Richey, J.K., Mulder, C.P.H., Winton, L.M. and Stanosz, G. (2011). Physiological performance of an Alaskan shrub (*Alnus fruticosa*) in response to disease (*Valsa melanodiscus*) and water stress. *New Phytologist* 189: 295–307.

Roller, E. and Poehling, H.F. (2004). *Ecological Infrastructures: Ideal Book on Functional Biodiversity at the Farm Level.* IOBC, OILB, Mattenbach AG, Winterthur, Switzerland.

Romies, J., Shanower, T.G. and Zebitz, C.P.W. (1999). *Trichogramma* egg parasitism of *Helicoverpa armigera* on pigeonpea and sorghum in southern India. *Entomologia Experimentalis et Applicata* 90: 69-81.

Roos, J., Hopkins, R., Kvarnheden, A. and Dixelius, C. (2011). The impact of global warming on plant diseases and insect vectors in Sweden. *Eur. J. Plant Pathol.* 129: 9–19.

Root, T.L., Price, J.T., Hall, K.R. *et al.* (2003). Finger prints of global warming on wild animals and plants. *Nature* 421: 57-60.

Rosenzweig, C., Iglesias, A., Yang X.B. *et al.* (2000). *Implications of Climate Change for U.S. Agriculture: Extreme Weather Events, Plant Diseases, and Pests.* Centre for Health and the Global Environment, Harvard Medical School, Cambridge, Massachusetts, 56 pp.

Rosenzweig, C. and Parry, M.L. (1994). Potential impact of climate change on world food supply. *Nature* 367: 133-138.

Rosenzweig, C., Yang, X.B., Anderson, P. *et al.* (2005). Agriculture: climate change, crop pests and diseases. In: P. Epstein and E. Mills (eds.) *Climate Change Futures: Health, Ecolovical and Economic Dimensions.* Centre for Health and the Global Environment at Harvard Medical School, Boston, MA, USA, pp. 70–77.

Rossi, M., Goggin, F.L., Milligan, S.B. *et al.* (1998). The nematode resistance gene Mi of tomato confers resistance against the potato aphid. *Proceedings of National Acadamy of Science USA* 95: 9750–9754.

Rossi, V. and Battilani, P. (1991). CERCOPRI: A Forecasting model for primary infections of *Cercospora* leaf spot of sugar beet. *EPPO Bull.* 21: 527–531.

Rosso, M.N., Hussey, R.S., Davis, E.L. *et al.* (2012). Nematode effector proteins: targets and functions in plant parasitism. *In:* F. Martin and S. Kamoun (éds.) *Effectors in Plant–Microbe Interactions.* John Wiley & Sons Inc., New York, USA, pp. 327-354.

Roth, S.K. and Lindroth, R.L. (1995). Elevated atmospheric CO_2 effects on phytochemistry, insect performance and insect–parasitoid interactions. *Global Change Biol.* 1: 173–182.

Rowntree, P.R. (1990). Estimate of future climatic change over Britain. *Weather* 45: 79-88.

Roy, B.A., Güsewell, S. and Harte, J. (2004). Response of plant pathogens and herbivores to a warming experiment. *Ecology* 85: 2570–2581.

Roy, D.B., Rothery, P., Moss, D. *et al.* (2001). Butterfly numbers and weather: predicting historical trends in abundance and the future effects of climate change. *Journal of Applied Ecology* 25: 819-828.

Roy, D.B. and Sparks, T.H. (2000) Phenology of British butterflies and climate change. *Global Change Biology* 6: 407–416.

Roy, H.E., Peyton, J., Aldridge, D.C. *et al.* (2014). Horizon scanning for invasive alien species with the potential to threaten biodiversity in Great Britain. *Glob Chang Biol.* 20: 3859–3871.

Roy, K.W., Rupe, J.C., Hershman, D.E. and Abney, T.S. (1997). Sudden death syndrome. *Plant Disease* 81: 1100-1111.

Runion, G.B., Curl, E.A., Rogers, H.H. *et al.* (1994). Effects of free-air CO_2 enrichment on microbial populations in the rhizosphere and phyllosphere of cotton. *Agricultural and Forest Meteorology* 70: 117-130.

Ruttan, V.W. (1999). The transition to agricultural sustainability. *Proc Natl Acad Sci USA* 96: 5960–5967.

Sachs, E.S., Benedict, J.H., Stelly, D.M. *et al.* (1998). Expression and segregation of genes encoding Cry1A insecticidal proteins in cotton. *Crop Science* 38: 1-11.

Sachs, J. (2008). *Common Wealth: Economics for a Crowded Planet.* Allen Lane, London.

Salam, M.U., MacLeod, W.J., Salam, K.P. *et al.* (2011). Impact of climate change in relation to *Ascochyta* blight on field pea in Western Australia. *Austral Plant Pathol* 40: 397-406.

Salinari, F., Giosue, S., Rossi, V. *et al.* (2007). Downy mildew outbreaks on grapevine under climate change: elaboration and application of an empirical-statistical model. *OEPP/EPPO Bulletin* 37: 317-326.

Salinari, F., Giosue, S., Tubiello, F.N. *et al.* (2006). Downy mildew (*Plasmopara viticola*) epidemics on grapevine under climate change. *Global Change Biology* 12: 1299-1307.

Salinger, M.J., Stigter, C.J. and Das H.P. (2000). Agrometeorological adaptation strategies to increasing climate variability and climate change. *Agric Forest Meteorol* 103:167–184.

Samietz, S., Stoeckli, S., Hirschi, M. *et al.* (2014). Modelling the impact of climate change on sustainable management of the codling moth (*Cydia pomonella* L.) as key pest in apple. *Acta Hort.* 18(4): 1087-1996.

Samra, J.S. (2003). Impact of climate and weather on Indian agriculture. *Indian Soc. Soil Sci.* 51: 418-430.

Samra, J.S. and Singh, G. (2004). *Heat Wave of March 2004: Impact on Agriculture.* Indian Council of Agricultural Research, New Delhi, 32 pp.

Sandermann Jr., H. (2000). Ozone/biotic disease interactions: Molecular biomarkers as a new experimental tool. *Environmental Pollution* 108: 327- 332.

Sankoh, O.A. (1999). Environmental assessment and management. *Afric. J. of Environ. Managment* 1(1): 3-10

Saracino, F., Scardicchio, D., Faretra, F. *et al.* (1995). SEMM: a multimedia expert system for integrated and crop protection. *Difesa Piante* 18(2): 143–153.

Sarfraz, M., Keddie, A.B. and Dosdall, L.M. (2005). Biological control of the diamondback moth, *Plutella xylostella*: a review. *Biocontrol Sci Technol* 15: 763–789.

Sasser, J.N. and Freckman, D.W. (1987). A world perspective on Nematology: The role of the Society. In J.A. Veech and D.W. Dickson (eds.) *Vistas of Nematology*. Society of Nematologists, Hyattsville, Maryland, pp. 7-14.

Savary, S., Mila, A., Willocquet, L. *et al.* (2011a). Risk factors for crop health under global change and agricultural shifts: a framework of analyses using rice in tropical and subtropical Asia as a model. *Phytopathology* 101: 696–709.

Savary, S., Nelson, A., Sparks, A.H. *et al.* (2011b). International agricultural research tackling the effects of global and climate changes on plant diseases in the developing world. *Plant Disease* 95: 1204–1216.

Savile, D.B.O. and Urban, Z. (1982). Evolution and ecology of *Puccinia graminis*. *Preslia* 54: 97-104.

Saxena, K. (1987). Indigeneous crop protection practices Sub-Saharan East Africa. *In: Database of Natural Crop Protectant Chemicals (DNCPC).* Available at http: www.ipmafrica. htpm.

Sayre, R.M., Patrick, Z.A. and Thorpe, H.J. (1965). Identification of a selective nematicidal component in extracts of plant residues decomposing in soil. *Nematologica* 11: 263–268.

Schaafsma, A.W. and Hooker, D.C. (2007). Climatic models to predict occurrence of *Fusarium* toxins in wheat and maize. *International Journal of Food Microbiology* 119: 116–125.

Schepers, H.T.A.M. (1996). Effect of rain on efficacy of fungicides on potato against *Phytophthora infestans*. *Potato Research* 39: 541-550.

Scherm, H. (2000). Simulating uncertainty in climate-pest models with fuzzy numbers. *Environmental Pollution* 108: 373-379.

Scherm, H. (2004). Climate change: can we predict the impacts on plant pathology and pest management? *Can J Plant Pathol* 26: 267–273.

Scherm, H., Sutherst, R.W., Harrington, R. and Ingram, J.S.I. (2000). Global networking for assessment of impacts of global change on plant pests. *Environmental Pollution* 108: 333-341.

Scherm, H. and Van Bruggen, A.H.C. (1994). Global warming and non-linear growth: how important are changes in average temperature? *Phytopathology* 84: 1380-1384.

Scherm, H. and Yang, X.B. (1995). Interannual variations in wheat rust development in China and the United States in relation to the El Ni˜no/Southern Oscillation. *Phytopathology* 85: 970–976.

Schillhorn, T.W. (1999). Agricultural Pest Management at Crossroads: New Opportunities and New Risks. The World Bank, Washington DC, USA.

Schmidt, G.A., Hoffmann, G., Shindell, D.T. and Hu, Y. (2005). Modeling atmospheric stable water isotopes and the potential for constraining cloud processes and stratosphere-troposphere water exchange. *J. Geophys. Res.* 110: D21314. doi:10.1029/2005JD005790.

Schmidt, M.H., Thewes, U., Thies, C. and Tscharntke, T. (2004). Aphid suppression by natural enemies in mulched cereals. *Entomologia Experimentalis et Applicata* 113: 87–93.

Schmutterer, H. (1990). Properties and potentials of natural pesticides from the Neem tree. *Annual Review of Entomology* 35: 371-298.

Schoeneweiss, D.F. (1975). Predisposition, stress, and plant disease. *Annu. Rev. Phytopathol* 13: 193–211.

Schulze, E.D. and Mooney, H.A. (1994). *Biodiversity and Ecosystem Function.* Springer, New York, 531 pp.

Scott, J.J. and Kirkpatrick, J.B. (2005). Changes in Sub-Antarctic Heard Island vegetation at sites occupied by *Poa annua*, 1987–2000. *Arctic, Antarctic, and Alpine Research* 37: 366-371.

Seem, R.C., Magarey, R.D., Zack, J.W. and Russo, J.M. (2000). Weather forecasts at the whole plant level: implications for estimating disease risk with global climate models. *Environmental Pollution* 108: 389-395.

Senft, D. (1995). FACE-ing the future. *Agricultural Research* 43: 4-6.

Shabani, F. and Kumar, L. (2013). Risk levels of invasive *Fusarium oxysporum* f. sp. in areas suitable for date palm (*Phoenix dactylifera*) cultivation under various climate change projections. PLOS ONE 8, e83404 (DOI: 10.1371/journal.pone.0083404).

Shabani, F., Kumar, L. and Esmaeili, A. (2014). Future distributions of *Fusarium oxysporum* f. sp. in European, Middle Eastern and North African agricultural regions under climate change. *Agr Ecosyst Environ* 197: 96-105.

Shao, Y., Bao, W., Chen, D. *et al.* (2015). Using structural equation modeling to test established theory and develop novel hypotheses for the structuring forces in soil food webs. *Pedobiologia* 58: 137-145.

Sharma, H. (2010). Global warming and climate change: Impact on arthropod biodiversity, pest management and food security. In *Souvenir National Symposium on Perspectives and Challenges of IPM for Sustainable Agriculture*, Y.S. Paramar Univ. of Hort. and Forestry, Solan and Indian Society of Pest Management and Economic Zoology, pp. 1-14.

Sharma, H.C. (2002). Host plant resistance to insects: Principles and practices. In B.B. Sarath, K.S. Varaprasad, K. Anitha *et al.* (eds.) *Resources Management in Plant Protection. Volume 1.* Plant Protection Association of India, Rajendarnagar, Hyderabad, Andhra Pradesh, India, pp. 37-63.

Sharma, H.C. (2005). Heliothis/Helicoverpa *Management: Emerging Trends and Strategies for Future Research.* Oxford and IBH Science Publishers, USA, 469 pp.

Sharma, H.C. (2014). Climate change effects on insects: Implications for crop protection and food security. *Journal of Crop Improvement* 28: 229-259.

Sharma, H.C., Dhillon, M.K., Kibuka, J. and Mukuru, S.Z. (2005). Plant defense responses to sorghum spotted stem borer, *Chilo partellus* under irrigated and drought conditions. *International Sorghum and Millets Newsletter* 46: 49-52.

Sharma, H.C., Mukuru, S.Z., Manyasa, E and Were, J.W. (1999). Breakdown of resistance to sorghum midge, *Stenodiplosis sorghicola. Euphytica* 109(2): 131-140.

Sharma, H.C. and Ortiz, R. (2000). Transgenics, pest management, and the environment. *Current Science* 79: 421-437.

Sharma, H.C., Srivastava, C.P., Durairaj, C. and Gowda, C.L.L. (1991). Pest management in grain legumes and climate change. In S.S. Yadav, R.W. Sutherst, G.F. Maywald and W. Bottomly (eds.) Climate change and management of cool season grain legume crops from CLIMEX to PESKY, a generic expert system for risk assessment. *EPPO Bulletin* 21: 595–608.

Sharma, H.C., Srivastava, C.P., Durairaj, C. and Gowda, C.L.L. (2010). Pest management in grain legumes and climate change. In S.S. Yadav, D.L. McNeil, R. Redden and S.A. Patil, (eds.) *Climate Change and Management of Cool Season Grain Legume Crops*. Business Media, Springer Science, Dordrecht, The Netherlands, pp. 115-140.

Sharma, M., Mangala, U.N., Krishnamurthy, M. et al. (2010). Drought and dry root of chickpea. In Proceedings of the 5th International Food Legumes Research Conference (IFLRC V) and 7th European Conference on Grain Legumes (AEP VII), Antalya, Turkey.

Sharma, S.D. and Singh, M. (2001). Environmental factors affecting absorption and bio-efficacy of glyphosate in Florida beggar weed (*Desmodium tortuosum*). *Crop Prot.* 20: 511–516.

Shaw, M.R., Zavaleta, E.S., Chiariello, N.R. *et al.* (2002). Grassland responses to global environmental changes suppressed by elevated CO_2. *Science* 298: 1987–1990.

Shaw, M.W. (2009). Preparing for changes in plant disease due to climate change. *Plant Protec Sci* 45(Special Issue): S3-S10.

Shaw, M.W. and Osborne, T.M. (2011). Geographic distribution of plant pathogens in response to climate change. *Plant Pathol* 60: 31–43.

Shelton, A.M., Wilsey, W.R. and Soderlund, D.M. (2001). *Classification of Insecticides and Acaricides for Resistance Management*. Dept. of Entomology, NYSAES, Geneva, NY.

Sidhu, A.K. and Mehta, H.S. (2008). Role of butterflies in the natural ecosystem with special reference to high altitude (Pangi Valley, Himachal Pradesh). In *Proceedings of International Conference on Climate Change, Biodiversity and Food Security in the South Asian Region*. Punjab State Council for Science and Technology, Chandigarh and United Nations Educational, Scientific and Cultural Organization, New Delhi, 36 pp.

Siebold, M. and Tiedemann, A.V. (2012). Potential effects of global warming on oilseed rape pathogens in Northern Germany. *Fung Ecol* 5: 62-72.

Siebold, M. and Tiedemann, A.V. (2013). Effects of experimental warming on fungal disease progress in oilseed rape. *Global Change Biol* 6: 1736-1747.

Sikora, R.A. (1992). Management of the antagonistic potential in agricultural ecosystems for the biological control of plant parasitic nematodes. *Ann. Rev. Phytopathol.* 30: 245-247.

Silc, U., Vrbnicanin, S., Bozi, C.D. *et al.* (2009). Weed vegetation in the north-western Balkans: diversity and species composition. *Weed Res* 49: 602–612.

Singer, A., Travis, J.M.J. and Johst, K. (2013). Interspecific interactions affect species and community responses to climate shifts. *Oikos* 122: 358–366.

Singh, D.P., Moore, C.A., Gilliland, A. and Carr, J.P. 2004. Activation of multiple antiviral defence mechanisms by salicylic acid. *Mol. Plant Pathol.* 5: 57-63.

Singer, S.F. (1996). Climate change and consensus. *Science* 271: 581–582.

Singh, R.P., Singh, R.K. and Singh, M.K. (2011). Impact of climate and carbon dioxide change on weeds and their management–A Review. *Indian J. Weed Sci.* 43(1 & 2): 1-11.

Sipes, B.S. and Arakaki, A.S. (1997). Root-knot nematodes management in dryland taro with tropical cover crops. *Journal of Nematology Supplement* 29: 721-724.

Skelsey, P. and Newton, A.C. (2015). Future environmental and geographic risks of Fusarium head blight of wheat in Scotland. *Eur J Plant Pathol* 142: 133-147.

Skirvin, D.J., Perry, J.N. and Harrington, R. (1997). The effect of climate change on an aphid–coccinellid interaction. *Global Change Biology* 3: 1–11.

Skov, F. and Svenning, J.C. (2004). Potential impact of climatic change on the distribution of forest herbs in Europe. *Ecography* 27: 366–380.

Šmatas, R., Semaškienė, R. and Lazauskas, S. (2008). The impact of the changing climate conditions on the occurrence of the Colorado potato beetle (*Leptinotarsa decemlineata*). *Zemdirbyste-Agriculture* 95(3): 235–241.

Smit, A.L. and Vamerali, T. (1998). The influence of potato cyst nematodes (*Globodera pallida*) and drought on rooting dynamics of potato (*Solanum tuberosum* L.). *European Journal of Agronomy* 9: 137-146.

Smith, C., van Klinken, R.D., Seabrook, L. and McAlpine, C. (2011). Estimating the influence of land management change on weed invasion potential using expert knowledge. *Divers Distrib* 1: 1-14.

Smith, K.P. and Goodman R.M. (1999). Host variation for interactions with beneficial plant-associated microbes. *Ann. Rev. Phytopathol.* 37: 473-491.

Smith, P. (2004). Engineered biological sinks on land. In C.B. Field and M.R. Raupach (eds.) *The Global Carbon Cycle. Integrating Humans, Climate, and the Natural World SCOPE 62.* Island Press, Washington DC, pp. 479-491.

Smith, P., Martino D., Cai, Z. *et al.* (2007). Agriculture. In B. Metz, O.R. Davidson, P.R. Bosch *et al.* (eds.) *Climate Change 2007: Mitigation.* Contribution of Working Group III to the Fourth Assessment Report of the IPCC, Cambridge University Press, Cambridge, United Kingdom and New York, USA.

Smith, R.F. and Doutt, R.L. (1971). The pesticide syndrome – diagnosis and suggested prophylaxis. In C.B. Huffaker (ed.) *Biological Control. AAAS Symposium Proceedings on Biological Control,* Boston, December 1969, Plenum Press, New York, pp. 331-345

Smith, S.D., Huxman, T.E., Zitzer, S.F. *et al.* (2000). Elevated CO_2 increases productivity and invasive species success in an arid ecosystem. *Nature* 408: 79–82.

Somasekhar, N. and Prasad, J.S. (2012). Plant–nematode interactions: consequences of climate change. In B. Venkateswarlu, A.K. Shanker, C. Shanker and M. Maheswari (éds.) *Crop Stress and its Management: Perspectives and Strategies.* Springer, Houten, The Netherlands, pp. 547-564.

Sonnemann, L. and Wolters, V. (2005). The micro-food web of grassland soils respond to a moderate increase in atmospheric CO_2. *Global Change Biology* 11: 1148-1155.

Soussana, J.F., Graux, A.I. and Tubiello, F.N. (2010). Improving the use of modelling for projections of climate change impacts on crops and pastures. *Journal of Experimental Botany* 61: 2217–2228.

Soussana, J.F. and Hartwig, U.A. (1996). The effect of elevated CO_2 on symbiotic N_2 fixation: a link between the carbon and nitrogen cycles. *Plant Soil* 187: 321–332.

Sparks, A.H., Forbes, G.A., Hijmans, R.J. and Garrett, K.A. (2014). Climate change may have limited effect on global risk of potato late blight. *Global Change Biol* 20: 3621-3631.

Srikanth, J. (2007). World and Indian scenario of sugarcane woolly aphid. In N. Mukunthan, J. Srikanth, B. Singaravelu *et al.* (eds.) *Woolly Aphid Management in Sugarcane.* Extension Publication, Sugarcane Breeding Institute, Coimbatore, pp. 1-12.

Srinivasa Rao, M., Dharma Reddy, K. and Singh, T.V.K. (2003). Impact of intercropping on predators of insect pests of pigeonpea in rainy and post rainy seasons. *Indian Journal of Entomology* 65: 245-251.

Srinivasa Rao, M., Ranga Rao, G.V. and Venkateswarlu, B. (2010). Impact of climate change on insect pests. *National Symposium on Climate Change and Rainfed Agriculture,* CRIDA, Hyderabad, India, pp. 43-54.

Srinivasa Rao, M., Srinivas, K., Vanaja, M. *et al.* (2008). *Impact of Elevated CO_2 on Insect Herbivore-Host Interactions.* Research Bulletin, Central Research Institute for Dryland Agriculture (CRIDA), Hyderabad, India, 36 pp.

Srinivasan, K. and Krishna Moorthy, P.N. (1991). Indian mustard as a trap crop for management of major lepidopterous pests on cabbage. *Trop Pest Mangmt* 37: 26-32.

Srinivasan, K., Krishna Moorthy, P.N. and Raviprasad, T.N. (1994). African marigold as a trap crop for the management of the fruit borer *Helicoverpa armigera* on tomato. *Int J Pest Mangmt* 40: 56-63.

Srivastava, A., Kumar, S.N. and Aggarwal, P.K. (2010). Assessment of vulnerability of sorghum to climate change in India. *Agric Ecosyst Environ* 138: 160-169.

Stacey, D.A. and Fellows, M.D.E. (2002). Influence of temperature on pea aphid *Acythosiphan pisum* (Hemiptera: Aphidae) resistence to natural enemy attack. *Bull. Entomol. Res.* 92: 351-7.

Staddon, P.L. and Fitter, A.H. (1998). Does elevated carbon dioxide affect arbuscular mycorrhizas? *Trends in Ecology and Evolution* 13: 455-458.

Standards Australia. (2006). *National Post-Border Weed Risk Management Protocol.* CRC Australian Weed Management, Adelaide and Standards Australia International Ltd, Sydney.

Starr, J.L. and Roberts, P.A. (2004). Resistance to plant-parasitic nematodes. In: Z.X.Chen, S.Y. Chen and D.W. Dickson (eds.) *Nematology, Advances and Perspective, Vol. 2.* CAB International, Wallingford, UK, pp. 879-907.

Stephens, A.E.A., Kriticos, D.J. and Leriche, A. (2007). The current and future potential geographical distribution of the oriental fruit fly, *Bactrocera dorsalis* (Diptera: Tephritidae). *Bull Entomol Res* 97: 369–378.

Sticht, C., Schrader, S., Giesemann, A. and Weigel, H.J. (2009). Sensitivity of nematode feeding types in arable soil to free air CO_2 enrichment (FACE) is crop specific. *Pedobiologia* 52: 337-349.

Stivers, L. (1999). *Crop Profile for Corn (Sweet) in New York.* http://pestdata.ncsu.edu/cropprofiles/docs/nycorn-sweet.html.

Stoeckli, S., Hirschi, M., Spirig, C. *et al.* (2012). *Impact of Climate Change on Voltinism and Prospective Diapause Induction of a Global Pest Insect*—Cydia pomonella (L.). PLoS One 7:e35723. doi:10.1371/journal.pone.0035723PubMedCentralPubMedGoogle Scholar

Stohlgren, T.J., Binkley, D., Chong, G.W. *et al.* (1999). Exotic plant species invade hot spots of native plant diversity. *Ecol Monogr* 69: 25–46.

Stonard, J.F., Latunde-Dada, A.O., Huang, Y.J. *et al.* (2010). Geographic variation in severity of phoma stem canker and *Leptosphaeria masculans/L. biglobosa* populations on UK winter oilseed rape (*Brassica napus*). *European Journal of Plant Pathology* 126: 97–109.

Stonard, J.F., Marchant, B., Latunde-Dada, A.O. *et al.* (2008). Effect of climate on the proportions of *Leptosphaeria maculans* and *L. biglobosa* in phoma stem cankers on oilseed rape in England. *Asp Appl Biol* 88: 161-164.

Stone, A.C., Gehring, C.A. and Whitham, T.G. (2010). Drought negatively affects communities on a foundation tree: growth rings predict diversity. *Oecologia* 164: 751–761.

Strand, J.F. (2000). Some agro-meteorological aspects of pest and disease management for the 21st century. *Agricultural and Forest Meteorology* 103: 73–82.

Strange, R.N. (1993). *Plant Disease Control: Towards Environmentally Acceptable Methods*. Chapman and Hall, London.

Strange, R.N. and Scott, P.R. (2005). Plant disease: a threat to global food security. *Annual Review of Phytopathology* 43: 83–116.

Stratonovitch, P., Storkey, J. and Semenov, M.A. (2012). A process-based approach to modelling impacts of climate change on the damage niche of an agricultural weed. *Global Chang Biol* 18: 2071–2080.

Sturrock, R.N., Frankel, S.J., Brown, A.V. *et al.* (2011). Climate change and forest diseases. *Plant Pathol.* 60: 133–149.

Sun, Y., Cao, H., Yin, J. *et al.* (2010). Elevated CO_2 changes the interactions between nematode and tomato genotypes differing in the JA pathway. *Plant Cell Environ.* 33: 729-739.

Sun, Y., Ding, J. and Frye, M.J. (2010). Effects of resource availability on tolerance of herbivory in the invasive *Alternanthera philoxeroides* and the native *Alternanthera sessilis*. *Weed Research* 50: 527-536.

Sun, Y., Yin J., Cao H. *et al.* (2011). Elevated CO_2 influences nematode-induced defense responses of tomato genotypes differing in the JA pathway. *PLoS ONE* 6: e19751.

Sun, Y., Yin, J., Chen, F. and Ge, F. (2011). How does atmospheric elevated CO_2 affect crop pests and their natural enemies? Case histories from China. *Insect Science* 18(4): 393-400.

Suppiah, R., Hennessy, K.J., Whetton, P.H. *et al.* (2007). Australian climate change projections derived from simulations performed for the IPCC 4th Assessment Report. *Aust. Meteorol. Mag.* 56: 131–52.

Sutherst, R.W. (1998). Implications of global change and climate variability for vector-borne diseases: generic approaches to impact assessments. *International Journal for Parasitology* 28: 935–945.

Sutherst, R.W. (2000). Climate change and invasive species – a conceptual framework. In H.A. Mooney and R.J. Hobbs (eds.) *Invasive Species in a Changing World.* Island Press, Washington, DC, pp. 211–240.

Sutherst, R.W. (2001). The vulnerability of animal and human health to parasites under global change. *International Journal for Parasitology* 31: 933–948.

Sutherst, R.W., Baker, R.H.A., Coakley, S.M. *et al.* (2007). Pests under global change - meeting your future landlords? In: J.G. Canadell, D.E. Pataki and L.F. Pitelka (eds.) *Terrestrial Ecosystems in a Changing World.* Springer, New York, pp. 211-223.

Sutherst, R.W. and Maywald, G.F. (1985). A computerised system for matching climates in ecology. *Agricultural Ecosystem and Environment* 13: 281– 299.

Sutherst, R.W., Maywald, G.F. and Bottomley, W. (1991). From CLIMEX to PESKY, a generic expert system for pest risk assessment. *EPPO Bulletin* 21: 595–608.

Sutherst, R.W., Maywald, G.F. and Russell, B.L. (2000). Estimating vulnerability under global change: modular modelling of pests. *Agriculture Ecosystems and Environment* 82: 303–319.

Sutherst, R.W., Maywald, G.F. and Skarratt, D.B. (1995). Predicting insect distributions in a changed climate. In R. Harrington and N.E. Stork (eds.) *Insects in a Changing Environment.* Academic Press, London, pp. 59–91.

Sutherst, R.W., Maywald, G.F., Yonow, T. and Stevens, P.M. (1999). *CLIMEX: Predicting the Effects of Climate on Plants and Animals.* CD-ROM and User guide. CSIRO, Melbourne, 90 pp.

Sutherst, R.W., Yonow T., Chakraborty, S. *et al.* (1996). A generic approach to defining impacts of climate change on pests, weeds and diseases in Australasia. In W.J. Bouma, G.I. Pearman and M.R. Manning (eds.) *Greenhouse: Coping with Climate Change.* CSIRO Publishing, Collingwood, Victoria, Australia, pp. 281-307.

Suzuki, N., Rivero, R.M., Shulaev, V. *et al.* (2014). Abiotic and biotic stress combinations. *New Phytol.* 203: 32-43.

Swaminathan, M.S. (1995). Foreword. In S. Nagarajan and K. Muralidharan (eds.) *Dynamics of Plant Diseases.* Allied Publishers Ltd., New Delhi, p.v.

Swanton, C.J., Huang, J.Z., Deen, W. *et al.* (1999). Effects of temperature and photoperiod on *Setaria viridis. Weed Science* 47: 446–453.

Swezey, O.H. (1925). Records of introduction of beneficial insects into the Hawaiian Islands. *Hawaiian Planter' Record* 29: 369–376.

Sysoeva, M.I., Lavrova V.V., Matveeva E.M. *et al.* (2011). Cross adaptation of potato plants to low temperatures and potato cyst nematode infestation. *Russ. J. Plant Physiol.* 58: 999-1004.

Sysoeva, M.I., Matveeva E.M., Lavrova V.V. and Sherudilo E.G. (2012). Potato plant responses to temperature drop and phytonematode infestation under continuous lighting. *Acta Hortic.* 956: 621-625.

Syvertsen, J. and Levy, Y. (2005). Salinity interactions with other abiotic and biotic stresses in citrus. *HortTechnology* 15: 100-103.

Tahiri, A., Sekkat, A., Bennani, A. *et al.* (2006). Distribution of tomato-infecting begomoviruses and *Bemisia tabaci* biotypes in Morocco. *Ann Appl Biol* 149: 175–186.

Talbot Trotter III, R., Cobb, N.S. and Whitham, T.G. (2008). Arthropod community diversity and trophic structure: a comparison between extremes of plant stress. *Ecological Entomology* 33: 1–11.

Tang'an, B.N.S., Adebitan, S.A., Adbul, S.D. and Ogidi, J.A. (2002). Fungicide activity of some plant extracts on *Cercosporea sojina* Hara: the causal agent of frogeye leaf spot of soybeans. *Paper presented at the 30th Annual Conference of the Nigerian Society for Plant Protection*, University of Agriculture, Abeokuta, pp. 1-4.

Tanwar, R.K., Jeyakumar, P. and Monga, D. (2007). *Mealybugs and their Management.* National Centre for Integrated Pest Management, New Delhi.

Tanwar, R.K., Jeyakumar, P. and Vennila, S. (2010). *Papaya Mealybug and its Management Strategies.* Technical bulletin, National Centre for Integrated Pest Management, New Delhi, 26 pp.

Teng, P.S., Heong, K.L., Kropff, M.J. *et al.* (1996). Linked pest-crop models under global change. In B. Walker and W. Steffen (eds.), *Global Change and Terrestrial Ecosystems.* University Press, Cambridge, UK, pp. 291–316.

Teng, P.S. and Yang, X.B. (1993). Biological impact and risk assessment in plant pathology. *Annual Review of Phytopathology* 31: 495– 521.

Thaler, J.S. and Bostock R.M. (2004). Interactions between abscisic acid-mediated responses and plant resistance to pathogens and insects. *Ecology* 85: 48-58.

Theilert, W. (2006). A unique product: The story of the imidacloprid stress shield. *Pflanzenschutz-Nachrichten Science Forum Bayer* 59: 73-86.

Thielert, W., Metzlaff, M. and De Block, M. (2006). *Increase of Stress Tolerance by Application of Neonicotinoids on Plants Engineered to be Stress Tolerant.* Patent No. European Patent Application EP1731037.

Thies, J.A. and Fery, R.L. (1998). Modified expression of the N gene for southern root-knot nematode resistance in pepper at high soil temperatures. *Journal of the American Society for Horticultural Science* 123: 1012–1015.

Thies, J.A., Mueller, J.D. and Fery, R.L. (1998). Use of a resistant pepper as a rotational crop to manage southern root-knot nematode. *HortScience* 33: 716-718.

Thind, T.S. (2012). Fungicides in crop health security. *Indian Phytopathol.* 65(2): 109–115.

Thomas, A. (2008). Agricultural irrigation demand under present and future climate scenarios in China. *Glob. Planet Change* 60: 306–326.

Thomas, C. and Cottage, A. (2006). Genetic engineering for resistance. In R.N. Perry and M. Moens (eds.) *Plant Nematology*. CABI, King's Lynn, UK, pp. 255–272.

Thomas, C.D., Cameron, A., Green, R.E. *et al.* (2004). Extinction risk from climate change. *Nature* 427: 145-148.

Thompson, G.B., Brown, J.K.M. and Woodward, F.I. (1993). The effects of host carbon dioxide, nitrogen and water supply on the infection of wheat by powdery mildew and aphids. *Plant, Cell and Environment* 16: 687-694.

Thompson, S., Alvarez-Loayza, P., Terborgh, J. and Katul, G. (2010). The effects of plant pathogens on tree recruitment in the Western Amazon under a projected future climate: a dynamical systems analysis. *Journal of Ecology* 98: 1434–1446.

Thomson, L.J. and Hoffmann, A.A. (2007). Effects of ground cover (straw and compost) on the abundance of natural enemies and soil macro invertebrates in vineyards. *Agricultural and Forest Entomology* 9: 173–179.

Thomson, L.J., Macfadyen, S. and Hoffmann, A.A. (2010). Predicting the effects of climate change on natural enemies of agricultural pests. *Biological Control* 52(3): 296-306.

Thuiller, W., Albert, C., Araújo, M.B. *et al.* (2008). Predicting global change impacts on plant species' distributions: future challenges. *Perspect Plant Ecol* 9: 137–152.

Tilman, D., Cassman, K.G., Matson, P.A. *et al.* (2002). Agricultural sustainability and intensive production practices. *Nature* 418: 671–677.

Tilman, D., Fargione, J., Wolff, B. *et al.* (2001). Forecasting agriculturally driven global environmental change. *Science* 292: 281-284.

Timoney, K.P. (2003). The changing disturbance regime of the boreal forest of the Canadian Prairie Provinces. *Forest Chronicle* 79: 502-516.

Tixier, P., Mal´ezieux, E., Dorel, M. and Wery, J. (2004b). SIMBA: a comprehensive model for agro-ecological assessment and prototyping of banana based cropping systems. An application to the banana cropping systems in French West Indies. In: *Proceedings of the Fourth International Crop Science Congress*, Brisbane.

Tixier, P., Risede, J.M., Dorel, M. and Malezieux, E. (2006a). Modelling population dynamics of banana plant-parasitic nematodes: A contribution to the design of sustainable cropping systems. *Ecological Modelling* 198: 321–331.

Tokatlidis, I.S. (2013). Adapting maize crop to climate change. *Agron Sustain Dev* 33:63–79.

Torriani, D.S., Calanca, P., Schmid, S. *et al.* (2007). Potential effects of changes in mean climate and climate variability on the yield of winter and spring crops in Switzerland. *Climate Research* 34: 59–69.

Trivedi, P.C. and Barker, K.R. (1986). Management of nematodes by cultural practices.

Nematropica 16: 213–236.

Trnka, M., Muska, F., Semeradova, D. *et al.* (2007). European corn borer life stage model: Regional estimates of pest development and spatial distribution under present and future climate. *Ecological Modelling* 207: 61–84.

Trudgill, D.L. (1991). Resistance to and tolerance of plant parasitic nematodes in plants. *Annual Review of Phytopathology* 29: 167–193.

Trumble, J.T. and Butler, C.D. (2009). Climate change will exacerbate California's insect pest problems. *Cal Agric* 63(2): 73–78.

Tsado, E.K. and Tanko, M.A. (2002). Evaluation of three local plants for the control of storage insect pests of maize and cowpea. *Paper presented at the 30th Annual Conference of the Nigerian Society for Plant Protection*, University of Agriculture, Abeokuta, pp. 1-4.

Tubiello, F.N., Donatelli, M., Rosenzweig, C. and Stockle, C.O. (2000). Effects of climate change and elevated CO_2 on cropping systems: model predictions at two Italian locations. *Eur J Agron* 13:179–189.

Tubiello, F.N., Soussana, J.F. and Howden, S.M. (2007). Crop and pasture response to climate change. *Proc Natl Acad Sci* 104: 19686–19690.

Tungate, K.D., Israel, D.W., Watson, D.M. and Rufty, T.W. (2007). Potential changes in weed competitiveness in an agroecological system with elevated temperatures. *Environ Exp Bot* 60: 42–49.

Tylka, G.L. (2012). *Soybean Cyst Nematode Field Guide, Second Edition.* Integrated Crop Management News 139. http://lib.dr.iastate.edu/cropnews/139.

UN-HABITAT (United Nations Human Settlements Programme). (2004). *The State of the World's Cities 2004/2005- Globalization and Urban Culture.* Earthscan, London, 198 pp.

USDA. (1999). *1999 Drought in the U.S.* (Available at http://www.ers.usda.gov).

USDA. (1999). *Economic Research Service of the U.S. Department of Agriculture Database.* http://usda.mannlib.cornell.edu/

Vaccari, F., Miglictta, F., Magliulo, V. *et al.* (2000) Free air CO_2 enrichment of potato: leaf consumption and larval growth of Colorado potato bettle *(Leptinotarsa decemlineata Say). Ital J Agron* 4: 37–41.

Van der Waals, J.E., Krüger, K., Franke, A.C. *et al.* (2013). Climate change and potato production in contrasting South African agro-ecosystems 3. Effects on relative development rates of selected pathogens and pests. *Potato Res* 56: 67-84.

van Overbeek, L.S., Van Doorn, J., Wichers, J.H. *et al.* (2014). The arable ecosystem as battle ground for emergence of new human pathogens. *Front Microbiol* 5: 1–17.

Váry, Z., Mullins, E., McElwain, J.C. and Doohan, F.M. (2015). The severity of wheat diseases increases when plants and pathogens are acclimatized to elevated carbon dioxide. *Global Change Biol* 21: 2661-2669.

Vásquez, L. (2011). *Climate Change, Pest Incidence and Resilient Agroecological Practices.* National Institute of Agricultural Sciences (INCA), Havana, 242 pp.

Vaughan, M.M., Huffaker, A., Schmelz, E.A. *et al.* (2014). Effects of elevated CO_2 on maize defence against mycotoxigenic *Fusarium verticillioides*. *Plant Cell Environ* 37: 2691-2706.

Vayssieres, J.F., Sinzogan, A., Korie, S. *et al.* (2009). Effectiveness of spinosad bait sprays (GF-120) in controlling mango-infesting fruit flies (Diptera: Tephritidae) in Benin. *J Econ Entomol* 102: 515–521.

Venter, C., DeWaele, D. and Meyer, A.J. (1992). Minimizing damage by *Ditylenchus destructor* to peanut seed with early harvest. *J. Nematol.* 24: 528–532.

Venugopal Rao, N., Raja Sekhar, P., Venkataiah, M. and Rajasri, M. (1995). Influence of habitat on *Helicoverpa armigera* (Hubner) in cotton ecosystem. *Indian Journal of Plant Protection* 23: 122-125.

Verdejo-Lucas, S., Blanco M., Cortada L. and Sorribas F.J. (2013). Resistance of tomato rootstocks to *Meloidogyne arenaria* and *Meloidogyne javanica* under intermittent elevated soil temperatures above 28° C. *Crop Prot.* 46: 57-62.

Veteli, T.O., Kuokkanen, K., Julkunen-Tiitto, R. *et al.* (2002). Effects of elevated CO_2 and temperature on plant growth and herbivore defensive chemistry. *Global Change Biology* 8(12): 1240-1252.

Vilà, M., Espinar, J.L., Hejda, M. *et al.* (2011). Ecological impacts of invasive alien plants: a meta-analysis of their effects on species, communities and ecosystems. *Ecol Lett* 14: 702–708.

Vincent, C., Hallman, G., Panneton, B. and Fleurat, F. 2003. Management of agricultural insects with physical control methods. *Ann. Rev. Entomol.* 48: 261-281.

Volk, T., Epke, K., Gerstner, V. *et al.* (2010). Klimawandel in Nordrhein- Westfalen – Auswirkungen auf Schädlinge und Pilzkrankheiten wichtiger Ackerbaukulturen. Abschlussbericht, pro- Plant GmbH, Münster.

von Tiedemann, A. and Firsching, K.H. (2000). Interactive effects of elevated ozone and carbon dioxide on growth and yield of leaf rust-infected versus non-infected wheat. *Environ. Pollut* 108: 357–363.

von Tiedemann, A. and Ulber, U. (2008). Verändertes Auftreten von Krankheiten und Schädlingen durch Klimaschwankungen. In: A.V. Tiedemann, R. Heitefuss and F. Feldmann (eds.) *Pflanzenproduktion im Wandel – Wandel im Pflanzenschutz*. DPG Selbstverlag, Braunschweig, Germany, pp. 79–89.

Wagenet, R.J., Hutson, J.L., Watson, D.G. *et al.* (1994). Computer simulation models as an aid in estimating the probability of pesticide leaching. *Computers in Agriculture 1994*, pp. 786-794.

Walck, J.L., Hidayati, S.N., Dixon, K.W. *et al.* (2011). Climate change and plant regeneration from seed. *Glob. Change Biol.* 17: 2145–2161.

Walker, P.A. and Cocks, K.D. (1991). HABITAT: a procedure for modelling a disjoint environmental envelope for a plant or animal species. *Global Ecology and Biogeography Letters* 1: 108-118.

Wallin, J.R. and Waggoner, P.E. (1950). The influence of climate on the development and spread of *Phytophthora infestans* in artificially inoculated potato plots. *Plant Disease Reporter Supplement* 190: 19-33.

Walter, J.M. (1967). Hereditary resistance to disease in tomato. *Annual Review of Phytopathology* 5: 131-162.

Walther, G., Berger, S. and Sykes, M. (2005). An ecological 'footprint' of climate change. *Proceedings of the Royal Society of London Series B, Biological Sciences* 272: 1427–1432.

Walther, G.R., Post, E., Convey, P. *et al.* (2002). Ecological responses to recent climate change. *Nature* 416: 389–395.

Wang, G., Peng, D., Gao, B. *et al.* (2014). *Comparative Transcriptome Analysis of Two Races of* Heterodera glycines *at Different Developmental Stages*. PLoS ONE 9 [En ligne], doi: 10.1371/journal.pone.0091634.

Wang, J., Niblack, T., Tremain, J.A. *et al.* (2003). Soybean cyst nematode reduces soybean yield without causing the above-ground symptoms. *Plant Dis.* 87: 623-628.

Wang, R.L. and Dekker, J. (1995). Weedy adaptation in *Setaria* sp.: III. Variation in herbicide resistance in *Setaria* sp. *Pestic Biochem Physiol* 51: 99–116.

Wang, W., Vinocur, B. and Altman, A. (2003). Plant responses to drought, salinity and extreme temperatures: Towards genetic engineering for stress tolerance. *Planta* 218: 1–14.

Ward, N. and Masters, G. 2007. Linking climate change and species invasion: an illustration using insect herbivores. *Glob. Chang. Biol.* 13: 1605–1615.

Ward, T.J., Clear, R.M., Rooney, A.P. *et al.* (2008). An adaptive evolutionary shift in Fusarium head blight pathogen populations is driving the rapid spread of more toxigenic *Fusarium graminearum* in North America. *Fungal Genet Biol* 45: 473–484.

Wardle, D.A., Yeates, G.W., Watson, R.N. and Nicholson, K.S. (1995). The detritus food web and the diversity of soil fauna as indicators of disturbance regimes in agroecosystems. *Plant Soil* 170: 35–43.

Warwick, S.I. (1990). Allozyme and life history variation in five northwardly colonizing North American weedy species. *Plant Systematics and Evolution* 169: 41–54.

Warwick, S.I., Phillips, D. and Andrews, C. (1986). Rhizome depth: the critical factor in winter survival of *Sorghum halepense* (L.) Pers. (Johnson grass). *Weed Research* 26: 381–387.

Warwick, S.I., Thompson, B.K. and Black, L.D. (1987). Life history variation in populations of the weed species *Setaria faberi. Canadian Journal of Botany* 65: 1396–1402.

Watson, G.W. and Malumphy, C.P. (2004) *Icerya purchasi* Maskell, cottony cushion scale (Hemiptera: Margarodidae), causing damage to ornamental plants growing outdoors in London. *British Journal of Entomology and Natural History* 17: 105-109.

Watson, R.T. (2001). *Climate Change 2001: Synthesis Report.* A Contribution of Working Groups I, II, and III to the Third Assessment Report of the Intergovernmental Panel on Climate Change, Cambridge, UK: Cambridge.

Watt, M.S., Ganley, R.J., Kriticos, D.J. and Manning, L.K. (2011). *Dothistroma* needle blight and pitch canker: the current and future potential distribution of two important diseases of *Pinus* species. *Canadian Journal of Forest Research* 41: 412–424.

Wayne, P.S., Foster, S., Connolly, J. *et al.* (2002). Production of allergenic pollen by ragweed (*Ambrosia artemissiifolia* L.) is increased in CO_2-enriched atmospheres. *Annal of Allergy, Asthma and Immunology* 80: 669-679.

Weaver, D.B., Rodriguez-Kabana R. and Carden E.L. (1995). Comparison of crop rotation and fallow for management of *Heterodera glycines* and *Meloidogyne* sp. in soybean. *Journal of Nematology Supplement* 27: 585-591.

Weaver, S.E., Dirks, V.A. and Warwick, S.I. (1985). Variation and climatic adaptation in northern populations of *Datura stramonium. Canadian Journal of Botany* 63: 1303–1308.

Web, E. and Gut, D. (2005). A survey of weeds that are increasingly spreading in Europe. *Agron Sustain Dev* 25: 109–121.

Webb, K.M., Ona, I., Bai, J. *et al.* (2011). A benefit of high temperature: increased effectiveness of a rice bacterial blight disease resistance gene. *New Phytol* 185: 568–576.

Webber, B.L., Yates, C.J., LeMaitre, D.C. *et al.* (2011). Modelling horses for novel climate courses: insights from projecting potential distributions of native and alien Australian acacias with correlative and mechanistic models. *Divers Distrib* 17: 978–1000.

Webber, J.F. (2008). Status of *Phytophthora ramorum* and *P. kernoviae* in Europe. In: S.J. Frankel, J.T. Kliejunas and K.M. Palmieri (eds.) *Proceedings of the Sudden Oak Death Third Science Symposium.* US Department of Agriculture, Forest Service, Pacific Southwest Research Station: General Technical Report PSW-GTR-214, Albany, CA, USA, pp 19–26.

Webber, J.F., Mullet, M. and Brasier, C.M. (2010). Dieback and mortality of plantation Japanese larch (*Larix kaempferi*) associated with infection by *Phytophthora ramorum*. *New Dis Rep* 22: 19.

Weber, E. and Gut, D. (2005). A survey of weeds that are increasingly spreading in Europe. *Agron. Sustain. Dev.* 25: 109–121.

Wei, G., Kloepper, J.W. and Tuzun, S. (1996). Induced systemic resistance to cucumber diseases and increased growth by plant growth-promoting rhizobacteria under field conditions. *Phytopathology* 86: 221–224.

Weiher, E. and van der Werf, A., Thompson, K. *et al.* (1999). Challenging Theophrastus: a common core list of plant traits for functional ecology. *J Veg Sci* 10: 609–620.

Welk, E. (2004). Constraints in range predictions of invasive plant species due to non-equilibrium distribution patterns: purple loosestrife (*Lythrum salicaria*) in North America. *Ecol Model* 179: 551–567.

Weller, D.M., Raaijmakers, J.M., Mcspadden Gardner, B.B. and Thomashow, L.S. (2002). Microbial populations responsible for specific soil suppressiveness to plant pathogens. *Annu. Rev. Phytopathol.* 40: 309-348.

Wellings, C.R. (2011). Global status of stripe rust: a review of historical and current threats. *Euphytica* 179: 1–13.

Weltzien, H.C. (1972). Geophytopathology. *Annual Review of Phytopathology* 10: 277–298.

Werner, T., Karlsson, G., Rydström, G. and Hallden, C. (1995). Breeding for beet cyst nematode (*Heterodera schachtii*) resistance. *Congress–International Institute of Beet Research* 58: 163–164.

Wessolek, G. and Asseng, S. (2006). Trade-off between wheat yield and drainage under current and climate change conditions in northeast Germany. *European Journal of Agronomy* 24: 333-342.

West, J.S., Kharbanda, P.D., Barbetti, M.J. and Fitt, B.D.L. (2001). Epidemiology and management of *Leptosphaeria maculans* (phoma stem canker) on oilseed rape in Australia, Canada and Europe. *Plant Pathol.* 50: 10–27.

West, J.S., Townsend, J.A., Stevens, M. and Fitt, B.D.L. (2012). Comparative biology of different palnt pathogens to estimate effects of climate change of crop diseases in Europe. *Eur J Plant Pathol* 133: 315–331.

Whitney, S., Whalen, J., VanGessel, M. and Mulrooney, B. (2000). *Crop Profile for Corn (Sweet) in Delaware.* http://www.ipmcenters.org/CropProfiles/docs/DEcorn-sweet.html.

Whittaker, J.B. (1999). Impacts and responses at population level of herbivores insects to elevated CO_2. *European Journal Entomology* 96: 149-156.

Wick, R.L. and Dicklow, B. (2001). Occurrence of *Pasteuria*-infected *Tylenchorhynchus* in golf course putting greens. *Phytopathology* 91: S198.

Wijnands, F.G. and Kroonen-Backbier, B.M.A. (1993). Management of farming systems to reduce pesticide inputs: the integrated approach. In J.C. Zadoks (ed.) *Modern Crop Protection: Developments and Perspectives.* Wageningen Publishers, Wageningen, The Netherlands, pp. 227–234.

Wilcove, D.S., Rothstein, D., Dubow, J. *et al.* (1998). Quantifying threats to imperiled species in the United States. *Bioscience* 48: 607–615.

Williams, K.J., Fisher, J.M. and Langridge, P. (1996). Development of a PCR-based allele-specific assay from an RFLP probe linked to resistance to cereal cyst nematode in wheat. *Genome* 39: 798–801.

Williams, R.S., Lincoln, D.E. and Norby, R.J. (2003). Development of gypsy moth larvae feeding on red maple saplings at elevated CO_2 and temperature. *Oecologia* 137: 114-122.

Williams, R.S., Norby, R.J. and Lincoln, D.E. (2000). Effects of elevated CO_2 and temperature-grown red and sugar maple on gypsy moth performance. *Global Change Biology* 6: 685-695.

Williamson, V.M., Ho, J.Y. and Ma, H.M. (1992). Molecular transfer of nematode resistance genes. *J. Nematol.* 24: 234–241.

Willis, C.G., Ruhfel, B., Primack, R.B. *et al.* (2008). Phylogenetic patterns of species loss in Thoreau's woods are driven by climate change. *Proceedings, National Academy of Sciences, USA* 105: 17029-17033.

Willmer, P., Stone, G. and Johnston, I. (2000). *Environmental Physiology of Animals 2nd Edition.* Blackwell Publishing, Malden, MA, USA, 768 pp.

Wilsey, B.J. (2001). Effects of elevated CO_2 on the response *Phlenum pratense* and *Poa pratensis* to above-ground defoliation and root-feeding nematodes. *International Journal of Plant Science* 162: 1275-1282.

Wilson, J.B. and Agnew, A.D.Q. (1992). Positive-feedback switches in plant communities. *Adv Ecol Res* 23: 263–336.

Wilson, J.R., Deinum, B. and Engels, F.M. (1991). Temperature effects on anatomy and digestibility of leaf and stem of tropical and temperate forage species. *Netherlands Journal of Agricultural Science* 39: 31-48.

Witzell, J., Berglund, M. and Rönnberg, J. (2011). Does temperature regime govern the establishment of *Heterobasidion annosum* in Scandinavia? *International Journal of Biometeorology* 55(3): 275-284.

Wolf, P.F.J. and Verreet, J.A. (2005). Factors affecting the onset of *Cercospora* leaf spot epidemics in sugar beet and establishment of disease-monitoring thresholds. *Phytopathology* 95: 269–274.

Wolfe, D.W. (1995). Physiological and growth responses to atmospheric carbon dioxide concentration. In: M. Pessarakli (ed.) *Handbook of Plant and Crop Physiology.* Marcel Dekker, New York, pp. 223-242.

Wolfe, D.W., Ziska, L., Petzoldt, C. *et al.* (2008). Projected change in climate thresholds in the Northeastern US: implications for crops, pests, livestock, and farmers. *Mitig Adapt Strat Glob Change* 13: 555-575.

Wolfe, M.S. (1985). The current status and prospects of multiline cultivars and variety mixtures for disease management. *Annual Review of Phytopathology* 23: 251–273.

Woodford, J.A.T. (1998). The impact of cropping policy on methods to control potato leafroll virus. *Asp Appl Biol* 17: 163–171.

Woodward, F.I. and Cramer, W. (1996). Plant functional types and climatic changes: introduction. *J Veg Sci* 7: 306–308.

Wright, K.M., Chapman, S., McGeachy, K. *et al.* (2013). The endophytic lifestyle of *Escherichia coli* O157:H7: quantification and internal localization in roots. *Phytopathology* 103: 333–340.

Wu, G., Chen, F.J. and Ge, F. (2006a). Direct effects of elevated CO_2 on growth, development and reproduction of cotton bollworm, *Helicoverpa armigera* Hubner [J]. *Acta Ecol Sin* 25 (6): 1732–1738.

Wu, G., Chen, F.J. and Ge, F. (2006b). Response of multiple generations of cotton bollworm *Helicoverpa armigera* Hubner, feeding on spring wheat, to elevated CO_2. *J Appl Entomol.*130 (1): 2–9.

Wu, Y.X. and Tiedemann, A.V. (2001). Physiological effects of azoxystrobin and epoxiconazole on senescence and the oxidative status of wheat. *Pesticide Biochemistry and Physiology* 71: 1–10.

Xannepuccia, A.D., Ciccihino, A., Escalante, A. *et al.* (2009). Differential responses of marsh arthropods to rainfall-induced habitat loss. *Zoological Studies* 48: 173-183.

Xiang-Qian, L., Jun-Zhi, W., Anderson, T. and Raffi, V.A. (2007). Bt Cry protein targets plant-parasitic nematode resistance to root-knot nematode in tomato roots expressing a nematicidal *Bacillus thuringiensis* crystal protein. *Plant Biotechnology Journal* 5: 455–464.

Xu, X.M. and Nicholson, P. (2009). Community ecology of fungal pathogens causing wheat head blight. *Ann Rev Phytopathol* 47: 83-103.

Xu, Y., Zhang, Y., Lin, E. *et al.* (2006). Analyses on the climate change responses over China under SRES B2 scenario using PRECIS. *Chin. Sci. Bull.* 51(18): 2260–2267.

Yadav, B.C., Veluthambi, K. and Subramaniam, K. (2006). Host-generated double stranded RNA induces RNAi in plant-parasitic nematodes and protects the host from infection. *Molecular Biochemical Parasitology* 148: 219-222.

Yadav, D.S., Subhash, C. and Selvraj, K. (2010). Agroecological zoning of brown plant hopper, [*Nilaparvata lugens* (Stal.)] incidence on rice [*Oryza sativa* (L.)]. *Journal of Scientific and Industrial Research* 69: 818-822.

Yadav, J.S., Ogwok, E., Wagaba, H. *et al.* (2011). RNAi-mediated resistance to cassava brown streak Uganda virus in transgenic cassava. *Molecular Plant Pathology* 12: 677–687.

Yamamura, K. and Kiritani, K. (1998). A simple method to estimate the potential increase in the number of generations under global warming in temperate zones. *Applied Entomology and Zoology* 33: 289-298.

Yang, X.B., Sun, P. and Hu, B.H. (1998). Decadal change of plant diseases as affected by climate in Chinese agroecosystems (Abstr.). *Proceedings of 7th International Congress on Plant Pathology*, Edinburgh, Scotland 3: 4.2.1.

Ye, L., Fu, X. and Ge, F. (2011). Enhanced sensitivity to higher ozone in a pathogen-resistant tobacco cultivar. *Journal of Experimental Botany* 63(3): 1341-1347.

Yeates, G.W. and Bongers, T. (1999). Nematode diversity in agricultural ecosystems. *Agricultural Ecosystems Environment* 74: 113-135.

Yeates, G.W., Bongers, T., De Goede, R.G.M. *et al.* (1993). Feeding habits in soil nematode families and genera an outline for soil ecologists. *Journal of Nematology* 25: 315-331.

Yeates, G.W., Newton, P.C.D. and Ross, D.J. (1999). Response of soil nematode fauna to naturally elevated CO_2 levels influenced by soil pattern. *Nematology* 1: 285-293.

Yeates, G.W., Newton, P.C.D. and Ross, D.J. (2003). Significant changes in soil microfauna in grazed pasture under elevated carbon dioxide. *Biology and Fertility of Soils* 38: 319-326.

Yeates, G.W. and Orchard, V.A. (1993). Response of pasture soil faunal populations and decomposition processes to elevated carbon dioxide and temperature: a climate chamber experiment. *Australian Grassland Invertebrate Ecology Conference* 6: 148-154.

Yeates, G.W., Tate, K.R. and Newton, P.C.D. (1997). Response of the fauna of a grassland soil to doubling of atmospheric carbon dioxide concentration. *Biology & Fertility of Soils* 25: 307-315.

Yi, Y.H. (1997). *Randomly Amplified Polymorphic DNA (RAPD) Analysis of Cultivated Tobacco* (Nicotiana tabacum L.). Ph D thesis. NC State Univ., Raleigh, 121 pp.

Yonow, T. and Sutherst, R.W. (1998) The geographical distribution of the Queensland fruit fly, *Bactrocera* (*Dacus*) *tryoni*, in relation to climate. *Australian Journal of Agricultural Research* 49: 935–953.

Yonow, T., Zalucki, M.P., Sutherst, R.W. *et al.* (2004). Modelling the population dynamics of the Queensland fruit fly, *Bactrocera (Dacus) tryoni*: a cohort based approach incorporating the effects of weather. *Ecological Modelling* 173: 9-30.

Young, A.M. (1982). *Population Biology of Tropical Insects*. Plenum Press, New York, USA.

Yu, L., Qin, X.Y., Du, J. *et al.* (2008). Bacterial leaf spot of tobacco caused by *Pseudomonas aeruginosa* in China. *New Dis Rep* 17: 3.

Yuan, C.P., Li, Y.H., Liu, Z.X. *et al.* (2012). DNA sequence polymorphism of the Rhg4 candidate gene conferring resistance to soybean cyst nematode in Chinese domesticated and wild soybeans. *Mol. Breed.* 30: 1155-1162.

Zadoks, J.C. (1985). On the conceptual basis of crop loss assessment: the threshold theory. *Ann. Rev. Phytopathol.* 23: 455-473.

Zanetti, S., Hartwig, U.A., Lüscher, A. *et al.* (1996). Stimulation of symbiotic N_2 fixation in *Trifolium repens* L. under elevated atmospheric CO_2 in a grassland ecosystem. *Plant Physiol.* 112: 575–583.

Zappalà, L., Biondi, A., Alma, A. *et al.* (2013). Natural enemies of the South American moth, *Tuta absoluta*, in Europe, North Africa and Middle-East, and their potential use in pest control strategies. *J Pest Sci* 86: 635–647.

Zavala, J.A., Casteel, C.L., DeLucia, E.H. and Berenbaum, M.R. (2008). Anthropogenic increase in carbon dioxide compromises plant defense against invasive insects. *Proc Natl Acad Sci USA* 105: 5129–5133.

Zavala, J.A., Casteel, C.L., Nabity, P.D. *et al.* (2009). Role of cysteine proteinase inhibitors in preference of Japanese beetles (*Popillia japonica*) for soybean (*Glycine max*) leaves of different ages and grown under elevated CO_2. *Oecologia* 161: 35–41.

Zawislak, K. and Tyburski, J. (1992). The tolerance of root, industrial and fodder crops to continuous cultivation. *Acta Acad. Agric. Tech Olst.* 55: 149–160.

Zhang, X., Halder, J., White, R.P. *et al.* (2014). Climate change increases risk of Fusarium ear blight on wheat in central China. *Ann Appl Biol* 164: 384-395.

Zheng, J.W., Li, Y.H. and Chen, S.Y. (2006) Characterization of the virulence phenotypes of *Heterodera glycines* in Minnesota. *J. Nematol.* 38: 383-390.

Zhou, X.L., Harrington, R., Woiwod, I.P. *et al.* (1995). Effects of temperature on aphid phenology. *Global Chang Biol* 1: 303–313.

Zhu, Y., Chen, H., Fan, J. *et al.* (2000). Genetic diversity and disease control in rice. Nature 406: 718–722.

Ziska, L.H. (1998). The influence of root zone temperature on photosynthetic acclimation to elevated carbon dioxide concentrations. *Ann. Bot.* 81: 717-721.

Ziska, L.H. (2000). The impact of elevated CO_2 on yield loss from a C3 and C4 weed in field-grown soybean. *Glob. Change Biol.* 6: 899–905.

Ziska, L.H. (2001). My View: Weed science and public health. *Weed Science* 49: 437-438.

Ziska, L.W. (2001). Changes in competitive ability between a C4 crop and a C3 weed with elevated carbon dioxide. *Weed Science* 49: 622-627.

Ziska, L.W. (2003). Evaluation of yield losses in field sorghum from a C3 and C4 weeds with increasing CO_2. *Weed Science* 51: 914-918.

Ziska, L.H. (2010). Elevated carbon dioxide alters chemical management of Canada thistle in no-till soybean. *Field Crop Res* 199: 299–303.

Ziska, L.H. (2011). Global climate change and carbon dioxide: assessing weed biology and management. In: D. Hillel (ed.) *Handbook on Climate Change and Agroecosystems: Impacts, Adaptation, and Mitigation.* Imperial College Press, London, pp. 191–208.

Ziska, L.H. and Bunce, J.A. (1993). Inhibition of whole plant respiration by elevated CO_2 as modified by growth temperature. *Physiologia Plantarum* 87: 459-466.

Ziska, L.H. and Caulfield, F.A. (2000). Rising carbon dioxide and pollen production of common ragweed, a known allergy-inducing species: implications for public health. *Aust J Plant Physiol.* 27: 893–898.

Ziska, L.H. and Dukes, J.S. (2011). *Weed Biology and Climate Change.* Blackwell Publishing Ltd, Ames, IA, pp. 68–205.

Ziska, L.H., Faulkner, S.S. and Lydon, J. (2004). Changes in biomass and root: shoot ratio of field- grown Canada thistle (*Cirsium arvense*), a noxious, invasive weed, with elevated CO_2: implications for control with glyphosate. *Weed Science* 52: 584-588.

Ziska, L.H., Gealy, D.R., Tomecek, M.B. *et al.* (2012). Recent and projected increases in atmospheric CO_2 concentration can enhance gene flow between wild and genetically altered rice (*Oryza sativa*). PLoS One 7:e37522. doi:10.1371/journal.pone.0037522 PubMedCentral PubMedGoogle Scholar

Ziska, L.H., Gebhard, D.E., Frenz, D.A. *et al.* (2003). Cities as harbingers of climate change: Common ragweed, urbanization, and public health. *Journal of Allergy and Clinical Immunology* 111: 290-295.

Ziska, L.H. and George, K. (2004). Rising carbon dioxide and invasive, noxious plants: potential threats and consequences. *World Res. Rev.* 16: 427–447.

Ziska, L.H. and Goins, E.W. (2006). Elevated atmospheric carbon dioxide and weed populations in glyphosate treated soybean. *Crop Sci.* 46: 1354-1359.

Ziska, L.H. and McConnell, L.L. (2015). Climate change, carbon dioxide, and pest biology: monitor, mitigate, management. *J. Agric. Food Chem.* 64: 6–12.

Ziska, L.H., Morris, C.F. and Goins, E.W. (2005). Quantitative and qualitative evaluation of selected wheat varieties released since 1903 to increasing atmospheric carbon dioxide: can yield sensitivity to carbon dioxide be a factor in wheat performance? *Glob Chang Biol* 10: 1810–1819.

Ziska, L.H., Runion, G.B. (2007). Future weed, pest, and disease problems for plants. In P.C.D. Newton, G. Edwards, A. Carran and P. Niklaus (eds.) *Agroecosystems in a Changing Climate. Advances in Agroecology series, Vol 12.* CRC Press, Boca Raton.

Ziska, L.H. and Teasdale, J.R. (2000). Sustained growth and increased tolerance to glyphosate observed in a C3 perennial weed, quackgrass (*Elytrigia repens* (L.) Nevski), grown at elevated carbon dioxide. *Australian Journal of Plant Physiology* 27: 159-164.

Ziska, L.H., Teasdale, J.R. and Bunce, J.A. (1999). Future atmospheric carbon dioxide concentrations may increase tolerance to glyphosate. *Weed Science* 47: 608-615.

Ziska, L.H., Tomecek, M.B. and Gealy, D.R. (2010). Competitive interactions between cultivated and red rice as a function of recent and projected increases in atmospheric carbon dioxide. *Agron. J.* 102: 118–123.

Ziter, C., Robinson, E.A. and Newman, J.A. (2012). Climate change and voltinism in Californian insect pest species: sensitivity to location, scenario and climate model choice. *Global Chang Biol* 18: 2771–2780.

Zocca, A., Zanini, C., Aimi, A. *et al.* (2008). Spread of plant pathogens and insect vectors at the northern range margin of cypress in Italy. *Acta Oecologica* 33: 307–313.

Zvereva, E.L. and Kozlov, M.V. (2006). Consequences of simultaneous elevation of carbon dioxide and temperature for plant–herbivore interactions: A metaanalysis. *Glob. Change Biol.* 12: 27–41.

Zvereva, E.L. and Kozlov, M.V. (2009). Responses of terrestrial arthropods to air pollution: A meta-analysis. *Environmental Science and Pollution Research* 17(2): 297-311.

Glossary

Abiotic stress: Abiotic stress is defined as the negative impact of non-living factors on the living organisms in a specific environment. Abiotic stress is essentially unavoidable.

Adaptation: Adjustments to current or expected climate variability and changing average climate conditions. This can serve to moderate harm and exploit beneficial opportunities. Various types of adaptation can be distinguished, including anticipatory and reactive adaptation, private and public adaptation, and autonomous and planned adaptation.

Adaptation benefits: Avoided damage costs or accrued benefits following the adoption and implementation of adaptation measures.

Adaptation strategies: A type of adjustment or response to climate change that is based on adapting to changing conditions in the environment usually through some type of technological on institutional innovation. An example of a technological innovation would be the development of new crop varieties or farming techniques to manage pests.

Agro-ecology: An ecological approach to agriculture that views agricultural areas as ecosystems and is concerned with the ecological impact of agricultural practices.

Agro-ecosystem: The organisms and environment of an agricultural area considered as an ecosystem.

Anthropogenic climate change: Climate change with the presumption of human influence, usually warming.

Anthropogenic global warming (AGW): Global warming with the presumption of human influence.

Agro-biodiversity: It is the result of natural selection processes and the careful selection and inventive developments of farmers over millennia. Agro-biodiversity is a vital sub-set of biodiversity.

Anthropogenic greenhouse emissions: Greenhouse-gas emissions resulting from human activities.

Biodiversity: The total diversity of all organisms and ecosystems at various spatial scales (from genes to entire biomass).

Bt-Cotton: Genetically modified cotton, which produced the toxin of *Bacillus thuringiensis* (*Bt*) against Lepidopteron larvae.

C3 plant species: Plant species with the C3 photosynthetic pathway (the first product in their biochemical sequence of reactions has three carbon atoms) use up some of the solar energy they absorb in a process known as photorespiration, in which a significant fraction of the CO_2 initially fixed into carbohydrates is reoxidised back to CO_2. C3 species tend to respond positively to increased CO_2 because it tends to suppress rates of photorespiration. Some of the current major food staples, such as wheat, rice and soybean, are C3 plants. In total, 16 of the world's 20 most important food crops would benefit from increased carbon-dioxide levels.

C4 plant species: In C4 plants (those in which the first product has four carbon atoms) CO_2 is first trapped inside the leaf and then concentrated in the cells which perform the photosynthesis. Although more efficient photosynthetically under current levels of CO_2, these plants are less responsive to increased CO_2 levels than C3 plants. The major C4 staples are maize, sorghum, sugarcane and millet.

Carbon dioxide (CO_2): A greenhouse gas whose atmospheric concentrations have been continually increasing from its pre-industrial (1750-1800) levels of 280 parts per million (ppm). It is currently increasing at a rate of 1.3-1.6 ppm per year, with a concentration (1995) ranging from 356-360 ppm, depending on location. There is a natural seasonal cycle in CO_2 levels in the atmosphere; CO_2 decreases in summer time when plant productivity consumes CO_2, and an increase in winter when biota are less active and respiration exceeds photosynthesis. The main source of CO_2 increase in the atmosphere has been fossil fuel consumption, with biomass burning becoming more significant over the past few decades, currently contributing approximately 30% as much as fossil fuel emissions.

CO_2 fertilization effect: The theory that forests and vegetation will experience enhanced growth or increased net primary productivity under elevated atmospheric CO_2 levels.

Climate: Climate should more accurately be the term applied to the average weather conditions over longer periods of years to decades.

Climate change: Climate change refers to any long-term trends in climate over many years or decades, around which climate variability may be evident year on year. Hence, a single warmer or cooler year on its own is not sufficient evidence to assert that climate is changing, but systematic changes in average conditions over many years do provide evidence of a changing climate.

Climate extreme (extreme weather or climate event): The occurrence of a value of a weather or climate variable above (or below) a threshold value near the upper (or lower) ends of the range of observed values of the variable. For simplicity, both extreme weather events and extreme climate events are referred to collectively as 'climate extremes.'

Climate-resilient agriculture (CRA): Agriculture that sustainably increases productivity, resilience (adaptation), reduces/removes greenhouse gases (mitigation), and enhances the achievement of national food security and development goals.

Crop diversification: Species diversification through varied crop associations and/or rotations (involving annual and/or perennial crops including trees). Crop diversification is intended to give a wider choice in the production of a variety of crops in a given area so as to expand production and protection related activities on various crops and also to lessen risks.

Diapause: A temporary pause in the growth and development of an organism due to adverse environmental conditions.

Drought: The phenomenon that exists when precipitation is significantly below normal recorded levels, causing serious hydrological imbalances that often adversely affect land resources and production systems.

Ecosystem: The interactive system formed from all living organisms and their abiotic (physical and chemical) environment within a given area. Ecosystems cover a hierarchy of spatial scales and can comprise the entire globe, biomes at the continental scale.

Entomopathogens: Disease which infect insects.

Farmer field school (FFS): The FFS approach is an innovative, participatory and interactive learning approach that emphasizes problem solving and discovery based learning. FFS aims to build farmers' capacity to analyze their production systems, identify problems, test possible solutions, and eventually encourage the participants to adopt the practices most suitable to their farming systems.

Food and nutrition security: This exists when all people, at all times, have physical, social and economic access to sufficient, safe and nutritious food that meets their dietary needs and food preferences for an active and healthy life.

General circulation model (GCM): A general circulation model is a generic term used to describe a computer model that simulates how climatic variables such as temperature and precipitation change over time. These models range in complexity from 0-dimensional models to 3-dimensional models, and are typically used to address the issue of global warming potential due to increasing atmospheric concentration of greenhouse gases.

Genetic resources: This includes any material of plant, microbial or other origin containing functional units of heredity that is of actual or potential value for food and agriculture. Genetic resources for food and agriculture includes the diversity present in agricultural, pastoral, and forest production systems or of importance to them: the variety and variability of plants and microorganisms at the genetic, species and ecosystem levels that sustain the structure, functions and processes of production systems. This diversity is often the result of the work of farmers, pastoralists, and forest dwellers over many hundreds of generations and reflects the diversity of both human activities and natural processes.

Geographical information system (GIS): A system of computing tools and procedures designed for capturing, managing, analyzing, modelling, and displaying spatially-referenced data.

Global circulation model: Numerical models that represent physical processes in the atmosphere, ocean, cryosphere and land surface, and are the most advanced tools currently available for simulating the response of the global climate system to increasing greenhouse gas concentrations.

Global warming potential (GWP): An index representing the combined effect of the differing times greenhouse gases remain in the atmosphere and their relative effectiveness in absorbing outgoing infrared radiation.

Greenhouse gases: Those gaseous constituents of the atmosphere, both natural and anthropogenic, that absorb and emit radiation at specific wavelengths within the spectrum of infrared radiation emitted by the Earth's surface, the atmosphere, and clouds. This property causes the greenhouse effect. Water vapor (H_2O), carbon dioxide (CO_2), nitrous oxide (N_2O), methane (CH_4), and ozone (O_3) are the primary greenhouse gases in the Earth's atmosphere. Moreover, there are a number of entirely human-made greenhouse gases in the atmosphere, such as the halocarbons and other chlorine and bromine-containing substances, dealt with under the Montreal Protocol. Besides CO_2, N_2O, and CH_4, the Kyoto Protocol deals with the greenhouse gases sulfur hexafluoride (SF_6), hydrofluorocarbons (HFCs), and perfluorocarbons (PFCs).

Intergovernmental Panel on Climate Change (IPCC): Established in 1988 by the World Meteorological Organization and the UN Environment Programme, the IPCC surveys world-wide scientific and technical literature and publishes assessment reports that are widely recognized as the most credible existing sources of information on climate change. The IPCC also works on methodologies and responds to specific requests from the Convention's subsidiary bodies. The IPCC is independent of the Convention.

Integrated pest management: An ecosystem approach to crop production and protection that combines different management strategies and practices to grow healthy crops and minimize the use of pesticides.

Landscape: Agricultural landscapes can be described in terms of the three elements: (i) structure – the interaction between environmental features, land use patterns and man-made objects; (ii) functions – the provision of landscape functions for farmers and for society (environmental services); and (iii) value – concerning the value the society places on agricultural landscape and the costs of maintaining and enhancing landscape provisions by agriculture. Because the underlying human and natural processes are subject to change and evolution, landscapes are dynamic systems.

Land-use change: A greenhouse gas inventory sector that covers emissions and removals of greenhouse gases resulting from direct human-induced land use, land-use change.

Mitigation: Technological change and substitution that reduces resource inputs and emissions per unit of output. Although several social, economic and technological policies would produce an emission reduction, with respect to climate change, mitigation means implementing policies to reduce GHG emissions and enhance sinks.

Multivoltine: Refers to organisms having more than two generations per year.

Mycorrhiza: This is a class of different types of fungi that symbiotically feed on plant roots. This symbiosis provides a jointly beneficial relationship between the fungus colony and its host.

Mulching: Application of mulch improves the water use efficiency and helps in water saving by reducing the ET losses, control weeds and increased yields of a number of field crops during summer months.

No regret options: Options that are valid whether climate change will occur as expected or not. In general, they are aimed at increasing the resilience of rural population and reducing their vulnerability to water-related shocks.

Overwinter: Passing through or waiting out the winter.

Parasitoids: Organisms that are parasitic during part of their life cycle, especially the ones that eventually kills their hosts.

Pathogens: Disease causing organisms.

Phenology: Studies that describes the periods of plant life cycle events related to climate.

Pheromones: A chemical secreted by an animal, especially an insect that influences the behaviour or development of others of the same species.

Resurgence: Abnormal increase in pest populations often exceeding the economic threshold level, following the insecticide application.

Transgenic plants: Transgenic plants are plants possessing a single or multiple genes, transferred from a different species.

Univoltine: Refers to organisms having one generation per year.

Voltinism: Indicates the number of generations of an organism per year.

Water vapour (H_2O): An important greenhouse gas in the troposphere that also plays a role in ozone depletion chemistry in the stratosphere.

Weather: Weather is the brief, rapidly changing condition of the atmosphere at a given place and time, influenced by the movement of air masses.

Acronyms

AICRIP. All India Coordinated Rice Improvement Project

AMF. Arbuscular mycorrhizal fungi

ATEAM. Advanced terrestrial ecosystem analysis and modelling

BCAs. Biological control agents

BPH. Brown plant hopper of rice

Bt. *Bacillus thuringiensis*

BYDV. Barley yellow dwarf virus

C3 plants. Produce a three carbon compound during photosynthesis

C4 plants. Produce a four carbon compound during photosynthesis

CBD. Convention on Biological Diversity

CC. Climate change

CIKS. Centre for Indian Knowledge Systems

CIP. International Potato Centre, Lima, Peru

CH_4. Methane

CMV. Cucumber mosaic virus

C: N Ratio. Carbon: Nitrogen ratio

CO_2. Carbon dioxide

CO_2e. Carbon dioxide equivalent

CPB. Colorado potato beetle

DNA. Deoxyribo nucleic acid

dsRNA. Double-stranded ribonucleic acid

eCO_2. Elevated CO_2

EPA. Environment Protection Agency

EPPO. European and Mediterranean Plant Protection Organization

ET. Ethylene

ETLs. Economic threshold levels

FACE. Free air CO_2 enrichment

FAO. Food and Agriculture Organisation, Rome

FFS. Farmer Field Schools

GCMs. Global circulation models

FHB. Fusarium head blight

GHGs. Greenhouse gases

GIS. Geographic information systems

GLM. Groundnut leaf miner

GPS. Global positioning system

GWP. Global warming potential

CEM. Corn earworm

CRA. Climate-resilient agriculture

HFCs. Hydro-fluorocarbons

IARI. Indian Agricultural Research Institute, New Delhi

ICAR. Indian Council Agricultural Research, New Delhi

ICIPE. International Centre of Insect Physiology and Ecology, Nairobi, Kenya

IDM. Integrated disease management

IFPRI. International Food Policy Research Institute

IITA. International Institute of Tropical Agriculture, Ibadan

ILCYM. Insect Life Cycle Modelling software

IMD. Indian Meteorological Department

INM. Integrated nematode management

IPCC. Intergovernmental Panel on Climate Change

IPM. Integrated pest management

IRRI. International Rice Research Institute, Philippines

ITK. Indigenous technical knowledge

IUCN. International Union for Conservation of Nature, Gland, Switzerland

IWM. Integrated weed management

JA. Jasmonic acid

MfE. Ministry for the Environment, Wellington

NACCAP. National Agriculture and Climate Change Action Plan

NARI. Nimbkar Agricultural Research Institute, Phaltan, Maharashtra

N_2O. Nitrous oxide

NSKE. Neem seed kernel extract

NPV. Nucleopolyhedrosis virus

O_3. Ozone

PDM. Probability distribution maps

PepMV. Pepino mosaic virus

PFCs. Perfluorocarbons

PLS. Peanut leaf spot

PLB. Potato late blight

PPPs. Plant protection products

PRA. Pest Risk Analysis

PRSV. Papaya ringspot virus

PTGS. Post transcriptional gene silencing

PVY. Potato virus Y

RCMs. Regional climate models

RKN. Root-knot nematodes

RNAi. Ribonucleic acid interference

SA. Salicylic acid

SAR. Systemic acquired resistance

SCN. Soybean cyst nematode

SDMs. Species distribution models

SF_6. Sulphur hexafluoride

siRNA. Silencing ribonucleic acid

SO_2. Sulphur dioxide

SRI. System of Rice Intensification

TYLCV. Tomato yellow leaf curl virus

UNFCCC. United Nations Framework Convention on Climate Change, New York, USA

USDA. United States Department of Agriculture

UV-B. Ultraviolet B

WMV2. Watermelon mosaic virus 2

ZYMV. Zucchini yellow mosaic virus

Subject Index

■